Introduction to Electronic Materials and Devices

Sergio M. Rezende

Introduction to Electronic Materials and Devices

 Springer

Sergio M. Rezende
Departamento de Física
Universidade Federal de Pernambuco (UFPE)
Recife, Pernambuco, Brazil

ISBN 978-3-030-81774-9   ISBN 978-3-030-81772-5   (eBook)
https://doi.org/10.1007/978-3-030-81772-5

Preface

The development of electronics and related technologies has had a major impact on the customs of modern society for their key role in telecommunication and information processing. Electronic technologies make use of active electronic components, such as vacuum tubes, transistors, diodes, integrated circuits, optoelectronic and magnetic devices, as well as associated passive electric components. The operation of all electronic components and devices is governed by the properties of the constituent materials and the laws of physics.

This book is intended to be used as a text for a course on electronic materials and devices, and taught for junior and senior undergraduate students in electrical, electronic, and computer engineering, physics, and materials science. The initial chapters present the basic concepts of waves and quantum mechanics at a level accessible to students that have not had courses on electromagnetism and modern physics. The book has an introductory character and does not go into the more specific technical details of the devices and methods of materials manufacturing. The emphasis is on the physical concepts of the properties of materials and the basic principles of device operation.

The material is suitable for one or two traditional semesters of classes. Chapters 1–3 provide the basic introduction of materials for electronics and the physical concepts necessary to understand the phenomena underlying the operation of devices. In this part, the wave concept is widely explored, since it plays a fundamental role in quantum mechanics and, therefore, in the properties of electrons in the atoms and ions. Chapter 4 is dedicated to the study of the main properties of electrons in materials and therefore is also basic for the following chapters, devoted to specific materials and devices.

Chapter 5 presents the main characteristics of semiconductors. Chapters 6 and 7 are dedicated to the principles of operation of devices manufactured with these materials, namely diodes, transistors, integrated circuits, and related devices, which today exist in a wide variety of types and categories. The junction diode and the junction transistor are studied in greater detail, since their operation can be fully deduced from the basic laws and equations presented in the initial chapters.

Chapter 8 is devoted to the basic properties of the interaction of light with matter and a variety of devices used to convert light into electric current or vice versa. These devices are responsible for making optoelectronics feasible and its applications in several areas of science, medicine, and engineering. In this category are photodetectors, such as photodiodes and solar cells, light-emitting diodes (LED), and lasers. The basic principles of semiconductor lasers and optical fibers are studied in more detail, due to their importance in optical communications.

Chapter 9 is dedicated to magnetic materials and devices, which play a key role in electronics and are not usually treated in introductory books. Special emphasis is placed on magnetic recording processes, since they are very important in computers and in countless applications of daily life. The basic concepts of spintronics are also presented, since this area of physics and technology is becoming increasingly important in several applications. Ferrite devices for use in microwave systems are also addressed in this chapter.

Finally, Chapter 10 presents a variety of materials with specific applications, but very important in the growing range of electronic devices. Among them are piezoelectric materials, dielectrics, and electrets used in electronics and in photonics, as well as materials employed in the manufacture of video displays, phosphorescent ceramics, liquid crystals, and organic conductors. In the last section, basic properties of the ever-exciting superconducting materials are emphasized. Notwithstanding the long search for a higher transition temperature, the exotic superconductivity property is bound for important real-world utilization, from Josephson junctions to maglev trains, from transmission cables to microwave frequency mixers, from magnetoencephalograms to MRI in medical applications, both resulting from the scientific demand for high magnetic fields in NMR research, for instance.

I am pleased to thank the collaboration of several colleagues at the Physics Department of Universidade Federal de Pernambuco and other Brazilian universities in bringing out suggestions, reviewing some texts, and pointing mistakes in the original Portuguese version. I am particularly grateful to Anderson Gomes, Antônio Azevedo, Belita Koiller, Celso Melo, Cid Araújo, Fernando Machado, Fernando Parisio, Ricardo Emmanuel de Souza, and Sergio Bampi. I have no words to express my gratitude to my colleague Flávio Aguiar, for carefully reviewing the whole manuscript and making many suggestions, and to Caio Nascimento for his work on the figures. My research activities, and therefore the conditions for writing this book, would not be possible without the financial support of CNPq, FINEP, CAPES, FACEPE, and UFPE. I would like to thank in advance all those who, in the future, send me criticisms and suggestions for improving the book.

I also express my deep gratitude to Leo and Elsa, my parents, who educated me with love and always stimulated my studies, and to Cláudia, Isabel, and Marta, my daughters, who as adults understood why I did not dedicate them more time when they were children.

Finally, my deepest thanks go to Adélia, my wife, who has always encouraged me in this endeavor, helped me to solve countless orthographic doubts, and followed with great interest each stage of the work on the three Portuguese editions of the book and on this new one in English.

Recife, Brazil   Sergio M. Rezende
October 2021

About This Book

Introduction to Electronic Materials and Devices is intended to be used as a text for a course on electronic materials and devices, and taught for junior and senior undergraduate students in electrical, electronic, and computer engineering, physics, and materials science. The initial chapters present the basic concepts of waves and quantum mechanics at a level accessible to students that have not had courses on electromagnetism and modern physics. The book has an introductory character and does not go into the more specific technical details of the devices and methods of materials manufacturing. The emphasis is on the physical concepts of the properties of materials and the basic principles of device operation. The material is suitable for one or two traditional semesters of classes. The first three chapters introduce the basic introduction of materials for electronics and the physical concepts necessary to understand the phenomena underlying the operation of devices. In this part, the wave concept is widely explored, since it plays a fundamental role in quantum mechanics and, therefore, in the properties of electrons in the atoms and ions. One chapter is dedicated to the study of the main properties of electrons in materials and therefore is also basic for the following chapters, devoted to specific materials and devices. More specific chapters present the basic properties and conduction mechanisms in semi-conductors and their use in diodes, transistors, and integrated circuits. One chapter is devoted to optoelectronic and photonic devices, including the light-emitting diode, solar cells, and various types of lasers. Another chapter is devoted to the magnetic properties of materials and their applications in magnetic and spintronic devices. The last chapter is dedicated to a variety of materials with specific applications in the growing range of electronic devices, such as dielectric materials used in electronics and photonics, liquid crystals, and organic conductors used in video displays, and superconducting materials. This breath of topics is not covered in any other single textbook for undergraduate engineering and science students. End-of-chapter problems are included.

Contents

About the Author

Sergio M. Rezende graduated in Electrical Engineering in Rio de Janeiro in 1963 and received the M.Sc. (1965) and Ph.D. (1967) degrees from the Massachusetts Institute of Technology, both in Electrical Engineering-Materials Science. He was one of the founders and first chairman of the Physics Department of the Federal University of Pernambuco (UFPE) (1972–1976), in Recife, where he is Emeritus Full Professor. He was twice Visiting Professor at the University of California at Santa Barbara (1975–1976 and 1982–1984). He has published over 300 scientific papers in international journals on dynamic excitation phenomena in magnetic materials, magnetic nanostructures, and spintronics, and has supervised over 40 M.Sc. and Ph.D. students. His scientific activities have never been interrupted by science managing positions he held, namely Dean of the Center for Exact Sciences of UFPE (1984–1988), Scientific Director of the Pernambuco Science Foundation (1990–1993), Secretary for Science and Technology of Pernambuco (1995–1998), President of FINEP, the main federal agency for funding of S&T in Brazil (2003–2005), and Minister for Science and Technology of Brazil (2005–2010), under President Luiz Inácio Lula da Silva. He is Member of several scientific societies and Honorary President of the Brazilian Society for the Advancement of Science (SBPC), and has received several prizes and awards in Brazil and abroad.

Chapter 1
Materials for Electronics

This first chapter presents the basic concepts that underly the formation of materials and how the atoms and ions are arranged in solids. Initially we describe qualitatively the various types of atomic bonding: ionic; covalent; molecular, or Van der Waals; and metallic. Then we present the 14 crystal lattices in three dimensions and the crystal structures of some important minerals and materials for electronics. The last section is devoted to a brief description of some features and preparation methods of important classes of materials that are employed in the fabrication of electronic devices.

1.1 Electronics and Condensed Matter Physics

Electronics was the most important technology of the twentieth century and continues to be so in this century. Its history dates back to 1904, when John Fleming invented the simplest vacuum tube, the diode, that has only two electrodes, the cathode and the anode. When heated, the cathode emits electrons that are collected by the anode, so that an electric current can flow in only one direction. Soon after, in 1907, Lee De Forest invented the triode, a vacuum tube that has a metallic grid between the cathode and the anode. In the triode, the electron flow from the cathode to the anode is controlled by the voltage between the grid and the cathode, making possible signal amplification. The origin of the name **electronics** lies in the fact that the operation of vacuum tubes is based on the control of the **electron** flow.

The main product of electronics in the first half of the twentieth century was the radio, which enabled the transmission of information at a distance and communication through voice and music. Later, the system for the transmission and reception of moving images, the television, was developed. Then came computers and also a wide variety of equipment for different purposes. However, electronics based on vacuum tubes had major limitations and disadvantages. The vacuum tubes are large, fragile, overheated, short-lived and expensive to manufacture, in addition to several technical

© The Author(s), under exclusive license to Springer Nature Switzerland AG 2022
S. M. Rezende, *Introduction to Electronic Materials and Devices*,
https://doi.org/10.1007/978-3-030-81772-5_1

drawbacks. For this reason, since before the Second World War, a solid-state device was sought that could replace the vacuum tubes in electronic equipment. The big step in this direction was taken in 1947 by J. Bardeen, W. Brattain and W. Shockley, three physicists at the Bell Telephone laboratories who studied properties of electronic conduction in semiconductors. That year they invented a three-element device that opened the possibility to control the electric current inside a piece of germanium, a semiconductor material. The device was called **transistor**, a name resulting from the contraction of the term *transresistance*, that had the potential of replacing the triode vacuum tube. For their invention, Shockley, Bardeen, and Brattain received the Physics Nobel Prize in 1954.

During the 1950s, the transistor was improved, becoming a reliable device, with applications in radios, television sets, computers and the most diverse electronic equipment and with increasingly lower manufacturing costs. Transistors revolutionized the field of electronics, and paved the way for smaller devices. In the 1960s we witnessed the miniaturization of electronics, with the development of the **integrated circuit** (IC) containing countless transistors and diodes, interconnected with resistors and capacitors, made up in the same semiconductor chip. The fabrication of integrated circuits with elements of dimensions on the order of a few micrometers gave rise to the technology of **microelectronics**. The increasing miniaturization of components and the development of the metal–oxide–semiconductor field-effect transistor (MOSFET) were essential for the birth of microcomputers. The production of integrated circuits and microprocessors with larger number of increasingly faster elements, together with the invention of devices for the visualization of information and with the developments in software engineering, led to the creation of a wide range of digital equipment that has produced a continuous evolution in electronics. This resulted in a tremendous change in the customs of society, provided by modern communication systems, the widespread use of computers, cell phones, tablets, watches, appliances used in our daily life, automation of industrial production, among others.

In addition to diodes, transistors, integrated circuits and microprocessors, whose operation is based on the electronic transport properties of semiconductors, there is a large number of other devices that give electronics a huge variety of applications. They are based on various properties of materials, electric conduction, optical, magnetic, thermal, among others, that will be presented in this book. The development of these devices was only possible thanks to the knowledge accumulated with the research activities in Solid State Physics. This is the area of physics that investigates the properties and phenomena that occur in solid materials, and that gained a great boost with the discovery of the transistor. Until the 1950s, work in this area was concentrated on crystalline solids, that have constituent atoms or ions with a periodic orderly arrangement. In these solids there are phenomena that do not exist in amorphous materials. Furthermore, since they have a crystalline structure with well-defined symmetry properties, many phenomena can be more easily interpreted by the laws of physics.

The progress in experimental and theoretical research in Solid State Physics and in Materials Science has made possible the discovery of more complex materials, such as conducting polymers, special amorphous alloys, liquid crystals, thin films and multilayers, and the development of new devices based on their properties. With broadened scope, the field of physics devoted to materials is now called Condensed Matter Physics, and is considered one of the largest and most versatile fields of study in physics, primarily due to the diversity of topics and phenomena that are available to study. Over 40% of the physicists currently work in this field worldwide, with new sub-fields of research continuing to emerge, driven by the discovery of new artificial materials, new properties and new phenomena. These, in turn, open the potential for the development of new devices that find applications in many technological segments, and whose economic interest drives basic and applied research. However, it was not only because of its technological importance that the new area developed quickly. The wide variety of phenomena that electrons and nuclei collectively exhibit in materials has given rise to exciting fundamental discoveries. This is one of the reasons for the fact that about 40% of the Physics Nobel Prizes in the past 50 years have been awarded to scientists working in this area.

The materials investigated in Condensed Matter Physics and used in electronic devices are generally not found in nature. They are produced artificially from chemical compounds with high degree of purity, through different physical chemical processes. Materials fabrication techniques have become increasingly sophisticated, making possible to obtain artificial structures not imaginable a few decades ago. It is possible, for example, to use very thin film fabrication processes to deposit individual atomic layers, one after the other, forming a multilayer or a crystalline superlattice. In addition, lithographic techniques can be used to define structures with nanometric lateral dimensions and desired shapes. The field of materials preparation and processing is therefore essential for the research in Condensed Matter Physics and Materials Science, as well as for the fabrication of electronic devices. In order to understand the phenomena that occur in solids and the operation of devices it is necessary to know several fundamental concepts that will be presented in the initial chapters of this book. We shall start by discussing a basic question: why and how do the atoms of the various elements bond together to form solid materials?

1.2 Atomic Bonding

Let us first consider the case of a simple solid, sodium chloride, NaCl. For reasons known from chemistry, and which are explained in detail by quantum mechanics, a chlorine atom, with its 17 electrons, tends to capture another extra electron to fill its $3p$ electronic shell and become stable. On the other hand, a sodium atom with 11 electrons tends to lose its single electron in the $3s$ shell so that the two inner shells form a closed nucleus. So, when a sodium atom is close to a chlorine atom, it transfers its electron to the chlorine atom, giving rise to two ions with opposite electric charges, which are attracted by means of the electrostatic interaction. In other words, the chlorine and sodium ions close together form a system that has **less energy** than if they are far apart. However, when the two ions are very close, the repulsion between their outermost electrons causes the energy to increase, preventing further

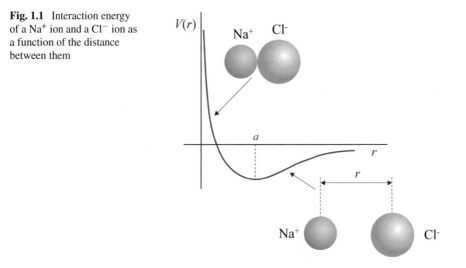

Fig. 1.1 Interaction energy of a Na$^+$ ion and a Cl$^-$ ion as a function of the distance between them

approximation. Figure 1.1 shows the variation of the interaction energy between the two ions as a function of the distance between them. When the ions are far apart, the electrostatic energy decreases, in absolute value, with increasing distance r, approximately as $(1/r)$. On the other hand, when the ions are very close, the energy grows exponentially as the distance decreases. Thus, there is a distance a at which the energy is minimum and the system can be in stable equilibrium. A similar situation occurs if there are 10^{23} sodium atoms "near" 10^{23} chlorine atoms, but now the ions tend to form a three-dimensional system, in the form of a solid crystal. This type of bonding is called **ionic**, and is the simplest to understand. There are three other types of bonds between atoms in materials: **covalent**, **molecular** and **metallic**. All of them result from the Coulomb interaction involving electrons and nuclei of the atoms. The type of bonding determines some properties of the material, as briefly presented below.

As we have seen, in **ionic solids** the bonding is due to the electrostatic attraction between ions of opposite charges, as illustrated schematically in two dimensions (2D) in Fig. 1.2a. This bond is very strong and therefore the melting point of the material is high. In other words, it takes a great deal of thermal agitation energy for the atoms to break loose of each other to form a liquid state. As electrons are strongly bound to atoms, these crystals generally have small electric and thermal conductivities, that is, they are good insulators. The absence of free electrons also results in good optical transparency over a large range of the electromagnetic spectrum. Some typical examples of ionic solids are alkaline halides (NaCl, KCl, NaBr, LiF, etc.), various oxides, sulfides, selenides, tellurides, and other compounds.

In the **covalent solids**, valence electrons are shared between neighboring atoms, as illustrated in Fig. 1.2b. In this case the attraction is due to the presence of electrons between the atoms, which simultaneously attract neighboring atoms that are left positive with their absence. Covalent solids generally have a lower melting point

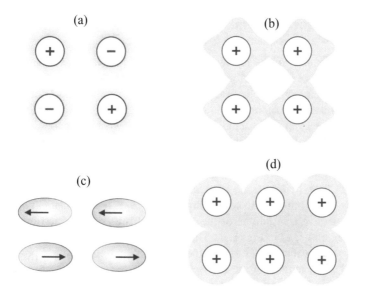

Fig. 1.2 Illustration in 2D of the main types of crystal bondings in solids. **A** Ionic; **b** covalent; **c** molecular; and **d** metallic

than ionic ones, but have greater hardness. Some of the important covalent materials are the semiconductors silicon, germanium, GaAs, InSb, GaN, etc.

The **molecular bonding** is much weaker than in the previous two cases. It is an attractive interaction between two molecules that results in a stable association in which the molecules are in close proximity to each other. It is also present in materials with neutral atoms that have closed electron shells, mediated by Van der Waals forces. These forces result from the attraction between electric dipoles formed in the atoms by a small displacement of the electronic shells relative to the nuclei, as in Fig. 1.2c. Solids with this type of bonding have very low melting point, generally less than 10 K, as is the case of crystals of solidified gases, such as oxygen, nitrogen, helium and other inert gases. As will be mentioned in Sect. 1.4.6, molecular bonding is also important in layered materials.

In some aspects, in **metals** the bonding can be considered ionic. These materials are formed by atoms that have few electrons outside their last filled shell and are therefore weakly bound to the atomic nuclei. When put together, these atoms release their last electrons that are free to move about, forming a "sea" of electrons. This negative sea of electrons tends to hold positive ions together due to the electrostatic attraction, as shown schematically in Fig. 1.2d. Thus, the bond is reasonably weak, which results in relatively low melting point, high malleability and ductility, and high thermal and electric conductivities, which are characteristic features of metals.

1.3   Crystalline Materials

Many materials used in the fabrication of electronic devices have the structure of crystalline solids or crystals. A perfect crystal is one that has a regular and periodic arrangement of atoms or ions, formed by the repetitive translation of a unit cell. The regular ordering of atoms or ions is the arrangement that minimizes the total electrostatic energy of the ensemble. For this reason, when a material is melted and then cooled slowly, the atoms or ions search for the lowest energy positions and tend to form a crystal.

Figure 1.3a shows the structure of a cesium chloride crystal. It is made of pairs of Cs^+ and Cl^- ions, that form the **base**, associated with each point of a crystal space lattice. The space lattice, also called Bravais lattice, is a mathematical abstraction, made of repetitive translations of the points of a **unit cell**, defined by three unit vectors, $\vec{a}$, $\vec{b}$, and $\vec{c}$, as in Fig. 1.3a. The lattice of cesium chloride is simple cubic, and the base consists of a Cl^- ion at position 000 and a Cs^+ ion at position ½ ½ ½, referred to the lengths of the unit vectors. Thus, the crystal structure of CsCl is obtained by translations of the unit cell, shown in Fig. 1.3b. Notice that in the center of the cube there is one ion of the base, but not a lattice point. For this reason the lattice is simple cubic.

1.3.1   Crystal Lattices

Although the number of crystal structures is very large, there are only 14 different types of Bravais lattices in three dimensions, that are shown in Fig. 1.4. The lattices

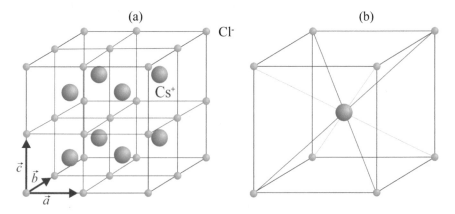

(a)   (b)

Cl⁻

Cs⁺

$\vec{c}$
$\vec{b}$
$\vec{a}$

Fig. 1.3 a Crystal of cesium chloride, CsCl. The crystal lattice is simple cubic. The base consists of a Cl^- ion at position 000 and a Cs^+ ion at position ½ ½ ½, referred to the length of the unit vectors. Note that the ions are represented by small spheres to facilitate the visualization. **b** Unit cell of CsCl

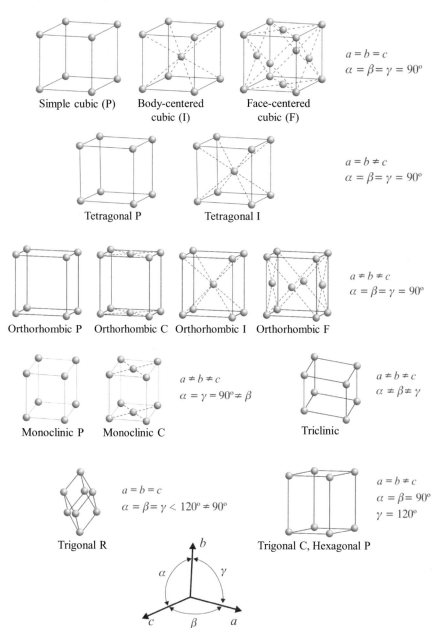

Fig. 1.4   Unit cells of the 14 crystal (Bravais) lattices in 3 dimensions

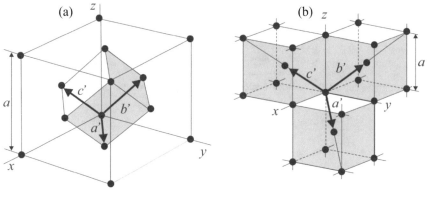

Face-centered cubic (fcc) lattice   Body-centered cubic (bcc) lattice

Fig. 1.5 Unit cell and primitive vectors of the face-centered (fcc) and body-centered (bcc) cubic lattices

are grouped into seven systems according to the type of the unit cell: cubic, tetragonal, orthorhombic, monoclinic, triclinic, trigonal, and hexagonal. Figure 1.4 also shows the relations between the angles α, β, γ, and between the lengths a, b, c of the edges of the unit cell. The lengths a, b, c are the **lattice parameters**. The unit cells shown in the figure are called conventional cells. They are the easiest to be viewed but not necessarily the smallest that can be used to reproduce the lattice by its repetitive translation. The smallest unit cell that reproduces the lattice is called **primitive cell**. Figure 1.5 shows the primitive vectors $\vec{a}'$, $\vec{b}'$ and $\vec{c}'$ of the face-centered cubic lattice (fcc) and body-centered cubic lattice (bcc).

The planes and axes that contain points of the crystalline lattice are represented by three digits that characterize their coordinates, called **Miller indices**. In order to obtain the Miller indices of a plane it is necessary first to determine its intersections with the axes x, y, z of the unit cell. The intersections are then represented by numbers p, q, r, expressing their coordinates pa, qb, rc in those axes. The Miller indices h, k, l are the smallest integers in proportion to $1/p$, $1/q$, $1/r$. To represent the plane, the indices are placed in parentheses. The axis perpendicular to the plane $(h\,k\,l)$ is represented by $[h\,k\,l]$.

Figure 1.6 shows the three most important planes and three main axes of a cubic lattice. Note that the plane parallel to the z axis that intersects axes x and y at the points $x = a$ and $y = a$, respectively, is characterized by intersections $p = 1$, $q = 1$, $r = \infty$. The inverses of these numbers give the Miller indices of the plane, which are denoted by (110). Note that since the cubic lattice is invariant under rotations of $90°$ around the z axis, the plane (110) is equivalent to the planes ($\bar{1}$10), ($1\bar{1}$0) and ($\bar{1}\bar{1}$0), where the bar above the index indicates the intersection on the negative side of the coordinate axis. These planes are also equivalent to the planes (101), (011) and their equivalents with negative indices. The set of equivalent planes is represented

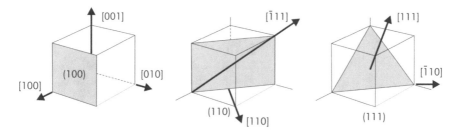

Fig. 1.6 Illustration of the three main crystal symmetry planes and main axes of the cubic lattices

by the symbol {110}, while the set of axes that can be obtained from the axis [110] by symmetry operations is represented by the symbol < 110 > .

1.3.2 Simple Crystalline Structures

In general, many different substances crystallize with the same crystalline structure. Some structures are simple and are characteristic of certain well- known materials, in general natural minerals. This section presents the main features of the crystal structures of some important materials.

The structure **of cesium chloride**, CsCl, shown in Fig. 1.3, is characterized by a simple cubic lattice with the base formed by two ions of opposite charges, Cl^- at position 000 and Cs^+ at position ½ ½ ½. Note that it is sufficient to specify an ion Cl^- at the base because all eight ions at the vertices of the unit cell are equivalent, that is, any of them can be obtained from the other by a translation by a unit vector. Since only 1/8 of each Cl^- ion is contained within the unit cell, for all purposes the cell contains only one Cl^- ion and one Cs^+ ion. The CsCl lattice parameter is $a =$ 4.11 Å. Other crystals with the same structure are TlBr (3.97 Å), CuZn (2.94 Å), which is the type β brass, AgMg (3.28 Å), and BeCu (2.70 Å).

The structure of **sodium chloride**, NaCl, is shown in Fig. 1.7a. It is formed by a face-centered cubic lattice with two ions at the base, one Na^+ and one Cl^-, separated by one-half the body diagonal of the unit cell cube. Note that the primitive cell, not shown in the figure, contains only one ion of each element. On the other hand, the unit cell contains four ions of each element (1/2 of the six ions at the faces and 1/8 of the eight ions on the vertices). Note also that the NaCl structure can be seen as formed by two face-centered cubic lattices, one with Na^+ ions and the other with Cl^- ions, displaced by one-half diagonal of the cube. The NaCl crystal has lattice parameter $a =$ 5.63 Å. Another crystal that has the NaCl structure is PbS (5.92 Å), known as galena. It is a semiconductor material and was widely used several decades ago to make detection diodes by metal contact in "galena radios". Today, PbS is used in detectors for infrared radiation. There are also several important materials for electronics that have the NaCl structure, such as MgO (4.20 Å), widely used in optical components, and NiO (4.18 Å), an antiferromagnetic material used in

(a)   (b)

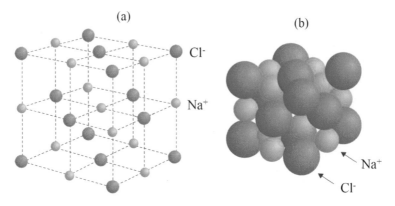

Cl⁻

Na⁺

Na⁺

Cl⁻

Fig. 1.7 a Crystal structure of sodium chloride, NaCl, formed by a face-centered cubic lattice with two ions at the base, one Na⁺ and one Cl⁻, separated by one-half the body diagonal of the unit cell cube. This structure can also be seen as formed by two face-centered cubic lattices, one with Na⁺ ions and the other with Cl⁻ ions, displaced by one-half of the cube diagonal. **b** View of the NaCl crystal, with the ions represented by spheres with sizes comparable to their distances

magnetic applications. Note that Fig. 1.7a is a simplified view of the NaCl structure, with the ions represented by small spheres. Actually, since the last electronic shells of neighboring ions are very close to each other, a more realistic view of the crystal structure is that shown in Fig. 1.7b. The apparent radius of each ion is called **ionic radius**. In the case of NaCl, the ionic radius of the Na⁺ ion is 1.220 Å and of the Cl⁻ ion is 1.595 Å. The sum of these two ionic radii is one-half the NaCl lattice parameter (5.63 Å).

The crystal structure of cubic **zinc sulfide**, ZnS, called zinc-blende, also has a face-centered cubic lattice, as shown in Fig. 1.8a. The base is formed by the atom of one of the elements at position 000 and an atom of the other element at position ¼ ¼ ¼. The structure can also be seen as formed by two face-centered cubic lattices, one with Zn atoms and the other with S atoms, displaced from each other by one-quarter of a body diagonal of the cube. Thus, as can be seen in Fig. 1.8a, each Zn atom has four S neighbors, and vice-versa, with a tetrahedral covalent bond between them. The ZnS lattice parameter is $a = 5.41$ Å. Other important semiconductors that crystallize with the zinc-blende structure are formed by elements of groups III and V of the periodic table and by elements of groups II and VI. Examples of III-V semiconductors are GaAs (5.65 Å), AlAs (5.66 Å), and InSb (6.49 Å), and examples of type II-VI semiconductors are CdS (5.82 Å) and CdTe (6.48 Å).

The last example of an important crystal structure is that of **diamond**. Its conventional unit cell, shown in Fig. 1.8b, is the same as in ZnS, but with all atoms of the same element. In the case of diamond, the element is carbon, C, with lattice parameter $a = 3.56$ Å. The diamond structure is characterized by covalent tetrahedral bonds between the neighbors. The important semiconductors silicon, Si (5.43 Å), and germanium, Ge (5.65 Å), also have crystals with this structure.

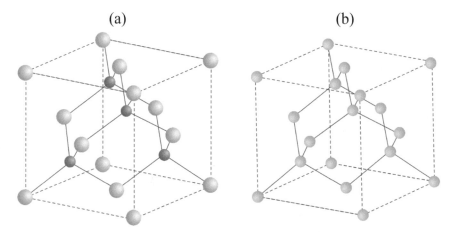

Fig. 1.8   a Unit cell of zinc sulfide, ZnS. The crystal structure is formed by a face-centered cubic lattice, having in the base the atom of one of the elements at position 000 and the atom of the other element at position ¼ ¼ ¼. The structure can also be seen as formed by two face-centered cubic lattices, one with Zn atoms and one with S atoms, displaced by one-quarter of the cube diagonal. **b** Unit cell of the diamond crystal, formed by C atoms only, and also of the semiconductors Si and Ge

1.4   Materials for Electronic Devices

Very often, books on Materials Science and Engineering classify materials in categories based on their mechanical properties. One common classification is: metals, ceramics, polymers, and composites. Other books include the category of semiconductors due to their key role in electronics. Another classification of materials used in electronics and adopted in this book is based on their main physical properties. The categories are: **Metals, semiconductors, insulators, optoelectronic, magnetic, dielectric,** and **superconductors**. In the remainder of this chapter we shall briefly present some characteristics of materials grouping them according to their microstructure or preparation process, as follows: single crystals; ceramics and glasses; polymers; liquid crystals; thin films and multilayers; graphene, carbon nanotubes, and 2D materials.

1.4.1   Single Crystals

A single crystal, also called simply a **crystal**, is a material that presents crystalline order throughout its volume, having typical dimensions that vary from a few millimeters to many centimeters. There are numerous methods for making single crystals, each suitable for certain classes of materials. In general, the crystal is produced from a liquid containing the elements that form the crystal structure. When a small piece of the desired crystal, the **seed**, is placed in the solution, if the conditions of concentration and temperature are adequate, its volume increases slowly forming a larger crystal. The essential factor in this method consists in allowing the atoms of the solution to bond to the atoms of the seed slowly. This bonding occurs in positions that minimize the total binding energy, that turn out to be those of the crystal structure. In some simple cases, one can use the liquid solution of the substance in a certain solvent. This is the case of NaCl, that can be diluted with water. By placing a small NaCl crystal in salted water, one can observe the crystal growth in a few hours. Most methods for growing crystals employ a solution obtained by melting a mixture containing the basic compounds of the desired crystal at high temperatures, producing a molten solution. The heating of the mixture is done within a container, called a **crucible**, using a resistive or radio frequency (RF) oven.

The two most used techniques for growing crystals from the molten solution are the Bridgman and the Czochralsky methods. In the first, illustrated in Fig. 1.9, the seed is placed at the bottom of the crucible containing the molten solution. The temperature of the crucible is slowly decreased while maintaining a gradient like the one in Fig. 1.9, so that the crystal grows from the bottom up. In the Czochralsky method, illustrated in Fig. 1.10, the seed is placed at the lower end of a rod, touching the surface of the molten solution. Then, as the rod is slowly pulled upwards and rotated simultaneously, by controlling the temperature gradients, rate of pulling, and speed of rotation, it is possible to extract a large, single-crystal, cylindrical ingot from the melt. Figure 1.11 shows a cylindrical ingot of single-crystal silicon grown by the Czochralsky method, with a diameter of 10.2 cm (4 inches). The discrete devices and integrated circuits used in microelectronics are manufactured on Si wafers obtained by cutting the rods, like the one in the figure. Currently, the microelectronics industry uses Si ingots with diameters of 20 cm and 30 cm.

One important technique used to grow single crystal films with thickness typically in the range 1–100 μm is the liquid phase epitaxy (LPE). It consists of inserting a thick wafer of some single crystal material, **the substrate**, in a mixture or molten solution having the elements of the desired film. The wafer can be static or made to rotate slowly in the same plane. Then, atoms in the molten solution gradually stick to the substrate, so that the single crystal film grows with one atomic layer after the

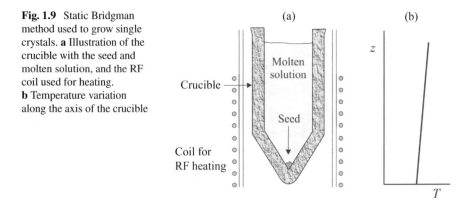

Fig. 1.9 Static Bridgman method used to grow single crystals. **a** Illustration of the crucible with the seed and molten solution, and the RF coil used for heating. **b** Temperature variation along the axis of the crucible

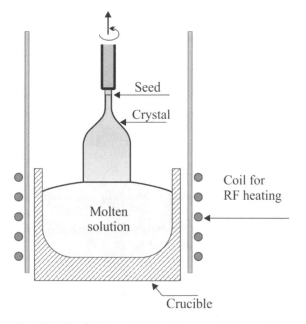

Fig. 1.10 Illustration of the Czochralsky method used to grow single crystals

other. Interestingly, if the wafer is made of the same material of the desired film, it is possible to use mixtures of two or more substances heated at temperatures below their melting points. This method was widely used to grow films of the important semiconductor GaAs in the early days of studies of diode lasers and light emitting diodes (LED). This is possible because the melting point of GaAs is 1238 °C, while the mixture of GaAs with Ga has a quite lower melting point. If a GaAs seed layer is immersed in a Ga + GaAs solution, melted at a temperature lower than 1238 °C,

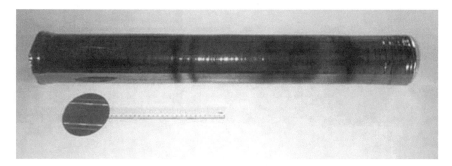

Fig. 1.11 Single crystal ingot of Si with diameter 10.2 cm (4 inches) grown by the Czochralsky method. The wafer in the pho to was obtained by cutting the rod with a diamond saw and processed to make a solar cell (Courtesy of Heliodinâmica Ltd.)

the seed layer remains solid while new crystalline layers are formed on top with atoms of As and Ga of the mixture. Nowadays, nanometric films of GaAs and other III-V and II-VI semiconductors are fabricated with thin film techniques presented in Sect. 1.4.5.

1.4.2 Ceramics and Glasses

The word **ceramics** comes from the Greek "*keramos*", which was the name of the clay used to make jars in ancient Greece. It is currently used to designate a variety of non-metallic inorganic compounds, usually hard, brittle and with a high melting point. They can be an amorphous solid or polycrystalline. To understand the difference between the two types we will consider the cases of silica (SiO_2) and alumina (Al_2O_3). The atomic bond in these materials has a mixed ionic and covalent character and, depending on the method of preparation, can result in amorphous or crystalline solids. If the cooling of the molten solution is slow, the material tends to become crystalline. In the case of silica, this occurs with a cubic or hexagonal lattice of oxygen atoms, with the Si ions between them having tetrahedral bonds, as illustrated in Fig. 1.12a. When crystallization is done from a seed, a single crystal of SiO_2, called quartz, is formed. Quartz is a natural mineral, abundant in some regions of the world, that is actually used to make the Si ingots used in electronics. If there is no seed in the preparation process, crystallization originates from many points in the material. In this case, randomly oriented crystalline grains are formed, producing a **polycrystal**, as illustrated in Fig. 1.12b. On the other hand, if the cooling is fast, the atoms will not have time to find the lowest energy positions and a crystalline structure is not formed. In this case, there is no long-range order and the material is **amorphous**, with atomic bonds as shown in Fig. 1.12c for silica, which is also called fused quartz. The case of Al_2O_3 is similar to that of silica. It can be found in the amorphous form, called alumina, or in the form of a crystal, called sapphire.

(a) (b) (c)

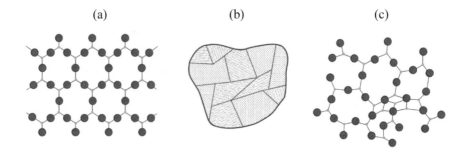

Fig. 1.12 **a** 2D view of the atomic bonds and crystalline arrangement of quartz, SiO_2. **b** Illustration of a polycrystalline material. **c** Atomic bonds and arrangement in silica, which is amorphous SiO_2

Ceramics can also be prepared by sintering. In this process, the constituents of the material in powder form are mixed and compacted to the desired final shape. The material is heated to near the melting point and then cooled, resulting in a ceramic made of polycrystalline grains with a strong adhesion to each other. This is the process used to make ceramic objects of daily use, such as pitchers, adornments, etc. When the raw material is of high quality and the processing is done under very controlled conditions, the so-called advanced ceramics are obtained, which find diverse applications in electronics and in other areas of technology. Currently it is possible to manufacture particles with dimensions of tens of nanometers (1 nm $= 10^{-9}$ m) with great uniformity of sizes, which when compacted and thermally processed result in ceramics with special properties for various applications.

Amorphous materials are also called **glasses** and are characterized by the absence of a well-defined melting temperature. When a glass is heated to the melting temperature, in contrast to crystals, it softens gradually to become a liquid without a sharp transition from the solid phase to the liquid phase. In fact, a glass can be seen as a liquid with very high viscosity, which, for practical purposes behaves as if the atoms were frozen in disorder. From the point of view of electric conductivity, amorphous materials, or glasses, can be metallic, insulating or semiconductors. In electronics they find many applications with any of these properties.

1.4.3 Polymers

Polymers consist of long chain molecular structures, usually organic, that result from the chemical combination of a large number (typically thousands) of simpler units, called monomers. The word **mer** originates from the Greek *meros*, which means part. A single mer is a monomer, while a polymer is made of monomers repeated on a regular or random manner. While natural polymers, such as rubber, have been known since immemorial times, only in the twentieth century, with the development of the chemical industry, it became possible to prepare large-scale synthetic polymers with a large variety of properties. Not only changes in the chemical nature of the monomers, but even simple structural differences in the type of chain organization, can lead to molecules with profoundly different physical and chemical properties. This is illustrated in Fig. 1.13 showing the chains of two widely used polymers: polyethylene and polyvinyl chloride (PVC). Polyethylene consists of monomers with one carbon atom and two hydrogen atoms. Replacing a hydrogen atom in ethylene

Fig. 1.13 Chains of two common polymers. **a** Polyethylene. **b** Polyvinyl chloride (PVC)

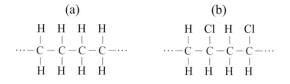

with a chlorine atom results in PVC, a completely different material. This example illustrates the enormous diversity of existing polymers.

The polymeric materials most used in electronics are "plastics" that serve as electrical insulators for covering wires, for encapsulating devices and for manufacturing parts with various functions. The intense research activities in the last decades has led to the discovery of polymers that can transport electric current as in metals or semiconductors, called conducting polymers. Some of them have optical properties similar to those in semiconductors, so that they can be used to make light emitting diodes or solar cells, with the advantage that a large number of cells can be fabricated in a flexible plastic sheet. Section 10.3 presents devices made of polymers that replace those made of traditional metals or semiconductors in some electronic and optoelectronic applications.

1.4.4 Liquid Crystals

Liquid crystals are materials that have a molecular structure with characteristics intermediate between the long-range orientational and positional order of crystals and the disorder typical of liquids and gases. Liquid crystals also have properties that are not found in liquids nor in solids, such as: formation of single crystals with the application of electric fields; optical activity much greater than typical solids and liquids, that are controllable by electric fields; large temperature sensitivity that can result in changes in their colors.

There are two major classes of liquid crystals: lyotropic and thermotropic. Lyotropic liquid crystals are generally obtained by dispersing a compound in a solvent. This is the case of several systems of biological importance, such as lipid-water, lipid-water-protein, etc. The liquid crystals relevant for electronics are thermotropic. They are formed by long molecules, usually of organic compounds, arranged in two types of structures, nematic or smectic. These structures are illustrated in Fig. 1.14, which also shows the random orientation of the molecules in an isotropic liquid. In nematic liquid crystals the molecules have parallel or almost parallel order, as in Fig. 1.14b. They are mobile in all three directions and therefore have positional disorder. In smectic liquid crystals, the molecules are also oriented parallel to each other, but have a structure stratified in layers. Within the same layer, the molecules occupy random positions, maintaining the same distance to the molecules of the neighboring layers. In smectic liquid crystals type A, the orientation of the molecules is perpendicular to the plane of the layers, whereas in type C they are tilted relative to the plane of the layers. In both the smectic A and smectic C types, the molecules randomly diffuse within each plane. No positional order exists within each plane.

Liquid crystals have wide application in electronics, especially for the manufacture of displays known as LCD (*Liquid Crystal Display*). This application is based on the fact that the orientation of the molecules can be controlled by applying an electric field, making it possible to vary the amount of light transmitted or reflected by the

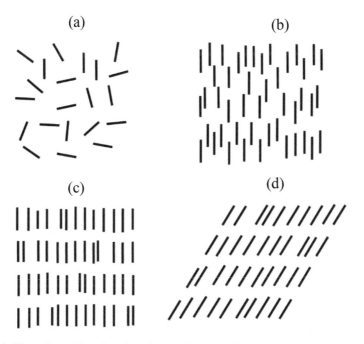

Fig. 1.14 Illustration of the orientation of molecules in the following systems: **a** isotropic liquid; **b** nematic liquid crystal; **c** smectic liquid crystal A; **d** smectic liquid crystal C

material. This can be done by means of low voltages and low power consumption, giving LCD displays a great advantage over other types, as presented in Sect. 10.3.

1.4.5 Thin Films and Multilayers

Many materials used in electronic devices are manufactured in the form of thin films, that is, layers with thicknesses that vary from few nanometers to tens of micrometers. The films are made with metals, insulators, semiconductors, or superconductors, depending on the desired application. They are used in numerous components, simple ones such as resistors, capacitors, and metal contacts in semiconductor devices, to sophisticated nanometric structures used in microelectronic, optoelectronic or spintronic devices. Thin films can be prepared by several different methods, depending on composition, structure, thickness and application. All of them are based on the gradual deposition of atoms or molecules of the desired material on the surface of another material that serves as support, called **substrate**. The most commonly used methods are grouped in two major categories, physical vapor and chemical vapor.

The notable progress in vacuum techniques in the last decades has made it possible to improve the deposition processes of very thin films. Currently, vacuum chambers with volumes on the order of 1 m³ can be routinely evacuated to pressures as low as

$10^{-11} - 10^{-9}$ Torr (1 Torr = 1 mm Hg). This enables the preparation of thin films by depositing individual layers of atoms or molecules, one on top of the other, using several different techniques. In all techniques the processing is done in a high vacuum chamber, and consists of three stages: in the first stage the substances that serve as raw materials are broken down into neutral atoms, ions or molecules, through the action of thermal sources, or a plasma, or a laser, or bombardment by accelerated electrons or ions; in the second stage, the physical vapor formed by the fragments of matter flows towards the substrate; finally, in the third stage, the fragments deposited on the substrate interact physically and chemically with each other, nucleating and forming larger portions of material, resulting in the desired film. The main differences between the different methods are in the first stage. The simplest method is thermal evaporation, in which the original substance is heated at high temperature until it evaporates. Heating is done by means of an electric current in a resistive metal that withstands high temperatures, such as tungsten, to melt the material to be deposited. This method is used to deposit simple films of metals or simple substances, to make mirrors or metallic contacts, for example.

One of the most sophisticated techniques to fabricate very thin films, multilayers and superlattices is the *Molecular Beam Epitaxy* (MBE), schematically illustrated in Fig. 1.15. The substances of the elements that form the desired material are heated separately in individual sources, inside a high vacuum chamber. Each source is made of a closed crucible, containing a small hole at one end. As the substance in the source is heated until it melts, it generates a vapor under pressure inside the crucible that is ejected in the vacuum through the hole, producing an atomic or molecular beam, which deposits on the substrate. Through the precise control of the evaporation rates and the movement of the shutters of each source it is possible to fabricate high quality crystalline films. With this method it is also possible to manufacture crystals with abrupt changes in composition forming a multilayer, or superlattice. A system of great

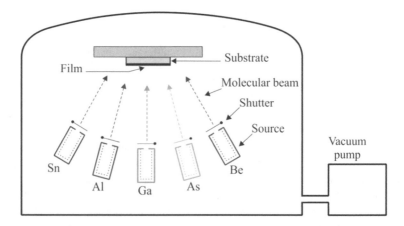

Fig. 1.15 Schematic illustration of the apparatus for molecular beam epitaxy (MBE), with the sources of the elements used to fabricate multilayers of GaAs and (GaAl)As, doped with impurities of Sn or Be

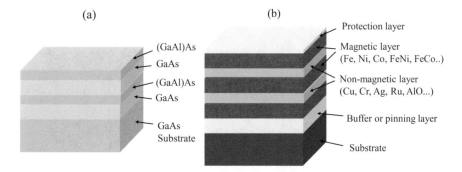

Fig. 1.16 Illustration of two types of important multilayers used in electronics: **a** multilayer of GaAs and (GaAl)As employed in semiconductor lasers. **b** Magnetic multilayer used in spin-valve reading heads in magnetic recording and other spintronic devices

technological interest is that formed by GaAs and AlAs, used in the fabrication of semiconductor lasers. The crystals of these substances have the ZnS crystal structure, with very similar lattice parameters, $a = 5.65$ Å. Because of this, it is possible to deposit epitaxially crystalline atomic layers of the ternary alloy $Ga_{1-x}Al_xAs$ on a single crystal substrate of GaAs, to artificially build layers, superlattices or "quantum wells", with chosen concentration x. Figure 1.16a illustrates a multilayer of GaAs and the alloy (GaAl)As used in semiconductor lasers. Figure 1.16b illustrates a magnetic multilayer, formed by several magnetic layers, intercalated by non-magnetic layers, metallic or insulating, used in magnetic recording and spintronic devices, described in Chap. 9.

Another important technique for the fabrication of thin films and multilayers that is widely employed in research laboratories and in industrial installations is the **sputtering** deposition. The main components of the sputtering equipment, illustrated in Fig. 1.17, are the vacuum chamber and the metallic supports of the substrate and the targets. The targets are made of the raw materials from which the atoms are pulled out to be deposited on the substrate. Before starting the deposition process, the chamber is evacuated for several hours to very low pressure ($10^{-11} - 10^{-8}$ torr), in order to eliminate residual gases. Then a noble gas (Ar, Ne) is injected into the chamber with a pressure of the order of 10^{-3} Torr, forming an inert atmosphere. A high voltage of the order of a few kV is then applied between the substrate and targets supports, ionizing the gas in the region and forming a plasma. The ions in the plasma that are accelerated by the voltage gain enough energy to pull the atoms or molecules from the target material producing a vapor that deposits on the substrate. The process usually employs several targets mounted on a wheel that can be rotated so as to have the substrate underneath a certain target during some time. By depositing different materials successively, one can fabricate a multilayer. Currently the sputtering systems have another important component, not shown in Fig. 1.17, namely a set of permanent magnets that creates a nonuniform magnetic field. Its purpose is to confine the plasma in the target region, so as to increase the efficiency of the process, that is then called *magnetron sputtering*. The high applied

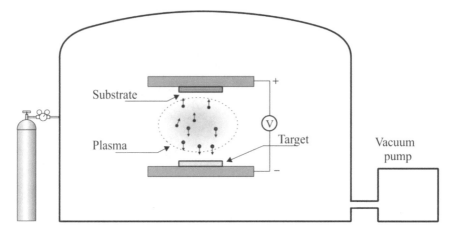

Fig. 1.17 Schematic illustration of the main components of a sputtering deposition system

voltage can be DC, used to vaporize metals, or RF, suitable for insulating materials. Recent improvements in sputtering deposition have made this technique increasingly powerful, contributing to spread its use in the processing of electronic and magnetic devices, both in research laboratories and in industrial plants.

1.4.6   Graphene, Carbon Nanotubes, and 2D Materials

Graphene is one of the most unusual artificial forms of matter. It is formed by a single plane of carbon atoms located at the vertices of the hexagons in a distribution that resembles a honeycomb, shown in Fig. 1.18a. While the diamond crystal is formed by carbon atoms in three dimensions, graphene is a carbon crystal in two dimensions. The natural mineral graphite is the amorphous form of carbon. In graphene each carbon atom is bonded to the other three atoms through covalent bonds, in which three of the four electrons in the $2s$ and $2p$ shells participate. Since the covalent bonds are very strong, graphene has a very good structural stability. The fourth electron of each C atom can be above or below the plane and has an enormous degree of freedom to move in two dimensions (2D). These almost free electrons can carry heat and electric current. Hence, high quality graphene has a very strong structure, it is lightweight, transparent and good conductor of heat and electricity. Suitably prepared graphene behaves like a semiconductor, that under the action of an external electric field has electrons and holes that move with average velocity proportional to the field intensity. As we shall see in Sect. 5.4, the mobility, defined as the ratio between the speed and the field intensity, is an important characteristic of semiconductors. At room temperature, electrons in graphene have mobility around 130 and 25 times larger than in silicon and gallium arsenide, respectively, which are the most used materials in the semiconductor industry. These properties make graphene a very promising material for the use in unique electronic devices.

Although some of its properties had been theoretically predicted in the 1960s, only in 2004 Andre Geim and Konstantin Novoselov at the University of Manchester were able to manufacture graphene samples and measure their exceptional properties. This caused an explosion of research in graphene and in other carbon materials, and Geim and Novoselov received the Physics Nobel Prize in 2010. The method they used to make graphene is very simple, exfoliation of atomic layers of pure carbon graphite with an adhesive tape and transfer to a SiO_2 film substrate on a silicon wafer. Graphene is currently fabricated in research laboratories by various techniques, such as *Chemical Vapor Deposition* (CVD). This technique can be used to manufacture a single graphene layer, or stacks of two or more layers with their bond directions parallel or twisted at some angle. By changing the twist angle it is possible to obtain unique electronic properties, such as superconductivity observed at a 'magic angle'. Research laboratories have used graphene in a variety of devices, such as high frequency transistors, logic transistors, integrated circuits, solar cells, optical modulators, etc., but to date they are not used in commercial products.

Another form of unusual arrangement of carbon atoms, discovered in the 1990s, is that of nanotubes. These consist of rolled graphene strips, as shown in Fig. 1.18b. Carbon nanotubes also have exceptional properties, characteristic of metals or semiconductors, depending on how these strips are arranged. Like graphene, carbon nanotubes have been used to make experimental electronic devices in research laboratories.

Several other materials can be made of a single atomic layer or stacked layers in which the electric properties are governed by the 2D potential acting on the electrons. These materials are made with atoms of the same element, like graphene, or with atoms of different elements. One example of single element 2D material is germanene, made with Ge atoms arranged in a buckled honeycomb structure, consisting of two hexagonal sublattices displaced by 0.2 Å from each other. Another example is silicene, made of Si atoms with a hexagonal honeycomb structure similar to that of graphene. An important class of two-element 2D compounds is that of

(a)   (b)

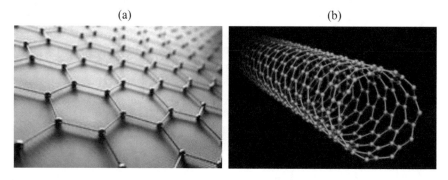

Fig. 1.18 Illustrations of two unique crystalline arrangements of carbon atoms. **a** Graphene, with the C atoms in a single plane, arranged as in a honeycomb. **b** Carbon nanotube, that can be seen as formed by a rolled graphene strip (Wikipedia)

transition metal dichalcogenides (TMD). One of the most studied is molybdenum disulfide, MoS_2, that has a unit layer structure consisting of one layer of Mo atoms covalently bonded to two layers of S atoms. These unit layers can also be stacked with bonding provided by the Van der Waals interaction. Since this interaction is relatively weak, MoS_2 layers can be obtained by exfoliation from the natural mineral molybdenite.

Another class of materials with unique properties that have potential applications in electronic and spintronic devices is that of **topological insulators**. They are materials that in the bulk have electronic energy bands with a gap, like an ordinary insulator, but have conducting states on their edges or surfaces with energy bands such that the electrons are free to move, like in a metal. The electrons at the edges or surfaces are said to have topologically protected 2D gapless states, resulting in very unique electronic and magnetic properties. Topological insulators are grouped in several classes depending on their symmetry and electronic properties. Examples of well-studied materials are HgTe, CaAs, Bi_xSb_{1-x}, Bi_2Se_3, Sb_2Te_3, and α-Sn. Actually, topological insulators are members of a more general family of material systems where quantum effects remain manifest over a wider range of energy and length scales. The so-called **quantum materials** include superconductors, graphene, topological insulators, Weyl semimetals, quantum spin liquids, and spin ices, among others.

Problems

1.1   Calculate the angle between the direction [111] and the plane (001) in a cubic crystal lattice.

1.2   Calculate the director cosines of the axis [1$\bar{2}$2].

1.3   Show, with a clear drawing, which are the primitive vectors of a simple tetragonal 3D lattice. Show why there is no face-centered tetragonal lattice.

1.4   Silicon, the most important semiconductor in electronics, crystallizes in the diamond structure, whose unit cell is shown in Fig. 1.8. At room temperature the lattice parameter is 5.42 Å. Since Si belongs to group IV of the periodic table, its atom has four valence electrons. Calculate the total number of Si valence electrons per unit volume, in cm^{-3}.

1.5   Like Si, germanium also crystallizes in the diamond structure, with a lattice parameter of 5.65 Å. Knowing that the atomic mass of Ge is 72.59 (referred to the mass of H), calculate the specific mass of Ge in g/cm^3 and compare it with the value at the table in Appendix C.

1.6   The alloy Al_xGa_{1-x} As is an important semiconductor used to manufacture optoelectronic devices. In the crystalline phase, it has the crystal structure of GaAs, in which Ga atoms in a fraction x are randomly replaced by Al atoms. Knowing that GaAs and AlAs crystallize in the ZnS structure, as in Fig. 1.8a, calculate the number of atoms per cm^3 and the specific mass of $Al_{0.3}Ga_{0.7}$ As.

1.7   A mathematical model for the total energy of an ionic bond crystal lattice is $U = N\left(\gamma e^{-R/\rho} - \alpha q^2/R\right)$, where $2N$ is the number of ions in the lattice, q is the ionic charge, γ, ρ, and α are constants that depend on the crystal structure and the constituent atoms, and R is the distance between two nearest neighbors. For NaCl, which crystallizes in the fcc structure of Fig. 1.7, $\gamma = 1.05 \times 10^{-15}$ J, $\rho = 0.321$ Å, and $\alpha = 1.747/4\pi\varepsilon_0$.

(a) Make a plot of the two terms in the expression of the energy per molecule, U/N, as a function of distance R, and interpret the meaning of each term. Use a computer to make a nice quantitative plot. Note that the second term in the equation, which results from the attraction between the two ions with opposite charges, tends to $-\infty$ at $R = 0$. Actually, the energy expression is not valid for $R = 0$, because the ions are not point charges. To avoid the divergence of the second term as $R \to 0$, the suggestion is to consider the energy constant for $R \leq 1$ Å and equal to the value at $R = 1$ Å.

(b) Make a plot of the sum of the two terms, that is, the energy U/N. Suggestion: use for the horizontal axis the range 0–10 Å, and for the vertical axis a scale in units of joule divided by a power of 10, convenient to show the minimum energy, as in Fig. 1.1.

(c) Calculate the values of the equilibrium distance R_0 and the lattice parameter of the NaCl crystal. Compare the latter value with the one given in the text.

(d) Calculate the energy per molecule needed to dissolve the crystal, that is, so that the distance between neighbors is infinite.

1.8 A single crystal film of Fe is grown on the plane (100) with a certain deposition technique, at a rate of 1.4 Å per second. Knowing that Fe crystallizes in the bcc structure, with a lattice parameter of 2.8 Å, calculate the number of atoms deposited during 20 s on a substrate in the form of a disk, with diameter 1.0 cm.

Further Reading

W.D. Callister Jr., *Materials Science and Engineering, an Introduction* (John Wiley & Sons, New York, 2010)

P.J. Collings, J.W. Goodby, *Introduction to Liquid Crystals: Chemistry and Physics* (CRC Press, Boca Raton, 2019)

R.E. Hummel, *Electronic Properties of Materials* (Springer, Berlin, 2011)

C. Kittel, *Introduction to Solid State Physics* (John Wiley & Sons, New York, 2004)

E.J. Mittemeijer, *Fundamentals of Materials Science* (Springer, Berlin, Heidelberg, 2010)

J.F. Shackelford, *Introduction to Materials Science for Engineers* (Prentice-Hall, New Jersey, 2021)

B.J. Streetman, S. Banerjee, *Solid State Electronic Devices* (Prentice-Hall, New Jersey, 2015)

H.-S.P. Wong, D. Akinwande, *Carbon Nanotubes and Graphene Devices* (Cambridge University Press, Cambridge, 2011)

Chapter 2
Waves and Particles in Matter

The phenomenon of wave propagation plays an essential role in electronics and in condensed matter physics. In electronics, a very important use of electromagnetic waves in several frequency ranges is to carry audio, video, and data signals through cables, optical fibers, and in free space. These applications are the subject of books on electromagnetics. Here we concentrate on waves that propagate inside the materials and are at the root of several electronic phenomena. In this chapter we initially briefly review the properties of electromagnetic waves and then treat the vibrations of atoms in the crystal lattice in the form of elastic waves. Then we describe the photoelectric effect that one century ago revealed the quantization of light waves and launched the idea of the wave-particle duality of electrons. This discovery opened the path to the development of quantum mechanics that is key for the understanding of the functioning of electronic devices.

2.1 Electromagnetic Waves

To introduce some important concepts in wave propagation, we start this chapter by reviewing the main characteristics of electromagnetic waves. The evolution of electromagnetic fields in space and time is described by Maxwell's equations,

$$\nabla \cdot \vec{D} = \rho, \tag{2.1}$$

$$\nabla \cdot \vec{B} = 0, \tag{2.2}$$

$$\nabla \times \vec{E} = -\frac{\partial \vec{B}}{\partial t}, \tag{2.3}$$

$$\nabla \times \vec{H} = \vec{J} + \frac{\partial \vec{D}}{\partial t}, \tag{2.4}$$

© The Author(s), under exclusive license to Springer Nature Switzerland AG 2022
S. M. Rezende, *Introduction to Electronic Materials and Devices*,
https://doi.org/10.1007/978-3-030-81772-5_2

where $\vec{E}$ and $\vec{H}$ are the electric and magnetic field vectors, respectively, $\vec{B}$ is the magnetic induction vector, $\vec{D}$ is the electric displacement vector, ρ is the free charge density, and $\vec{J}$ is the electric current density. In a linear and isotropic material, $\vec{D} = \varepsilon\vec{E}$ and $\vec{B} = \mu\vec{H}$, where ε is the electric permittivity and μ is the magnetic permeability. If the material is insulating and has no free charges, $\rho = 0$ and $\vec{J} = 0$. In these conditions, replacing (2.4) in (2.3) and using (2.1), with known relations between differential operators, we obtain the equation that describes the evolution of the electric field (Problem 2.1),

$$\nabla^2 \vec{E}(\vec{r}, t) - \mu\varepsilon\frac{\partial^2 \vec{E}(\vec{r}, t)}{\partial t^2} = 0. \tag{2.5}$$

This is the **wave equation** for a vector field in three dimensions. It relates the spatial variation of the field to its temporal variation. For plane waves propagating in the direction of the x-axis of a coordinate system, Eq. (2.5) reduces to

$$\frac{\partial^2 \vec{E}(x, t)}{\partial t^2} - \frac{1}{v^2}\frac{\partial^2 \vec{E}(x, t)}{\partial t^2} = 0, \tag{2.6}$$

where $v = 1/\sqrt{\mu\varepsilon}$. One solution of Eq. (2.6) is (Problem 2.3),

$$\vec{E}(x, t) = \vec{E}_0 \cos(k\,x - \omega t), \tag{2.7}$$

where $\vec{E}_0$ is a constant vector. Substitution in (2.6) shows that (2.7) is a solution if $\omega = k\,v$. Using (2.1) it can be shown that $\vec{E}_0$ is necessarily perpendicular to the direction of propagation $\hat{x}$. Substituting Eq. (2.7) in (2.3) and using (2.2) we obtain the solution for the magnetic field

$$\vec{H}(x, t) = \vec{H}_0 \cos(k\,x - \omega t), \tag{2.8}$$

where $\vec{H}_0$ is perpendicular to the field $\vec{E}_0$ and to the propagation direction $\hat{x}$, and the two field amplitudes are related by $E_0 = \sqrt{\mu/\varepsilon}\, H_0$. Equations (2.7) and (2.8) show that at any point in space, with coordinate x_1, the fields $\vec{E}$ and $\vec{H}$ vary harmonically in time with angular frequency ω. This can be written as $\omega = 2\pi\, v$, where $v = 1/T$ is the frequency and T is the period of the oscillation. Equations (2.7) and (2.8) show that both fields $\vec{E}$ and $\vec{H}$ have the same behavior at all points in the plane $x = x_1$. For this reason, the planes perpendicular to the propagation axis are called **phase planes**. The vector perpendicular to these planes, $\vec{k} = \hat{x}\,k$, is the **wave vector**, and its interpretation is linked to the spatial behavior of the wave. To understand this, consider the variation of $\vec{E}$, or $\vec{H}$, in space at some instant t. As shown in Fig. 2.1, the field varies sinusoidally along the direction of propagation, having the phase repeated at each distance λ, called **wavelength**. Since the argument kx corresponding to a complete period is 2π, the relation between k and λ is

$$\lambda = \frac{2\pi}{k}. \tag{2.9}$$

The spatial variation of the field at a later time $t + \Delta t$ is given by the same wave function shifted in the coordinate x by a distance $\Delta x = \omega \, \Delta t / k$, as in Fig. 2.1. Then, as time goes on, the fields $\vec{E}$ and $\vec{H}$ vary as if the wave function translates along the positive x axis, with speed $\Delta x / \Delta t = \omega / k$. This ratio is called the **phase velocity** of the wave v_p, which in this case is

$$v_P = \frac{\omega}{k} = \frac{c}{n}, \tag{2.10}$$

where $n = c\sqrt{\mu\varepsilon}$ is the index of refraction of the material, $c = 1/\sqrt{\mu_0\varepsilon_0}$ is the phase velocity in vacuum, which is the **speed of light**, given approximately by $c = 3 \times 10^8$ m/s. It is not difficult to see that if the wave propagates in an arbitrary direction, the wavevector $\vec{k}$ is parallel to the direction of propagation, that is perpendicular to the phase planes, with modulus related to the wavelength by Eq. (2.9). In this general case, it can be shown that the solutions of Eq. (2.5) are given by

$$\vec{E}(\vec{r}, t) = \vec{E}_0 \cos(\vec{k} \cdot \vec{r} - \omega t + \phi), \tag{2.11}$$

$$\vec{H}(\vec{r}, t) = \vec{H}_0 \cos(\vec{k} \cdot \vec{r} - \omega t + \phi), \tag{2.12}$$

where

$$\vec{H}_0 = \frac{\sqrt{\varepsilon/\mu}}{k} \vec{k} \times \vec{E}_0. \tag{2.13}$$

Fig. 2.1 Variation of the electric field intensity in space at two instants of time, t and $t + \Delta t$

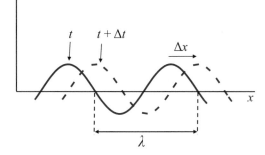

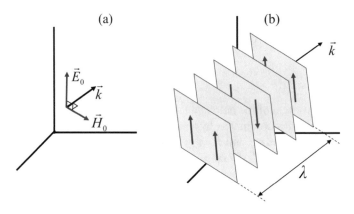

Fig. 2.2 Properties of an electromagnetic wave propagating in a generic direction in space. **a** Vectors $\vec{E}_0$, $\vec{H}_0$, and $\vec{k}$ in space. **b** Phase planes of the wave at a certain instant. The vectors in the planes represent the electric field. The distance between two consecutive planes at which the fields have the same phase is equal to the wavelength λ

In addition to the harmonic form (2.11), it is also very useful to represent the fields in complex form, using Euler's identity $e^{i\theta} = \cos\theta + i\sin\theta$. Thus, the electric field of Eq. (2.11) can be written as

$$\vec{E}(\vec{r}, t) = \text{Re}[\vec{E}_0 \, e^{i(\vec{k}\cdot\vec{r} - \omega t + \phi)}]. \tag{2.14}$$

This representation of time-harmonic fields is very useful for calculating products or divisions of quantities because the phase angles in the exponential functions simply add or subtract. Figure 2.2 shows the phase planes and the electric and magnetic fields of a wave propagating in a generic direction. The function $\omega(k)$ is called **dispersion relation**, and it contains important information about the behavior of the waves. One of them is the phase velocity $v_p = \omega/k$. As we have seen, in the case of electromagnetic waves, $\omega(k) = ck/n$, that is, the relation is linear, as shown in Fig. 2.3. For other types of waves in solids, however, this relation is a more complicated function of k. In the next section, we will see, for example, that elastic waves in solids have a nonlinear dispersion relation.

One way to generate electromagnetic waves is through moving electric charges. Harmonic waves of type (2.11) result from charges in oscillatory motion, or alternating currents (AC). The frequency of the motion, or the current, determines the frequency of the wave and therefore the type of radiation that is produced. AC currents with frequency in the range of 100 kHz (10^5 Hz) to 100 MHz (10^8 Hz), generated by transistor or vacuum tube oscillators, produce waves that are used to carry audio signals, called **radio waves**. The range from just below 100 MHz to 1000 MHz, or 1 GHz (10^9 Hz), is used to carry television signals. During the 1990s, there was a great evolution in mobile telephony, which started using frequencies in the range of hundreds of MHz, and nowadays uses a few GHz, in the microwave region. The various ranges of the electromagnetic spectrum are illustrated in Fig. 2.4, by means

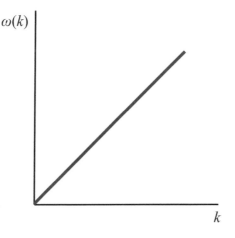

Fig. 2.3 Dispersion relation of electromagnetic waves in an isotropic, homogeneous, and linear medium

of logarithmic scales of wavelength λ in vacuum, the corresponding frequency ν, and energy E (this will be defined in Sect. 2.3). The radiation in the microwave range (1 GHz–300 GHz) is also produced by vacuum tube or transistor oscillators. In the infrared, visible and ultraviolet regions, the radiation is produced by lamps of incandescent filaments, by atomic transitions in electric discharge lamps, or gas

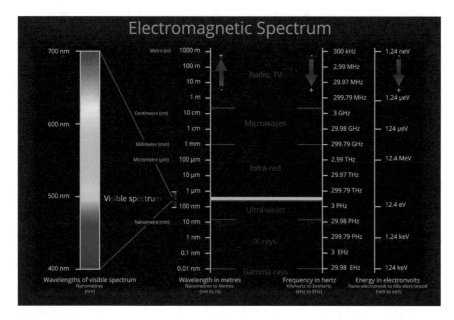

Fig. 2.4 Illustration of the electromagnetic spectrum in units of wavelength λ, frequency ν, and energy E (in eV) (courtesy of Lightcolourvision.org)

lasers, and also by electronic transitions in different materials or in semiconductor diodes. The visible region of the spectrum is shown in an expanded scale on the left side of Fig. 2.4.

The wave function described by Eq. (2.11) represents an electric field that fills the entire space, which, of course, represents an unreal situation. Despite this, it is very important in physics for several reasons. One is that any variation of the electric field that occurs in practice can be decomposed into a sum of plane waves of the type (2.11), using the Fourier transform technique. The Fourier transform allows to decompose any form of variation into plane waves of different frequencies and wave vectors. For example, let us consider an electric field that varies only in the x direction. At a given instant, say $t = 0$, we can write this field as

$$\vec{E}(x,0) = \int_{-\infty}^{\infty} \vec{E}_k \, e^{ikx} \, dk, \qquad (2.15)$$

where

$$\vec{E}_k = \int_{-\infty}^{\infty} \vec{E}(x,0) \, e^{-ikx} \, dx. \qquad (2.16)$$

Equation (2.15) means that the field is a superposition of several plane waves, each characterized by a wave vector $\vec{k} = \hat{x}\,k$ and amplitude $\vec{E}_k$. The value of $\vec{E}_k$ is given by the Fourier transform (2.16). Let us consider the case of an electromagnetic field confined to a small region of space, as shown in Fig. 2.5a, at a time $t = 0$. As time goes on, this pulse propagates in space. It can be shown that the Fourier transform of the wave packet also has the shape of a pulse, shown in Fig. 2.5b. In other words, the superposition of several plane waves, with wave vectors close to k_0 and with amplitude of the type represented in Fig. 2.5b, reproduces a spatial variation

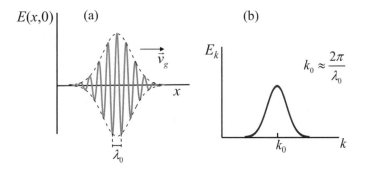

Fig. 2.5 a Electric field in an electromagnetic wave pulse in space. **b** Amplitude of the Fourier transform of the field pulse in (**a**)

in the shape of the pulse in Fig. 2.5a. It can also be shown that as time evolves, the field pulse propagates with the **group velocity**, given by

$$v_g = \left. \frac{\partial \omega}{\partial k} \right|_{k_0}.$$

(2.17)

This result is valid for any type of wave. In the case of electromagnetic waves in a vacuum or in isotropic, linear and homogeneous media, the group velocity is equal to the phase velocity (Problem 2.4). However, in other situations like the ones we will find later, this does not happen, the velocity of pulse propagation is different than the phase velocity.

2.2 Elastic Waves in Solids

In this section we study some properties of the simplest type of waves in crystals, the waves of lattice vibrations, called **elastic waves**. One of the reasons for the simplicity of this phenomenon is that its basic properties can be described with classical physics, since the atoms that form the crystal structure are relatively heavy.

The main properties of the crystal vibrations can be studied considering initially the simple case of two identical atoms, bound as explained in Sect. 1.2. Classically, in equilibrium, the distance between the two atoms corresponds to the minimum of the interaction energy, as illustrated in Fig. 2.6. Actually, this situation only occurs at a temperature of 0 K, i.e., in the absence thermal activation forces. In solids the typical equilibrium distance is a few Å. When the atoms are displaced from their equilibrium positions, they tend to oscillate about it due to the restoring forces. For small deviations, the variation of the interaction energy can be approximated by a parabolic well, so that the motion of the atoms is that of a harmonic oscillator.

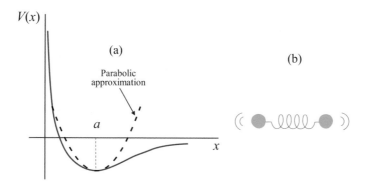

Fig. 2.6 a Effective interaction energy between two bound ions. The dashed curve represents a parabolic approximation for small deviations around the equilibrium distance a. **b** Equivalent system in the vicinity of $x = a$

Considering $u = x - a$ the deviation around the equilibrium distance, the first terms of the energy expansion in a Taylor series are

$$V(u) = V(0) + \frac{dV}{du}\Big|_0 u + \frac{1}{2}\frac{d^2V}{du^2}\Big|_0 u^2 + \ldots \ldots \tag{2.18}$$

In equilibrium, the force of interaction between the two atoms is zero, so that $dV/du|_0 = 0$. Thus, around the equilibrium we can write the energy as

$$V(u) \approx V(0) + \frac{1}{2}C\,u^2 \tag{2.19}$$

where $C = d^2V/du^2|_0$ is a constant characteristic of the atomic bonding. In this approximation, the interaction force between the atoms is proportional to the deviation from equilibrium

$$F(u) = -\frac{dV}{du} = -C\,u, \tag{2.20}$$

which is the same as in a simple harmonic oscillator. This result shows that two atoms bound by the electrostatic interaction behave like two masses connected by a spring.

In a crystal at $T = 0$, and without external disturbances, the atoms, or ions, are at their equilibrium positions. As the crystal temperature is increased, the atoms vibrate with increasing amplitude. This vibration is incoherent, in the sense that the motion of one atom has no correlation with that of another atom. However, the collective vibration of the atoms can be seen as a superposition of waves. In other words, the excitations of a crystal have a wave character. These waves of crystal vibrations are called **elastic waves**. The vibrations of a crystal with one atom in the primitive cell can be modelled by the motion of identical atoms in a linear infinite chain, as illustrated in Fig. 2.7. Consider a simplified model of the crystal lattice consisting of atoms with mass M, connected by springs with elastic constant C, as in Fig. 2.7a. Denoting by u_n the displacement of the nth-atom from its equilibrium position along the chain, since the interaction force between two atoms is given by Eq. (2.20), the force on the nth-atom exerted by its two neighbors is

$$F_n = C\left[(u_{n+1} - u_n) - (u_n - u_{n-1})\right] = C\,(u_{n+1} - 2u_n + u_{n-1}). \tag{2.21}$$

With Newton's law we can write the equation of motion for the atoms as

$$M\frac{d^2u_n}{dt^2} \equiv M\,\ddot{u}_n = C\,(u_{n+1} - 2u_n + u_{n-1}). \tag{2.22}$$

As expected, the motion of the nth-atom depends on the motion of the atoms $n \pm 1$, which in turn depend on the atoms $n \pm 2$, and so on. Thus, the motion of the chain is

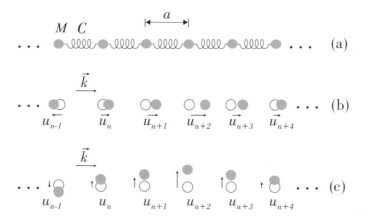

Fig. 2.7 **a** Monoatomic chain model in equilibrium. **b** Displacements of the atoms in the presence of a longitudinal vibration wave. **c** Displacements in a transverse wave

a collective one, described by an infinite number of coupled equations. To solve the system of equations, we use separation of variables and write the possible solution for $u_n(x, t)$ in the form of a wave

$$u_n(x, t) = u_k(t) \, e^{ikx}. \tag{2.23}$$

Substituting this function in the equation of motion (2.22), and considering that the equilibrium coordinate of the nth-atom is $x = na$, we obtain

$$M \, \ddot{u}_k = C u_k(e^{ika} - 2 + e^{-ika}) = 2C \, u_k(\cos ka - 1). \tag{2.24}$$

Thus, we obtain only one equation for the time dependence of the variable $u_k(t)$. Due to the collective nature of the motion, the variation in space is contained in Eq. (2.23). Note that (2.24) is the equation for a simple harmonic oscillator, whose solution is

$$u_k(t) = Ae^{-i\omega_k t}. \tag{2.25}$$

Substituting (2.25) in (2.24) we obtain the oscillation frequency of the chain as a function of the wave number

$$\omega_k = \left(\frac{2C}{M}\right)^{1/2} (1 - \cos ka)^{1/2}. \tag{2.26}$$

This result means that, when excited externally, the atoms in the chain oscillate collectively with frequency ω_k, in the form of elastic waves. The wave illustrated in Fig. 2.7b is called longitudinal, because the displacements of the atoms have the same

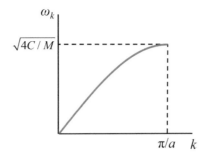

Fig. 2.8 Dispersion relation of elastic waves in a linear monoatomic chain

direction of the wave propagation. A similar calculation can be done for displacements perpendicular to the chain direction. In this case one obtains the equations for transverse waves, whose mode of vibration is illustrated in Fig. 2.7c.

Equation (2.26) is the **dispersion relation** of longitudinal elastic waves in the linear monoatomic chain. Since this relation is periodic in k and $\omega_k = \omega_{-k}$, we will only consider the values of ω_k between $k = 0$ and $k = \pi/a$, as in Fig. 2.8. This range of wavenumber contains all necessary information about ω_k. The region $-\pi/a < k < \pi/a$ is called the **first Brillouin Zone** of wave vector space.

Note that in (2.26), ka represents the phase angle between the motions of two neighboring atoms. For long wavelength waves, *i.e.* for $\lambda \gg a$, this angle is small and we can use the approximation $\cos ka \approx 1 - (ka)^2/2$, so that the dispersion relation becomes

$$\omega_k = \sqrt{C/M}\, ka. \tag{2.27}$$

Hence, for $ka \ll 1$ the dispersion relation is approximately linear, like in electromagnetic waves. In this case the phase and group velocities of the wave are equal, given by

$$v = \sqrt{C/M}\, a. \tag{2.28}$$

In general, this velocity is on the order of 10^4 m/s, that is, 10^4 times smaller than the speed of light. For long wavelength waves we can approximate the displacement function by a continuous function in x, $u(x, t)$. In this case, it is possible to show that the equation of $u(x, t)$ is equal to the wave equation for the electric field, Eq. (2.6) (Problem 2.5). On the other hand, when the wavelength is small, the discrete nature of the atomic chain becomes important. The wave with $\lambda = 2a$ has the maximum vibration frequency. Using $ka = \pi$ in Eq. (2.26) we see that the maximum value of ω is given by $\sqrt{4C/M}$. The value of this frequency varies from one material to another and is in the range of 1–10 THz (1 THz = 10^{12} Hz), which corresponds to the far infrared region of the electromagnetic spectrum (Problem 2.6).

In a 3D crystal there are two facts that make the problem of elastic waves more complex: the first is that the atoms, or ions, interact with their neighbors in three dimensions, generally with different elastic constants; the second is that the crystals

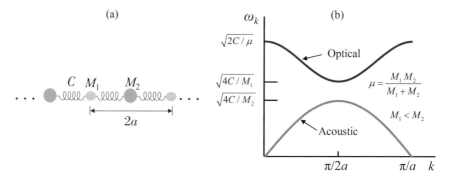

Fig. 2.9 **a** Illustration of a linear chain with two different atoms, or ions, in equilibrium. **b** Dispersion relation of elastic waves in the diatomic chain with acoustic and optical branches

usually contain different atoms or ions. In the case of cubic crystals with one atom per primitive cell, for waves propagating along the [100], [110], or [111] axes, the equations are basically the same as in the 1D chain. This is so because entire planes of atoms move in phase, with displacements either parallel or perpendicular to the propagation direction. Thus, the displacement of the nth plane can be described by a single variable u_n, just as in the case of the linear chain.

In the case the crystal is made of atoms, or ions, of different elements, the waves have new characteristics, which can also be qualitatively explained assuming a 1D model. Consider a linear chain with two types of atoms, with masses M_1 and M_2, in an alternating arrangement, as in Fig. 2.9a. Then, the motions of the atoms are described by two equations of the form Eq. (2.22), instead of just one, as in the case of identical atoms. We will then have two solutions for the vibration frequencies and, consequently, two branches in the dispersion relation, qualitatively shown in Fig. 2.9b. In this case, the possible vibration frequencies of the system form two **bands**, defined by the two branches of the dispersion relation. Between them there is a forbidden band, whose width depends on the difference between the masses. When the two masses are equal, the forbidden band disappears, that is, the lower branch in the region $0 < k < \pi/2a$ and the upper branch in the region $\pi/2a < k < \pi/a$ make up the dispersion relation of the monoatomic chain of Fig. 2.8.

In the so-called **acoustic modes**, with frequencies given by the lower branch of Fig. 2.9b, a wave with $ka \ll 1$ has two neighboring atoms moving in phase. In the **optical modes** (upper branch in Fig. 2.9b), a wave with $ka \ll 1$ has two neighboring atoms moving in opposite phases. Waves in the acoustic branch can be excited by forces that make neighboring atoms move in the same direction, as in a sound wave (hence its name, acoustic). On the other hand, waves in the optical branch are created when the excitation produces opposite motions of neighbors, as is the case of the electric field of infrared light acting on neighboring ions of opposite charges (hence its name, optical).

The complexity arising from the 3D character of the crystal results in the existence of a larger number of degrees of freedom in the system. In this case, the displacement

Fig. 2.10 Dispersion
relations for elastic waves in
a diatomic cubic crystal,
with wavevector in the
direction of one main axis (L
= longitudinal, T =
transverse, O = optical, and
A = acoustic)

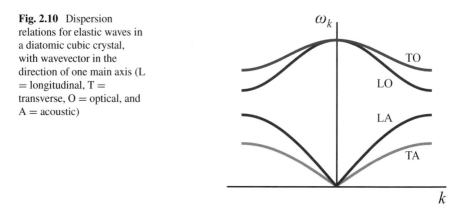

of an atom from its equilibrium position $\vec{r}$ is characterized by a vector $\vec{R}(\vec{r}, t)$.
Solutions of the general equations of motion lead to

$$\vec{R}_\lambda(\vec{r}, t) = \mathrm{Re}[A_k\, e^{i(\vec{k}\cdot\vec{r} - \omega_{\lambda k} t)}], \qquad (2.29)$$

where λ is an index that represents the type of vibration and the direction of displacement $\vec{R}_\lambda$, that is, it expresses the polarization of the wave and its type (optical or acoustic). For a given direction of the wavevector $\vec{k}$ there are three polarizations for each type of wave. For particular directions we can have one longitudinal and two transverse waves. The frequency $\omega_{\lambda k}$ depends on k and the type of wave. Figure 2.10 illustrates typical shapes of dispersion relations for elastic waves in a cubic crystal with two different atoms or ions per unit cell.

2.3 Photoelectric Effect: Waves and Particles

At the end of the nineteenth century, the first evidence emerged that, in some situations, an electromagnetic wave behaved with typical characteristics of particles. In 1886–87, Heinrich Hertz carried out several experiments that confirmed the existence of electromagnetic waves and Maxwell's theory. In one of these experiments, he observed that an electric discharge between two electrodes occurred more easily when one of the electrodes was illuminated with ultraviolet light. Later, Lenard showed that the discharge occurred more easily because the ultraviolet light facilitates the emission of electrons from the cathode surface. The ejection of electrons from a surface by the action of light was later called the **photoelectric effect**.

Figure 2.11 shows the basic apparatus used to study the photoelectric effect. It consists of an evacuated glass tube, with a flat quartz "window" through which the incident light passes. Monochromatic light is shed on the metal plate of the cathode C, causing it to release electrons. These so-called **photoelectrons** are attracted to the

Fig. 2.11 Illustration of the apparatus used by Millikan to study the photoelectric effect

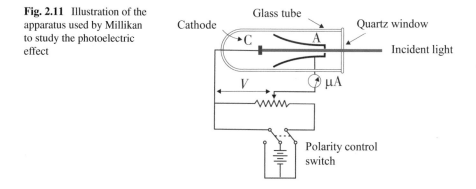

metal surface of anode A by means of the electric field due to a potential difference V between the two electrodes, producing an electric current that is measured by the microammeter μA. In a typical experiment, the current variation is measured as a function of the voltage V, which can be varied by means of a potentiometer. Such apparatus was used in 1914 by Robert Millikan in his pioneering studies of the photoelectric effect, for which he received the Physics Nobel Prize in 1923.

One of the measurements of Millikan is shown in Fig. 2.12. It demonstrates that the variation of the photoelectric current I with the applied voltage V depends on the incident light intensity. When V is positive and sufficiently large, the current saturates at a value I_a corresponding to a certain light intensity a. The current saturation occurs when all photoelectrons emitted by the cathode are collected by the anode. One of the important results of this experiment is obtained when the polarity of the voltage is reversed. The current does not go to zero immediately with the negative voltage, indicating that electrons are still emitted from C with a certain kinetic energy. However, when the voltage reaches a value $- V_0$, even the electrons ejected with highest energy are stopped, so that the current goes to zero. From this result one can conclude that the voltage V_0, called the **stopping potential**, or stopping voltage,

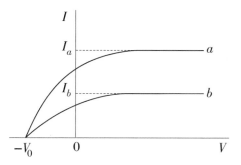

Fig. 2.12 Variation of the photoelectric current I with the applied voltage V, for two intensities of the incident light, a and b. The saturation current is directly proportional to the light intensity, but the stopping voltage V_0 is independent of the intensity

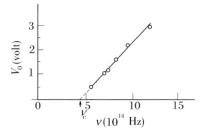

Fig. 2.13 Millikan's measurements of the stopping voltage in the photoelectric effect in sodium, as a function of the incident light frequency

corresponds to the kinetic energy T_{max} of the electrons that are emitted with the maximum energy. Thus, we have

$$T_{max} = e\, V_0, \tag{2.30}$$

where e is the absolute value of the electron charge. The second result of the experiments is obtained by measuring the I versus V curve with another light intensity b, which is one-half the intensity a. As shown by the two curves in Fig. 2.12, the maximum kinetic energy is independent of the light intensity of the incident light. Then, the question is what happens if the light frequency changes. Figure 2.13 shows the variation of the stopping potential V_0 as a function of the frequency of the incident light on a sodium cathode, measured by Millikan in 1914. These measurements show that there is a cutoff frequency v_c, below which there is no photoelectric effect. The value of this frequency varies from one material to another, and for sodium it is $v_c = 4.39 \times 10^{14}$ Hz ($\lambda \approx 683$ nm, which corresponds to red light, almost in the infrared).

The first observations of the photoelectric effect in the nineteenth century could not be explained with the classical wave theory of light, and for several years they were a challenge for physicists. Then, in 1905, Albert Einstein questioned the classical theory of light and used the ideas of quantization, initially proposed by Planck, to explain the photoelectric effect. Several years later, Millikan's measurements confirmed the quantum theory of light, and in 1921 Einstein received the Physics Nobel Prize. The rigorous formulation of the quantum model for the electromagnetic radiation was made possible many years later by the quantum field theory. One of the most important results of this theory is that an electromagnetic wave is **quantized in energy**. This means that if light has frequency v, it can only be generated with discrete energy values nhv, where n is an integer and h is Planck's constant ($h = 6.6262 \times 10^{-34}$ J s).

According to Einstein, the energy of electromagnetic radiation is quantized in the form of packets, called **photons**. When an electromagnetic wave has high energy, that is, much larger than hv, the number of photons is so large, that the discrete nature of the energy is not perceived. In this situation, the wave behaves classically. The energy of a radiation photon with frequency v, or angular frequency $\omega = 2\pi v$, is

$$E = h\,v = \hbar\,\omega, \tag{2.31}$$

where $\hbar = h/2\pi$. Photons have, in many situations, particle-like behavior. However, they are not common particles, since they only exist at the speed of light c, and have zero rest mass. The relation between energy and frequency, given by (2.31), allows the electromagnetic spectrum to be represented in energy units, such as eV. Using the values of Planck's constant and the electron charge, one can show that to convert Hz to eV it is necessary to multiply by 4.1357×10^{-15}. This conversion factor was used to make Fig. 2.4. Thus, the visible region of the spectrum has wavelength from 700 to 400 nm, frequency from 4.3×10^{14} Hz to 7.5×10^{14} Hz, and energy from 1.7 eV to 3.1 eV. Appendix B presents a conversion table between several units of energy.

We know from electromagnetic theory that the momentum p of a wave in a vacuum is related to its energy E by

$$p = \frac{E}{c}. \tag{2.32}$$

Using Eq. (2.10), with $n = 1$, and Eq. (2.31) in (2.32), we obtain the expression for the photon momentum,

$$\vec{p} = \hbar \vec{k}. \tag{2.33}$$

It follows, then, that in an electromagnetic wave of frequency ν and wave vector $\vec{k}$, both energy and momentum are quantized. It is important to draw attention to the fact that the theory does not impose a spatial quantization of the electromagnetic wave. In other words, there is nothing that limits the existence of a photon to a finite region of space. It is possible to have a flat electromagnetic wave, filling the entire space, corresponding to just one photon. Quantization is done only in terms of momentum and energy. It is possible, however, to have an electromagnetic wave confined to a limited region of space, as for example in the wave pulse in Fig. 2.5, containing only one photon. In this case, the photon is more like a particle, or a corpuscle.

In the photoelectric effect, photons are absorbed by a metal surface in an interaction process that results in the emission of electrons. Since there is energy conservation in the electron-photon interaction, when the electron is emitted from the metal surface, its kinetic energy is

$$T = h\nu - W, \tag{2.34}$$

where W is the work required to pull the electron out of the metal. Since there are electrons more attached to atoms than others, T varies from one electron to another. Electrons that are less bound are ejected from the surface with maximum kinetic energy. For these electrons we can write

$$T_{\max} = h\nu - W_0, \tag{2.35}$$

where W_0, a characteristic quantity of each metal called **work function**, is the minimum energy necessary for an electron to overcome the internal forces of attraction and cross the surface. Using the equations above, one can show that Einstein's theory explains the main observations of the photoelectric effect. First, we see that electrons "ejected" from the metal by the absorption of photons with energy

$$h \nu_c = W_0 \tag{2.36}$$

have zero kinetic energy. Therefore, as can be seen in Fig. 2.13, they correspond to $V_0 = 0$ and do not contribute to the photoelectric current. Thus, the value of the frequency ν_c in Eq. (2.36) is the cutoff frequency, which is independent of the incident light intensity. When $\nu > \nu_c$ and $V > -V_0$, there is a photoelectric current resulting from the emission of electrons. If the light intensity increases, the number of incident photons per unit of time increases, resulting in a proportional increase in the photoelectric current.

With Eqs. (2.30) and (2.35) we can obtain an expression for the stopping potential V_0 in terms of the photon frequency

$$e V_0 = h \nu - W_0. \tag{2.37}$$

Using (2.36), we obtain for $\nu \geq \nu_c$ the main result of Einsteins's theory for the photoelectric effect

$$V_0 = \frac{h}{e}(\nu - \nu_c), \tag{2.38}$$

which shows the linear variation of V_0 with ν, in agreement with Millikan's experimental data in Fig. 2.13.

The ideas of energy quantization and of the corpuscular nature of electromagnetic radiation had a profound impact on physics in the beginning of the twentieth century. Based on these ideas, several physicists began to look for effects of quantization and wave behavior in electrons. These efforts led to the formulation of quantum mechanics in 1926 by Schrödinger and independently by Heisenberg. The equations of quantum mechanics govern the behavior of electrons in atoms and solids and their knowledge is essential for the understanding of electronic phenomena that occur in different materials.

Example 2.1 In a photoelectric effect experiment, the material of the photo-cathode is lithium, whose work function is 2.3 eV, and the wavelength of the light used to illuminate the photocathode is 300 nm. Determine: (a) The cut-off frequency of lithium; (b) The value of the stopping potential.

(a) The relation between the work function and the cutoff frequency is given by Eq. (2.36). Thus, using the quantities above and the values of the physical constants given in Appendix B, in SI units, we have

$$\nu_c = \frac{W_0}{h} = \frac{2.3 \times 1.6 \times 10^{-19}}{6.63 \times 10^{-34}} \cong 5.5 \times 10^{14} \text{ Hz.}$$

(b) The stopping potential is related to the cutoff frequency and the light frequency by Eq. (2.38). We first need to calculate the light frequency

$$\nu = \frac{c}{\lambda} = \frac{3.0 \times 10^8}{300 \times 10^{-9}} = 10.0 \times 10^{14} \text{ Hz.}$$

Thus

$$V_0 = \frac{h}{e}(\nu - \nu_c) = \frac{6.63 \times 10^{-34}}{1.6 \times 10^{-19}} \times 4.5 \times 10^{14} = 1.86 \text{ V.}$$

2.4 The Electron as a Wave: Uncertainty Principle

As studied in the previous section, electromagnetic radiation is quantized in energy, acquiring, in certain situations, the nature of corpuscles, or particles. This concept was introduced in physics to explain an experimental result, the photoelectric effect, which could not be understood in a classic context. On the other hand, the concept that the electron, a particle in the classical sense, is also a wave, resulted from a theoretical hypothesis that was later confirmed experimentally. It was Louis de Broglie, in his doctoral thesis presented in 1924 to the University of Paris, who proposed the revolutionary idea of waves of matter. For his theory, de Broglie won the Physics Nobel Prize in 1929, after it was confirmed experimentally.

The hypothesis of de Broglie that the electron may behave as a particle and as a wave was inspired by the concept, already accepted at the time, that electromagnetic radiation has a particle-like behavior. He postulated that the electron is characterized by a frequency ν and wavelength λ, related to energy and momentum in exactly the same way as for photons. As in Eq. (2.31), the electron energy is expressed in the form

$$E = h\nu, \tag{2.39}$$

while the momentum is

$$p = h/\lambda. \tag{2.40}$$

These two equations lead to quite interesting results. Multiplying and dividing the right-hand side of Eq. (2.40) by 2π, and using the expression $k = 2\pi/\lambda$, we obtain the relation between momentum and wavevector

$$p = \hbar k, \tag{2.41}$$

which is the same as Eq. (2.33). Then comes an important question. If matter has a wave behavior, why we do not notice this in daily life? Consider an object of mass $m = 1.0$ kg moving with speed $v = 100$ m/s. With Eq. (2.40) we find that the corresponding wavelength is

$$\lambda = \frac{h}{p} = \frac{h}{mv} = \frac{6.6 \times 10^{-34}}{100} = 6.6 \times 10^{-36} \text{ m}. \tag{2.42}$$

Thus, the wavelength is very small compared to the typical dimensions of ordinary objects. Therefore, the diffraction and interference effects, which are characteristic of waves, are entirely negligible.

Now consider an electron with kinetic energy $T = 100$ eV. The corresponding wavelength is

$$\lambda = \frac{h}{p} = \frac{h}{\sqrt{2mT}} \approx 1.2 \times 10^{-10} \text{ m} = 1.2 \text{ \AA}. \tag{2.43}$$

This wavelength has the same order of magnitude as the size of the atoms and the distance between them in matter. For this reason, wave effects are very important on the atomic scale. The wave nature of electrons, proposed by de Broglie, was confirmed experimentally in 1927 by Clinton Davisson and his student Lester Germer, and independently by George Thomson. Both experiments consisted of making an electron fall on a thin metallic crystal film and observing the transmitted pattern. They found that the crystal, with its periodic atomic structure, behaves like a diffraction grating for the electron wave, producing interference fringes in the scattered electron beam, just like in optics experiments. Davisson and Thomson received the Physics Nobel Prize in 1939 for their pioneering studies.

The fact that electrons with energies of dozens of eV behave as waves with wavelength several orders of magnitude smaller than that of visible light, has an important practical application. When an electron beam falls on a material, the analysis of the scattered electrons makes it possible to observe much smaller details than it is possible with visible light in an optical microscope. This is the basic principle of operation of the electron microscopes. In the optical microscope, the observer sees the image of the object enlarged by means of glass lenses, which process the light scattered by the details of the analyzed material. As the minimum wavelength of visible light is around 300 nm, it is not possible to distinguish details with dimensions much smaller than this value. However, since electron microscopes use waves of electron beams, it is possible to observe details with dimensions of a few nanometers. In this case, the image of the object is formed by magnetic lenses (magnetic fields produced by coils with suitable shapes) and converted into electrical signals by means of detectors, that produce the images observed on computer displays.

Another important application of matter waves is in the study of crystalline solids by means of electron diffraction. Since the crystal lattice parameter is on the order of a few angstroms, diffraction only occurs with radiation of wavelength close to this value. It is then possible to use electron or neutron beams to study the atomic structure of materials or their elementary excitations, that will be described in the following section. One advantage of neutron scattering lies in the fact that since they are electrically neutral, their penetration into the solids is much larger than that of electrons. Another advantage is that neutrons have spin, so that they can be used to study the magnetic structure and magnetic excitations of materials.

Example 2.2 Calculate the energies and velocities of an electron beam and of a neutron beam, such that both have a wavelength of 2 Å.

From the relationship between energy and wavelength, given by Eq. (2.43), we obtain $T = h^2/(2m\lambda^2)$.

For the electron beam $m = 9.1 \times 10^{-31}$ kg, therefore

$$T = \frac{6.63^2 \times 10^{-68}}{2 \times 9.1 \times 10^{-31} \times 2^2 \times 10^{-20}} = 6.0 \times 10^{-18} \text{ J}.$$

The electron energy in eV is

$$T = \frac{6.0 \times 10^{-18}}{1.6 \times 10^{-19}} = 37.5 \text{ eV}.$$

To find the electron speed we use its relation to the kinetic energy $T = mv^2/2$. Thus

$$v = (2T/m)^{1/2} = \left(\frac{2 \times 6.0 \times 10^{-18}}{9.1 \times 10^{-31}}\right)^{1/2} = 3.6 \times 10^6 \text{ m/s}.$$

In the case of the neutron beam we use $m = 1.67 \times 10^{-27}$ kg

$$T = \frac{6.63^2 \times 10^{-68}}{2 \times 1.67 \times 10^{-27} \times 2^2 \times 10^{-20}} = 3.3 \times 10^{-21} \text{ J}$$

$$v = \left(\frac{2 \times 3.3 \times 10^{-21}}{1.67 \times 10^{-27}}\right)^{1/2} = 2.0 \times 10^3 \text{ m/s}.$$

The characteristics of an electron can be described quantitatively by means of a **wave function** Ψ, that will be formally introduced in the next chapter. If the electron

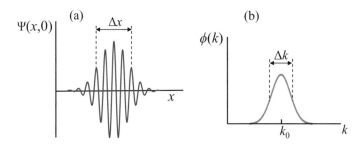

Fig. 2.14 **a** Wave packet that describes the state of a free particle located in a region of space. **b** Fourier transform of the wave packet shown in **a**

is a plane wave propagating in the x direction with a well-defined momentum p_0, it will have a wave number $k_0 = p_0/\hbar$, and its wave function can be written in the form

$$\Psi(x, t) = A \, e^{ik_0 x - i\omega t}, \tag{2.44}$$

where the angular frequency ω is related to the energy by $E = \hbar\omega$. The wave function (2.44) describes an electron that fills the entire space, and therefore has an infinitely large uncertainty Δx in its position. Clearly, it is very difficult to "produce" an electron with the wave function (2.44) throughout space. However, it is possible to have a more localized electron, with a wave function like the one in Fig. 2.14a. In this case, the electron does not have a well-defined wavevector k_0. It is described, say at $t = 0$, by a wave function $\Psi(x, 0)$ that is a superposition of plane waves with wave vectors close to k_0 and amplitudes $\phi(k)$ with maximum at $k = k_0$ and width Δk, as in Fig. 2.14b, in a similar way to the case of the electric field $E(x, 0)$ discussed in Sect. 2.1. An uncertainty in the determination of k implies an uncertainty in the electron momentum $\Delta p = \hbar \, \Delta k$. It is possible to show, by the Fourier transform of a function of the type in Fig. 2.14a, that $\Delta x \, \Delta k \approx 1$ (Problem 2.11). Thus, for an electron described by the wave function Ψ, this it means that

$$\Delta x \, \Delta p \approx \hbar. \tag{2.45}$$

The result in Eq. (2.45) has an important consequence in quantum physics. It means that in an experimental measurement of the position and momentum of an electron, there is an uncertainty Δx in the measured position, related to the uncertainty of the measured momentum Δp. This was postulated in 1927 by Heisenberg, and is known as the **uncertainty principle**. According to this principle, in an experiment it is not possible to determine the exact position value of the electron and its momentum simultaneously. There is a minimum uncertainty in the measurement process that is given by

$$\Delta x \, \Delta p \geq \hbar/2. \tag{2.46}$$

Notice that in the case of a plane wave function like that of Eq. (2.44), the momentum is well determined ($\Delta p = 0$), on the other hand $\Delta x \rightarrow \infty$.

There is another version of the uncertainty principle, related to the determination of the electron energy E and the time interval Δt necessary to measure it. According to Heisenberg, if the measurement is performed in a finite time interval Δt, there is uncertainty ΔE in determining E given by

$$\Delta E\,\Delta t \geq \hbar/2. \tag{2.47}$$

The uncertainty principle, represented by Eqs. (2.46) and (2.47), was proposed by Heisenberg at a time when the concept of the electron wave function was not yet known. It had a profound impact on physics and also generated many philosophical speculations. In fact, it is a natural consequence of the wave character of the particles of matter, whose formalization is given by quantum mechanics, which will be studied in the next chapter.

2.5   Phonons and Other Elementary Excitations in Solids

The quantization of the electromagnetic wave and of the electron wave are just two examples of a general phenomenon that occurs with any type of wave. This phenomenon is observed experimentally through several effects and has a rigorous explanation in quantum field theory. Any wave is formed by energy "packages", called quanta (plural of quantum) of energy. Therefore, the energy of a wave is discrete and has a value equal to a multiple of h. The quantum of a wave has both wave and particle behavior, having energy and momentum given by

$$E = \hbar\omega, \tag{2.48}$$

$$\vec{p} = \hbar\vec{k}, \tag{2.49}$$

which are identical relations to those seen previously for electromagnetic waves and for electrons. Excitations in a solid have a wave character and are therefore quantized. The quanta of the various waves are called elementary excitations. Thus, the quantum of an elastic wave in a crystal is called a **phonon**.

There are many other elementary excitations in solids, usually with names ending in **on**. The quantum of a spin wave in a magnetic material is called **magnon**. The quantum of a plasma wave in a metal or semiconductor is called **plasmon**. Other excitations that will not be presented in this book are excitons, polarons, polaritons, helicons, plasmaritons, rotons, etc.

Problems

2.1   Apply the rotational operator $(\nabla\times)$ to Eq. (2.3), use Eq. (2.4), together with the relationships between the fields and the vector identity

$$\nabla \times (\nabla \times \vec{A}) = \nabla(\nabla \cdot \vec{A}) - \nabla^2 \vec{A}$$

to show that in a medium without charges or currents, the electric field obeys the wave equation, Eq. (2.5).

2.2   Consider an electric field with amplitude that varies in time and space with a function $E(x,t)$.

   (a)   Show that this function will be a solution of the wave equation, Eq. (2.6), if the argument has the form $E(x,t) = f(x - vt) + g(x + vt)$, where f and g are any differentiable functions.

   (b)   Choose a function $f(x)$ at $t = 0$ which satisfies the wave equation and make a qualitative plot of its variation with x at any two instants of time $t > 0$. Interpret the result.

2.3   Consider the electric field $\vec{E}(x, t) = \hat{y}\, E_0 \cos(kx - \omega t)$ of an electromagnetic plane wave.

   (a)   Show that this field is a solution of the wave equation by the direct substitution in Eq. (2.6).

   (b)   Show that this function is a solution to the wave equation, since it is a particular case of the solution obtained in Problem 2.2.

   (c)   Make a qualitative plot of $E(x, t)$ as a function of x for $t = 0$, and obtain the relationship of the distance between two consecutive maxima of the wave with ω and k.

   (d)   Make the plot of $E(x, t)$ as a function of x for $t = \Delta t$ and relate the speed of a maximum, $\Delta x/\Delta t$, with ω and k.

2.4   Consider an electromagnetic wave pulse of Gaussian form at the instant $t = 0$

$$E(x, 0) = E_0\, e^{-x^2/2L^2} \cos k_0 x.$$

   (a)   Make a semi-quantitative plot of E as a function of x for $E_0 = 1$ (arbitrary units) and $L = 5 \times 2\pi/k_0$. Use a computer to make a quantitative plot using $k_0 = 1$.

   (b)   Determine the function that describes the pulse at an arbitrary instant t, by imposing that $E(x, t)$ satisfies Eq. (2.6).

   (c)   Repeat item (a) for the field obtained in item (b).

2.5   Show that in the limit of long wavelengths, $\lambda \gg a$, Eq. (2.22) reduces to a wave equation for a continuous variable u,

$$\frac{\partial^2 u}{\partial t^2} = v^2 \frac{\partial^2 u}{\partial x^2}.$$

2.6   The lattice vibrations of a certain crystal can be described by the one-dimensional chain model given by Eq. (2.22), with atoms of atomic weight 56, lattice parameter 2.9 Å, and elastic constant $C = 10^4$ g/s^2.

   (a)   Calculate the propagation velocity of the elastic wave in the chain in the limit of long wavelengths, $\lambda \gg a$ (or $ka \ll 1$), in cm/s, and compare it with the speed of light.
   (b)   Calculate the maximum value of the atomic vibration frequency in the chain in rad/s and in Hz.

2.7   From the measurements of the photoelectric effect shown in Fig. 2.13 calculate.

   (a)   The work function of sodium, in eV.
   (b)   The stopping potential V_0 of a photoelectric cell with a sodium photocathode, illuminated by light with wavelength $\lambda = 350$ nm.

2.8   A photoelectric effect measurement apparatus uses a cell with an aluminum photocathode, whose work function is 4.2 eV. The ultraviolet light used has a wavelength of 180 nm.

   (a)   What is the cutoff frequency of aluminum?
   (b)   What is the stopping potential of aluminum for this wavelength?
   (c)   Calculate the kinetic energy of the fastest electron emitted.
   (d)   What is the energy of an electron in aluminum that is ejected with zero speed?

2.9   A light-emitting diode made with the semiconductor GaP emits light with a wavelength of 549 nm, with a power 1 μW.

   (a)   What is the energy, in eV, of the photons emitted by the diode?
   (b)   How many photons per second are emitted by the diode?

2.10   In a photoelectric effect experiment with a laser, light with intensity of 1.0 W and a certain frequency is focused on a lithium photocathode, whose work function is 2.3 eV.

   (a)   What is the stopping potential for a frequency with value twice the cutoff frequency?
   (b)   Suppose that for every ten photons that reach the photocathode, one electron is emitted, and that the positive potential applied between anode and cathode is such that the current is saturated. Calculate the value of this current, in ampere.

2.11 An electron is described by a wave function in the form of a Gaussian packet, given at $t = 0$ by

$$\Psi(x, 0) = A\, e^{-x^2/2L^2}\, e^{ik_0 x}.$$

(a) Make a qualitative plot of $|\Psi(x, 0)|^2$ as a function of x.

(b) The width of the wave packet can be characterized by $\Delta x = \sqrt{\langle \Delta x^2 \rangle}$, where $\langle \Delta x^2 \rangle$ is the mean square deviation of the function relative to its average value x_m,

$$\langle \Delta x^2 \rangle = \int |\Psi(x, 0)|^2 (x - x_m)^2 dx.$$

Calculate Δx for the electron.

(c) Calculate the Fourier transform of the wave function $\phi(k)$, using the definition of Eq. (2.16), with E replaced by Ψ. Make a qualitative plot of $|\phi(k)|^2$ and calculate the width Δk of the packet in the same way as in item b).

(d) Calculate the product $\Delta x \, \Delta k$ and interpret the result. Hint, use

$$\int_{-\infty}^{\infty} e^{-y^2} dy = 2 \int_{-\infty}^{\infty} y^2 e^{-y^2} dy = \sqrt{\pi}.$$

2.12 An electron and a photon each have a wavelength of 3.0 Å. Calculate the energy and momentum of each one and interpret the result.

2.13 The maximum resolution of a microscope is limited by the wavelength of the radiation used, that is, the shortest distance that can be observed is approximately equal to the wavelength. What should be, in eV, the electron energy in an electron microscope so that its resolution is 10 Å?

2.14 Show that the uncertainty relation for a particle, in terms of the uncertainties in position Δx and in wavelength $\Delta \lambda$ that can be measured simultaneously, is given by $\Delta x \, \Delta \lambda \geq \lambda^2/4\pi$.

2.15 Consider that the uncertainty in the measurement of the photon wavelength is $\Delta \lambda / \lambda = 10^{-7}$. Calculate the uncertainty in the measurement of the photon position for the following wavelengths: 5×10^{-4} Å (γ-ray); 5 Å (X-ray); and 500 nm (visible light).

2.16 The atomic vibrations of certain diatomic crystal can be described by the one-dimensional model studied in Sect. 2.2, with atoms of atomic weights 39 and 80, and elastic constant $C = 10^4$ g/s^2.

(a) Calculate the value of the optical mode vibration frequency with $k = 0$, in rad/s and in Hz.

(b) What is the phonon energy corresponding to the vibration of item (a) in eV?

(c) For the phonon of item (b) to be excited resonantly by a photon of the same energy, what should be the wavelength of this photon and in which region of the electromagnetic spectrum is it situated?

Further Reading

J.J. Brehm, W.J. Mullins, *Introduction to the Structure of Matter—A Course in Modern Physics* (John Wiley & Sons, New York, 1989)

D. Halliday, R. Resnick, J. Walker, *Fundamentals of Physics* (John Wiley & Sons, New York, 2015)

C. Kittel, *Introduction to Solid State Physics* (John Wiley & Sons, New York, 2004)

S.M. Riad, I.M. Salama, *Electromagnetic Fields and Waves: Fundamentals of Engineering* (McGraw-Hill Book Co., New York, 2019)

F.G. Smith, T.A. King, D. Wilkins, *Optics and Photonics: An Introduction* (John Wiley & Sons, New York, 2007)

S. T. Thornton, A. Rex, *Modern Physics for Scientists and Engineers* (Brooks/Cole, Cengage Learning, Boston, 2013)

P.A. Tipler, R.A. Llewellyn, *Modern Physics* (W. H. Freeman & Co., New York, 2012)

S.M. Wentworth, *Fundamentals of Electromagnetics with Engineering Applications* (John Wiley & Sons, New York, 2013)

Chapter 3
Quantum Mechanics: Electrons in the Atom

This chapter presents the basic concepts of quantum mechanics that are necessary for the understanding of physical properties of materials and phenomena that underly the operation of electronic devices. It is designed mainly to engineering students with no background in modern physics. Initially we present postulates and arguments that lead to the Schrödinger equation, that is essential to calculate the properties of atoms and materials. Then we apply it to the "motion" of particles in simple potentials and show two important purely quantum effects, the energy quantization in potential wells and the tunnel effect. Finally, we present the solution of the Schrödinger equation for the hydrogen atom, that is key for the understanding of atoms with many electrons and the periodic table of the chemical elements.

3.1 The Postulates of Quantum Mechanics

A major step for the development of quantum mechanics was taken in 1913 when Niels Bohr explained the emission spectrum of the hydrogen atom. Inspired by the ideas of quantization of Planck and Einstein, Bohr proposed a model for the atom based on some assumptions: 1-An electron in an atom moves around the nucleus under the influence of the Coulomb attraction only in discrete circular orbits, such that the angular momentum is quantized, $L = n\hbar$, where n is an integer; 2-In each orbit the electron is in a stationary state with energy given by the electrostatic interaction with the nucleus; 3-The electron may undergo a transition from an initial state with energy E_i to another state with energy E_f by either absorbing ($E_f > E_i$) or emitting ($E_f < E_i$) a photon with energy $h\nu = |E_f - E_i|$.

The Bohr model was very successful in explaining quantitatively the puzzling experimental observations of the discrete radiation spectrum of the atoms in a hydrogen gas under electric discharge. Despite its success, the model left many physicists uneasy because of the ad-hoc imposed quantization. There was still skepticism with the quantization of the thermal oscillations proposed by Planck in 1900

© The Author(s), under exclusive license to Springer Nature Switzerland AG 2022
S. M. Rezende, *Introduction to Electronic Materials and Devices*,
https://doi.org/10.1007/978-3-030-81772-5_3

and of the light proposed by Einstein in 1905. However, Bohr's theory for the atom together with de Broglie's postulate for the wave nature of matter, paved the way for the development of quantum mechanics, announced in 1926 by Werner Heisenberg and independently by Erwin Schrödinger. The two formulations were different, but soon it was found that they were equivalent and led to identical results. Due to the success of quantum mechanics in explaining the atom, the Physics Nobel Prize was awarded to Bohr in 1922, to Heisenberg in 1932, and to Schrödinger in 1933.

Here we present only the formulation of Schrödinger which is based on an equation for a complex wave function. Some authors try to justify this equation using several plausibility arguments. Of course, some of them are useful in understanding several aspects of quantum mechanics. However, the Schrödinger equation is considered a fundamental equation of physics, that cannot be derived from classical laws. Its best justification is the fact that its results explain experimental observations and quantitative measurements. Notice that most results presented in this section can be rigorously derived by the linear vector space formalism which underpins the modern quantum theory and may be found in more advanced texts on the subject. It is particularly more adequate for the cases where there is no classical analog to genuinely quantum objects such as the *spin*, a key quantity in devices based on magnetic materials, discussed in Chap. 9. For our purposes, here we follow the more intuitive wave approach. Quantum mechanics is based on four postulates, presented in the following sub-sections.

3.1.1 The Wave Function

The state of an electron, or any material "particle", is characterized by a complex **wave function** $\Psi(x, t)$. In three dimensions, it is actually a function of $\vec{r}$, not of x, but we will keep it for simplicity. The function Ψ and its derivatives with respect to x and t are continuous, finite and unique. If at a certain instant t we make a measurement to determine the location of a particle with wave function $\Psi(x, t)$, the **probability** of finding the particle between x and $x + dx$ is given by $P(x, t)\, dx$, where

$$P(x, t) = \Psi^*(x, t)\, \Psi(x, t) = |\Psi(x, t)|^2. \tag{3.1}$$

As an example, consider a particle with a wave function in the form of a wave packet, such as in Fig. 2.14a. The probability of finding the particle within the region Δx is close to unity, so that the wave packet is a good wave function for a reasonably well localized particle.

For any wave function, the probability of finding the particle in the entire space is 1. Thus, we have

$$\int_{-\infty}^{\infty} P(x, t)\, dx = \int_{-\infty}^{\infty} \Psi^*(x, t)\, \Psi(x, t)\, dx = 1. \tag{3.2}$$

This condition is sufficient to determine the amplitude of the wave function with a known shape. We say that the wave function that satisfies Eq. (3.2) is **normalized**.

3.1.2 Quantum Operators

The wave function allows the calculation of the probability of "localization" of a particle at any time. In other words, the wave function alone only gives information on the position of a particle. Since in physics it is necessary to calculate other quantities related to the motion or properties of a particle, the next question is how to extract more information from the wave function. The answer is in the concept of **quantum operator**. Each physical quantity corresponds to a mathematical operator, which operates on the wave function.

The operator for the **linear momentum** in one dimension, say x, is

$$p_{op} = -i\hbar \frac{\partial}{\partial x}, \tag{3.3}$$

where i is the imaginary unit. Let us apply this operator to the plane wave function of a free electron given by Eq. (2.44). The result is

$$p_{op}\Psi(x,t) = -i\hbar \frac{\partial}{\partial x} A\, e^{ik_0 x - i\omega t} = \hbar k_0\, \Psi(x,t). \tag{3.4}$$

Thus, application of the momentum operator (3.3) to the wave function of a free electron gives the momentum proposed by de Broglie, Eq. (2.41), multiplied by the wave function. This is a strong indication of the consistency of Eq. (3.3) with the wave nature of the electron. It is important to note that when an operator is applied to a wave function, in general, the value of the associated physical quantity is not obtained directly as in Eq. (3.4). When an operator applied to Ψ reproduces the wave function multiplied by a constant, we say that Ψ is an **eigenfunction** of the operator. Thus, if

$$p_{op}\Psi(x,t) = p\,\Psi(x,t), \tag{3.5}$$

Ψ is an eigenfunction of p_{op}, and p is called the **eigenvalue**. In this case, the momentum of the particle is well determined, that is, its uncertainty is zero. This is the case for the electron described by the wave function in Eq. (2.44).

In the more general case of three dimensions, the momentum operator is defined by

$$\vec{p}_{op} = -i\hbar \nabla = -i\hbar \left(\hat{x} \frac{\partial}{\partial x} + \hat{y} \frac{\partial}{\partial y} + \hat{z} \frac{\partial}{\partial z} \right). \tag{3.6}$$

Another important operator is that of energy, given by

$$E_{op} = i\hbar \frac{\partial}{\partial t}. \tag{3.7}$$

Application to a free electron gives

$$E_{op}\Psi(x, t) = i\hbar \frac{\partial}{\partial t} A\, e^{ik_0 x - i\omega t} = \hbar\omega\,\Psi(x, t). \tag{3.8}$$

Thus, the wave function of a free electron is also an eigenfunction of the energy, with eigenvalue

$$E = \hbar\omega, \tag{3.9}$$

which is also in agreement with de Broglie's theory. From these operators it is possible to build others. For example, the kinetic energy operator is

$$T_{op} = \frac{1}{2m}\vec{p}_{op} \cdot \vec{p}_{op} = -\frac{\hbar^2}{2m}\nabla \cdot \nabla = -\frac{\hbar^2}{2m}\nabla^2 \tag{3.10}$$

where m is the particle mass and ∇^2 is the Laplacian operator. In Cartesian coordinates the Laplacian is

$$\nabla^2 = \frac{\partial^2}{\partial x^2} + \frac{\partial^2}{\partial y^2} + \frac{\partial^2}{\partial z^2}. \tag{3.11}$$

In the particular case of one-dimension x, the kinetic energy operator is

$$T_{op} = -\frac{\hbar^2}{2m}\frac{\partial^2}{\partial x^2} \tag{3.12}$$

Table 3.1 shows the quantum operators corresponding to some classical quantities. The important point is that having the wave function of a particle, several physical quantities associated with the particle motion or properties can be readily calculated. The operators for other quantities can be derived from the ones above. For instance, since the angular momentum of a particle is $\vec{r} \times \vec{p}$, the corresponding operator is given by the vector product of the position vector with the momentum operator in Eq. (3.6), as in Table 3.1.

Table 3.1 Correspondence between some quantum operators and classical quantities

Quantity	Classical quantity	Quantum operator
Position (1D)	x	x
Position (3D)	$\vec{r}$	$\vec{r}$
Linear momentum (1D)	p_x	$-i\hbar\,\partial/\partial x$
Linear momentum (3D)	$\vec{p}$	$-i\hbar\,\nabla$
Energy	E	$-i\hbar\,\partial/\partial t$
Kinetic energy	T	$-(\hbar^2/2m)\nabla^2$
Angular momentum	$\vec{L}$	$-i\hbar\,\vec{r}\times\nabla$

3.1.3 Expectation Value of a Quantity

The next important question in the formulation of quantum mechanics is how to extract a quantitative value for a variable associated with a particle described by a given wave function. Before answering this question, we note that quantum mechanics deals with probabilities, so that measurements also have to be done considering this aspect. This means that if a particle is described by a wave function $\Psi(x, t)$, the measurement of some of its properties should be done several times at the same value of t, and recording the observed values. Then, the *average* of the observed values is considered the result of the measurement. Correspondingly, a certain quantity associated with an operator acting on a wave function, should in general give an average value, that is called *expectation value*. This is the most probable value, that is an average value in the statistical sense. Thus, since the probability of finding a particle with wave function $\Psi(x, t)$, between x and $x + dx$ is given by $P(x, t)\,dx$, where $P(x, t)$ is given by Eq. (3.1), the expectation value $\langle Q \rangle$ of a quantity corresponding to the operator Q_{op} is, in 3D,

$$\langle Q \rangle = \int_{-\infty}^{\infty} \Psi^* Q_{op} \Psi \, dx \, dy \, dz, \qquad (3.13)$$

where we have simplified the notation for a triple integral. It is also common to represent the expectation value by $\bar{Q}$. Notice that in 3D, the condition of normalization is expressed by

$$\int_{-\infty}^{\infty} \Psi^*(\vec{r}, t)\, \Psi(\vec{r}, t)\, dx \, dy \, dz = 1. \qquad (3.14)$$

From Eqs. (3.13) and (3.14) we can see that if Ψ is an eigenfunction of Q_{op}, then the expectation value of Q_{op} is the eigenvalue itself. In this case, the value of the quantity can be determined precisely, with zero uncertainty.

3.1.4   The Schrödinger Equation

The evolution of the wave function of a particle in a physical system is determined by a differential equation proposed by Schrödinger. As stated earlier, this equation cannot be derived from the laws of classical physics, but it can be understood with plausibility arguments. The Schrödinger equation states that the total energy of a particle, or a system of particles, with wave function $\Psi(\vec{r}, t)$, expressed in terms of the energy operator acting on the wave function, is the sum of the kinetic and potential energies. The equation can be written as

$$(T_{op} + V_{op})\Psi(\vec{r}, t) = E_{op}\Psi(\vec{r}, t). \tag{3.15}$$

Using the forms of the operators in Eqs. (3.7) and (3.10), the Schrödinger equation for a particle of mass m becomes

$$-\frac{\hbar^2}{2m}\nabla^2\Psi(\vec{r}, t) + V_{op}\Psi(\vec{r}, t) = i\hbar\frac{\partial\Psi(\vec{r}, t)}{\partial t}. \tag{3.16}$$

where the operator V_{op} represents the interaction potential to which the particle is subjected in a given physical situation, that varies from one problem to another. If the particle motion is restricted to the coordinate x, Schrödinger equation reduces to

$$-\frac{\hbar^2}{2m}\frac{\partial^2\Psi(x, t)}{\partial x^2} + V(x, t)\Psi(x, t) = i\hbar\frac{\partial\Psi(x, t)}{\partial t}. \tag{3.17}$$

From now on we will no longer use the subscript "op" in the operator to simplify the notation. Equation (3.16) is a differential equation of partial derivatives that has, for each potential, an infinite number of solutions. The solutions for each problem are determined by the boundary conditions that Ψ and $\partial\Psi/\partial x$ must satisfy, as well as the normalization condition (3.2) that "ties" the amplitudes of the wave function. Equation (3.16) has another important characteristic, it is a **linear** differential equation, since the operators and the wave functions have power one. An important property of a linear equation is that the superposition of two or more solutions, is also a solution of the equation (see Problem 3.1).

3.2   The Time-Independent Schrödinger Equation

We shall now consider the case where the potential energy to which a particle is subjected to does not vary in time, which is the most common physical situation. In this case we can use a standard procedure for solving partial differential equations, namely, separation of variables. This consists in searching for solutions in the form of products of functions such that each one contains only one of the independent variables involved in the equation.

In the case of the Schrödinger equation, if V is a function of the particle position only, it is possible to find a solution for Eq. (3.16) in the form

$$\Psi(\vec{r}, t) = \psi(\vec{r})\, \phi(t), \tag{3.18}$$

where $\psi(\vec{r})$ and $\phi(t)$ are functions, respectively, of $\vec{r}$ and t only. Substitution of (3.18) into (3.16) leads to

$$-\frac{\hbar^2}{2m}\nabla^2 \psi(\vec{r})\, \phi(t) + V(\vec{r})\, \psi(\vec{r})\, \phi(t) = i\hbar\, \frac{\partial \phi(t)}{\partial t} \psi(\vec{r}). \tag{3.19}$$

Dividing both sides by the product $\psi(\vec{r})\, \phi(t)$ we obtain

$$\frac{1}{\psi(\vec{r})}\left[-\frac{\hbar^2}{2m}\nabla^2 \psi(\vec{r}) + V(\vec{r})\, \psi(\vec{r})\right] = \frac{1}{\phi(t)}\left[i\hbar\, \frac{\partial \phi(t)}{\partial t}\right]. \tag{3.20}$$

Notice that the left-hand side of (3.20) does not depend on t, while the right-hand side does not depend on $\vec{r}$. Thus, the common value of the two sides cannot depend on t or $\vec{r}$, and therefore it must be a constant. The equation obtained by equating the right-hand side of (3.20) to a constant E is

$$\frac{1}{\phi(t)}\left[i\hbar\, \frac{\partial \phi(t)}{\partial t}\right] = E.$$

This equation gives

$$\frac{d\phi(t)}{dt} = -i\frac{E}{\hbar}\phi(t), \tag{3.21}$$

where we have replaced the symbol of partial derivative by the one of total derivative, since $\phi(t)$ is a function of t only. The solution of (3.21) is

$$\phi(t) = \exp\left(-i\frac{E}{\hbar}t\right). \tag{3.22}$$

This shows that $\phi(t)$ is an oscillating function in time with angular frequency $\omega = E/\hbar$. Therefore, we can associate the constant introduced in the separation of variables with the **energy** of the state whose wave function is the solution of Eq. (3.16).

The equation obtained by equating the left-hand side of (3.20) to E is a differential equation in the spatial variables,

$$-\frac{\hbar^2}{2m}\nabla^2\psi(\vec{r}) + V(\vec{r})\,\psi(\vec{r}) = E\,\psi(\vec{r}). \qquad (3.23)$$

This is known as the **time-independent Schrödinger equation**. The total energy operator is also called the **Hamiltonian** of the system, so that Eq. (3.23) can be written in the form

$$H\,\psi(\vec{r}) = E\,\psi(\vec{r}), \qquad (3.24)$$

where

$$H = -\frac{\hbar^2}{2m}\nabla^2 + V(\vec{r}). \qquad (3.25)$$

is the Hamiltonian operator. Equation (3.24) is called an eigenvalue equation. Its solution gives the **eingenfunctions** of the Hamiltonian, as well as the corresponding energy eigenvalues. Hence, the complete solution of Eq. (3.16) is

$$\Psi(\vec{r}, t) = \psi(\vec{r})\,\exp\left(-i\frac{E}{\hbar}t\right), \qquad (3.26)$$

where $\psi(\vec{r})$ is the eigenfunction of Eq. (3.24) with energy E. Note that the probability density of finding the particle with the wave function (3.26) at the position $\vec{r}$ at an instant t, given by

$$P(\vec{r}, t) = \Psi^*(\vec{r}, t)\,\Psi(\vec{r}, t) = |\psi(\vec{r})|^2 \qquad (3.27)$$

is independent of time. This means that if the particle is initially in an eigenstate of the Hamiltonian, it remains indefinitely in that state. We say that the particle in this situation is in a **stationary state**. In the following subsections we shall use the time-independent Schrödinger equation to calculate the properties of a particle in some simple situations.

3.3   Simple Applications of Quantum Mechanics

3.3.1   Free Electron

The simplest example for application of the Schrödinger equation is that of a uniform potential, $V(r) = $ constant. Classically, a particle in such potential is subjected to a force $\vec{F} = -\nabla V = 0$. Therefore, it is a free particle and moves at a constant speed. Since the value of the constant potential does not influence the motion, we consider

$V = 0$. Assuming that the electron moves in one dimension along the x direction, Eq. (3.23) becomes

$$-\frac{\hbar^2}{2m}\frac{\partial^2\psi(x)}{\partial x^2} = E\,\psi(x). \qquad (3.28)$$

The solution of this equation can be written in the form

$$\psi(x) = A\,e^{ikx} + B\,e^{-ikx}. \qquad (3.29)$$

where A and B are constants to be determined by the boundary conditions and k is a wave number. Substituting (3.29) in (3.28) one can see that the energy is related to k by

$$E = \frac{\hbar^2 k^2}{2m}. \qquad (3.30)$$

Using for $\phi(t)$ the expression (3.22), we obtain with the first term in Eq. (3.29) a wave function like (2.14), corresponding to a plane wave propagating in the positive x direction, that is

$$\Psi(x, t) = A\,e^{ikx - i\omega t}. \qquad (3.31)$$

This is the wave function of a free electron, moving at a constant speed in the $+x$ direction, as had been anticipated in Eq. (2.44) obtained from de Broglie's hypothesis. Likewise, the wave function obtained with the second term in (3.29) corresponds to a free electron moving in the $-x$ direction. In both cases, the wave functions are eigenfunctions of the momentum operator, and the momentum eigenvalues are $p = \pm\hbar k$. In three dimensions the momentum is $\vec{p} = \hbar\vec{k}$, and is related to the energy by Eq. (3.30), which can be written as

$$E = \frac{p^2}{2m}. \qquad (3.32)$$

As expected, the energy is exactly the kinetic energy, because we considered a free particle in zero potential. Note that in this problem there is no condition that restricts the value of the energy E. Thus, E can vary continuously from 0 to infinity. Note that from Eq. (3.30) one can obtain the **dispersion relation** $\omega(k)$ for the free electron. Using $E = \hbar\omega$ we have

$$\omega(k) = \frac{\hbar k^2}{2m}, \qquad (3.33)$$

which is the parabolic function illustrated in Fig. 3.1. The particle with the wave function (3.31) behaves like a plane wave that fills the entire space, having wavelength

Fig. 3.1 Parabolic
dispersion relation for a free
electron

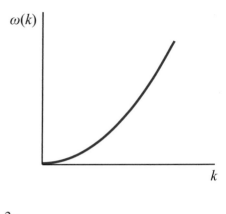

$$\lambda = \frac{2\pi}{k}, \tag{3.34}$$

where

$$k = \sqrt{2mE}/\hbar. \tag{3.35}$$

An important remark is that the results above are in complete agreement with
de Broglie's theory for the wave nature of the electron, studied in Chap. 2. Notice
that if the wave function (3.31) is normalized through Eq. (3.2) we obtain $A \rightarrow 0$.
Therefore, it does not make sense to normalize (3.31) across the whole space. In fact,
there is not much physical sense in considering a particle in the entire space. However,
as we saw in Sect. 2.1, plane waves of type (3.31) can be used mathematically to
build a wave packet like the one in Fig. 2.14a, which represents a particle confined
to a small region in space. So, let us consider a free particle represented by a wave
function in the form of a wave packet. As discussed in Sect. 2.1, the packet propagates
with group velocity

$$v_g = \left.\frac{\partial \omega}{\partial k}\right|_{k_0}. \tag{3.36}$$

Using the dispersion relation (3.33) we obtain the velocity of a particle represented
by a wave function as in Fig. 2.14

$$v_{part} = v_g = \frac{\hbar}{m} k_0. \tag{3.37}$$

The momentum of this particle is then

$$p = m v_{part} = \hbar k_0. \tag{3.38}$$

which is also in agreement with the concept introduced by de Broglie.

3.3.2 Particle in an Infinite Square-Well Potential

We shall now study the properties of a particle with mass m, that moves "freely" in one dimension, confined to a region in space by an infinite square well potential, illustrated in Fig. 3.2. This case is often called a **particle in a box**. The square well represents, qualitatively, the situation of a free electron confined to the interior of a metallic crystal, since the electron cannot simply jump out at the surfaces. In fact, as we know, the potential well in a crystal is not infinite, because the electron can be ejected from the solid, as in the photoelectric effect. The first task here is to solve the time-independent Schrödinger equation to find the stationary states. The eigenfunctions of Eq. (3.24) for this problem are determined in the same way as for a uniform potential, but in this case we have

$$V(x) = \begin{cases} 0 & 0 < x < L \\ \infty & x \le 0; \quad x \ge L \end{cases}. \qquad (3.39)$$

In the range $0 < x < L$, the equation is identical to that of the free electron, and therefore its solution is the same as in (3.29)

$$\psi(x) = A\,e^{ikx} + B\,e^{-ikx}, \quad (0 < x < L), \qquad (3.40)$$

and the energy eigenvalue is related to k as in (3.30), $E = (\hbar k)^2/2m$. For $x \le 0$ and $x \ge L$ the wave function must be zero, $\psi = 0$, because the infinitely large potential does not allow the electron to be in this region. Since the electron momentum, given by $-i\hbar d\psi/dx$, cannot be infinite, ψ must be a continuous function in x, and therefore

$$\psi(x = 0) = 0, \quad \text{and} \quad \psi(x = L) = 0. \qquad (3.41)$$

These are the boundary conditions that are used to find the coefficients in Eq. (3.40). Using the first condition in (3.41) in Eq. (3.40) we obtain $B = -A$, hence we can write the eigenfunctions of the infinite square-well potential as

Fig. 3.2 Infinite square-well potential

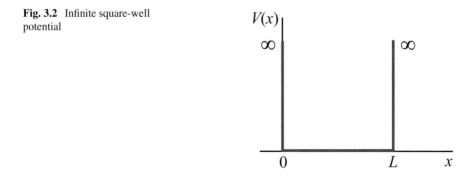

$$\psi_n(x) = A_n \sin k_n x. \tag{3.42}$$

The second condition in (3.41) imposed on (3.42) restricts the values of k_n to

$$k_n = n\frac{\pi}{L}, \tag{3.43}$$

where n is an integer ($n = 1, 2, 3, \ldots$). Thus, unlike the free electron, the electron in the infinite square well potential cannot have an arbitrary energy value. The energy, obtained using the condition (3.43) in $E = (\hbar k)^2/2m$, can only assume discrete values, given by

$$E_n = \frac{\hbar^2\pi^2}{2mL^2}n^2, \tag{3.44}$$

where n is called a **quantum number**, because it determines the **quantized values** of the energy. The values E_n are the energy eigenvalues, and the functions ψ_n are the eigenfunctions of Eq. (3.24) for the infinite square-well potential. Figure 3.3 shows a representation of the wave functions and the corresponding energies, for the first four values of the quantum number n.

Some results of this simple problem are, at least qualitatively, of very general validity for potential wells, regardless of their detailed form. They are:

1. Particles that have a motion confined to a limited region of space have stationary states with discrete energies, $i.e.$, they have **quantized** energies. Mathematically, this results from the boundary conditions imposed to the wave functions at the

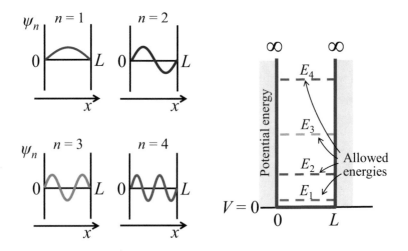

Fig. 3.3 Wave functions and corresponding energies of a particle in an infinite square well potential, for the first four values of the quantum number n

limits of the region. This is the same reason for which a string attached at the ends can only vibrate at certain discrete frequencies. The state with lowest energy is called **ground state**.

2. The wave function of a state confined to a region of space has a certain number of zeros, which increases with increasing energy.

Example 3.1 A particle is in the ground state in an infinite square well potential of width L. Calculate: (a) The expectation values of the position x and of the momentum p_x; (b) The mean square deviations of x and p_x.

(a) The wave function of the particle in the ground state is given by (3.42) and (3.43) with $n = 1$, $\psi = A \sin(\pi x/L)$. For the normalization of the wave function we use condition (3.1),

$$\int_0^L A^2 \sin^2(\frac{\pi}{L}x)\,dx = A^2 \frac{L}{\pi} \int_0^\pi \sin^2 \alpha \, d\alpha = 1,$$

where $\alpha = \pi x/L$. Using the relation $\sin^2 \alpha = (1 - \cos 2\alpha)/2$, this gives

$$A^2 \frac{L}{\pi} \int_0^\pi \frac{1}{2}\,d\alpha - A^2 \frac{L}{2\pi} \int_0^\pi \cos 2\alpha \, d\alpha = 1.$$

Since the second term is zero, this gives

$$A^2 \frac{L}{\pi}\frac{\pi}{2} = 1, \text{ so that } A = \sqrt{\frac{2}{L}}.$$

The expectation value of the position x is calculated with Eq. (3.14). Thus, we have

$$\bar{x} = \int_{-\infty}^{\infty} \psi^* x \psi \, dx = \int_0^L A^2 x \sin^2(\frac{\pi}{L}x)\,dx = \frac{2}{L}\int_0^L x \sin^2(\frac{\pi}{L}x)\,dx,$$

so that

$$\bar{x} = \frac{1}{L}\int_0^L x\,dx - \frac{1}{L}\int_0^L x \cos(\frac{2\pi}{L}x)\,dx.$$

The second term can be calculated with the following integration by parts

$$\int x \, \cos(ax) \, dx = \frac{1}{a^2} \cos(ax) + \frac{x}{a} \sin(ax),$$

where $a = 2\pi/L$. We can see that the integration from 0 to L in the second term vanishes. Thus

$$\bar{x} = \frac{1}{L} \left[\frac{x^2}{2} \right]_0^L = \frac{L}{2}.$$

This result is, in some way, expected, because a particle that moves freely between 0 and L has an average position at $L/2$.

The expectation value for the momentum is

$$\bar{p}_x = \int_{-\infty}^{\infty} \psi^* \, (-i\hbar) \frac{\partial \psi}{\partial x} \, dx = -i\hbar A^2 \int_0^L \sin(\frac{\pi}{L} x) \frac{\pi}{L} \cos(\frac{\pi}{L} x) dx$$

$$= -i\hbar A^2 \frac{\pi}{2L} \int_0^L \sin(\frac{2\pi}{L} x) \, dx = 0,$$

This result is also expected, since a particle that moves back and forth in a box with constant energy has zero average speed.

(b)   The mean square deviation of x is defined by.

$$\overline{\Delta x^2} = \langle x^2 - \bar{x}^2 \rangle$$

For the particle in a box

$$\overline{\Delta x^2} = A^2 \int_0^L (x^2 - L^2/4) \sin^2(\frac{\pi}{L} x) \, dx = \frac{A^2}{2} \int_0^L (x^2 - L^2/4) \, [1 - \cos(\frac{2\pi}{L} x)] \, dx.$$

Thus

$$\overline{\Delta x^2} = \frac{A^2}{2} \int_0^L [(x^2 - L^2/4) - x^2 \cos(\frac{2\pi}{L} x) + \frac{L^2}{4} \cos(\frac{2\pi}{L} x)] \, dx.$$

The integral of the first term is simple to do, the integral of the third term is zero, and to calculate the second term we use the expression

$$\int x^2 \cos(ax)\, dx = \frac{2x}{a^2}\cos(ax) + \frac{a^2x^2 - 2}{a^3}\sin(ax).$$

Using $A^2 = 2/L$ we have

$$\overline{\Delta x^2} = \frac{1}{L}\left[\frac{L^3}{3} - \frac{L^3}{4} - \frac{2L^3}{(2\pi)^2}\right] = \frac{L^2}{(2\pi)^2}\left[\frac{\pi^2 - 6}{3}\right] = 0.033\, L^2.$$

The mean quadratic deviation of the momentum can be calculated in a similar way. The result is

$$\overline{\Delta p_x^2} = \left(\frac{\hbar\pi}{L}\right)^2.$$

It is interesting to note that the uncertainties in the determination of the position and of the momentum can be considered as the square roots of the mean square deviations. Thus

$$\Delta x = \left(\overline{\Delta x^2}\right)^{1/2} = \sqrt{0.033}\, L = 0.18\, L,$$

$$\Delta p_x = \left(\overline{\Delta p_x^2}\right)^{1/2} = \hbar\frac{\pi}{L}.$$

The product of the two uncertainties gives

$$\Delta x\, \Delta p_x = 0.18\, \pi\, \hbar = 0.57\, \hbar.$$

This result is consistent with the uncertainty principle, that establishes the value $\hbar/2$ for the lower limit of the product of the two uncertainties.

3.3.3 Potential Barrier: Tunnel Effect

Consider a free electron with energy E, moving in the $+x$ direction in a region where the potential is zero for $x < 0$, and has at $x = 0$ a barrier, with V_0 higher than its energy, as illustrated in Fig. 3.4. This is also called a step potential. The task here is to find the wave functions in both regions, $x < 0$ and $x > 0$.

Since region 1 is semi-infinite, the electron energy is not quantized. The problem is to find what happens when the electron encounters the potential barrier. In region 1, as in the previous example, the electron wave function is given by

Fig. 3.4 Potential barrier

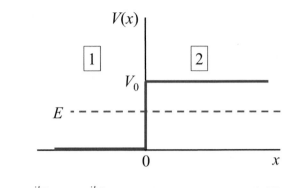

$$\psi_1(x) = A\,e^{ikx} + B\,e^{-ikx}, \quad (x < 0), \tag{3.45}$$

where the wavenumber is related to the energy E as in (3.35), $k = (2mE)^{1/2}/\hbar$.
   In region 2, Schrödinger equation leads to

$$\frac{\partial^2 \psi_2}{\partial t^2} = \frac{2m}{\hbar^2}(V_0 - E)\,\psi_2. \tag{3.46}$$

For $V_0 - E > 0$, the solution of Eq. (3.46) is

$$\psi_2(x) = C\,e^{\gamma x} + D\,e^{-\gamma x}, \quad (x > 0), \tag{3.47}$$

where

$$\gamma = [2m\,(V_0 - E)]^{1/2}/\hbar. \tag{3.48}$$

   Notice that in the case $E > V_0$, the exponent in (3.48) is imaginary, and the two
terms in (3.47) represent propagating waves (Problem 3.6). However, in the case $E <
V_0$, that will be considered here, γ is positive, so that the first term in Eq. (3.47) grows
exponentially with x, while the second term decreases exponentially. To determine
the four constants A, B, C, and D, that appear in (3.45) and in (3.47), it is necessary
to use the boundary conditions for the wave function. For $x \to \infty$, Eq. (3.47) shows
that if C is nonzero, $\psi_2 \to \infty$. Since the wave function cannot diverge, the constant
C must be null.
   At $x = 0$, the wave functions in the two regions must be equal, because ψ is
continuous throughout space. With $C = 0$ in (3.47), we obtain from $\psi_1(0) = \psi_2(0)$

$$A + B = D. \tag{3.49}$$

   At $x = 0$, the derivative of ψ with respect to x must also be continuous

$$\left.\frac{\partial \psi_1}{\partial x}\right|_{x=0} = \left.\frac{\partial \psi_2}{\partial x}\right|_{x=0}, \tag{3.50}$$

otherwise the kinetic energy, which is proportional to $d^2\psi/dx^2$, would be infinite at $x = 0$. Using (3.45) and (3.47) in Eq. (3.50), we obtain

$$ik(A - B) = -\gamma D. \tag{3.51}$$

With the two conditions (3.49) and (3.51) we can determine the amplitudes of the reflected and transmitted waves in terms of the incident wave amplitude

$$B = \frac{k - i\gamma}{k + i\gamma} A, \qquad D = \frac{2k}{k + i\gamma} A. \tag{3.52}$$

Note that the constants B and A have the same modulus, meaning that the amplitudes of the incident and reflected waves in region 1 are equal, so that the wave function in region 1 is a standing wave. Since $C = 0$, in region 2 the electron wave function is.

$$\psi_2(x) = D e^{-\gamma x}, \tag{3.53}$$

which shows that, despite the fact that the incident electron has an energy smaller than the potential of the barrier, there is a certain probability that the electron will be found in region 2. This is a purely quantum effect, because classically a particle would be fully reflected by a potential barrier larger than its energy. As illustrated in Fig. 3.5, the amplitude of ψ_2 decays exponentially with x but at some position $x = a$ the wave function ψ_2 can be nonzero. Thus, if the barrier has a finite thickness a, the probability that the electron crosses it is, approximately

$$|\psi_2(a)|^2 \approx e^{-2\gamma a}, \tag{3.54}$$

This is a quantum phenomenon, called **tunnel effect**, because classically the electron would only cross the potential barrier if there were a tunnel under the barrier. Note that the result (3.54) is approximate, because if we had considered the barrier width finite from the beginning, we could not have done $C = 0$. However, if $\exp(-2\gamma a)$ is sufficiently small, the amplitude C of the reflected wave at $x = a$ is negligible and the expression (3.54) is a good approximation for the rigorous result.

Fig. 3.5 Spatial behavior of the wave function for a particle with energy E, incident on a potential barrier with $V_0 > E$

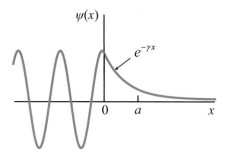

Example 3.2 Another important application of quantum mechanics is that of a particle with mass m, subjected to an interaction of a simple harmonic oscillator with potential energy

$$V(x) = \frac{1}{2} k x^2 = \frac{1}{2} m \omega_0^2 x^2,$$

where $\omega_0 = (k/m)^{1/2}$ is the natural frequency of the oscillator.

Show that the functions $\psi_0 = A_0 e^{-ax^2}$ and $\psi_1 = A_1 x e^{-ax^2}$ are eigenfunctions of the Schrödinger equation for the ground state and first excited state of the harmonic oscillator and determine their energies.

The Schrödinger equation for the harmonic oscillator has the form

$$-\frac{\hbar^2}{2m} \frac{d^2 \psi}{dx^2} + \frac{1}{2} m \omega_0^2 x^2 \, \psi = E \, \psi.$$

For the ground state we have the following derivatives of ψ_0

$$\frac{d\psi_0}{dx} = -2a \, x \, A_0 \, e^{-ax^2},$$

$$\frac{d^2 \psi_0}{dx^2} = -2a \, A_0 \, e^{-ax^2} + 4a^2 x^2 \, A_0 \, e^{-ax^2}.$$

Substitution of these derivatives in the Schrödinger equation above gives

$$-\frac{\hbar^2}{2m}(-2a \, A_0 \, e^{-ax^2} + 4a^2 x^2 \, A_0 \, e^{-ax^2}) + \frac{1}{2} m \omega_0^2 x^2 A_0 \, e^{-ax^2} = E \, A_0 \, e^{-ax^2}.$$

Dividing all terms by $A_0 \, e^{-ax^2}$ we obtain

$$\frac{\hbar^2}{m} a - 2\frac{\hbar^2}{m} a^2 x^2 + \frac{1}{2} m \omega_0^2 x^2 = E.$$

For this equation to be satisfied for any value of x, it is necessary that the two terms in x^2 cancel out. With this condition, we obtain for the parameter a

$$a = \frac{m \, \omega_0}{2\hbar}.$$

Substituting this expression in the previous equation, we obtain the energy of the ground state.

$$E_0 = \frac{\hbar^2 a}{m} = \frac{1}{2}\hbar\omega_0.$$

We follow the same procedure to obtain the energy of the first excited state with wave function ψ_1. Calculating the second derivative with respect to x, replacing in Schrödinger equation and diving all terms by the common factor we obtain

$$-\frac{\hbar^2}{2m}(-2a\,x - 4a\,x + 4a^2\,x^3) + \frac{1}{2}m\,\omega_0^2 x^3 = E\,x.$$

In this case it is necessary to equate to zero separately all terms with the same power in x. The term in x^3 gives for the parameter a the same expression obtained for the ground state, while the term in x gives

$$E_1 = 3\frac{\hbar^2 a}{m} = \frac{3}{2}\hbar\omega_0.$$

This is the energy of the first excited state, whose wave function is precisely ψ_1. The general solution of the Schrödinger equation for the harmonic oscillator, which is presented in detail in the books on quantum mechanics, is given by functions of the type,

$$\psi_n(x) = (c_0 + c_1 x + c_2 x^2 + \dots c_n x^n)\,e^{-ax^2},$$

where the function in parentheses is known as the Hermite polynomial. The demonstration that this expression is an eigenfunction of the Schrödinger equation for the harmonic oscillator is done in an analogous manner to what we did for $n = 0$ and $n = 1$, that correspond to the two lowest energy states. The general solution shows that the energy of the excited state of order n is given by

$$E_n = (n + 1/2)\,\hbar\omega_0.$$

This is an important result that shows that the energy levels of the harmonic oscillator states are equally spaced, with a difference between two consecutive levels of $\hbar\omega_0$.

3.4   Electron in the Hydrogen Atom

One of the most important simple applications of quantum mechanics is in the hydrogen atom. This was one of the first problems to which Erwin Schrödinger addressed in the studies that led to his equation. The agreement he obtained with the experiments and with the energy eigenvalues of the Bohr model was the first important test of the validity of his theory.

The hydrogen atom is the simplest of all atoms, it has only one electron with charge $-e$ around a proton with charge $+e$. The potential energy that acts on the electron due to the electrostatic interaction is

$$V(r) = -\frac{e^2}{4\pi\varepsilon_0}\frac{1}{r}, \tag{3.55}$$

where r is the distance between the electron and the proton. Despite the simplicity of this potential, the solution of the Schrödinger equation is reasonably complicated because of its three-dimensional nature. To solve the equation more easily, we use a system of spherical coordinates, illustrated in Fig. 3.6 The electron position relative to the proton is characterized by the coordinates r, θ, and φ. In spherical coordinates the Laplacian operator that appears in the Schrödinger equation has the following form.

$$\nabla^2 = \frac{1}{r^2}\frac{\partial}{\partial r}\left(r^2\frac{\partial}{\partial r}\right) + \frac{1}{r^2\sin^2\theta}\frac{\partial^2}{\partial\varphi^2} + \frac{1}{r^2\sin\theta}\frac{\partial}{\partial\theta}\left(\sin\theta\frac{\partial}{\partial\theta}\right). \tag{3.56}$$

To solve the Schrödinger Eq. (3.23) with the Laplacian operator (3.56) and the potential $V(r)$ in (3.55), we shall assume that the proton mass is infinitely larger than the electron mass. This corresponds to say that the electron moves around a fixed proton, so that the problem of two particles is reduced to just one. Thus, in Eq. (3.23) we can neglect the kinetic energy of the proton, which would not be possible if its mass were not very large. Since the potential (3.55) depends only on the variable r, it is possible to find solutions of the Schrödinger equation in the form

Fig. 3.6 Spherical coordinates r, θ, φ of point P in the Cartesian coordinate system (x, y, z)

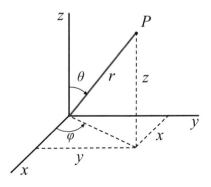

$$\Psi(r, \theta, \varphi) = R(r)\,\Theta(\theta)\,\Phi(\varphi). \tag{3.57}$$

With the function (3.57) it is possible to separate the partial differential equation with three variables into three ordinary differential equations in the coordinates r, θ, φ, by the method of separation of variables used to treat Eq. (3.19). Substituting the function (3.57) into Eq. (3.23), and using the Laplacian operator in (3.56), we obtain

$$-\frac{\hbar^2}{2m}\left[\frac{1}{r^2}\frac{\partial}{\partial r}\left(r^2\frac{\partial R\Theta\Phi}{\partial r}\right) + \frac{1}{r^2\sin^2\theta}\frac{\partial^2 R\Theta\Phi}{\partial\varphi^2} + \frac{1}{r^2\sin\theta}\frac{\partial}{\partial\theta}\left(\sin\theta\frac{\partial R\Theta\Phi}{\partial\theta}\right)\right]$$
$$+ V(r)\,R\Theta\Phi = E\,R\Theta\Phi.$$

Operating with the partial derivatives it follows that

$$-\frac{\hbar^2}{2m}\left[\frac{\Theta\Phi}{r^2}\frac{d}{dr}\left(r^2\frac{dR}{dr}\right) + \frac{R\Theta}{r^2\sin^2\theta}\frac{d^2\Phi}{d\varphi^2} + \frac{R\Phi}{r^2\sin\theta}\frac{d}{d\theta}\left(\sin\theta\frac{d\Theta}{d\theta}\right)\right]$$
$$+ V(r)\,R\Theta\Phi = E\,R\Theta\Phi \quad .$$

Notice that in this equation we have replaced the symbols of the partial derivatives by those of total derivatives because the three functions depend only on one variable each. Multiplying all terms by $-2mr^2\sin^2\theta/(R\Theta\Phi\,\hbar^2)$ and rearranging the terms we have

$$\frac{1}{\Phi}\frac{d^2\Phi}{d\varphi^2} = -\frac{\sin^2\theta}{R}\frac{d}{dr}\left(r^2\frac{dR}{dr}\right) - \frac{\sin\theta}{\Theta}\frac{d}{d\theta}\left(\sin\theta\frac{d\Theta}{d\theta}\right)$$
$$- \frac{2m}{\hbar^2}r^2\sin^2\theta[E - V(r)]. \tag{3.58}$$

Now, since the left-hand side of this equation does not depend on r or θ, while the right-hand side does not depend on φ, their common value must be a constant, which we will designate by $-m_l^2$. Thus, we obtain two equations

$$\frac{d^2\Phi}{d\varphi^2} = -m_l^2\,\Phi, \tag{3.59}$$

$$\frac{1}{R}\frac{d}{dr}\left(r^2\frac{dR}{dr}\right) + \frac{1}{\Theta\sin\theta}\frac{d}{d\theta}\left(\sin\theta\frac{d\Theta}{d\theta}\right) + \frac{2m}{\hbar^2}r^2[E - V(r)] = \frac{m_l^2}{\sin^2\theta}. \tag{3.60}$$

Equation (3.59) can be solved with a function of φ, while Eq. (3.60) can be rewritten in the form

$$\frac{1}{R}\frac{d}{dr}\left(r^2\frac{dR}{dr}\right) + \frac{2m}{\hbar^2}r^2[E - V(r)] = \frac{m_l^2}{\sin^2\theta} - \frac{1}{\Theta\sin\theta}\frac{d}{d\theta}\left(\sin\theta\frac{d\Theta}{d\theta}\right),$$

which can also be separated in the variables r and θ. Using for separation constant l $(l + 1)$, we obtain two equations in the variables r and θ

$$-\frac{1}{\sin\theta}\frac{d}{d\theta}\left(\sin\theta\frac{d\,\Theta}{d\theta}\right) + \frac{m_l^2\,\Theta}{\sin^2\theta} = l(l+1)\,\Theta, \tag{3.61}$$

$$\frac{1}{r^2}\frac{d}{dr}\left(r^2\frac{dR}{dr}\right) + \frac{2m}{\hbar^2}[E - V(r)]\,R = l(l+1)\frac{R}{r^2}. \tag{3.62}$$

Equations (3.59), (3.61) and (3.62) are now independent of each other and can be solved separately. The complete solution for the electron wave function is the product of the three solutions of those equations.

Let us first consider Eq. (3.59) for φ. Its solution is

$$\Phi(\varphi) = e^{im_l\varphi}. \tag{3.63}$$

Mathematically this function is a solution of Eq. (3.59) for any value of m_l. However, physically the wave function of the electron wave must have for $\varphi = 0$ the same value as for $\varphi = 2\pi$, 4π, 6π, etc. This requires m_l to have only the following values

$$|m_l| = 0, 1, 2, 3, \ldots, \tag{3.64}$$

that is, m_l must be an integer, positive or negative, or zero. It is a **quantum number**. The solutions of Eq. (3.61) and (3.62) are much more complex. However, they are well-known equations, studied extensively in text books of advanced calculus for physics and engineering students. The solutions of (3.61) are the so-called associated Legendre polynomials, which are finite only if l is a positive integer number, limited by

$$|m_l| \leq l. \tag{3.65}$$

The solutions of the radial Eq. (3.62) are the Laguerre polynomials, which are finite if the constant E is given by

$$E = -\frac{m\,e^4}{2\hbar^2(4\pi\varepsilon_0)^2\,n^2}, \tag{3.66}$$

where n is also an integer, which satisfies the relation

$$0 \leq l \leq n - 1. \tag{3.67}$$

The constant E in Eq. (3.66) is the energy eigenvalue of the wave function in the hydrogen atom. This result means that the electron energy in the hydrogen atom is quantized (discrete), similarly to what occurs in the infinite potential well studied in Sect. 3.3. Substituting the physical constants in (3.66), we can express the energy in eV as

$$E = -\frac{13.6}{n^2} \text{ eV} \tag{3.68}$$

Figure 3.7 illustrates the energy levels of the infinite potential well and the Coulomb electron well in the atom ($V = -A/r$). Note that in both cases the lowest energy value is not the potential at the bottom of the well. Instead, it is a value above the minimum, which is called called zero-point energy, or **ground state** energy.

The general solution of the Schrödinger equation for the electron in the hydrogen atom is given by the product of the three functions in the variables r, θ, φ, solutions of Eqs. (3.59), (3.61) and (3.62), which can be written as

$$\Psi_{n l m_l}(r, \theta, \varphi) = R_{n l}(r) \, \Theta_{l m_l}(\theta) \, \Phi_{m_l}(\varphi). \tag{3.69}$$

where

$$\Phi_{m_l}(\varphi) = e^{i \, m_l \, \varphi},$$

$$\Theta_{l m_l}(\theta) = \sin^{|m_l|} \theta \times (\text{polynomial in } \cos\theta),$$

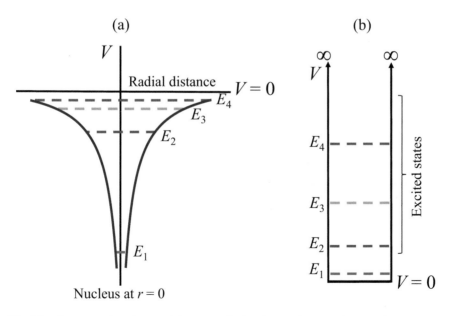

Fig. 3.7 a Representation of a potential energy well of an electron due to the electrostatic interaction with the nucleus of an atom, indicating the positions of the energy eigenvalues. **b** Energy levels in an infinite square-well potential

$$R_{nl}(r) = e^{-Cr/n} r^l \times (\text{polynomial in } r),$$

where C is a constant. Although the energy eigenvalues of the electron in the atom with Coulomb potential depend only on the quantum number n, the wave functions also depend on l and on m_l. The fact that there are three quantum numbers, instead of just one, as in the square well potential studied in Sect. 3.3, is a consequence of the fact that the Schrödinger equation for the atom contains three independent variables. Grouping the conditions (3.64), (3.65) and (3.67), we can write the relationships between the quantum numbers in the form.

Principal quantum number: $n = 1, 2, 3, \ldots$

Azimuthal quantum number: $l = 0, 1, 2, \ldots, n - 1$

Magnetic quantum number: $m_l = -l, -l + 1, ..0, \ldots, l - 1, l$

The reason for l to be called the azimuthal quantum number is because it determines the angular variation of the electron wave function. On the other hand, m_l is called the magnetic quantum number, because it defines the separation of the energy levels when the atom is placed in a magnetic field. Table 3.2 shows the normalized eigenfunctions corresponding to the first three values of n for an atom with the nucleus charge of $+ Ze$ (Z is the atomic number), and only one electron. The wave function Ψ_{100} corresponds to the state with minimum energy, the ground state.

Notice that the wave functions Ψ_{200}, Ψ_{210}, and $\Psi_{21\pm1}$ are quite different from each other, but they have the same energy, since they all have the same principal quantum number, $n = 2$. States with different wave functions that have the same energy are called **degenerate**. It is common to find solutions to the Schrödinger equation that are degenerate states.

To understand the meaning of the electron eigenfunctions in the hydrogen atom, let us calculate some of their associated quantities. The first is the probability density function $\Psi * \Psi$. It is not simple to plot this function in three coordinates simultaneously, so we will consider each one separately. Initially we consider the dependence on r. Since the probability of finding the electron in the elementary volume d^3r is $\Psi * \Psi d^3r$, it does not make much sense to study the behavior of $\Psi^* \Psi$ only, because in spherical coordinates $d^3r = r^2 \sin\theta \, dr d\theta d\varphi$ also depends on r. Thus, we consider the **radial probability density** $P(r)$, defined as the probability of finding the electron with the radial coordinate between r and $r + dr$. For the wave function Ψ_{nlm_l}, this probability density is given by (Problem 3.16)

$$P_{nl}(r) = r^2 R_{nl}^*(r) R_{nl}(r), \tag{3.70}$$

where the factor r^2 is due to the volume of the region between the spheres of radii r and $r + dr$. Note that the quantum number m_l does not influence the radial density because the function $\exp(im_l\varphi)$ disappears in the product with its complex conjugate. Figure 3.8 represents the radial probability density of the electron in the hydrogen

Table 3.2 Eigenfunctions of an atom with one electron around a nucleus with Z protons, for the first values of the principal quantum number n

Quantum numbers			Eigenfunction
n	l	m_l	
1	0	0	$\psi_{100} = \frac{1}{\sqrt{\pi}}\left(\frac{Z}{a_0}\right)^{3/2} e^{-Zr/a_0}$
2	0	0	$\psi_{200} = \frac{1}{4\sqrt{2\pi}}\left(\frac{Z}{a_0}\right)^{3/2}\left(2 - \frac{Zr}{a_0}\right)e^{-Zr/2a_0}$
2	1	0	$\psi_{210} = \frac{1}{4\sqrt{2\pi}}\left(\frac{Z}{a_0}\right)^{3/2}\frac{Zr}{a_0}e^{-Zr/2a_0}\cos\theta$
2	1	± 1	$\psi_{21\pm1} = \frac{1}{8\sqrt{\pi}}\left(\frac{Z}{a_0}\right)^{3/2}\frac{Zr}{a_0}e^{-Zr/2a_0}\sin\theta\, e^{\pm i\varphi}$
3	0	0	$\psi_{300} = \frac{1}{81\sqrt{3\pi}}\left(\frac{Z}{a_0}\right)^{3/2}\left(27 - 18\frac{Zr}{a_0} + 2\frac{Z^2r^2}{a_0^2}\right)e^{-Zr/3a_0}$
3	1	0	$\psi_{310} = \frac{\sqrt{2}}{81\sqrt{\pi}}\left(\frac{Z}{a_0}\right)^{3/2}\left(6 - \frac{Zr}{a_0}\right)\frac{Zr}{a_0}e^{-Zr/3a_0}\cos\theta$
3	1	± 1	$\psi_{31\pm1} = \frac{1}{81\sqrt{\pi}}\left(\frac{Z}{a_0}\right)^{3/2}\left(6 - \frac{Zr}{a_0}\right)\frac{Zr}{a_0}e^{-Zr/3a_0}\sin\theta\, e^{\pm i\varphi}$
3	2	0	$\psi_{320} = \frac{1}{81\sqrt{6\pi}}\left(\frac{Z}{a_0}\right)^{3/2}\frac{Z^2r^2}{a_0^2}e^{-Zr/3a_0}\left(3\cos^2\theta - 1\right)$
3	2	± 1	$\psi_{32\pm1} = \frac{1}{81\sqrt{\pi}}\left(\frac{Z}{a_0}\right)^{3/2}\frac{Z^2r^2}{a_0^2}e^{-Zr/3a_0}\sin\theta\cos\theta\, e^{\pm i\varphi}$
3	2	± 2	$\psi_{32\pm2} = \frac{1}{162\sqrt{\pi}}\left(\frac{Z}{a_0}\right)^{3/2}\frac{Z^2r^2}{a_0^2}e^{-Zr/3a_0}\sin^2\theta\, e^{\pm 2i\varphi}$

The parameter $a_0 = 4\pi\varepsilon_0\hbar^2/me^2$ is the Bohr radius

atom with the eigenfunctions for $n = 1$, 2 and 3, using dimensionless quantities in both axes. The figure shows clearly that the electrons are not particles with well-defined orbits, as predicted in the Bohr model. In fact, each electron occupies a large region around the nucleus, with a distribution in space such that it has a maximum probability to be found at a certain radius, with a value that increases with n. Let us calculate the radius of maximum probability density for the ground state ($n = 1$), that is, the state of minimum energy. Using in Eq. (3.70) the radial function $R_{10}(r)$ of the eigenfunction Ψ_{100} in Table 3.1, with $Z = 1$, we have

$$P_{10}(r) = e^{-2r/a_0}r^2, \tag{3.71}$$

which is proportional to the function represented by the curve in the upper panel of Fig. 3.8. The maximum of this function is obtained with

$$\frac{dP_{10}(r)}{dr} = 2r\, e^{-2r/a_0}(1 - r/a_0) = 0. \tag{3.72}$$

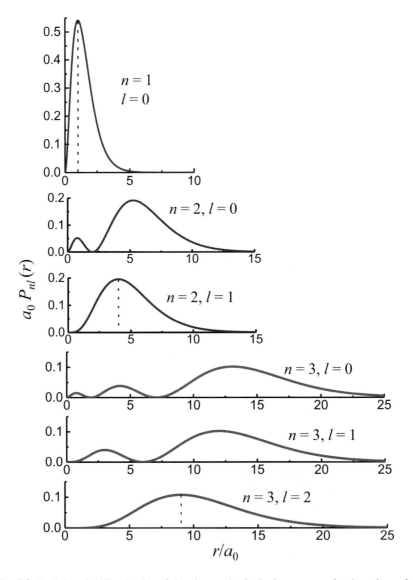

Fig. 3.8 Radial probability density of the electron in the hydrogen atom for the values of the quantum numbers n and l indicated. Note that in the three cases for which $l = l_{max} = n - 1$, the maximum occurs at a radius $r/a_0 = n^2$, indicated by the dashed line

Thus, the radius of maximum probability density of finding the electron in the ground state is

$$r = a_0 = \frac{4\pi\varepsilon_0\hbar^2}{me^2} \approx 0.53 \ \overset{\circ}{A} \tag{3.73}$$

which is exactly the Bohr radius. This is another strong argument for demonstrating the consistency of the Schrödinger formulation of quantum mechanics with the Bohr model for the atom.

The angular variation of the probability density can be represented in several different ways. One of them uses a polar plot, in which the amplitude of the probability density of finding the electron at the position x, y, z is represented by the distance of this point to the origin. Figure 3.9 shows the polar plots corresponding to the quantum numbers $l = 0$ and $l = 1$. These plots provide a view of the electronic shell in each state. They clearly show that the electron is not characterized by an orbit in the classical sense, but by a probability density of being found at each position. The variation in the probability density with angular position suggests the name **orbital** to designate the atomic wave functions. Since the quantum number l determines the form of the angular variation of the orbital, it is a very important number that is designated by letters used to denote the radiation emission spectral lines of the hydrogen atom. The orbitals with $l = 0, 1, 2, 3, 4, \ldots$ are designated by the letters s, p, d, f, g, ...

Figure 3.10 shows another way of representing the angular and radial variations of the electron probability densities, corresponding to the first four energy levels in the hydrogen atom. This representation takes into account the dependencies of the states with the angle and also with the radial distance to the nucleus.

Before closing this section, it is necessary to mention two important facts related to the hydrogen atom: The first is that the electron has, in addition to mass m and charge $-e$, another property, the **spin**. As the names implies, classically the spin would correspond to a rotation of the electron around itself, analogous to the rotation of the planet Earth around itself. However, the electron is not a particle in the classical sense, so there is not much sense in speaking of a rotation around itself. The spin is

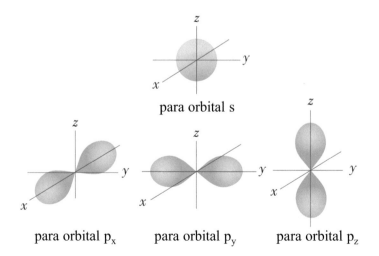

para orbital s

para orbital p_x \quad para orbital p_y \quad para orbital p_z

Fig. 3.9 Illustration of the function $|\Theta(\theta)\Phi(\varphi)|^2$, that represents the electronic probability densities in the hydrogen atom for $l = 0$ and $l = 1$

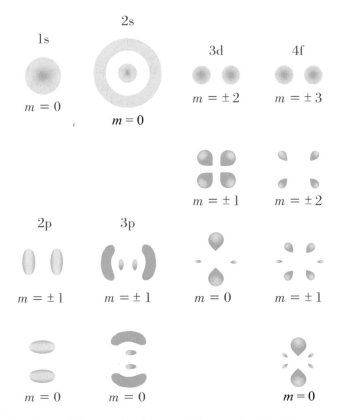

Fig. 3.10 Illustration in different scales of the probability densities of the electron states in the first four energy levels in a hydrogen atom

a quantum entity, with no classical analog, which arises naturally from a relativistic quantum theory. The electron spin is characterized by a fourth quantum number m_s, that can have two values, $+1/2$ or $-1/2$, each one would correspond to a rotation in one sense around an axis, or in the opposite sense. In the classical view, the rotation of a charged particle around itself also gives rise to a magnetic moment. Thus, in addition to the spin, the electron also has a magnetic dipole moment. For this reason, the potential that appears in the Schrödinger equation for the hydrogen atom is actually more complex than Eq. (3.55). Consequently, its solution is more complex than the one presented here. However, since the effect of the spin is relatively small, the calculation of the energy levels of the hydrogen atom can be treated approximately with **perturbation theory**. The main result is that the energy of the electron does not depend only on the quantum number n, it also depends on the orbital quantum number l. However, the separation between the energy levels of states with the same n and different ls is small compared to the separation of states with different ns.

The other important note is about transitions of the electron between states. When the electron is "placed" in a certain state characterized by an eigenfunction of the

Schrödinger equation, if there is no disturbance in the atom the electron remains indefinitely in that state. A possible disturbance is that of electromagnetic radiation, which contributes to the Schrödinger equation with a time-varying potential. As we shall see in Chap. 8, quantum theory shows that the electron may undergo a transition to a state of higher energy by absorbing a photon of frequency v, provided that the difference between the energies of the final state f and the initial state i is equal to the energy of the photon, that is

$$E_f - E_i = h\,v. \tag{3.74}$$

This expression is just a result of the energy conservation equation. The electron can also go from a higher to a lower energy state through the emission of photons with frequency given by $v = \Delta E/h$, where ΔE is the energy difference between the two levels. The measurements of the absorption and emission spectra of light in the beginning of the twentieth century were very important to show that a new theory for the hydrogen atom was needed to explain the observations. Later, the comparison of the experimental results with the theoretical calculations were decisive for the acceptance of the quantum theory. Even today, optical spectroscopy techniques are widely used for the study and identification of atoms, molecules and solids.

Figure 3.11 shows several transitions between some of the lower energy levels in the hydrogen atom. Quantum theory shows that transitions between states with emission or absorption of photons can only occur if the orbital quantum numbers of the initial and final states differ by 1, $\Delta l = \pm 1$. This is called a **selection rule** for the transitions in an atom. Only when two states have orbital numbers differing by $\Delta l = \pm 1$, the electric field of the radiation manages to induce an electric-dipole transition between them. For this reason, the lines representing the transitions in

Fig. 3.11 Representation of the electronic transitions in a hydrogen atom with absorption or emission of photons. The diagonal lines represent the transitions allowed by the electric dipole selection rules. The corresponding wavelengths are indicated in nanometers

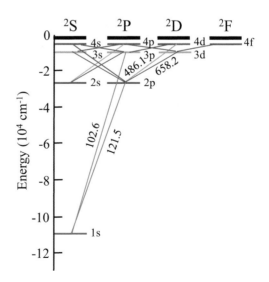

Fig. 3.11 are diagonal, not vertical. If the electric field is linearly polarized, another selection rule is $\Delta m_l = 0$. However, if the electric field is circularly polarized, the selection rule is $\Delta m_l = \pm 1$.

3.5 Atoms with Many Electrons

In atoms with more than one electron the potential that enters in the Schrödinger equation is much more complicated than in the hydrogen atom. This is due to the fact that each electron interacts not only with the nucleus, but also with the other electrons. Thus, the "motion" of an electron, and therefore its wave function, affects all other electrons. It is possible to write the Schrödinger equation for all electrons, but it cannot be solved analytically. The solutions can only be obtained approximately, and with the use of computer numerical calculations.

There are several approximate methods to solve Schrödinger equation for an atom or a crystal. One of the simplest is the **mean field** method, proposed by D. R. Hartree, that is essentially the following: the Schrödinger equation is written for a certain electron taking into account the interaction with the nucleus and with the other electrons. However, the interaction potential with the other electrons is considered only on an average, and the problem is solved in steps. In the first step the wave function for a certain electron is calculated assuming a tentative average interaction potential due to the other electrons. In the following steps the wave function for each electron is calculated sequentially, and the average potential is modified with the electronic densities calculated in the previous steps. This process is repeated several times, until the differences between the wave functions obtained in consecutive cycles are negligible. At the end one obtains a consistent set of eigenfunctions, as well as the corresponding electronic energies.

Once the atomic orbitals are obtained, the next question is: how are the electrons of the atom distributed in these orbitals, that is, in the energy levels? The distribution is based on two fundamental principles: the first is that electrons must occupy the lowest possible energy states. However, they cannot all go to the ground state, due to the **Pauli exclusion principle**. According to this principle, two electrons cannot occupy exactly the same state. Since each electron can have a spin $m_s = \pm 1/2$, the distribution in the atom is done by filling the lower energy states, starting from the ground state, successively with two electrons each. Thus, the lowest energy configuration of an atom with atomic number Z, has two electrons with quantum numbers $n = 1, l = 0, m_l = 0$, two electrons with $n = 2, l = 0, m_l = 0$, two with $n = 2, l = 0, m_l = -1$, two with $n = 2, l = 0, m_l = +1$, and so on. Since an orbital l can have $m_l = 0, \pm 1, ... \pm l$, it can accommodate $2(2l + 1)$ electrons. To facilitate the notation, the orbitals are represented by letters corresponding to the values of the quantum number l. The orbitals corresponding to $l = 0, 1, 2, 3, 4 \ldots$ are denoted by s, p, d, f, g, … respectively. Likewise, the quantum number n is designated by a letter that represents the "shell", with the letters K, L, M, N, O, … associated respectively to $n = 1, 2, 3, 4, 5, \ldots$.

The element whose atom has one electron is hydrogen. In the ground state the electron of the H atom has one orbital represented by 1s. In the element with two electrons, helium, both electrons have 1s orbitals, one with spin $+ 1/2$ and one with $- 1/2$. Thus, the ground state is represented by $1s^2$, where the superscript denotes the number of electrons in the orbital. Since in the K shell, of orbital 1s, only two electrons can fit, the helium atom is formed by a "closed shell" with zero spin. This fact gives He a great chemical stability, which is the reason for it to be called a **noble gas**. The next atom is the one of lithium, with three electrons and therefore represented by the notation $1s^2$ 2s. Since the two electrons $1s^2$ form a closed shell, the third electrons can easily get loose in solid Li and move about the atoms, so that Li is a metal. Thus, for each element the electrons successively fill in the orbital states with lower energies and give the elements their own chemical characteristics. It is important to note, however, that several elements have similar chemical properties, since there is a periodic repetition in the formation of shells. For example, the argon atom has ten electrons, with the configuration $1s^2$ $2s^2 2p^6$. Thus, Ar has two closed shells, K and L, and has properties similar to helium. On the other hand, sodium, with eleven electrons, has configuration $1s^2 2s^2 2p^6 3s$, and has properties similar to those of lithium. This periodicity of behavior with the atomic number is the reason for the name Periodic Table, in which the elements are organized, as in Table 3.3. This Table shows the atomic number of each element, as well as the number of electrons and the corresponding orbitals of the last occupied shells. The Periodic Table in Appendix C contains other important properties of the elements.

Table 3.3 Periodic Table of the elements

The spectroscopic notation indicates the number of electrons and the corresponding orbitals of the last occupied shells

Problems

3.1   Show that the sum of the wave functions $\psi_1 = A\,e^{ikx-i\omega t}$ and $\psi_2 = B\,e^{-ikx-i\omega t}$ is a solution of Schrödinger Eq. (3.17) for a particle with mass m in a constant potential, $V = V_0$, and obtain the relation between k and ω.

3.2   Calculate the constants A_1 and A_2, defined in Eq. (3.42), in order to normalize the two lowest energy eigenfunctions of a particle in an infinite square well potential.

3.3   Calculate the difference in energies, in eV, of the two lowest energy states of an electron in an infinite square well potential of width: (a) $L = 30$ Å; (b) $L = 1$ cm.

3.4   Consider an electron in an infinite square well of width L, in the first excited state. Calculate.

(a)   The expectation value of the electron position x.
(b)   The mean square deviation of the position $\overline{\Delta x^2} = \left\langle x^2 - \bar{x}^2 \right\rangle$.
(c)   The expectation value of the momentum p_x.
(d)   The mean square deviation of the momentum.
(e)   The product of the two uncertainties $\Delta x\,\Delta p_x$.

3.5   An electron moves at a constant speed towards a potential barrier, like the one illustrated in Fig. 3.4. Considering the electron energy E larger that the height of the barrier V_0, calculate the probabilities for the electron to be reflected and to be transmitted into the barrier, as a function of E and V_0. Obtain the numerical values for $E = 2\,V_0$.

3.6   An electron moves with energy E in a multilayer formed by two thick films of a semiconductor A, separated by a thin film of thickness d of another semiconductor B. In a first approximation, the potential seen by the electron is as shown in the figure below.

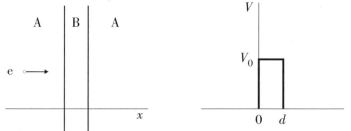

(a)   For $E > V_0$, calculate the probability of the electron, initially in the left layer A, to cross layer B perpendicularly and to reach the right layer A.
(b)   What is the condition for the probability calculated in a) to be equal to 1?
(c)   Calculate the thickness d so that the probability for the electron to cross the barrier is 1, for $E = 1.0$ eV and $V_0 = 0.8$ eV.
(d)   For $E = 1.0$ eV, $V_0 = 1.2$ eV, and the value of d obtained in c), calculate the probability for the electron to go through layer B.

3.7   Consider the situation of Problem 3.6 with $E < V_0$. Calculate the probability that the electron will go through layer B for $E = 1.0$ eV, $V_0 = 1.2$ eV and $d = 5$ Å (tunnel effect).

3.8   Obtain the wave function and the energy of the second excited state of a simple harmonic oscillator. (Suggestion: use the polynomial function at the end of Example 3.2 with $n = 2$).

3.9   Calculate the energies, in eV, of the states of the hydrogen atom with $n = 1, 2, 3$, and obtain the frequencies, in Hz, of all possible transitions between these levels.

3.10   (a) Show that the wavelength of the photon absorbed, or emitted, in a transition between the levels n_1 and n_2 in a hydrogen atom is, in Angstroms,

$$\lambda(\overset{\circ}{A}) = \frac{911\, n_1^2 n_2^2}{n_2^2 - n_1^2}.$$

(b) Compare the result obtained in Problem 3.9 for the levels $n = 2$ and $n = 3$ with this expression. In which region of the electromagnetic spectrum is located the radiation involved in this transition?

3.11   The attraction of an electron by a hole in a semiconductor can be described by the coulomb potential

$$U(r) = -\frac{e^2}{4\pi\,\varepsilon\,r},$$

where ε is the electric permissivity of the material and r the distance between the electron and the hole. In contrast to hydrogen, where the proton is much heavier than the electron, here the energy levels depend on the reduced mass μ of the electron–hole pair, and are given by

$$E_n = E_c - \frac{\mu\, e^4}{2\hbar^2 (4\pi\,\varepsilon)^2 n^2},$$

where E_c is the energy of the conduction band. For Cu_2O, which has $\varepsilon = 10\varepsilon_0$, the frequencies of the corresponding transitions obtained experimentally can be described by

$$\nu(\mathrm{cm}^{-1}) = 17\,508 - \frac{800}{n^2}.$$

(a)   From the results above, determine the reduced mass of the electron–hole pair;

(b)   Determine also the average radius of the orbital for the state ψ_{100};

(c)   Draw the energy levels relative to the energy of the conduction band.

3.12   Check, by direct substitution, that the eigenfunction ψ_{100} for the hydrogen atom, given in Table 3.2, is a solution of the time-independent Schrödinger equation, and obtain the values of the constants a_0 and E.

3.13   Show that the eigenfunctions ψ_{100} and ψ_{211} given in Table 3.2 are normalized.

3.14   From the expression of the angular momentum operator given in Table 3.1, it can be shown that in Cartesian coordinates its component z is given by

$$L_{zop} = -i\hbar\, \partial/\partial\varphi,$$

and its module squared is

$$L_{op}^2 = -\hbar^2\left[\frac{1}{\sin\theta}\frac{\partial}{\partial\theta}\left(\sin\theta\frac{\partial}{\partial\theta}\right) + \frac{1}{\sin^2\theta}\frac{\partial^2}{\partial\varphi^2}\right].$$

(a)   Show that the eigenfunctions ψ_{nlm_l} of the hydrogen atom are also eigenfunctions of L_{zop}, and give an interpretation for the quantum number m_l.

(b)   Show that they are also eigenfunctions of L_{op}^2, and interpret the quantum number l [Suggestion: use the expression above combined with Eq. (3.61)].

3.15   An electron in the hydrogen atom has a wave function

$$\psi = A(6 - r/a_0)\frac{r}{a_0}e^{-r/3a_0}\sin\theta\, e^{i\varphi}.$$

Calculate the electron energy by substituting this function into the Schrödinger equation.

3.16   Calculate the integral of the probability density $\Psi^*\Psi$ in the volume between the spheres of radii r and $r + dr$, for the wave function of the hydrogen atom given by Eq. (3.69), and show that the radial probability density is given by Eq. (3.70).

Further Reading

J.J. Brehm, W.J. Mullins, *Introduction to the Structure of Matter—A Course in Modern Physics* (Wiley, New York, 1989)

R. Eisberg, R. Resnick, *Quantum Physics of Atoms, Molecules, Solids, Nuclei, and Particles* (Wiley, New York, 1985)

D.J. Griffiths, D.F. Schroeter, *Introduction to Quantum Mechanics*, 3rd edn. (Cambridge University Press, New York, 2018)

S.T. Thornton, A. Rex, *Modern Physics for Scientists and Engineers* (Brooks/Cole, Cengage Learning, Boston, 2013)

P.A. Tipler, R.A. Llewellyn, *Modern Physics* (W. H. Freeman & Co., New York, 2012)

Chapter 4
Electrons in Crystals

In this chapter we use quantum mechanics to introduce the concept of energy bands in solids, that is essential to explain why, in terms of the electric properties, materials behave as metals, insulators, or semiconductors. We then present the meaning of the electron effective mass and study the behavior of electrons based on the statistical Fermi–Dirac distribution. Finally, we discuss the mechanism of the electric current in conducting materials and calculate the electric conductivity in terms of material parameters.

4.1 Energy Bands in Crystals

In this Chapter we study some basic properties of electrons in crystals, which are essential for the understanding of the mechanisms responsible for the electric current in materials and, therefore, for its use in Electronics.

As we saw in the previous chapter, an electron in an isolated atom has stationary quantum states characterized by discrete and quantized energy levels, corresponding to the atomic orbitals designated by 1s, 2s, 2p, 3s, 3p, 3d, etc. In an atom with many electrons, the ground state is obtained by distributing the various electrons in the lowest possible energy levels, obeying the Pauli Exclusion Principle. Since the electron has spin, each orbital state may accommodate two electrons with opposite spins. The question we now ask is: how are the electronic states modified when a large number of atoms approach each other to make a crystal?

The problem of finding the quantum states in a crystal is much more complicated than in an isolated atom, since the outer electrons of each atom also interact with the electrons of neighboring atoms. A crude explanation of what happens is as follows: When two isolated atoms are brought into close proximity, the energy levels of a certain state in the two atoms are slightly disturbed by the presence of the neighbor, resulting in two closely spaced levels. This is analogous to what happens when two identical oscillators are weakly coupled, they may oscillate in two close frequencies.

S. M. Rezende, *Introduction to Electronic Materials and Devices*,
https://doi.org/10.1007/978-3-030-81772-5_4

Thus, if a large number of atoms are brought into close proximity, the energy levels of the electrons in the same state in different atoms form an almost continuous energy band. This is illustrated in Fig. 4.1, showing the variation of the energies of the electronic states with the interatomic distance for N sodium atoms, that have electronic configuration $1s^2\ 2s^2\ 2p^6\ 3s$. For very large distances the energy levels of equivalent states coincide, and are the same as those of an isolated atom. As the distance decreases, the levels split due to the interaction with the neighbors, giving rise to several energy bands. At the equilibrium separation distance a, there are four bands, each corresponding to an orbital state. Since each shell with quantum orbital number l has $2(2\,l + 1)$ states, the number of states in a band is $2(2\,l + 1)N$. This description of the origin of the energy bands in crystals is very crude and hides some essential features of the electronic states. Actually, it is the wave nature of the electrons in crystals that gives rise to the energy bands, in a manner analogous to the formation of the various branches in the dispersion relation of elastic waves, such as those in Fig. 2.10.

The quantum calculation of electronic states and energies in a solid is quite complex, and can only be done with several approximations to the problem. The first is to assume that the nuclei of the atoms are fixed and with known positions in the crystal structure. Another one is to consider that the problem involves a single electron (one-electron model), and that all other electrons are considered an integral part of the ions that create a periodic potential. This is illustrated in Fig. 4.2, which shows qualitatively the potential seen by an electron along an axis in the crystal. The periodic potential to which the electron is subjected leads to solutions of the Schrödinger equation with energies that form bands. Since the solution for a periodic potential, even the simplest ones, is complex, let us understand what happens with an approximate model. In the case of alkali metals, such as sodium, the electron 3s of the last shell "sees" a potential of the nucleus that is shielded by the electrons

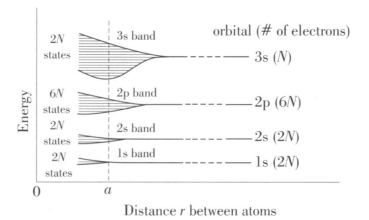

Fig. 4.1 Illustration of the formation of bands of energy levels due to the proximity of N sodium atoms

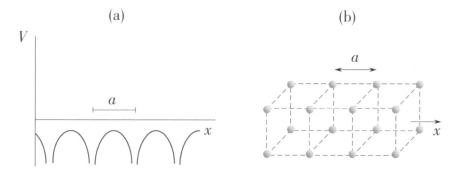

Fig. 4.2 **a** Potential energy V for an electron moving along the x axis of the crystal shown in (**b**)

of the inner shells, so that it is almost free. For this electron we can assume, in a first approximation, that the potential is a well with infinite walls at the surfaces of the crystal and constant inside it, as in Fig. 3.2. In this case, as we saw in Sect. 3.2, the electron wave functions are of the type

$$\Psi(\vec{r}, t) = A\, e^{i\vec{k}\cdot\vec{r}-i\omega t}, \tag{4.1}$$

where its energy is

$$E = \hbar\omega = \frac{\hbar^2 k^2}{2m}, \tag{4.2}$$

and the wave number k has discrete values as in Eq. (3.43), $k_n = n\,\pi/L$, where L is the dimension of the crystal in the direction of propagation. The dispersion relation (4.2) is represented by the dashed curve in Fig. 4.3. However, since the potential is not uniform inside the well, its small periodic variation has an effect on the electron wave function and consequently on the dispersion relation (4.2). This effect is noticeable in periodic ranges in k-space, and can be understood with an analogy to the effect of a diffraction grating on light. Considering the periodicity of the crystal in one dimension, the waves most affected are those that have a wave number satisfying the Bragg condition for diffraction

$$2a\,\sin\theta = m\lambda = m\,2\pi/k. \tag{4.3}$$

Electron waves that satisfy the condition (4.3) are reflected by the crystal, resulting in standing waves. Depending on the spatial configuration of the standing wave relative to the lattice, it can have two energy values. Thus, at the points $k = m\pi/a$, where m is a positive or negative integer, the dispersion curve separates in two branches. This gives rise to the solid lines in Fig. 4.3, which represent the electron dispersion relation in the periodic potential. The separation of the curves results in

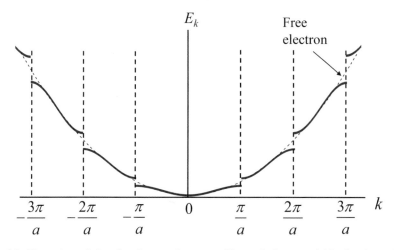

Fig. 4.3 Dispersion relation for electrons in a crystalline periodic potential in the almost free-electron model

energy bands for the electronic states. This means that electrons can only occupy states with energy in one of the bands in Fig. 4.3.

The model of a solid as a potential well with almost free electrons is a reasonable approximation for a metal like sodium. However, more generally the electron wave functions do not have the simple plane wave form of Eq. (4.1). Despite this, the problem can still be treated with plane waves because of a general result of great importance, the Bloch's theorem, that applies to systems that are invariant under translations. Let us suppose that the interactions on an electron can be represented by an effective one-electron potential $U(\vec{r})$, that is constant in time. Since the atoms, or ions, in a crystal are arranged in a regular periodic array, this potential is invariant under translations by a multiple of a primitive cell unit vector, i.e.,

$$U(\vec{r} + n\vec{a}) = U(\vec{r}), \tag{4.4}$$

where n is an integer. In this case, it is possible to show that the time-independent Schrödinger equation with a potential that has translation symmetry, as in (4.4), has solutions of the type

$$\psi(\vec{r}) = e^{i\vec{k}\cdot\vec{r}} u_k(\vec{r}), \tag{4.5}$$

where $u_k(\vec{r})$ is a function with the same periodicity as $U(\vec{r})$. This wave function represents a plane wave, whose amplitude is modulated by a periodic function that reflects the effect of the crystalline potential. The functions (4.5) are called **Bloch's functions**. This result, that can be demonstrated by replacing the function (4.5) in the Schrödinger equation, is of great importance in crystals, as it applies to any type of excitation. The elastic waves studied in Chap. 2 are an example that (4.5) applies to a periodic atomic chain. In the case of electrons, the important consequence of (4.5) is that in a crystal they are described by waves, characterized by a wave vector $\vec{k}$ and an energy E_k. The energy is a function, not only of the modulus of the wave vector, but also of its direction in the crystal. Since the wave vector can have any direction in k-space, in general the variation of E_k with k is represented for the main symmetry directions in the crystals. Thus, Fig. 4.3 can represent the variation of energy with the wave vector in the direction [100] of a cubic crystal. This way of representing the energy of the electronic states is called the extended zone scheme.

Another more useful way of representing the energy bands is in the so-called reduced zone scheme. Consider initially a simple one-dimensional model, as in Fig. 4.3. The points in a one-dimensional wave vector space with coordinates $n2\pi/a$, where n is an integer, form a periodic array called **reciprocal lattice**. In three dimensions each space lattice as in Fig. 1.4 has a different reciprocal lattice. The primitive cell of each reciprocal lattice is called **first Brillouin zone** (BZ). In the one-dimensional model, the first Brillouin zone corresponds to wave numbers in the range $-\pi/a < k < \pi/a$, while the second Brillouin zone corresponds to $-2\pi/a < k < -\pi/a$, etc. Thus, as shown in Fig. 4.4a, an electron state with wave number k' in the range $\pi/a < k' < 2\pi/a$ is in the second BZ, and has energy in another band. But then, if we subtract from k' a wave number $G = 2\pi/a$, this results in a wave function that is identical to that of $k = k' - G$, because of the result (4.5). Thus, it is possible to translate the bands in the momentum space by a multiple of G, that is, $n2\pi/a$, in order to represent all bands in the first Brillouin zone. This operation, shown in Fig. 4.4 for the first bands, results in the **reduced scheme to the first zone**. In this scheme it is evident that there are no electronic states between the energy bands. For this reason, the regions between the bands are called forbidden-energy gaps, or just **energy gaps**.

In the case of a three-dimensional crystal, the representation of the bands is a little more complicated. Figure 4.5a shows the surfaces that limit the first Brillouin

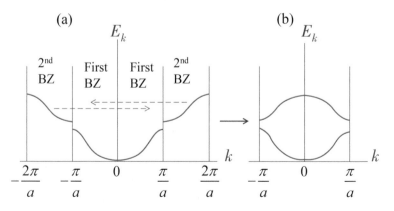

Fig. 4.4 a Illustration of the displacement of the two bands in the second Brillouin zone by $\pm 2\pi/a$. **b** Reduced first-zone scheme resulting from the displacement

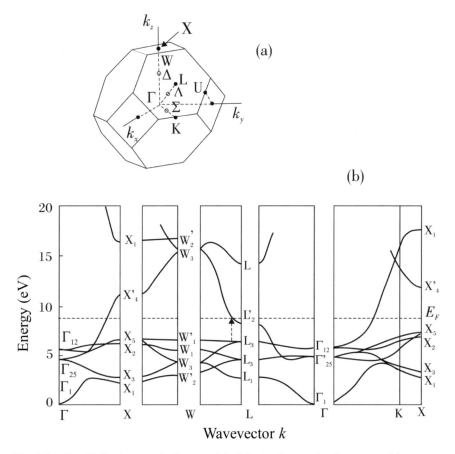

Fig. 4.5 a First Brillouin zone of a fcc crystal. **b** Calculated energy band structure of fcc copper. The Fermi energy E_F will be defined in Sect. 4.4. Reprinted with permission from B. Segal, Phys. Rev. **125**, 109 (1962). Copyright (1962) by the American Physical Society

zone of a face-centered cubic lattice. Figure 4.5b shows the energy band structure of crystalline copper with the fcc lattice. Evidently, in the three-dimensional crystal it is not possible to represent the variation of the energy in all directions of $\vec{k}$. Then, one chooses the main directions of $\vec{k}$ in the first Brillouin zone, shown in Fig. 4.5a. The horizontal axis is segmented and in each section one represents the energy bands for one main symmetry direction, indicated by the letters that designate the characteristic points of the Brillouin zone. Note that for each wave vector $\vec{k}$ the electron can have several wave functions, each with an energy in a different band.

To close this section, we note that due to the boundary conditions on the crystal, k cannot assume an arbitrary value. The allowed values are discrete, as in the infinite square well potential, so that the number of states in each band is finite. If the number of primitive cells in the crystal is N, **each band** contains $2N$ electronic states, where the factor 2 is due to the two possible spin states. This result comes from Eq. (3.43) generalized for three dimensions. In one dimension, k can assume the values $k = m\pi/N_1 a$, where N_1 is the number of primitive cells of size a along the length of the crystal. Since m is a positive integer, in the range between 0 and π/a there are N_1 different values for k, and therefore, the number of electronic states in each band is $2N_1$. Note that if we allow k to have positive or negative values, as it is more appropriate for a travelling wave like (4.1), it is necessary to change the boundary conditions so that $k = m2\pi/N_1 a$, where m is a positive or negative integer. But this does not change the number of electronic states in each band.

The formation of the ground state of the crystal is done by filling the discrete levels of lower energy states with electrons, analogously to what occurs in an atom. As we will see in the next section, the result of this filling process determines whether the solid is an electric insulator or a conductor.

4.2 Conductors, Insulators and Semiconductors

In a crystal with n electrons, the ground state is obtained by filling the lowest energy levels with only one electron in each state, so as to satisfy the Pauli exclusion principle. Since there are $2N$ states in each band, the number of occupied bands in the ground state is $n/2N$. Considering that n/N is the number of electrons per primitive cell, this is an integer number, and therefore $n/2N$ is an integer or half-integer. Thus, in a crystal at 0 K, there are several bands completely filled with electrons, and the last (upper) one is necessarily filled completely or in half. The conducting properties of the crystal depend essentially on whether the last band is full or not. The reason is that the wave vector can have any direction in k-space and the bands are symmetrical, so that

$$\sum_{\substack{all\ states \\ in\ a\ band}} \vec{k} = 0. \qquad (4.6)$$

The electric **insulators**, which are materials that do not conduct electric current, are made of crystals that have the last band completely filled. In these crystals, the application of an external electric field cannot change the total electron momentum, which is null, since all available states are occupied. In this case there is no net electron flow when the field is applied and thus no electric current. So, the necessary condition for a crystal to be insulating is that it has an even number of electrons per unit cell (the condition is not sufficient, as we shall see below). Figure 4.6a shows a possible distribution of the last energy bands in an insulating crystal and their occupation by electrons. The energy level above which there are no occupied states at a temperature $T = 0$ K is called the **Fermi level**, E_F. In Sect. 4.4 we shall discuss in more detail the important role that the Fermi level plays in the properties of solids.

Conductors, also called metals, are materials in which at $T = 0$ K the electrons flow under the action of an electric field producing an electric current. The condition

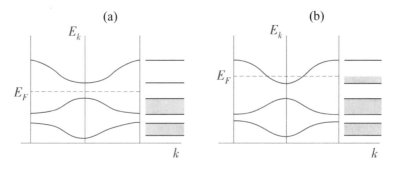

Fig. 4.6 Energy band structure and occupation of the electron states in insulators (**a**) and in metals (**b**). The colored regions represent the occupied bands

to be a conductor is to have the last band partially filled. This occurs whenever the number of electrons per primitive cell is odd. In this case it is possible to change the states of the electrons with an electric field, resulting in an electric current. In this category are the alkali metals (Li^3, Na^{11}, K^{19}, etc.) and the noble metals (Cu^{29}, Ag^{47}, Au^{79}), that have an odd number of electrons. In this case, as shown in Fig. 4.6b, the energy band above the last full band is only partially filled. As shown in Fig. 4.6b, in metals the Fermi level is located in the middle of the last occupied energy band.

It is also possible to have a metal formed by atoms with an even number of electrons in the unit cell, such as the alkaline earth metals (Be^4, Mg^{12}, Ca^{20}, Sr^{38}, Ba^{56}). In these metals, the distribution of bands is not as simple as those in Fig. 4.6. As shown in Fig. 4.7, in these materials band 1, which would normally be the last full one, has its maximum above the minimum of the next band 2. Since the electrons occupy the lowest energy states, the electrons that would be at the top of band 1 actually occupy states in band 2, so that both bands are partially filled. In these materials, the application of an external electric field causes electrons to change states, which results in an electric current. Therefore, they are also conductors, but not as good as the alkali metals. For this reason, they are also called **semimetals**.

In an insulating crystal, only at a temperature $T = 0$ K, the last occupied band, called **valence band**, is completely filled. When the temperature is above 0 K, electrons in the valence band can gain enough thermal energy to jump to the next band,

Fig. 4.7 Energy band structure and occupation of states in a semimetal

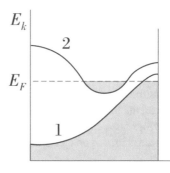

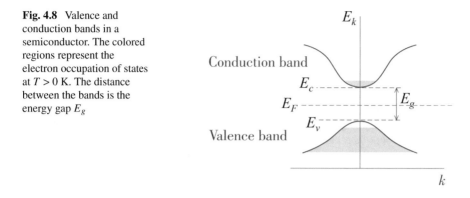

Fig. 4.8 Valence and conduction bands in a semiconductor. The colored regions represent the electron occupation of states at $T > 0$ K. The distance between the bands is the energy gap E_g

called **conduction band**, which was empty at $T = 0$ K. The transfer of electrons to the conduction band leaves in the valence band vacant states that behave as carriers of positive electric charge, called **holes**. Electrons in the conduction band and holes in the valence band produce electric current under the action of an external electric field. The conductivity of the material depends on the number of electrons that are excited into the conduction band, which can be calculated probabilistically, as we shall see in the next section. This number increases with increasing temperature and also with decreasing in the energy separation of the two bands. The difference between the minimum of the conduction band (E_c) and the maximum of the valence band (E_v) is called energy gap, represented by E_g. Materials that are insulating at 0 K but have a relatively small E_g, on the order of 1 eV or less at room temperature, have significant conductivity and are called **semiconductors**. Figure 4.8 illustrates the occupation of the valence and conduction bands in a semiconductor at $T > 0$. In these materials the number of electrons in the conduction band can be significant compared to an insulator, but it is still much smaller than the number of free electrons in a metal. Therefore, the conductivity of semiconductors is much smaller than in metals. The main difference between an insulator and a semiconductor is in the value of E_g. For example, silicon has $E_g = 1.1$ eV and is a semiconductor, while diamond, which has the same structure as Si, formed by C atoms, has $E_g = 5$ eV, and is an excellent insulator. Silicon oxide, SiO, has $E_g = 8$ eV and is also an insulator. The difference in the values of E_g may not seem so big to produce such a radical change in conductivity. However, as we will see later, the occupation of the conduction band decreases exponentially with the increase of the ratio E_g/k_BT.

4.3 Effective Mass

For the study of the electric properties of metals and semiconductors, it will be necessary first to understand how an electron behaves in a crystal under the action of an external electric field. As discussed in Sect. 3.3.1, the electron is described by a wave packet that propagates with group velocity $v_g = \partial\omega/\partial k$. Since the electron energy is $E = \hbar\omega$, we can write

$$\frac{\partial E}{\partial k} = \hbar v_g. \tag{4.7}$$

If the electron is subjected to a force F, for instance due to an electric field, its energy varies in a displacement dx, such that $dE = F\,dx$. Using (4.7) we see that the electron velocity is related to the force by,

$$F\,dx = \hbar v_g dk.$$

Using $dx = v_g dt$ this gives

$$F = \hbar \frac{dk}{dt}. \tag{4.8}$$

This result was somehow expected, because as the electron momentum is $\hbar k$, Eq. (4.8) is nothing more than Newton's second law. However, it is also somewhat surprising, since we might have expected that the potential of the crystal lattice would have a more drastic effect on the electron motion. Equation (4.8) means that the lattice does not change the momentum variation, what changes is the dependence of the energy on momentum, which corresponds to changing the electron mass. To show this we use Eq. (4.7) to express the electron acceleration as a function of E and k

$$a = \frac{\partial v_g}{\partial t} = \hbar^{-1}\frac{\partial^2 E}{\partial k \partial t} = \hbar^{-1}\frac{\partial^2 E}{\partial k^2}\frac{dk}{dt}. \tag{4.9}$$

Substituting in this equation the expression for dk/dt from (4.8) we obtain

$$F = \frac{\hbar^2}{\partial^2 E/\partial k^2} a. \tag{4.10}$$

Recalling that $F = ma$, we see that under the action of an external force, the electron in the crystal behaves similarly to a free electron, but with an **effective mass** given by

$$m^* = \frac{\hbar^2}{\partial^2 E/\partial k^2}. \tag{4.11}$$

This result also applies to a free electron. In this case, using the dispersion relation (3.30) we obtain $m^* = m$, that is, the effective mass is the electron mass itself.

Equation (4.11) was obtained assuming that the energy only depends on the modulus of the wave vector. Actually, as shown in Fig. 4.5, the energy also depends on the direction of $\vec{k}$. This means that the effective mass depends on the direction of $\vec{k}$. In the most general definition, the effective mass is not a scalar, it is a tensor quantity represented by a matrix, whose element $\alpha\beta$ is given by

$$m^*_{\alpha\beta} = \frac{\hbar^2}{\partial^2 E / \partial k_\alpha \partial k_\beta}. \tag{4.12}$$

This definition applies to electrons in metals and in semiconductors.

4.4 Electron Behavior at $T > 0$: The Fermi-Dirac Distribution

As we have seen, at $T = 0$ K electrons in a crystal occupy the states with the lowest lying allowed energy levels so as to fill, one by one, all states up to a certain energy E_F, the Fermi energy, or Fermi level. At temperatures above 0 K electrons with energy close to the Fermi level are thermally excited to states above E_F. The thermal equilibrium distribution is calculated in statistical mechanics and takes into account that the electrons are particles indistinguishable from each other and that they obey the Pauli exclusion principle. Here we will simply state the known result and will use it to calculate quantities of interest. The probability of finding the states with energy in the range from E to $E + dE$, occupied with electrons, at a temperature T, is given by $f(E)\,dE$, where

$$f(E) = \frac{1}{1 + e^{(E-E_F)/k_B T}} \tag{4.13}$$

is the Fermi–Dirac distribution function. In this expression E_F is the Fermi level and k_B is the Boltzmann constant ($k_B = 1.38 \times 10^{-23}$ J/ K). The shape of $f(E)$ is shown in Fig. 4.9 for various temperatures. Note that at $T = 0$ the function is discontinuous at $E = E_F$, that is $f(E < E_F) = 1$ and $f(E > E_F) = 0$. This means that

Fig. 4.9 Fermi–Dirac distribution function at various temperatures

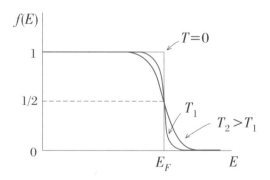

the states with $E < E_F$ are occupied by electrons while those with energies $E > E_F$ are empty. At temperatures above 0 K the Fermi–Dirac distribution changes mainly in the vicinity of E_F. Due to thermal excitation, the probability f (E) of states with $E > E_F$ to be occupied is nonzero. The value of f (E) decreases exponentially with the distance $E - E_F$ and increases exponentially with temperature. The probability f (E) of occupation of states with $E > E_F$ is the same as the probability $1 - f$ (E) of the states with $E < E_F$ to be empty. This is so because the Fermi–Dirac function f (E) is symmetrical about E_F, where its value is f $(E_F) = \frac{1}{2}$. In each material the value of E_F depends on the shape of the bands and the number of electrons.

The Fermi–Dirac distribution f (E) represents the probability of occupation of a state with energy E. To calculate the number of electrons in a given energy range, it is also necessary to know the number of states in that range. This number is given by the density of states D (E), which can be calculated from the relation E (k). Let us consider the simple model for a metal like the one in Sect. 4.1, in which free electrons are confined by an infinite potential well. In this case, the electron energy is characterized by a parabolic function given by Eq. (4.2)

$$E = \frac{\hbar^2 k^2}{2m}.$$

Actually, this result was demonstrated for a one-dimensional potential well. However, it also applies to a potential well in three dimensions with infinite walls. In this case, the wave number k is replaced by a wave vector $\vec{k}$ with three components k_x, k_y, k_z, so that the energy becomes

$$E = \frac{\hbar^2}{2m}(k_x^2 + k_y^2 + k_z^2). \tag{4.14}$$

Analogously to the problem in one dimension, the three components of the wave vector can only assume discrete values, that are determined by the boundary conditions on the crystal surfaces. Assuming that the crystal is a cube with sides of length L, the values are

$$k_x = n_x \frac{2\pi}{L}, \qquad k_y = n_y \frac{2\pi}{L}, \qquad k_z = n_z \frac{2\pi}{L}, \tag{4.15}$$

where n_x, n_y, and n_z are positive or negative integers. This result is a generalization of Eq. (3.43) for three dimensions and for travelling waves. Due to the spin of the electron, for each set of quantum numbers (n_x, n_y, n_z), and therefore in each volume $(2\pi/L)^3$ in space, there are **two** electronic states. The density of states $D(E)$ is a quantity that expresses the number of states with energy E. By definition, $VD(E)dE$ is the number of states with energy between E and $E + dE$, where V is the volume of the crystal. In wave vector space the surfaces of constant energy are spheres of radius k. Therefore, the number of states with energy in the range $(E, E + dE)$ is the volume between the spheres of radii k_E and k_{E+dE}, multiplied by number of states

Fig. 4.10 Density of
electronic states $D(E)$ for a
parabolic energy band

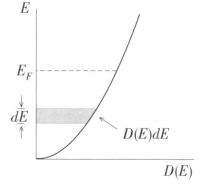

per unit volume in k-space. Since this is given by $2(2\pi/L)^3$, we have

$$V\,D(E)\,dE = 2(L/2\pi)^3\,4\pi k_E^2\,dk,\qquad(4.16)$$

where k_E is the modulus of the wave vector corresponding to energy E. From Eq. (4.2)
we have

$$k_E^2\,dk = \frac{1}{2}\left(\frac{2m}{\hbar^2}\right)^{3/2} E^{1/2}\,dE,$$

which, substituted in (4.16), gives for the density of states

$$D(E) = \frac{1}{2\pi^2}\left(\frac{2m}{\hbar^2}\right)^{3/2} E^{1/2}.\qquad(4.17)$$

Figure 4.10 shows a plot of the density of states for electrons with a parabolic
energy band. Note that the energy is placed on the vertical axis in order to facilitate
the visualization of the electron filling of the lower energy states. At $T = 0$, all states
with energy below the Fermi level E_F are filled. If there are N electrons in the band,
per unit volume, the condition that determines E_F is

$$\int_0^{E_F} D(E)\,dE = N.\qquad(4.18)$$

Using Eq. (4.17), this relation gives.

$$\frac{1}{3\pi^2}\left(\frac{2m}{\hbar^2}\right)^{3/2} E_F^{3/2} = N.\qquad(4.19)$$

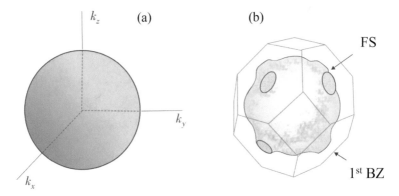

Fig. 4.11 **a** Fermi surface for a system of free electrons. **b** Fermi surface (FS) and the first Brillouin zone of fcc coper

From (4.19) we obtain an expression for the Fermi level of a system with N electrons with a parabolic energy band, at $T = 0$,

$$E_F = (3\pi^2 N)^{2/3} \frac{\hbar^2}{2m}. \tag{4.20}$$

At $T = 0$ all states with energy $E \leq E_F$ are occupied. These states are characterized by wave vectors with modulus $k \leq k_F$, where k_F, given by

$$k_F^2 = \frac{2m \, E_F}{\hbar^2}, \tag{4.21}$$

is called the **Fermi wave vector**, or **Fermi radius**. The surface in $\vec{k}$-space for which all states inside are occupied $T = 0$, is called **Fermi surface**. In an electron system with a parabolic energy band, this surface is a sphere of radius k_F given by (4.21), as illustrated in Fig. 4.11a. The parabolic band (4.14) is valid only for free electrons. In crystals, the variation of energy with $\vec{k}$ is more complicated, as illustrated by the copper bands in Fig. 4.5. In this case, the Fermi surface is not a sphere, it has a more complex shape. Figure 4.11b shows the Fermi surface of copper, that is contained in the first Brillouin zone.

Example 4.1 Sodium crystallizes with the bcc structure, with two atoms per unit cell, each with one electron 3s. Considering that the lattice parameter of sodium at $T = 0$ is 4.225 Å, calculate: (a) The Fermi energy; (b) The velocity of electrons with energy at the Fermi level, called **Fermi velocity** v_F.

(a)   To obtain the Fermi energy with Eq. (4.20), it is necessary first to calculate the number of free electrons per unit volume. Since there are two electrons per unit cell with lattice parameter a, we have

$$N = \frac{2}{a^3} = \frac{2}{4.225^3 \times 10^{-30}} = 2.65 \times 10^{28} \text{ m}^{-3}$$

The Fermi energy is related with N by Eq. (4.20)

$$E_F = (3\pi^2 N)^{2/3} \left(\frac{\hbar^2}{2m} \right).$$

In a first approximation we can consider the mass of the free electrons in sodium as the electron mass in vacuum, $m = 9.1 \times 10^{-31}$ kg. Thus.
$E_F = (3 \times 3.14^2 \times 2.65 \times 10^{28})^{2/3} \frac{1.05^2 \times 10^{-68}}{2 \times 9.1 \times 10^{-31}} = 5.15 \times 10^{-19}$ J or

$$E_F = \frac{5.15 \times 10^{-19}}{1.6 \times 10^{-19}} = 3.22 \text{ eV}$$

(b) Since the energy of the free electrons is a kinetic energy, the velocity can be calculated using the relation

$$E_F = \frac{1}{2} m v_F^2.$$

Thus

$$v_F = \left(\frac{2E_F}{m} \right)^{1/2} = \left(\frac{2 \times 5.15 \times 10^{-19}}{9.1 \times 10^{-31}} \right)^{1/2}$$

$$v_F = 1.06 \times 10^6 \text{ m/s} = 1.06 \times 10^8 \text{ cm/s}$$

At temperatures above zero, the probability of occupation of the electron states is given by $f(E)$ in Eq. (4.13), so that the number of electrons, per unit volume in the energy range between E and $E + dE$ is

$$dN = f(E) D(E) dE. \tag{4.22}$$

Figure 4.12 illustrates the product of the functions $f(E)$ and $D(E)$ and shows the area element corresponding to dN. Note that the electrons excited to states above the Fermi level are those mainly in states with energies below and close to E_F. This result is quite general. Whenever there is a disturbance in the electron system, states with energy close to E_F are the most affected. This disturbance may be due to thermal excitation, or to excitation produced by external fields. In the next section, we shall study the effect of an applied electric field.

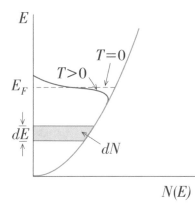

Fig. 4.12 Electron population $N(E) = f(E) D(E)$ for a parabolic energy band at $T > 0$. The area dN represents the number of electrons in the energy range dE

Example 4.2 Calculate the total energy of free electrons in a sample of sodium with volume 1 cm³, at $T = 0$.

The electron energy per unit volume is the sum of the free electron energies, which can be calculated using (4.22).

$$\frac{U}{V} = \int E \, dN = \int E \, f(E) \, D(E) \, dE.$$

At $T = 0$ K, the Fermi–Dirac distribution has a value 1 for $E < E_F$ and 0 for $E > E_F$, therefore, with Eq. (4.17) we have

$$\frac{U}{V} = \int_0^{E_F} E \, D(E) \, dE = \frac{1}{2\pi^2} \left(\frac{2m}{\hbar^2} \right)^{3/2} \int_0^{E_F} E^{3/2} \, dE = \frac{1}{2\pi^2} \left(\frac{2m}{\hbar^2} \right)^{3/2} \frac{2}{5} E_F^{5/2}.$$

Using (4.20) we can express this result in the form

$$\frac{U}{V} = \frac{3}{5} N E_F.$$

Thus, using the results of Example 4.1, we obtain

$$\frac{U}{V} = \frac{3}{5} \times 2.65 \times 10^{28} \times 5.15 \times 10^{-19} = 8.19 \times 10^9 \text{ J/m}^3.$$

So, the energy of the electrons in a sample of sodium with volume 1 cm³ is

$$U = 8.19 \times 10^9 \times 10^{-6} = 8.19 \times 10^3 \text{ J}.$$

4.5 The Mechanism of Electric Current in Metals

The electric current in a metal results from transport of electric charges by electrons. To understand the mechanism of the electric current we will have to use classical results combined with quantum concepts. When an external electric field is applied to the metal, the free electrons suffer the effect of this field superimposed on that of the crystalline potential. The effect of the latter is basically expressed by the electron effective mass m^*. The external field E exerts a force $\vec{F} = -e\vec{E}$ on the electron, so that its acceleration given by (4.10) and (4.11) is

$$a = \frac{dv}{dt} = -\frac{e}{m^*} E. \tag{4.23}$$

This result means that in a perfect crystal, a constant electric field E produces a constant acceleration on the electron and hence a velocity that increases linearly in time, $v = a\,t$. Equation (4.23) also implies that, even without an external field, electrons can have a constant nonzero velocity. This results from the fact that the steady state of the electron in a crystal without an external field is a plane wave, given by Eq. (4.5). This wave has a momentum $\hbar k$, that corresponds to a constant velocity. It turns out that the electron can only be in a plane wave steady state if the crystal is perfect, and at $T = 0$ K. At nonzero temperatures the crystal is not perfect because of the thermal lattice vibrations. Also, the crystal may have defects or impurities. To understand the electron behavior in a crystal we make a simple analogy with the motion of a skater along a row of regularly spaced obstacles, as illustrated in Fig. 4.13a. If the skater is well trained, he can perform this "zig-zag" motion without hitting the obstacles, and with constant average speed along the row. This motion is analogous to that of the electron in the perfect crystal, described by a plane wave with amplitude modulated by the periodic potential of the lattice, as in Eq. (4.5). However, if an obstacle is displaced from its normal position, or if there is an extra obstacle in the row, the skater will likely collide with it, as illustrated in Fig. 4.13b. What happens to an electron in the solid is somewhat analogous. If the regularity of the crystal lattice is disturbed, the electron remains in a steady state only for a certain period of time. The disturbance causes electron collisions, or scattering, resulting in a transition to another state. The two main sources of disturbances in the regularity of the lattice are the atomic vibrations due to thermal agitation, crystal defects, or the presence of atoms or ions of impurities. The collision with the lattice in

(a)   (b)

Fig. 4.13 a Illustration of a zig-zag motion of a skater along a row of regularly spaced obstacles. **b** Collision with an obstacle displaced from its regular position

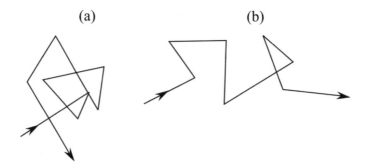

Fig. 4.14 Illustration of the electron motion in a solid. **a** Without an external electric field, the average velocity is zero. **b** In the presence of an electric field, in addition to the fast and random motion, there is a continuous displacement in the direction opposite to the field resulting in an electric current

thermal motion corresponds to the scattering of electrons by phonons. This process is similar to the collision between particles, in which there is conservation of energy and momentum. Due to the scattering processes, in the absence of an external electric field the average electron velocity is zero, as illustrated in Fig. 4.14a.

When an electric field is applied to the material, in addition to the fast and random motion of the electrons caused by the collisions, there is a continuous displacement in the direction opposite to the electric field. This displacement results in a transport of electric charge, that is, an electric current, called **drift current**, or **conduction current**.

In the quantum description of the behavior of electrons, it is necessary to consider that at $T = 0$ and without an external field, all states in k-space inside the Fermi surface are occupied. This is illustrated in Fig. 4.15a by a section in the plane (k_x,

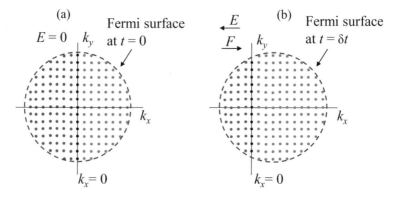

Fig. 4.15 **a** View of the Fermi surface in the k_x, k_y plane in the absence of an external electric field at $t = 0$. The dots represent the occupied electron states. **b** Fermi surface at an instant δt after the application of an electric field in the $- x$ direction. The occupied states are displaced by δk_x given by Eq. (4.24), so that the total momentum in the x direction is positive

k_y) of the Fermi sphere, which applies to the case of free electrons. Since all states with $k < k_F$ are filled, to each state $+\vec{k}$ occupied, another state $-\vec{k}$ is also occupied. Thus, $\sum \vec{k} = 0$ so that the electric current is zero. However, if an electric field E is applied in the direction $-x$ at the instant $t = 0$, the states $\vec{k}$ of the electrons change according to Eq. (4.8). Since the force on the electrons is $F_x = (-e)(-E) = eE$, the variation of $\vec{k}$ in the time interval δt is

$$\delta k_x = \frac{eE}{\hbar} \delta t. \tag{4.24}$$

As a consequence of Eq. (4.24), each electron in state $\vec{k}$ goes to another state $\vec{k} + \hat{x} \delta k_x$ after an interval δt, resulting in the occupation of states shown in Fig. 4.15b. The net result is a total momentum per unit of volume given by $N\,\delta k_x$, where N is electron concentration in the band. This results in an electric current in the direction $+ x$. Note that although all electrons have their states changed by the action of the electric field, only the states close to the Fermi surface contribute to make the vectorial sum of the velocities different than zero. Due to collisions, the displacement of the Fermi sphere stops after an average time interval τ, called **collision time**. The resulting average velocity can be obtained from Eq. (4.23), or directly from Eq. (4.24) using the relation $\vec{v} = \hbar \vec{k}/m^*$. This average velocity, called **drift velocity**, is then

$$v_x = \frac{eE\,\tau}{m^*}. \tag{4.25}$$

Considering that there are N free electrons per unit of volume, we obtain for the electric current density

$$J_x = (-e)N\,v_x = -N\,e^2\tau E/m^*. \tag{4.26}$$

This equation has the form of Ohm's law that relates the applied voltage V, the electric current intensity I, and the resistance R

$$I = V/R. \tag{4.27}$$

Considering that the resistance of a conductor of length L and cross section area A is

$$R = \frac{1}{\sigma}\frac{L}{A}, \tag{4.28}$$

where $\sigma = 1/\rho$ is the conductivity and ρ is the resistivity, we can use Eq. (4.28) in (4.27), together with the relations $J = I/A$ and $V = E L$, to write Ohm's law in the form

$$J = \sigma\,E. \tag{4.29}$$

Fig. 4.16 Measured
variation of the resistivity of
two samples of potassium at
low temperatures. The upper
curve corresponds to a
sample with larger amount of
impurities than the one of the
lower curve. Reprinted with
kind permission from D. K.
MacDonald and K.
Mendelssohn, Proc. Roy.
Soc. (London) A202, 103
(1950)

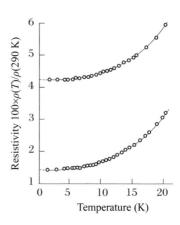

Substituting (4.29) in (4.26) we obtain the conductivity of a metal in terms of material parameters

$$\sigma = \frac{N e^2 \tau}{m^*}. \tag{4.30}$$

In a conductor with a perfect crystal lattice at $T = 0$, the collision time is infinite and therefore the conductivity is also infinite. In a real crystal, the collision time is limited because of the scattering of electrons by phonons and by the impurities and imperfections of the lattice. As thermal agitation increases with temperature, the collision time due to scattering by phonons decreases with increasing temperature. On the other hand, the contributions of impurities and imperfections do not vary with temperature and exist even at $T = 0$. This is illustrated in Fig. 4.16 which shows the measured variation of the resistivity ρ of potassium with the temperature. The increase in ρ with temperature is due to scattering by phonons, while the contribution at $T = 0$ comes from impurities and imperfections. The upper curve corresponds to a sample of a potassium with a larger amount of impurities than the lower one and, therefore, has a larger value of ρ.

Figure 4.17 shows the conductivity at room temperature for a variety of materials. It varies from 10^{-18} Ω^{-1} m^{-1} in quartz, which is a very good insulator, to about 10^8 Ω^{-1} m^{-1} in copper, which is a good conductor. This range of variation of 10^{26} is

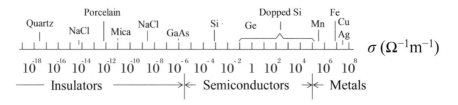

Fig. 4.17 Conductivity in Ω^{-1} m^{-1} for a variety of materials at room temperature

the largest known for the same physical quantity. In fact, the range of variation of σ is even larger since superconducting materials have conductivity several orders of magnitude larger than copper.

To close this chapter, let us give numerical estimates for some important quantities involved in the mechanism of electric current. Consider the case of copper, which at room temperature has conductivity of $\sigma \sim 10^8 \ \Omega^{-1} \ m^{-1}$. Since the number of free electrons is $N \sim 10^{23} \ cm^{-3}$, using Eq. (4.30) and the values for the mass and charge of the electron (Appendix B), we obtain for the collision time $\tau \sim 10^{-13}$ s at room temperature. The average distance that the electron travels between two collisions is called **mean free path**, denoted by $\bar{l}$. Since the electrons involved in the current are in states close to the Fermi surface, the mean free path is

$$\bar{l} = v_F \, \tau, \tag{4.31}$$

where v_F is the **Fermi velocity**, related to the radius of the Fermi sphere by the relation $v_F = \hbar k_F / m^*$. Using this expression and Eqs. (4.20) and (4.21), we obtain for copper $v_F \sim 10^6$ m/s. With Eq. (4.31) we obtain the mean free path for copper at room temperature $\bar{l} \sim 10^{-7}$ m $= 10^3$ Å, which corresponds to the length of hundreds of Cu unit cells in the crystal. It is quite surprising that in copper, at room temperature, an electron travels through hundreds of unit cells without colliding with the atoms.

From Eq. (4.25) we can also estimate the electron drift velocity. Considering that a voltage of 1 V is applied to the ends of a 1 m long copper wire, the electric field in the wire is $E = 1$ V/m. Using $\tau \sim 10^{-13}$ s we obtain with Eq. (4.25) $v_x \sim 10^{-2}$ m/s. This shows that the drift velocity is several orders of magnitude smaller than the velocity of the electrons between two collisions. In other words, the drift motion is much slower than the random motion of the electrons between two collisions.

Example 4.3 Considering that at room temperature free electrons in silver have collision time of $\tau = 3.8 \times 10^{-14}$ s, and concentration $N = 5.86 \times 10^{22}$ cm^{-3}, calculate: (a) The resistance of a silver wire with cross-section area 0.1 mm^2 and length 100 m; (b) The electric current in the wire when a voltage of 1.6 V is applied to its ends; (c) The drift velocity of electrons in the conditions of item (b).

(a)   For the calculation of the resistance it is necessary first to obtain the conductivity. With Eq. (4.30) we have

$$\sigma = \frac{N \, e^2 \, \tau}{m^*} = \frac{5.86 \times 10^{22} \times 10^6 \times 1.6^2 \times 10^{-38} \times 3.8 \times 10^{-14}}{9.1 \times 10^{-31}}$$
$$= 6.26 \times 10^7 \ \Omega^{-1} \ m^{-1}.$$

The wire resistance is then

$$R = \frac{1}{\sigma}\frac{L}{A} = \frac{100}{6.26 \times 10^7 \times 1 \times 10^{-7}} = 16\,\Omega.$$

(b)   The electric current is

$$I = \frac{V}{R} = \frac{1.6}{16} = 0.1A.$$

(c)   The drift velocity, related to the current density by Eq. (4.26) is

$$v_x = \frac{J}{Ne} = \frac{I}{NeA} = \frac{0.1}{5.86 \times 10^{28} \times 1.6 \times 10^{-19} \times 10^{-7}} = 1.7 \times 10^{-4}\,\text{m/s}.$$

Problems

4.1   Silver crystallizes in the fcc structure, with lattice parameter 4.086 Å, and four atoms per unit cell, each with a 5s electron. Calculate the concentration of free electrons in silver in cm^{-3}.

4.2   In a first approximation the electrons in silver have a 5s parabolic energy band. Calculate the Fermi level, E_F, in eV, assuming that the free electron mass is equal to the mass of electrons in a vacuum.

4.3   Using the results of Problems 4.1 and 4.2 to calculate for silver.

(a)   The Fermi velocity of the electrons.
(b)   The wavelength of a free electron moving with the Fermi velocity and compare it with the distance between the atoms (~ 4 Å).
(c)   At what temperature the probability of finding electrons with energy $E = E_F + 0.1$ eV is 10%?

4.4   A certain metal has a Fermi level $E_F = 1$ eV. Plot, preferably on a computer, the Fermi–Dirac distribution function for $T = 10$ K and 300 K.

4.5   Show that the probability that an electronic state with energy $E = E_F + \Delta E$ is occupied is equal to the probability that the state with energy $E = E_F - \Delta E$ is empty.

4.6   In a copper wire of cross-section area 1 mm^2 there is a current of 10 A. Considering that the concentration of free electrons is $N = 8.5 \times 10^{22}$ cm^{-3}, calculate:

(a)   The Fermi level E_F, using the same approximations of Problem 4.2;
(b)   The Fermi velocity;
(c)   The electron drift velocity, and compare it with the Fermi velocity.

4.7 Consider the resistivity of copper at room temperature 1.7×10^{-8} Ωm and use the data and results from the previous problem to calculate:

(a) The average electron collision time;
(b) The mean free-path of electrons.

Further Reading

R.E. Hummel, *Electronic Properties of Materials* (Springer, Berlin, 2011)
D. Jiles, *Introduction to the Electronic Properties of Materials* (CRC Press. Boca Raton, FA, 2017)
C. Kittel, *Introduction to Solid State Physics* (Wiley, New York, 2004)
D.J. Roulston, *An Introduction to the Physics of Semiconductor Devices* (Oxford University Press, Oxford, 1999)
L. Solymar, D. Walsh, *Lectures on the Electrical Properties of Materials* (Oxford University Press, Oxford, 2009)
F.F.Y. Wang, *Introduction to Solid State Electronics* (North-Holland, Amsterdam, 1989)

Chapter 5
Semiconductor Materials

In this chapter we present several basic concepts and properties of semiconductors that are essential for understanding the operation of devices. We begin with the introduction of the concept of holes, that together with electrons are the charge carriers in electric conduction processes. Then we show the key role played by crystal impurities used in the preparation of two types of extrinsic semiconductors, p and n, that make possible the fabrication of devices. One section is devoted to the calculation of the concentrations of electrons and holes in both types of semiconductors, that strongly depend on the position of the Fermi level relative to the valence and conduction energy bands. Finally, we study the various mechanisms by which electrons and holes carry electric current in semiconductors.

5.1 Semiconductors

As studied in the previous chapter, semiconductors are characterized by a full valence band and an empty conduction band at $T = 0$, separated by a relatively small energy gap, E_g, on the order of 1 eV or less. Due to the small gap, at room temperature the number of electrons in the conduction band is appreciable, although much smaller than the number of free electrons in metals. This results in conductivities with values intermediate between those of insulators and of metals, as illustrated in Fig. 4.17. This is the reason for the name **semiconductor**. The concentration of electrons in the conduction band of a pure semiconductor varies exponentially with temperature, which makes its conductivity strongly dependent on temperature. This is one of the reasons why pure semiconductors, also called **intrinsic**, are not much used in devices.

The conductivity of semiconductors can also be drastically changed with the presence of impurities, that is, atoms of elements different from those in the pure semiconductor crystal. It is this property that makes it possible to manufacture a variety of electronic devices from the same semiconductor material. The process of placing

© The Author(s), under exclusive license to Springer Nature Switzerland AG 2022
S. M. Rezende, *Introduction to Electronic Materials and Devices*,
https://doi.org/10.1007/978-3-030-81772-5_5

impurities from known elements in a semiconductor is called doping. Semiconductors with impurities are called **doped** or **extrinsic**.

The most important semiconductor for electronics is silicon. It has the same crystalline structure as diamond, shown in Fig. 1.8b, formed only by atoms of the element Si, that belongs to group IV of the periodic table. Figure 5.1 shows the electronic energy band structure of silicon. The maximum of the valence band occurs at $k = 0$, the Γ point of the Brillouin zone. The top of the valence band is taken as the reference for the energy scale, that is $E = 0$. The minimum of the conduction band occurs at a nonzero wave vector along the direction [100], close to point X on the egde of the Brillouin zone, with energy 1.12 eV. This is the value of the energy gap of Si at $T = 300$ K, $E_g = 1.12$ eV. Actually, the value of the gap varies with temperature. In Si at $T = 0$, the gap is 1.16 eV and decreases with increasing temperature. Another important semiconductor is germanium, also formed by an element of group IV, Ge, which also has the crystalline structure of diamond. Ge has a band structure similar to that of Si, but with a smaller gap, $E_g = 0.66$ eV at room temperature. This makes its electrical properties more sensitive to changes in temperature than in Si.

In Ge and Si the valence and conduction bands result from electronic states s and p that overlap. Since there are two s and six p states, there are eight hybrid bands s + p, which are separated into two sets of four bands each. The four lower energy bands can accommodate 4 N electrons. Since Si and Ge have four valence electrons per atom, the four s + p bands with smaller energy are completely filled, constituting the valence bands, shown in Fig. 5.1 for Si.

One of the most important semiconductors for application in opto-electronics is gallium arsenide, GaAs. It is formed by the elements Ga and As, from groups III and V respectively, and crystallizes in the *zinc-blende* structure of Fig. 1.8a. In the formation of GaAs, the atom of As loses an electron for a neighbor Ga atom, leaving both with four electrons in the $4s^2 \ 4p^2$ shells. Similar to Si and Ge, at $T = 0$, GaAs has a completely filled valence band and an empty conduction band. The

Fig. 5.1 Calculated electronic energy band structure of silicon. Reprinted with permission from J. R. Chelikowsky and M. L. Cohen, Phys. Rev. B 14, 556 (1976). Copyright (1976) by the American Physical Society

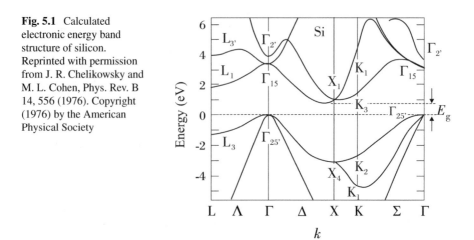

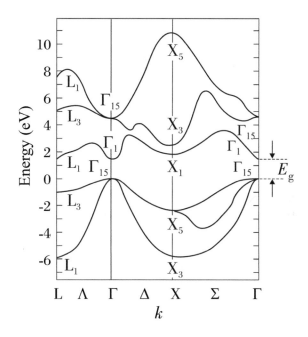

Fig. 5.2 Calculated electronic energy band structure of GaAs. Reprinted with permission from F. Herman and W. E. Spicer, Phys. Rev. **174**, 906 (1968). Copyright (1976) by the American Physical Society

band structure of gallium arsenide is shown in Fig. 5.2. Note that, in this case, the minimum of the conduction band occurs at the same wave vector as the maximum of the valence band, $k = 0$, with an energy gap $E_g = 1.43$ eV. There are several other semiconductors formed by elements of groups III and V, called III-V compounds, such as InSb ($E_g = 0.18$ eV), InP (1.35 eV), GaP (2.26 eV), and GaN (3.38 eV), for example. There are also important semiconductor compounds with elements of groups II and IV, such as CdS (2.42 eV), PbS (0.35 eV), PbTe (0.30 eV) and CdTe (1.45 eV), among others.

The conduction properties of semiconductors are mainly determined by the number of electrons in the conduction band. So, they depend heavily on the ratio E_g/k_BT, and therefore on the value of the energy gap, but they are not very much influenced by the shape of the bands. On the other hand, the optical properties strongly depend on the shape of the energy bands. As will be shown in Chap. 8, electronic transitions accompanied by the emission or absorption of photons in a crystal must conserve energy and momentum, i. e.

$$E_f - E_i = \pm \hbar\omega, \tag{5.1}$$

$$\vec{k}_f - \vec{k}_i = \pm \vec{k}, \tag{5.2}$$

where E_f and E_i are the electron energies in the final and initial states, respectively, $\vec{k}_f$, $\vec{k}_i$ the corresponding wave vectors, while ω and $\vec{k}$ are the frequency and

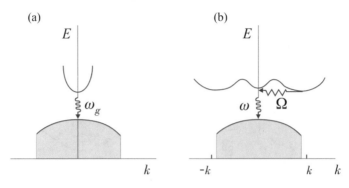

Fig. 5.3 Illustration of the interband electronic transitions in semiconductors. **a** Direct gap: electronic transition with emission of photons only. **b** Indirect gap: transition with emission of photons and phonons

wave vector of the photon absorbed ($E_f > E_i$) or emitted ($E_f < E_i$) in the transition. In the case of gallium arsenide, the transition from one electron at the minimum of the conduction band to the maximum of the valence band is accompanied by the emission of a photon with energy $\hbar\omega = E_g = 1.43$ eV and wave number $k = 2\pi/\lambda = 7.2 \times 10^4$ cm^{-1}. Since this value is much smaller than the wave number at the Brillouin zone boundary (ZB), $k_{ZB} \sim \pi/a \sim 10^8$ cm^{-1}, it is negligible on the scale of Fig. 5.2. Thus, momentum is conserved in the photon emission and the transition is allowed. This transition, illustrated in Fig. 5.3a, is called a direct emission process. Correspondingly, the material is called **direct gap** semiconductor.

In the case of silicon or germanium, it is not possible to have a transition between the minimum of the conduction band and the top of the valence band, with emission or absorption of photons only. The reason is that the photon with energy E_g has $k \ll k_{ZB}$ and this transition requires a variation of the wave vector of the order of k_{ZB} to conserve momentum. As we saw in Chap. 2, phonons have energy $\hbar\Omega \ll E_g$ and wave vector in the range $0 \leq k \leq k_{ZB}$. It is possible, then, to have a transition through the gap, with the emission or absorption of a photon, as long as it is accompanied by the emission or absorption of a phonon. This transition, illustrated in Fig. 5.3b, is called an indirect process. For this reason, Si and Ge are called **indirect gap** semiconductors. Since the transition in indirect gap semiconductors involves phonons and photons, the probability of emission or absorption of photons is much smaller than in direct gap semiconductors. For this reason, lasers and light emitting diodes (LED) are made with direct gap semiconductors. The most important ones are GaAs, InSb, InAs, InP, PbS, CdS, CdTe, and GaN. Not all compounds in group III-V have a direct gap. GaP and AlSb, for example, have indirect gap.

5.2 Electrons and Holes in Intrinsic Semiconductors

5.2.1 *Effective Mass of Electrons and Holes*

In a semiconductor at a finite temperature, electrons in the top of the valence band are thermally excited and can go to the conduction band. In this process they leave empty states in the valence band. Thus, if an external electric field is applied to the semiconductor, electrons in both partially filled bands contribute to the electric current. The electrons of the conduction band, under the action of the field $\vec{E}$, are subject to a force $\vec{F} = -e\vec{E}$ and move according to Newton's law, with effective mass given by Eq. (4.11). Since the electrons occupy states near the minimum of the conduction band, they all have approximately the same effective mass

$$m_e^* = \frac{\hbar^2}{(\partial^2 E/\partial k^2)_{k=k_{mc}}},\tag{5.3}$$

where k_{mc} corresponds to the minimum of the conduction band. Since the curvature of the conduction band is upward, the effective mass of the electrons is positive, so that they accelerate in the opposite direction to the field.

The behavior of the electrons in the valence band is quite different. Notice first, that the electrons near the top of the valence band have negative effective mass, because of the curvature of the function $E(k)$. To understand the electron behavior, let us assume that there is only one empty state at the top of the band. Figure 5.4 illustrates the behavior of this state when an electric field E_x is applied to the crystal in the direction $+ x$. Before application of the field, the empty state must be at the top, as in the energy diagram in Fig. 5.4a, so that the algebraic sum of the momenta of all electrons is zero. After the field is applied, all electrons tend to move in space in the $- x$ direction, because by Eq. (4.8)

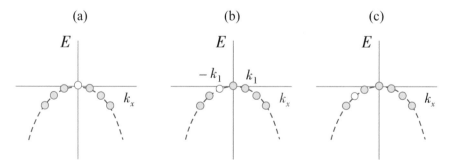

Fig. 5.4 Electron occupation of states at the top of the valence band. **a** Without external electric field. **b** and **c** With an external electric field in the direction $+ x$

$$\hbar \frac{dk_x}{dt} = -e\,E_x. \tag{5.4}$$

Thus, at some later instants, the empty state will be at the positions shown in Fig. 5.4b, c in the $E(k)$ diagram. The displacement of all electrons in the band in the negative k_x direction results in the displacement of the empty state in the same direction in k-space. Since all other states are occupied, the existence of an empty state (absence of electron) with momentum $-\hbar k_1$ implies that the total momentum of the system is $+\hbar k_1$. Therefore, the system behaves as if it were formed by a **hole** with wave vector

$$\vec{k}_h = -\vec{k}_e. \tag{5.5}$$

In this case, the force equation can then be written as

$$\vec{F}_e = \hbar \frac{d\vec{k}_e}{dt} = -\hbar \frac{d\vec{k}_h}{dt}. \tag{5.6}$$

Since the force on the electron due to an electric field is $\vec{F}_e = -e\vec{E}$, Eq. (5.6) gives

$$+e\vec{E} = \hbar \frac{d\vec{k}_h}{dt}.$$

This shows that the hole behaves as a particle with positive electric charge. A development analogous to that of Eqs. (4.9)–(4.11) shows that the effective mass of the hole is

$$m_h^* = -\frac{\hbar^2}{(\partial^2 E/\partial k^2)_{k=k_{mv}}}, \tag{5.7}$$

where k_{mv} corresponds to the maximum of the valence band. Since the curvature of the valence band is downward, the denominator is negative, $\partial^2 E/\partial k^2 < 0$, so that the hole has a positive effective mass. This is consistent with the fact that if an electric field is applied in the $+x$ direction, the holes have momentum $k_x > 0$ and therefore move in the $+x$ direction in real space.

Equations (5.6) and (5.7) lead to the conclusion that the empty states at the top of the valence band, behave as states of elementary excitations of positive charge, with modulus equal to that of the electron charge, and positive effective mass given by Eq. (5.7). They are the states of **holes**. Since the values of the curvatures of the valence and conduction bands are not the same, the effective masses of electrons and holes are different. In addition, it is possible to have crystals with more than one conduction or valence band, and also curvatures that vary with the direction of the wave vector, so there are several masses of electrons and holes. In Figs. 5.1 and 5.2 we see that both Si and GaAs have two valence bands degenerate at $k = 0$. The

Table 5.1 Energy gaps and effective masses of important semiconductors at 300 K

Crystal	E_g (eV)	m_e^*/m_0	m_h^*/m_0
Ge	0.66	$m_c^* = 0.55$	$m_v^* = 0.31$
		$m_e^* = 0.12$	$m_h^* = 0.23$
Si	1.12	$m_c^* = 1.10$	$m_v^* = 0.56$
		$m_e^* = 0.26$	$m_h^* = 0.38$
GaAs	1.43	0.068	0.5
GaN	3.38	0.13	0.74
InSb	0.18	0.013	0.6
InP	1.29	0.07	0.4

m_0 is the electron rest mass

In Si and Ge, m_c^* and m_v^* are the effective masses used to calculate the densities of state in the conduction and valence bands, respectively, while m_e^* and m_h^* are the masses used to calculate the motion of electrons and holes

holes in the band of larger curvature (larger modulus of $\partial^2 E/\partial k^2$) have smaller effective mass, are therefore called **light holes**, while the ones in the band of smaller curvature are called **heavy holes**. Because of the plurality of effective masses, and also the differences in the experimental measurements, the values of the electron and hole effective masses found in the literature vary from one source to another, even in the cases of the most studied semiconductors, such as Si, Ge and GaAs.

Table 5.1 shows the effective masses of some important semiconductors for applications in electronics, as well as the values of E_g at $T = 300$ K. Note that in the case of silicon and germanium there are two effective masses of electrons and two of holes. In this case, m_c^* and m_v^* are geometric averages of the effective masses used to calculate the densities of state in the conduction and valence bands, respectively, while m_e^* and m_h^* are the average masses used to calculate the motion of electrons and holes.

We note that electrons and holes in semiconductors are not simple particles like the free electron: they incorporate the effect of the crystal potential, as their masses are related to the curvature of the bands. Although the absolute charge of holes is the same as the electron charge, its sign is positive. This led to the concept of quasi-particles to define "particles-like" electrons and holes in semiconductors.

5.2.2 Creation and Recombination of Electron-Hole Pairs

In a pure semiconductor crystal at $T = 0$ and without any external disturbance, there are no electrons in the conduction band or holes in the valence band. In other words, there are no electric charge carriers and the material is an electric insulator. There are several processes for exciting electrons from the valence to the conduction band. The most common is thermal excitation, by which a number of electrons from the top of the valence band go to the first levels of the conduction band when $T > 0$. The

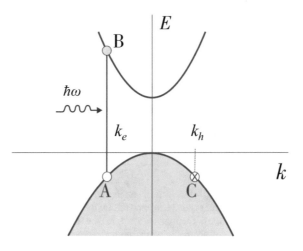

Fig. 5.5 Absorption of a photon with energy $\hbar\omega$ and negligible wave number with the creation of an electron–hole pair in a direct gap semiconductor

concentration of electrons and holes due to thermal excitation will be calculated in the next section. The point to note here is that, in an intrinsic semiconductor, the excitation of an electron to the conduction band always corresponds to the creation of a hole in the valence band, that is, electrons and holes are created in pairs.

Electrons and holes are also created in pairs by other processes, such as optical absorption. As illustrated in Fig. 5.5, when a photon with energy $\hbar\omega$ is absorbed in a semiconductor, an electron goes from the valence to the conduction band. Since the photon wave vector is negligible, the electron created in the conduction band has the same wave vector $\vec{k}_e$ of the electron removed from the valence band. This corresponds to the creation of a hole with wave vector $\vec{k}_h = -\vec{k}_e$. In other words, the absorption of the photon is accompanied by the creation of two quasi-particles: an electron and a hole. As they have momenta $\hbar\vec{k}_e$ and $-\hbar\vec{k}_e$, the total momentum before and after the photon absorption is zero, that is, momentum is conserved, as it should.

If n is the concentration of electrons per unit volume in the conduction band of an intrinsic semiconductor, and p is the concentration of holes in the valence band, we may state that $n = p$. In thermal equilibrium we then have

$$n = p = n_i, \tag{5.8}$$

where n_i is the concentration of carriers in the intrinsic semiconductor, which will be calculated in the next section. In any mechanism for the creation of electron–hole pairs, the process is not static, it is dynamic. Electrons go to the conduction band, leaving holes in the valence band, with a certain rate g that represents the number of pairs generated per unit volume and per unit time. Simultaneously, electrons **recombine** with holes at a recombination rate r. This is clear in the case of thermal excitation. In the optically induced process this is also true, because while the absorption of photons results in the creation of pairs, the recombination involves emission of photons. The fact is that, in the stationary regime, the number of pairs is constant. This requires that, for each pair generation and recombination mechanism, the creation and recombination rates are equal, that is

$$r = g. \tag{5.9}$$

This result is called the **principle of detailed balance**.

Example 5.1 A laser beam with wavelength 515.5 nm, with area 1 mm^2 and power 10 mW, is incident on a semiconductor and is fully absorbed in a length of 100 μm due to a process of generation of electron–hole pairs. Assuming that the conversion efficiency of the process is 10%, calculate the rate of creation of electron–hole pairs in cm^{-3} s^{-1}.

Initially, it is necessary to calculate the number of photons per unit time in the laser beam. Using Eq. (2.31) we can determine the energy of each photon:

$$E = h\nu = h\frac{c}{\lambda} = 6.63 \times 10^{-34} \frac{3 \times 10^8}{515.5 \times 10^{-9}} = 3.86 \times 10^{-19} \text{ J}.$$

The number of photons per unit time is the ratio of the laser power by the photon energy

$$\frac{P}{h\nu} = \frac{10 \times 10^{-3}}{3.86 \times 10^{-19}} = 2.59 \times 10^{16} \text{ s}^{-1}.$$

Since one electron–hole pair is created for 10 incident photons, the generation rate per unit volume is (with length unit in cm)

$$r = \frac{1}{10} \frac{2.59 \times 10^{16}}{10^{-2} \times 100 \times 10^{-4}} = 2.59 \times 10^{19} \text{ cm}^{-3}\text{s}^{-1}.$$

5.2.3 Concentrations of Carriers in Thermal Equilibrium

Several properties of semiconductors, such as the conductivity, depend strongly on the concentrations of the charge carriers, which are determined by the number of states available to be occupied and the probability of occupation of each one. Here

Fig. 5.6 Parabolic energy bands in a semiconductor used to calculate the density of states

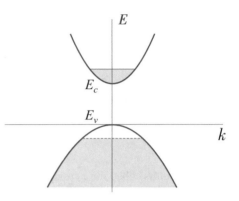

we shall calculate the concentrations in an intrinsic semiconductor at a temperature T using concepts presented in Chap. 4. The probability for electrons to occupy a state with energy E is given by the Fermi–Dirac function $f(E)$ given by Eq. (4.13). In semiconductors there is an additional difficulty compared to metals because the Fermi level is not known, a priori, as we will see below.

Let us consider a semiconductor with energy bands as in Fig. 5.6. The top of the valence band has energy E_v and the minimum of the conduction band is E_c, so that the energy gap is $E_g = E_c - E_v$. At $T = 0$, the valence band is full and the conduction band is empty. Then, the Fermi level is located between the two bands, $E_v < E_F < E_c$, but its exact position in the gap depends on the shape of the bands. Due to the symmetry of $f(E)$ and the fact that, at $T > 0$, the number of electrons in the conduction band is equal to the number of holes in the valence band, if the bands are symmetrical, E_F is located exactly in the middle of the gap. However, if the bands are not symmetrical, E_F is close but not exactly in the middle. In fact, the determination of E_F is made in the calculation of the concentrations of the carriers.

To calculate the concentration of carriers in the semiconductor, it is also necessary to know the number of electronic states available for occupation in the energy bands, which depends on the shape of the bands. Since the states involved in the conduction process are close to the ends of the two bands, as in Fig. 5.6, we can use a parabolic approximation for both. Assuming that the energy does not vary with the direction of k we can write for the conduction band and for the valence band.

$$E - E_c = \frac{\hbar^2 k^2}{2m_c^*}, \tag{5.10}$$

$$E_v - E = \frac{\hbar^2 k^2}{2m_v^*}, \tag{5.11}$$

where m_c^* and m_v^* are, respectively, the effective masses in the conduction and in the valence bands.

Except for the displacement of the reference, the above expressions are equal to Eq. (4.14). Thus, the wave vectors of the states that can be occupied are discrete and given by Eq. (4.15). Therefore, the density of electronic states in the conduction band is given by (4.17), with E replaced by $E - E_c$, and m replaced by m_c^*. So

$$D(E) = \frac{1}{2\pi^2} \left(\frac{2m_c^*}{\hbar^2} \right)^{3/2} (E - E_c)^{1/2}. \tag{5.12}$$

Likewise, the density of hole states in the valence band is

$$D(E) = \frac{1}{2\pi^2} \left(\frac{2m_v^*}{\hbar^2} \right)^{3/2} (E_v - E)^{1/2}. \tag{5.13}$$

From these results we can obtain the concentrations of electrons and holes in thermal equilibrium in the semiconductor. The concentration (number per unit volume) of electrons in the conduction band is obtained by the integral of the product of the density of states $D(E)$ with the occupation probability $f(E)$,

$$n = \int_{E_c}^{\infty} D(E) f(E) \, dE. \tag{5.14}$$

In this equation we made the upper limit infinite because the contribution of states with energy far above E_c is negligible, due to the fact that $f(E)$ falls exponentially with increasing E. To facilitate integration, we shall use an approximate expression for the Fermi–Dirac function. At a temperature $T = 290$ K, the Boltzmann factor is $k_B T = 0.025$ eV. As E_F is near the middle of the gap and E_g is of the order of 1 eV, we can consider $E - E_F \gg k_B T$. Therefore, Eq. (4.13) can be approximated by

$$f(E) \approx e^{-(E - E_F)/k_B T}. \tag{5.15}$$

Replacing Eqs. (5.12) and (5.15) in (5.14) we have

$$\begin{aligned}
n &= \frac{1}{2\pi^2} \left(\frac{2m_c^*}{\hbar^2} \right)^{3/2} \int_{E_c}^{\infty} (E - E_c)^{1/2} e^{-(E - E_F)/k_B T} \, dE \\
&= \frac{1}{2\pi^2} \left(\frac{2m_c^*}{\hbar^2} \right)^{3/2} e^{-(E_c - E_F)/k_B T} \int_0^{\infty} x^{1/2} e^{-x/a} \, dx
\end{aligned}$$

where $x = (E - E_c)$ and $a = k_B T$. The definite integral can be calculated analytically, and its value is $a^{3/2} \pi^{1/2}/2$. Thus, the concentration of electrons in the conduction band can be written as

$$n = N_c \, e^{-(E_c - E_F)/k_B T}, \tag{5.16}$$

where

$$N_c = 2\left(\frac{m_c^* k_B T}{2\pi \hbar^2}\right)^{3/2}. \tag{5.17}$$

The concentration N_c has two useful interpretations. Note that Eq. (5.16) would be obtained from (5.14) immediately if the density of states were a Dirac delta function at $E = E_c$, in the form

$$D(E) = N_c \, \delta(E - E_c). \tag{5.18}$$

This equation means that N_c plays the role of a concentration of states that, if totally located at the energy E_c, would give the electron population in the conduction band. Also, one can see the electron concentration n as given, approximately, by an effective concentration of states with a constant value N_c between E_c and $E_c + k_B T$, and zero outside this range.

In a similar way, we can obtain the hole concentration in the valence band. Since the number of holes is given by the absence of electrons in the valence band, we have

$$p = \int_{-\infty}^{E_v} [1 - f(E)] D(E) \, dE. \tag{5.19}$$

Considering $E_F - E \gg k_B T$, we can use the approximation

$$1 - f(E) \approx e^{(E - E_F)/k_B T}.$$

With this approximation, the integral (5.19) can be calculated in a manner analogous to that of n, leading to the following result for the concentration of holes

$$p = N_v \, e^{-(E_F - E_v)/k_B T}, \tag{5.20}$$

where N_v is the effective concentration of states with energy at the top of the valence band, given by

$$N_v = 2\left(\frac{m_v^* k_B T}{2\pi \hbar^2}\right)^{3/2}. \tag{5.21}$$

The calculation of n and p is illustrated graphically in Fig. 5.7 for the case of an intrinsic semiconductor with approximately symmetric bands. In this case, the Fermi level is approximately in the middle of the gap. In fact, since the Fermi–Dirac function can be approximated by the expression (5.15), Eqs. (5.16)–(5.21) are valid for intrinsic or extrinsic semiconductors. The difference between the two cases is the position of the Fermi level, which has not yet been calculated.

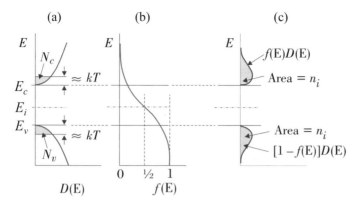

Fig. 5.7 Graphic illustration of the calculation of the charge carrier concentrations in an intrinsic semiconductor with symmetric energy bands. **a** The solid lines represent the density of states $D(E)$ in the two bands. **b** Representation of the Fermi–Dirac distribution $f(E)$. **c** The colored areas represent the concentrations of carriers in the two bands for $T > 0$

Example 5.2 Calculate the probability of occupation $f(E)$ of a state with energy E above the Fermi level, $E = E_f + 0.2$ eV, at a temperature $T = 290$ K, using the approximate expression above and also with the exact result (5.15).

Initially we need to calculate the thermal energy in eV,

$$k_B T = \frac{1.38 \times 10^{-23} \times 290}{1.6 \times 10^{-19}} = 0.025 \text{ eV}.$$

Thus

$$e^{(E-E_F)/k_B T} = e^{0.2/0.025} = e^8 = 2980.96.$$

The probability of occupation, given by the Fermi–Dirac distribution is

$$f(E) = \frac{1}{1 + e^{(E-E_F)/k_B T}} = \frac{1}{1 + 2980.96} = 3.3535 \times 10^{-4}.$$

The value calculated with Eq. (5.15) is

$$f(E) = \frac{1}{2980.96} = 3.3546 \times 10^{-4},$$

that is practically the same calculated with the exact expression.

In order to determine the Fermi level E_F, it is necessary to use the condition of conservation of the number of electrons. In the case of the intrinsic semiconductor, this imposes the condition that the number of electrons in the conduction band is equal to the number of holes in the valence band, $n = p = n_i$. Equating (5.16) and (5.20), making $E_F = E_i$ (the Fermi level in the intrinsic semiconductor), and using Eqs. (5.17) and (5.21), we obtain the Fermi energy in the intrinsic material (Problem 5.2),

$$E_i = \frac{1}{2}(E_c + E_v) + \frac{3}{4} k_B T \ln (m_v^*/m_c^*). \tag{5.22}$$

This equation clearly shows that only if $T = 0$, or if the effective masses of electrons and holes are equal, the Fermi level in the intrinsic semiconductor is exactly in the middle of the gap. At $T > 0$, in the general case where $m_v^* \neq m_c^*$ (non-symmetric bands), the Fermi level is not exactly in the middle of the gap and its position depends on the temperature. However, since at room temperature $E_g \gg k_B T$, this correction is very small in Si, Ge and GaAs.

Once the Fermi energy E_i for the intrinsic semiconductor is known, we can immediately calculate the concentrations of electrons in the conduction band and of holes in the valence band. Using $E_F = E_i$ in Eqs. (5.16) and (5.20) we obtain.

$$n_i = N_c \, e^{-(E_c - E_i)/k_B T}, \tag{5.23}$$

$$p_i = N_v \, e^{-(E_i - E_v)/k_B T}, \tag{5.24}$$

where n_i and p_i denote the concentrations of electrons and holes in intrinsic semiconductors. The product of these two quantities and the condition $n_i = p_i$ give

$$n_i = p_i = \sqrt{n_i p_i} = (N_c N_v)^{1/2} \, e^{-E_g/2k_B T}. \tag{5.25}$$

This important result shows that the number of carriers in the intrinsic semiconductor varies exponentially with $E_g/k_B T$. Figure 5.8 shows the variation with temperature of n_i for Ge, Si, and GaAs, calculated with Eq. (5.25), using Eqs. (5.17) and (5.21) for the effective concentrations. The calculation was made with the parameters in Table 5.1, assuming that E_g does not vary with temperature in the range of the figure. The strong variation of n_i with T is mainly due to the exponential factor in Eq. (5.25), but it also contains a contribution from the term $(N_c N_v)^{1/2}$ (Problem 5.3).

Table 5.2 presents the values of carrier concentrations and other important quantities for Ge, Si, and GaAs. The mobility and the diffusion coefficient will be defined in Sect. 5.5. Note that in all of them the intrinsic carrier concentration is in the range $10^7 - 10^{13}$ cm^{-3}. These values are extremely small compared to the number of free electrons in metals, on the order of 10^{22} cm^{-3}, and result from the fact that $E_g \gg k_B T$. Note that the values of N_c, N_v, and n_i given in Table 5.2 were obtained through independent measurements. For this reason, there is a small discrepancy between them and the values calculated with Eq. (5.25) (Problem 5.3).

Fig. 5.8 Variation with temperature of the intrinsic carrier concentrations in Ge, Si, and GaAs, calculated with Eqs. (5.17), (5.21), and (5.25), with the parameters given in Table 5.1

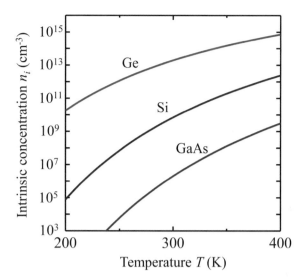

Table 5.2 Values of important quantities in Ge, Si, and GaAs at $T = 300$ K, as given in Sze and Lee, and in Streetman

Quantity	Ge	Si	GaAs
Atoms or molecules (10^{22}/cm^3)	4.42	5.0	2.21
Lattice parameter a (Å)	5.658	5.431	5.654
Dielectric constant $\varepsilon/\varepsilon_0$	16.0	11.8	10.9
Energy gap E_g (eV)	0.66	1.12	1.43
Intrinsic concentration n_i (cm^{-3})	2.5×10^{13}	1.5×10^{10}	10^7
Effective concentration N_c (cm^{-3})	1.04×10^{19}	2.8×10^{19}	4.7×10^7
Effective concentration N_v (cm^{-3})	6.1×10^{18}	1.02×10^{19}	7.0×10^{18}
Mobility μ_n (cm^2/Vs)	3900	1350	8600
Mobility μ_p (cm^2/Vs)	1900	480	400
Diffusion coefficient D_n (cm^2/s)	100	35	220
Diffusion coefficient D_p (cm^2/s)	50	12.5	10

Example 5.3 Obtain an expression for the concentration of electrons in the conduction band for a hypothetical intrinsic semiconductor with $m_v^* = m_c^* = m_0$, and calculate its value for $E_g = 1.0$ eV and $T = 300$ K.

For $m_v^* = m_c^* = m_0$, Eqs. (5.17) and (5.21) give, in SI units

$$N_c = N_v = 2 \left(\frac{m_0 k_B}{2\pi \, \hbar^2} \right)^{3/2} T^{3/2} = 2 \left(\frac{9.1 \times 10^{-31} \times 1.38 \times 10^{-23}}{2 \times 3.14 \times 1.05^2 \times 10^{-68}} \right)^{3/2} T^{3/2}.$$

That gives $N_c = N_v = 4.83 \times 10^{21} \, T^{3/2} \, \text{m}^{-3} \text{K}^{-3/2}$.

Using this result in Eq. (5.25), we have for the intrinsic carrier concentration in cm^{-3}

$$n_i = 4.83 \times 10^{15} \, T^{3/2} \, e^{-E_g/2k_B T} \, \text{cm}^{-3} \text{K}^{-3/2}.$$

To obtain the numerical value for n_i, it remains to calculate the value of the exponential factor. For this we first find the value of the thermal energy in eV at 300 K

$$k_B T = \frac{1.38 \times 10^{-23} \times 300}{1.6 \times 10^{-19}} = 0.026 \, \text{eV}.$$

Thus, finally

$$n_i = 4.83 \times 10^{15} \times 300^{3/2} \, e^{-1.0/0.052} = 1.12 \times 10^{10} \, \text{cm}^{-3}.$$

5.3   Extrinsic Semiconductors

Intrinsic semiconductors are rarely used in devices, among other reasons because they have small conductivity and that is strongly dependent on temperature. In general, semiconductors are used with a certain amount of atoms of elements different from the ones in the intrinsic semiconductor, called **impurities**. Semiconductors with impurities are called **extrinsic**, or **doped**. As we shall study in this section, by suitably doping a semiconductor it is possible to control the number of electrons and holes, and thus the value of the conductivity, with a small temperature dependence. Semiconductors with much larger concentration of electrons than of holes are called **type n** (negative), while semiconductors with larger concentration of holes are called **type p** (positive). The control of the properties of semiconductors by means of the doping makes it possible to use these materials to manufacture a huge variety of electronic devices.

The most common method for doping semiconductors is high temperature diffusion. The atoms of the desired impurity come from a gas, as AsH in the case of As, and diffuse into the material through its surface. This process is carried out in an oven where the material and the gas that supplies the impurity are heated to a temperature in the range of 400–700 °C. The depth of the surface layer that becomes doped and the concentration of impurities depend on the temperature and exposure time.

In the diffusion process, the boundary between the doped layer and the pure material is not well defined. Due to the thermal nature of the process, the concentration of impurities varies gradually at the border. Another method that allows to obtain doped regions with better defined borders is ion implantation. In this process, a beam of accelerated ions with energy in the range 10–100 keV bombards the surface of the material and penetrates into the interior. Impurity layers with well-defined and controlled borders can be produced by this process with thicknesses up to 1 μm.

5.3.1 Impurity Energy Levels in a Crystal

The presence of defects or impurities in a crystal changes the electrostatic potential in its vicinity, breaking the translation symmetry of the periodic potential. This disturbance can produce electronic wave functions that are localized in the vicinity of the impurity, and are not propagating throughout the crystal. The energies of these wave functions are obtained by solving the Schrödinger equation for the impurity potential. These energies appear in the form of discrete levels that can be located between the bands of the perfect crystal. Figure 5.9 illustrates possible energy levels of impurities between the bands of a doped crystal. In a first approximation, these energy levels can be calculated with a simple model. Let us consider, for example, the case of semiconductors such as germanium or silicon, which have a uniform covalent bond.

The elements of group V of the periodic table, such as P, As or Sb, have inner electronic shells like the ones in Si or Ge, but have five valence electrons instead of four. In small quantities, atoms of these elements can easily enter into the crystal replacing Ge or Si atoms, becoming substitutional impurities, as illustrated in Fig. 5.10. Doping can also be done with elements of group III, such as B, Al, Ga, and In, that have three valence electrons.

In the case of group V impurities, such as As, four of its five valence electrons are used in covalent bonding with neighboring atoms of Ge or Si. The fifth electron is weakly bound to the atom, and can be thermally ionized at relatively low temperatures, typically above 50 K. With ionization, the fifth electron is free to move about in the crystal, that is, it goes to the conduction band. This implies that the energy

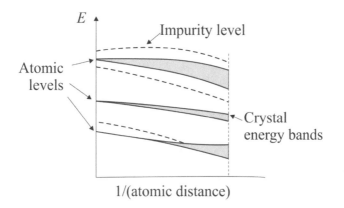

Fig. 5.9 Perturbation in the energy diagram of a crystal due to defects and impurities. Some energy levels are located in the gaps between the energy bands

Fig. 5.10 Schematic model
of a crystal of Si or Ge doped
with substitutional
impurities, such as Ga
(acceptor) and As (donor).
The white circles represent
the atoms of Si or Ge.

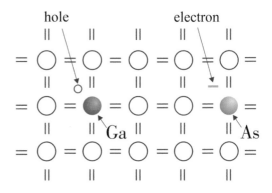

level of the As impurity is close to the conduction band. Thus, impurities of As and
other elements of group V are **donors**, since they donate electrons to the conduction band, as illustrated in Fig. 5.11a. Semiconductors with donor impurities have a
higher concentration of electrons than holes and are therefore called **type n**.

In the case of impurities of group III, such as Ga, there is one electron less than
the four that are necessary for the full covalent bond with the neighbors. At temperatures of 50 to 100 K, electrons from the valence band of the crystal are "captured"
and become part of the covalent bonding, leaving holes in the valence band. Thus,
impurities of elements of group III are called **acceptors** and form semiconductors
of **type p**. As illustrated in Fig. 5.11b, they have an electronic energy level close to
the valence band. The energy levels of impurities in the gap can be calculated using
a simple model of the hydrogen atom. Let us consider initially the case of a donor
impurity in a Ge crystal. The calculation assumes that the almost free electron with
effective mass due to the periodic crystal potential moves around the positive impurity ion. The energy of the impurity level is given by the equation for the ionization
energy of the hydrogen atom in the ground state, Eq. (3.66) with $n = 1$,

$$E = \frac{m_e^* \, e^4}{2(4\pi\varepsilon)^2 \, \hbar^2} = \frac{m_e^*}{m_0}\left(\frac{\varepsilon_0}{\varepsilon}\right)^2 E_H, \tag{5.26}$$

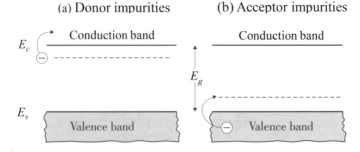

Fig. 5.11 Schematic illustration of the impurity energy levels in the gap of doped semiconductors.
E_c and E_v denote, respectively, the minimum of the conduction band and the maximum of the valence
band. Note that this diagram represents the energies along a physical distance in the semiconductor

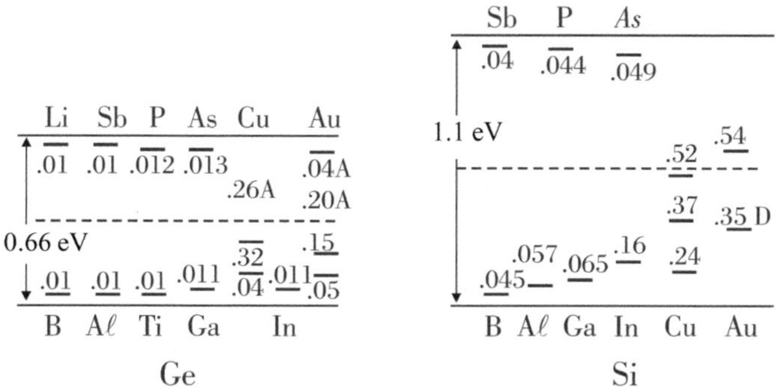

Li Sb P As Cu Au
.01 .01 .012 .013 .04A
.26A .20A
0.66 eV
.32 .15
.01 .01 .01 .011 .011 .04 .05
B Aℓ Ti Ga In
Ge

Sb P As
.04 .044 .049
1.1 eV
.54
.52
.37 .35 D
.057 .065 .16 .24
.045
B Aℓ Ga In Cu Au
Si

Fig. 5.12 Ionization energies of some impurity elements in Ge and Si at $T = 300$ K. The numbers indicate the distances in eV to the minimum of the conduction band for levels above the middle of the gap, or to the maximum of the valence band for levels below the middle of the gap. Note that Cu and Au have several impurity levels, both donors and acceptors [Sze and Lee]

where ε is the permittivity of the crystal, m_e^* is the electron effective mass, and E_H is the absolute value of the ground state energy of the hydrogen atom, 13.6 eV. In germanium, the effective mass in the conduction band, given in Table 5.1, is $m_e^* = 0.12\, m_0$, so that the ionization energy of the donor impurity is

$$E_1 = 13.6 \times (0.12/16^2) = 0.006\,\text{eV}.$$

Silicon has $\varepsilon = 12\varepsilon_0$ and larger effective mass than Ge, so that the energy calculated with this model is larger, 0.025 eV. This energy represents the distance between the impurity level and the minimum of the conduction band. Notice that this simple model, which does not take into account the detailed nature of the impurity atom, gives only approximate results. Figure 5.12 shows the energy levels of various impurities in Ge and Si. The impurities commonly used to make type n semiconductors, such as Sb, P and As, have levels close to the conduction band. On the other hand, impurities of elements used in type p semiconductors, such as B, Al, Ga, and In, have levels close to the valence band. In the case of Cu and Au, there are several impurity levels in the gap of Si and Ge. Some levels are far from the bands and are called deep levels. These levels are used to increase the rate of recombination of electron–hole pairs. Typically, the concentrations used for doping semiconductors vary from 10^{14} cm^{-3} (1 part in 10^8, considering 10^{22} atoms per cm^3), to 10^{20} cm^{-3} (1 part in 10^2, which is very strong).

5.3.2 Carrier Concentrations in Extrinsic Semiconductors

Equations (5.14) and (5.19) for the carrier concentrations are, of course, not restricted to intrinsic semiconductors. They also apply to extrinsic semiconductors, either with donor or acceptor impurities. Therefore, the results (5.16) and (5.20) also apply to extrinsic semiconductors, as long as the approximation (5.15) is valid. Denoting by n_0

and p_0 the thermal equilibrium concentrations of electrons in the conduction band and holes in the valence band, in a extrinsic semiconductor, we can then write

$$n_0 = N_c \, e^{-(E_c - E_F)/k_B T}, \tag{5.27}$$

$$p_0 = N_v \, e^{-(E_F - E_v)/k_B T}. \tag{5.28}$$

The calculation of n_0 and p_0 in a type n semiconductor is illustrated in Fig. 5.13. The main difference between extrinsic and intrinsic semiconductors is in the position of the Fermi level. For example, in type n semiconductors with donor impurity energy E_d close to the conduction band, at $T = 0$ the states with energy E_d are filled while those with energy $E > E_c$ are empty. Thus, at $T = 0$, the Fermi level E_F is between E_d and E_c. At $T > 0$ the Fermi level can be below E_d, but will not be far from this level. Since E_F is close to E_c, at room temperature the exponential in (5.27) is much larger than that in (5.28), so that the number of electrons is much larger than of holes. Physically what happens is that n_0 in a type n semiconductor increases relative to n_i because of the ionization of donor impurities. On the other hand, the number of holes decreases because there are more electrons to recombine with them. The product of the concentrations of electrons and holes, obtained from (5.27) and (5.28) is

$$n_0 p_0 = N_c N_v \, e^{-E_g/k_B T}. \tag{5.29}$$

Comparing this result with Eq. (5.25) we see that

$$n_0 p_0 = n_i^2. \tag{5.30}$$

In this way, the product $n_0 p_0$ is constant and independent of the type and concentration of impurities. This result, known as the **law of mass action**, is very important and will be used frequently later. Using Eqs. (5.23) and (5.24) we can rewrite (5.27) and (5.28) in a convenient way

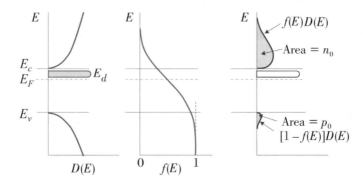

Fig. 5.13 Graphic illustration of the calculation of the charge carrier concentrations in a type n semiconductor

$$n_0 = n_i \, e^{(E_F - E_i)/k_B T}, \tag{5.31}$$

$$p_0 = n_i \, e^{(E_i - E_F)/k_B T}. \tag{5.32}$$

These expressions show clearly that $n_0 = p_0 = n_i$ when $E_F = E_i$, and that the concentrations vary exponentially when E_F departs from E_i.

In type n semiconductors the Fermi level E_F is close to the conduction band, so that $(E_F - E_i)/k_B T \gg 1$. In this case, $n_0 \gg n_i$ and $p_0 \ll n_i$, and for this reason electrons are called **majority carriers**, while holes are **minority carriers**. On the other hand, in type p semiconductors, E_F is close to the valence band, so that $(E_i - E_F)/k_B T \gg 1$, and $p_0 \gg n_i$ and $n_0 \ll n_i$. In this case, holes are the majority carriers while electrons are minority carriers.

Another important relation between carrier concentrations results from charge neutrality. Denoting by N_d^+ the concentration of ionized donor impurities (impurities that give electrons to the conduction band and are positively charged), and N_a^- the concentration of ionized acceptor impurities (that receive electrons from the valence band and are negative), the condition for the material to be **electrically neutral** is

$$n_0 + N_a^- = p_0 + N_d^+. \tag{5.33}$$

This is the equation of charge neutrality. For a given semiconductor with known concentrations of impurities, one can calculate the Fermi level and the concentrations of electrons and holes with the set of Eqs. (5.27)–(5.33).

Let us consider the case of a type n semiconductor with N_d donor impurities per unit volume, at a temperature such that almost all of them are ionized, that is $N_d^+ \approx N_d$. In this case Eq. (5.33) gives

$$n_0 \approx p_0 + N_d. \tag{5.34}$$

Using the law of mass action (5.30) in this equation, we obtain

$$n_0 = \frac{N_d}{2} + \left[\left(\frac{N_d}{2} \right)^2 + n_i^2 \right]^{1/2}, \tag{5.35}$$

$$p_0 = -\frac{N_d}{2} + \left[\left(\frac{N_d}{2} \right)^2 + n_i^2 \right]^{1/2}, \tag{5.36}$$

Typically, in a doped semiconductor, the concentration of impurities is much larger than the intrinsic concentration, $N_d \gg n_i$. In this case, neglecting n_i in (5.35) we obtain

$$n_0 \approx N_d, \tag{5.37}$$

as was expected. On the other hand, we cannot neglect n_i in (5.36) completely, since this would lead to $p_0 = 0$. Using the binomial approximation for the square root in Eq. (5.36) we obtain

$$p_0 \approx \frac{n_i^2}{N_d}. \tag{5.38}$$

which is consistent with Eqs. (5.30) and (5.37). With the expressions (5.37) and (5.38) for the carrier concentrations, the Fermi level can be determined with Eq. (5.27) or Eq. (5.31). For example, substituting (5.37) in (5.27) we have

$$E_F = E_c - k_B T \ln \frac{N_c}{N_d}. \tag{5.39}$$

Or else, substituting (5.37) in (5.31), we obtain another useful expression for E_F

$$E_F = E_i + k_B T \ln \frac{N_d}{n_i}. \tag{5.40}$$

It is important to note that these expressions for E_F are valid only for type n semiconductors, in the condition $N_d \gg n_i$.

Example 5.4 Calculate the concentrations of electrons and holes and the position of the Fermi level in a silicon crystal doped with $N_d = 10^{16}$ cm^{-3} atoms of arsenic, at $T = 290$ K.

From Table 5.2 we have $n_i = 1.5 \times 10^{10}$ cm^{-3}. Using Eqs. (5.37) and (5.38) we have

$$n_0 \approx N_d^+ \approx N_d = 10^{16}\, \text{cm}^{-3},$$

$$p_0 \approx \frac{n_i^2}{N_d} \approx 2.25 \times 10^4\, \text{cm}^{-3}.$$

Using $k_B T = 0.025$ eV and $N_c = 2.8 \times 10^{19}$ cm^{-3} in (5.39), it follows that

$$E_c - E_F = 0.025 \times \ln(2.8 \times 10^3) = 0.20\, \text{eV}.$$

Comparing this result with the energy given in Fig. 5.12, it can be seen that in this case the Fermi level is close and slightly below the level of the As impurity in silicon. On the other hand, with (5.40) we obtain

$$E_F = E_i + 0.34\, \text{eV}.$$

The energy diagram corresponding to the semiconductor of Example 5.4 is shown in Fig. 5.14a. This diagram is typical of type n semiconductors, in which the Fermi level is close to the conduction band. It is important to note that when the concentration of impurities is large, that is, comparable to N_c (2.8×10^{19} cm^{-3} in Si), the Fermi level approaches E_c. In this case, the result (5.39) is not valid

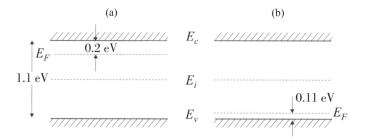

Fig. 5.14 Energy diagrams for doped silicon: **a** type n, with $N_d = 10^{16}$ cm^{-3} donor impurities; **b** type p, with $N_a = 10^{17}$ cm^{-3} acceptor impurities

because Eq. (5.15) is not a good approximation for $f(E)$. A semiconductor with N_d comparable to N_c is called **degenerate**, and it has $E_F \approx E_c$.

It is easy to see, by analogy with the development of Eqs. (5.34)–(5.40), that in type p semiconductors, doped with acceptor impurities with concentration N_a, the expressions for the concentrations of electrons and holes, and for the Fermi level are (Problem 5.6)

$$n_0 \approx \frac{n_i^2}{N_a}, \tag{5.41}$$

$$p_0 \approx N_a, \tag{5.42}$$

$$E_F = E_v + k_B T \ln \frac{N_v}{N_a}, \tag{5.43}$$

$$E_F = E_i - k_B T \ln \frac{N_a}{n_i}. \tag{5.44}$$

Example 5.5 Calculate the concentrations of electrons and holes and the position of the Fermi level in a silicon crystal doped with $N_a = 10^{17}$ cm^{-3} atoms of gallium, at $T = 290$ K.
   From Table 5.2 we have $n_i = 1.5 \times 10^{10}$ cm^{-3}. Using Eqs. (5.41) and (5.42) we have

$$p_0 \approx N_a = 10^{17} \text{ cm}^{-3},$$

$$n_0 \approx \frac{n_i^2}{N_a} \approx 2.25 \times 10^3 \text{ cm}^{-3}.$$

Using $k_B T = 0.025$ eV and $N_v = 1.02 \times 10^{19}$ cm^{-3} in (5.43), we have

$$E_F = E_v + 0.025 \times \ln(1.02 \times 10^2) = E_v + 0.11 \text{ eV}.$$

Fig. 5.15 Concentration of electrons as a function of temperature in type n silicon, with $N_d = 10^{16}$ cm^{-3} [Yang]

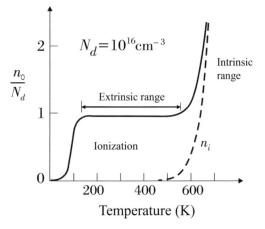

The energy diagram corresponding to Example 5.5, illustrated in Fig. 5.14b, is typical of a type p semiconductor, in which the Fermi level is close to and above the impurity level, that is close to the top of the valence band.

To conclude this section, we remark that the calculations presented here consider that $N_d^+ \approx N_d$ and $N_a^- \approx N_a$, that are valid only above a certain temperature, which in the case of silicon is about 100 K. Below this temperature the impurities are not all ionized and the numbers of carriers vary with temperature (see Problem 5.8). However, since in the range of 100 K to 500 K the impurities are practically all ionized, the concentrations are almost independent of the temperature, as illustrated in Fig. 5.15. Above 500 K the intrinsic concentration, which grows exponentially with T, becomes important and eventually dominates the extrinsic one.

5.4   Dynamics of Electron and Holes in Semiconductors

The operation of semiconductor devices is based on the dynamics of the electric charge carriers, electrons and holes. The main dynamic processes are the creation of electron–hole pairs, the recombination of pairs, and the collective motion of these carriers. The collective motion of charges results in an electric current, which is the main mechanism for the transmission of information in devices. There are two basic types of collective motion that we shall study in the section: the **drift motion** in an electric field and the **diffusion of charges** due to a spatial gradient in the concentrations of the carriers.

5.4.1 Conduction Current

The conduction current, or **drift current**, results from the slow average displacement of charge carriers produced by an external electric field, simultaneous with the rapid and random motion characteristic of particles in thermal agitation. While in a metal this current is produced by the motion of electrons only, in semiconductors it is formed by both electrons and holes.

When an electric field is applied to a semiconductor, electrons and holes drift in opposite directions. However, since they have opposite charges, the intensities of the electric currents of the two types of carriers add up. As we studied in Sect. 4.5, the electron current density is related to the electric field E by

$$J_n = \sigma_n E, \tag{5.45}$$

where σ_n is the conductivity due to electrons, that according to Eq. (4.25) is

$$\sigma_n = \frac{e^2 n_0 \tau_e}{m_e^*}, \tag{5.46}$$

where τ_e is the average time between two electron collisions, that we refer simply by electron collision time. In this expression we use the electron equilibrium concentration n_0 because the application of the electric field has negligible effect on the value of the carrier concentration. Since the conductivity results from the average motion of the ensemble of electrons, it is useful to introduce a new quantity, that describes the response of a single electron to the action of the external field. This quantity is the **mobility**, defined by the ratio between the drift velocity and the electric field

$$\mu = \frac{v}{E}. \tag{5.47}$$

The mobility is an important parameter because it describes how fast an electron drifts under the action of an electric field. Comparing Eqs. (4.25), (5.46) and (5.47) we see that the conductivity can be written as

$$\sigma_n = e \, n_0 \, \mu_n, \tag{5.48}$$

where μ_n is the electron mobility, given by

$$\mu_n = \frac{e \, \tau_e}{m_e^*}. \tag{5.49}$$

This result shows that the mobility involves only intrinsic parameters of the material. It depends on the concentration of impurities because this has a direct effect on the collision time. Figure 5.16 shows the variation of the electron mobility with temperature in type n silicon, for various concentrations of donor impurities. Note that the

Fig. 5.16 Electron mobility as a function of temperature in type n silicon, for several values of the impurity concentration N_d [Beadle, Tsai, and Plummer]

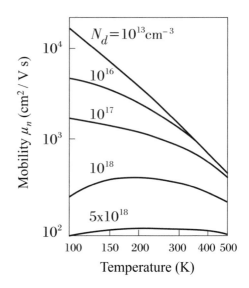

mobility decreases with increasing impurity concentration, due to the decrease of τ_e resulting from electron collisions with impurities. The decrease of the mobility with increasing temperature is due to electron collisions with thermal lattice vibrations.

Following the same procedure for electrons, we see that the current density due to holes is given by

$$J_p = \sigma_p E, \tag{5.50}$$

where σ_p is the conductivity due to holes, that according to Eq. (4.25) is

$$\sigma_p = e\, p_0\, \mu_p = \frac{e^2\, p_0\, \tau_p}{m_h^*}, \tag{5.51}$$

where τ_p is the collision time, p_0 the concentration, and μ_p the mobility of holes. The sum of Eqs. (5.45) and (5.50) gives the total current density $J = \sigma\, E$, where the conductivity is

$$\sigma = \sigma_n + \sigma_p = e\, (n_0\, \mu_n + p_0\, \mu_p). \tag{5.52}$$

At each temperature, the total conductivity can be calculated using the values of the concentrations of electrons and holes, obtained as in Sect. 5.3, and of the mobilities. Figure 5.17 shows the variation of μ_n and μ_p with the impurity concentration in silicon and gallium arsenide at $T = 300$ K. Note that the electron mobility in GaAs is about five times higher than in Si, due mainly to the smaller effective mass of electrons in GaAs. Of course, in a type n semiconductor the current is essentially due to electrons, while in type p it is due to holes.

Fig. 5.17 Mobilities of electrons and holes in Si and GaAs as a function of impurity concentration at $T = 300$ K [Beadle, Tsai, and Plummer]

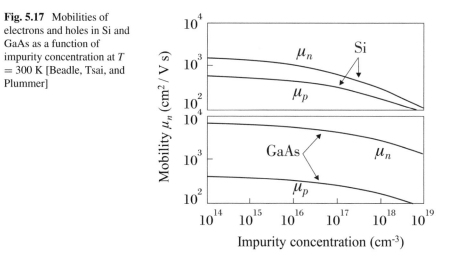

The electric current in a bar of a semiconductor material to which an external electric field is applied results from the mobility of electrons and holes. This field can be created by a potential difference applied at the ends of the bar by n external circuit, such as the one Fig. 5.18. The current in the semiconductor is the sum of the contributions of the two types of charge carriers, since electrons and holes move in opposite directions. Of course, in the metallic wire that provides the voltage, the current is entirely due to electrons. Thus, we need to ask what happens to the holes at the ends of the bar. Since the current in the wire is equal to the current in the bar, the number of electrons that go through the cross section of the wire per unit time is the sum of the numbers of electrons and holes drifting in the semiconductor, because their charges have the same absolute value. What happens is that at the interface between the metal and the semiconductor at the end A, there is a process of generation of electron–hole pairs. The electrons created at interface A pass to the wire, while the holes move in the bar towards the end B. At interface B, on the other hand,

Fig. 5.18 Illustration of the motion of electrons and holes in a semiconductor and in the external electric circuit

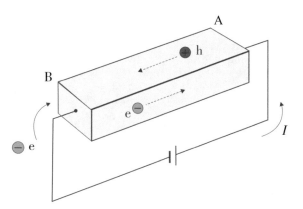
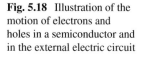

holes recombine with the excess electrons coming from the metallic wire, in such a way that the number of electrons in the processes of creation and recombination of pairs at the interfaces require that they behave as perfect sources or sinks of electrons and holes, without any tendency to favor one of the two charge carriers. A metal–semiconductor contact with these characteristics is called **ohmic**. In an ohmic contact the resistance is the same regardless of the direction of the current used to measure it. Actually, in real circuits, the contact between a metal and a semiconductor is never perfectly ohmic. The properties of metal–semiconductor contacts will be studied in Sect. 6.3.1.

Example 5.6 Calculate the resistivity of silicon at $T = 300$ K in two situations: (a) Intrinsic; (b) Doped with arsenic impurities with concentration $N_d = 2 \times 10^{16}$ cm^{-3}.

(a) The total conductivity is calculated with Eq. (5.52), considering $n_0 = p_0 = n_i$, and using the parameters of Table 5.2 for intrinsic Si. We have

$$\sigma = e\,n_i(\mu_n + \mu_p) = 1.9 \times 10^{-19} \times 1.5 \times 10^{10} \times (1350 + 480)\,\text{C cm}^{-3}\text{cm}^2/\text{Vs}$$

$$\sigma = 4.39 \times 10^{-6}\,(\Omega\,\text{cm})^{-1}.$$

The resistivity is the inverse of the conductivity, so

$$\rho = \frac{1}{\sigma} = \frac{1}{4.39 \times 10^{-6}} = 2.28 \times 10^5\,\Omega\,\text{cm} = 2.28 \times 10^3\,\Omega\,\text{m}$$

(b) In Si with donor impurities with $N_d \gg n_i$, the electron concentration is given by (5.37),

$$n_0 \approx N_d = 2 \times 10^{16}\,\text{cm}^{-3}.$$

Since $p_0 \ll n_i$, the conductivity is

$$\sigma \approx e\,n_0\,\mu_n.$$

Using the value of μ_n given in Fig. 5.17, we have

$$\sigma \approx 1.9 \times 10^{-19} \times 2 \times 10^{16} \times 10^3 = 3.2\,(\Omega\,\text{cm})^{-1}.$$

and

$$\rho = \frac{1}{3.2} = 0.31\,\Omega\,\text{cm}.$$

Note that a relatively weak doping (1 part out of 10^6) decreases the resistivity of the silicon by six orders of magnitude.

5.4.2 Motion in a Magnetic Field: Hall Effect

Consider a semiconductor with the shape of a bar, traversed by an electric current produced by an external circuit. If a static magnetic field is applied to the semiconductor, perpendicularly to the direction of motion of the carriers, they tend to be deflected laterally, resulting in a charge accumulation that produces a potential difference across the bar. Let us consider the geometry shown in Fig. 5.19, in which the z direction of the coordinate system is chosen as the direction of the magnetic field B, x is the direction of the current, and y is the transverse direction. The force exerted by the magnetic field on a charge q is given by

$$\vec{F} = q\vec{v} \times \vec{B}. \tag{5.53}$$

Let us assume a type p semiconductor, so that the current is essentially due to the holes. As they move in the $+x$ direction and have positive charge, the force on them has the $-y$ direction. This force deflects the holes and results in the accumulation of positive charges on the side $y = -d/2$ of the bar, thus leaving negative charges on the side $y = +d/2$. These charges create an electric field in the $+y$ direction that, after an initial transient, prevents the motion of the holes in the y direction to continue. The value of the transverse electric field can be calculated considering that the total force on a hole is given by

Fig. 5.19 Illustration of the Hall effect in a semiconductor. The application of a magnetic field in a bar with an electric current generates a transverse potential difference V_H that provides a measurement of the charge carrier concentration

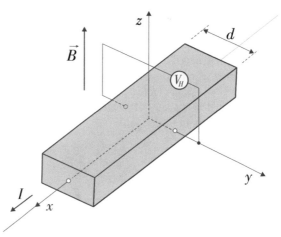

$$\vec{F} = q\,(\vec{E} + \vec{v} \times \vec{B}). \tag{5.54}$$

In steady state the y component of this force must be zero. Thus, the y component of the electric field is

$$E_y = -\,(\vec{v} \times \vec{B})_y = v_x B_z = v\,B. \tag{5.55}$$

The generation of this transverse electric field is known as the **Hall effect**, in honor of Edwin Herbert Hall, who observed this phenomenon in metals in 1879. The transverse voltage that appears in the bar, $V_H = E_y d$, is called Hall voltage. Using the relation between the hole current density and the drift velocity, $J_p = e\,p_0\,v$, we have

$$E_y = \frac{J_p}{e\,p_0} B_z \equiv R_H\,J_p\,B, \tag{5.56}$$

where $R_H = (e\,p_0)^{-1}$ is the **Hall coefficient**. In semiconductors, an important application of the Hall effect is that the measurement of the Hall voltage can be used to determine the concentration p_0 of holes with good precision. Actually, the voltage gives information about the difference between the concentrations of electrons and holes. Note that in the case of the current produced by electrons, the velocity v_x is negative and, therefore, by Eq. (5.55) the electric field and the Hall voltage have the opposite direction to that of holes. Thus, from the sign of the Hall voltage one can determine which are the majority carriers in the semiconductor. In case the concentrations of electrons and holes are comparable, the value of the Hall voltage allows to determine the difference $(p_0 - n_0)$.

Although it was discovered more than a century ago, the Hall effect is still an important technique for investigating the conduction properties of materials. This technique was used by Klaus von Klitzing to study the motion of electrons confined to two dimensions in a semiconductor. He discovered that the Hall voltage varies with the intensity of the magnetic field in steps. This is a quantum effect resulting from the quantization of electron energy levels in the magnetic field. For the discovery of the quantum Hall effect, von Klitzing earned the 1985 Physics Nobel Prize. The Hall effect also has several applications. One of the most important ones is in the measurement of magnetic fields. The Hall sensor is made up of a small semiconductor bar, traversed by a certain electric current. When placed in a magnetic field whose intensity one wants to measure, the value of the Hall voltage across the sensor provides a direct measurement of the field.

Example 5.7 A bar of type p silicon, of thickness $d = 0.5$ mm, with impurity concentration $N_a = 10^{14}$ cm^{-3}, is used as a Hall sensor. Calculate the Hall voltage for a probe current of 100 mA and a magnetic field perpendicular to the plane of $B = 0.1$ T.

The Hall voltage is given by $V_H = E_y w$, and for a current with intensity I the current density is $J = I/(wd)$, where w and d are, respectively, the width and thickness of the bar. Since $N_a \gg n_i$, we have $p_0 \approx N_a \gg n_0$, so that the current is dominated by holes. Using Eq. (5.56), we have

$$V_H = \frac{I/(w\,d)}{e\,p_0} B\,w = \frac{I\,B}{e\,p_0\,d},$$

and using SI units, the voltage is

$$V_H = \frac{10^{-1} \times 0.1}{1.6 \times 10^{-19} \times 10^{14} \times 10^6 \times 0.5 \times 10^{-3}} = 1.25\,\text{V}.$$

This example shows that for magnetic fields with intensities typical of those used in laboratories, the Hall voltage has a relatively high value for electronic circuits. This does not happen in metals, because the concentration of free electrons ($\sim 10^{22}$ cm^{-3}) is much larger than in semiconductors, and thus the voltage is quite small.

5.4.3 Diffusion Current

The conduction current results from the motion of charges produced by an electric field, that is, by the gradient of an electric potential. This is not the only gradient that produces electric current in a semiconductor. When particles are distributed non-uniformly in a medium, the concentration gradient produces a motion called diffusion. In semiconductors, diffusion occurs when charge carriers are created in a certain region, and flow to regions of smaller gradients. Since the carriers have electric charge, their diffusion motion results in an electric current, called **diffusion current.**

The diffusion motion is very common in physics. It is by diffusion that a drop of color ink, placed in a glass of water, spreads in the water so that after some time it becomes uniformly colored. The diffusion of the ink molecules in the water results from their random motion of thermal agitation. In this process, each molecule, both of water and of ink, moves in an arbitrary direction until it collides with another molecule. After the collision, the molecule moves in another direction, resulting in a completely random motion. In this way, the ink molecules, which were initially concentrated in a certain region, after a certain time are completely diffused in the water. In the case of a semiconductor, the diffusion of excess charge carriers, initially concentrated in a certain region, results from their random motion in the crystal lattice of the material.

To obtain the equation that describes the diffusion motion, we shall initially consider a simple model, in which holes move in one dimension, say the x direction. The concentration of holes in excess of equilibrium is described by the function $p(x)$. Consider $\bar{l}_p$ the average distance traveled by a hole between two collisions, the **mean free path**, and τ_p the average time between two collisions. Consider two planes perpendicular to x, with coordinates x and $x + \Delta x$, where $\Delta x = \bar{l}_p$, as in Fig. 5.20. In the random motion that characterizes diffusion, the holes that are between the planes x and $x + \Delta x$ are equally likely to move in the $+ x$ or $- x$ direction. Likewise, the holes between the planes $x - \Delta x$ and x can move in either direction with equal probability. If the concentration of holes is the same to the left or to the right of x, the net number of holes that cross the plane is zero, and the electric current is also zero. However, if there is a gradient of hole concentration, the current in the plane x will not be zero, it will be proportional to the difference in concentrations to the left and to the right of x. As half of the holes between and $x - \Delta x$ and x cross the plane x in the sense $+ x$, over a period of time τ_p, the current due to these holes in a cross section of area A is approximately

$$\frac{1}{2} e \bar{l}_p A p(x - \Delta x/2) \times \frac{1}{\tau_p},$$

because the current is the ratio between the total charge that crosses the section and the time interval. To obtain the current density in the plane x, it is necessary to subtract the contribution of the holes that are between x and $x + \Delta x$ crossing the plane x in the $- x$ direction, and dividing the difference by the area. The result is

$$\frac{1}{2\tau_p} e \bar{l}_p \left[p(x - \frac{\Delta x}{2}) - (x + \frac{\Delta x}{2}) \right].$$

Assuming that the variation of $p(x)$ with x occurs over distances much larger than Δx, we can consider Δx very small, so that the expression inside the brackets is $\Delta x\, dp/dx$. Thus, the **diffusion current** density of holes in the direction $+ x$ is given by

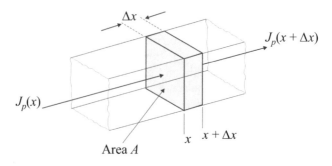

Fig. 5.20 Illustration of the currents flowing in and out of a region with charge of volume $A\, \Delta x$

$$J_p^{diff} = -e\,D_p\,\frac{dp(x)}{dx}\,, \tag{5.57}$$

where $D_p = \bar{l}_p^2/2\tau_p$ is the hole **diffusion coefficient**. The diffusion current of electrons can be obtained in a similar manner. Since the electron charge is negative, its diffusion current is given by

$$J_n^{diff} = +e\,D_n\,\frac{dn(x)}{dx}\,, \tag{5.58}$$

where D_n is the diffusion coefficient and $n(x)$ the concentration of the electrons. Both Eqs. (5.57) and (5.58) show that, as expected, the diffusion current will be zero if there is no spatial variation in the concentration of carriers. These equations, obtained assuming that the concentrations only vary in the x direction, represent the x components of the diffusion currents. In the general case of variation in three dimensions, the components y and z are given by similar expressions with derivatives of the concentrations with respect to y and z. Thus, generalization of Eqs. (5.57) and (5.58) leads to two equations involving the gradient operator

$$J_p^{diff} = -e\,D_p\,\nabla p\,, \tag{5.59}$$

$$J_n^{diff} = +e\,D_n\,\nabla n. \tag{5.60}$$

With Eqs. (5.59) and (5.60) one can calculate the diffusion currents of holes and electrons from the gradients in their concentrations. In most situations, however, these are not known a priori, they need to be calculated. To obtain the equations that provide the evolution of the concentrations, it is necessary to have another independent relation between the diffusion current and the concentration. To obtain this relation, we shall initially consider the one-dimensional model in Fig. 5.20 to relate the current density to the temporal variation of the density. Let us also assume, initially, that the phenomenon of generation and recombination of electron–hole pairs is negligible. Note that the net current I that enters the volume indicated in the figure, divided by the volume, is the difference of the current densities in x and in $x + \Delta x$, divided by Δx,

$$\frac{I}{A\,\Delta x} = \frac{J(x) - J(x + \Delta x)}{\Delta x}.$$

Since $I = dq/dt$, this result leads, in the limit $\Delta x \to 0$, to the following differential equation,

$$\frac{\partial \rho}{\partial t} = -\frac{\partial J(x)}{\partial x}\,, \tag{5.61}$$

where $\rho = q/(A\Delta x)$ is the volume charge density. This is the **equation of charge continuity**, that expresses the fact that the total charge is conserved. If the current density also has components y and z, Eq. (5.61) can be generalized to three dimensions

$$\frac{\partial \rho}{\partial t} = -\left(\frac{\partial J_x}{\partial x} + \frac{\partial J_y}{\partial y} + \frac{\partial J_z}{\partial z}\right),$$

or

$$\nabla \cdot \vec{J} = -\frac{\partial \rho}{\partial t}. \tag{5.62}$$

This is the equation of charge continuity in three dimensions. This equation is valid whatever the source of the current is. Notice that it is contained in Maxwell's equations studied in Chap. 2. Since $\nabla \cdot \nabla \times \vec{A} = 0$ for any vector field $\vec{A}$, the operation $\nabla\cdot$ in Eq. (2.4), together with Eq. (2.1), reproduce the continuity Eq. (5.62).

Notice that in a semiconductor, the charge density ρ is related to the concentrations of electrons and holes by

$$\rho = e\,(p - n). \tag{5.63}$$

To obtain the equation for the evolution of the concentration, let us suppose, to simplify, a semiconductor type n, that is, only with electrons in excess of equilibrium. From Eqs. (5.62) and (5.63) we have

$$\nabla \cdot \vec{J} = e\,\frac{\partial n}{\partial t}. \tag{5.64}$$

Substituting this result in Eq. (5.60), and operating on both sides with the operator $\nabla\cdot$ we obtain

$$D_n \nabla^2 n - \frac{\partial n}{\partial t} = 0. \tag{5.65}$$

This is the **diffusion equation**, with which we can calculate the spatial and temporal evolutions of the concentration of excess electrons, subject only to thermal agitation. An identical equation holds for the concentration p of holes, with the corresponding diffusion coefficient D_p, and also for the concentration of blue ink molecules in the glass of water. The diffusion equation shows that as long as there is a spatial variation in the concentration, there will also be variation in time. Figure 5.21 shows the evolution of the electron concentration $n(x)$ after the production of an electron pulse in the position $x = 0$ at an instant of time $t = 0$. If at $t = 0$, electrons are concentrated at $x = 0$, the concentration is represented by the **Dirac delta function**,

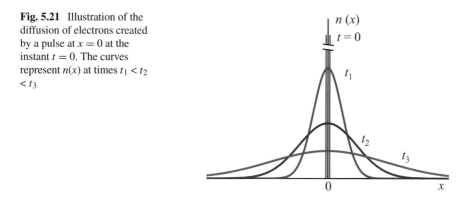

Fig. 5.21 Illustration of the diffusion of electrons created by a pulse at $x = 0$ at the instant $t = 0$. The curves represent $n(x)$ at times $t_1 < t_2 < t_3$

$n(x) = \delta(x)$. Then, at $t > 0$, the electrons diffuse to regions of lower concentration. The solution of Eq. (5.65) is a Gaussian function (Problem 5.17), which gradually expands with time and decreases in amplitude, as shown in Fig. 5.21. Since the total number of electrons is conserved, the area under the curve does not vary over time. After a long time, $n(x)$ is uniform, so that $\nabla^2 n = 0$, and hence $n(x)$ is uniform and constant.

If, in addition to the concentration gradients, there is an electric field $\vec{E}$ applied to the semiconductor, the electron and hole current densities will have conduction and diffusion components, and can be written as

$$\vec{J}_n = -e\,\mu_n\,n\,\vec{E} + e\,D_n\,\nabla n, \tag{5.66}$$

$$\vec{J}_p = e\,\mu_p\,p\,\vec{E} - e\,D_p\,\nabla p, \tag{5.67}$$

and the total current density is

$$\vec{J} = \vec{J}_n + \vec{J}_p. \tag{5.68}$$

As we shall study in the next chapter, all components of the current are relevant for the operation of semiconductor devices. It is the electric field and the concentration gradients of the carriers in the region of a junction of two types of semiconductors that determine the relation between voltage and current in a junction device and, therefore, its functioning.

To conclude this section, let us obtain an important relation between the diffusion coefficient and the mobility. When the semiconductor is in thermal equilibrium, with no external electric field, both the electron and the hole currents must be zero. In this situation if, due to the thermal motion, there is a variation in the concentrations of the charges, the electric field created by the charges produces a drift current that cancels the diffusion current. The relation between this internal field $\vec{E}$ and the equilibrium concentration gradient can be obtained from Eqs. (5.66) and (5.67) with $\vec{J}_n = \vec{J}_p = 0$. Since the electric field is the gradient of the electric potential

$$\vec{E} = -\nabla\phi, \tag{5.69}$$

from Eq. (5.67) with $\vec{J}_p = 0$ we obtain

$$\frac{\mu_p}{D_p}\nabla\phi = -\frac{1}{p_0}\nabla p_0, \tag{5.70}$$

where p_0 is the hole equilibrium concentration. A relationship analogous to (5.70) holds for electrons. Substituting in (5.70) the expression of p_0 given by (5.32) we obtain

$$\frac{\mu_p}{D_p}\nabla\phi = -\frac{1}{k_B T}\nabla(E_i - E_F). \tag{5.71}$$

The Fermi level cannot vary with the position since the system is in equilibrium, therefore $\nabla E_F = 0$. On the other hand, the energy of an electron in the electric potential ϕ is $E = -e\phi$. This means that if the electric potential varies in space, the electron energy levels and the energy bands accompany the electric potential, that is, $\nabla E_i = -e\nabla\phi$. Using this relation in Eq. (5.71) one obtains

$$\frac{D_p}{\mu_p} = \frac{k_B T}{e}. \tag{5.72}$$

Since the relation obtained for electrons is identical, we can write

$$\frac{D_p}{\mu_p} = \frac{D_n}{\mu_n} = \frac{k_B T}{e}. \tag{5.73}$$

This important result, known as the **Einstein relation**, makes possible to calculate the diffusion coefficient from the values of the measured mobilities, or vice versa. Notice in Table 5.2, that the ratio D/μ for Ge, Si, and GaAs, is close to 0.026 eV, which is the value of $k_B T$ at $T = 300$ K.

5.4.4 Injection of Carriers: Diffusion with Recombination

A very important process in the operation of devices is one in which carriers in excess of equilibrium are introduced into a region of the semiconductor by some external mechanism. This is called **injection of carriers**. This occurs, for example, when electrons, that are the majority carriers in type n semiconductor, flow into the p-side in a p–n junction. We say that in the region of the junction, electrons from the n-side are **injected** into the p-side.

In the carrier injection process, the mechanism of electron–hole pair recombination cannot be neglected, as we did in the previous section. Since the injected carriers are in excess of the equilibrium concentration, the recombination process results in a decrease of the concentration towards equilibrium. Consider, for example, holes injected into a semiconductor so that at a certain instant the concentration is

$$p = p_0 + \delta p. \tag{5.74}$$

The recombination of holes in excess of equilibrium with electrons in the semiconductor occurs at a rate that increases with increasing δp. In a first approach, the process can be described by

$$\frac{\partial \delta p}{\partial t} = -\frac{\delta p}{\tau_p}, \tag{5.75}$$

where τ_p is the hole **recombination time**. Note that if there is no other mechanism acting for the evolution of δp, the solution of Eq. (5.75) is

$$\delta p(t) = A\, e^{-t/\tau_p}, \tag{5.76}$$

where A is the value of δp at the instant $t = 0$. This result shows that the recombination acts to make the excess of carriers to decay exponentially in time, with a characteristic time τ_p. In the case of electrons, the excess concentration is described by an equation similar to Eq. (5.75), with a recombination time τ_n,

$$\frac{\partial \delta n}{\partial t} = -\frac{\delta n}{\tau_n}, \tag{5.77}$$

The carriers injected into a certain region of the semiconductor produce a concentration gradient that, in turn, results in a diffusion current. Thus, in the injection process, the spatial and temporal evolution of the carrier concentration is determined by the diffusion and recombination processes. To obtain the equation that describes both processes, we subtract from the time derivative of the concentration in the diffusion Eq. (5.65), the term that describes the recombination given by Eq. (5.77). Combining (5.65) with (5.77) and taking into account that $\partial n_0/\partial t = 0$, since the equilibrium concentration is constant, we obtain for electrons

$$\frac{\partial \delta n}{\partial t} = D_n \nabla^2 \delta n - \frac{\delta n}{\tau_n}, \tag{5.78}$$

A similar development for holes leads to

$$\frac{\partial \delta p}{\partial t} = D_p \nabla^2 \delta p - \frac{\delta p}{\tau_p}. \tag{5.79}$$

These are the **diffusion equations with recombination** for electrons and holes. They make possible to calculate the evolution in space and time of the concentrations of carriers injected in a region of the semiconductor. If a pulse in the concentration of electrons is produced in the position $x = 0$ at $t = 0$, the evolution of the pulse in time is similar to that of Fig. 5.21. The difference to the case described in the previous section, proposed in Problem 5.17, is that the area under the curve now decreases with time. This is due to the pair recombination process, which causes the electron concentration in excess of the equilibrium to decay exponentially with the characteristic time τ_n. If, in addition, there is an electric field along the bar, as the pulse concentration widens and decreases in area, it shifts due to the effect of the electron drift.

To conclude this Chapter, we shall apply the diffusion equation with recombination to the case of injection in steady state. This is what happens, for example, when a light beam of falls on a region of a semiconductor with constant intensity. The photons produce electron–hole pairs in the illuminated region. If the beam intensity is constant, after the transient that occurs when the light begins to fall, the process enters a steady state regime. In this situation, the rate of creation of pairs is constant and the time derivative is zero. This is also what happens with a constant electric current through a p–n junction. When majority carriers on one side arrive at the junction, they are injected into the other side at a constant rate. In steady state $\partial/\partial t = 0$, and from Eqs. (5.78) and (5.79) we obtain

$$\nabla^2 \delta n = \frac{\delta n}{L_n^2}, \tag{5.80}$$

$$\nabla^2 \delta p = \frac{\delta p}{L_p^2}, \tag{5.81}$$

where $L_n = \sqrt{D_n \tau_n}$ and $L_p = \sqrt{D_p \tau_p}$ are the **diffusion lengths** of electrons and holes, respectively. The reason for this name becomes clear with the following example: Consider a semi-infinite semiconductor bar, in which holes are injected uniformly in $x = 0$ at a constant rate, so that the excess concentration is kept constant at this point, $\delta p(x = 0) = \Delta p$. The injected holes diffuse along the bar and recombine with electrons. This results in a distribution of excess concentration along the bar, characterized by the function $\delta p(x)$. To obtain this function we use Eq. (5.81) and consider in the Laplacian operator $\partial^2/\partial y^2 = \partial^2/\partial z^2 = 0$, so that

$$\frac{d^2 \delta p(x)}{dx^2} = \frac{\delta p}{L_p^2}. \tag{5.82}$$

The solution of this equation is

$$\delta p(x) = C_1\, e^{-x/L_p} + C_2\, e^{x/L_p}. \tag{5.83}$$

where C_1 and C_2 are constants determined by the boundary conditions. Due to the recombination along the bar, δp should tend to zero at $x \to \infty$. Thus, we must have $C_2 = 0$. Since $\delta p(x = 0) = \Delta p$, we have $C_1 = \Delta p$. Therefore

$$\delta p(x) = \Delta p\, e^{-x/L_p}. \tag{5.84}$$

Fig. 5.22 Variation of the hole concentration with x due to diffusion, produced by injection at $x = 0$ with a constant rate

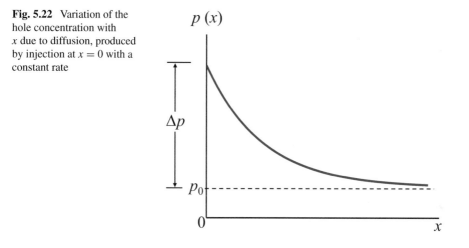

This function is shown in Fig. 5.22. The holes injected in $x = 0$ at a constant rate in time, produce a concentration in excess the equilibrium that falls exponentially with x. The characteristic length of this exponential is L_p, the hole diffusion length.

The value of diffusion length depends on the type of carrier, on the semiconductor material, and on the impurity concentration. The dependence on the carrier and on the material is due both to the diffusion coefficient D (see Table 5.2) and the recombination time τ. The dependence on the impurity concentration is also due to τ. Larger impurity concentrations result in shorter the recombination times. Since D varies in the range 10–200 cm^2/s, and τ is in the range 10^{-7}–10^{-6} s, the diffusion length $L = (D\,\tau)^{1/2}$ is typically in the range of 10^{-3}–10^{-2} cm, or 10–100 μm.

The results of this section will be extensively used in the next two chapters, since the diffusion motion of electrons and holes play a key role in the operation of semiconductor junction devices.

To close this chapter, we remark that we have treated the effects of impurities in the transport properties of semiconductors in a very simplified way. We have considered that the only role of the impurities is to increase the number of electrons in the conduction band, in the case of n-type, or to increase the number of holes in the valence band, in the case of p-type semiconductors. In both cases electrons and holes are treated as quasi-particles described by Bloch's functions as in Eq. (5.4.5), having effective masses determined by the curvatures of the conduction and valence bands, as in Eqs. (5.3) and (5.7). Actually, besides the introduction of charge carriers into the conduction and valence bands, the impurities create disorder in the crystal lattice, because their distribution is random. When the impurity concentration is very low, the Bohr radius of a single impurity wavefunction, as studied in Sect. 5.3.1, is much less than the average distance between impurities. In this case, the impurity levels are the same as for a single impurity, as assumed here, because the dopants do not feel the effects of the neighbors. Also, it is a good approximation to consider that the crystal potential remains the same as in the pure material, and the bands $E(k)$ keep the same dispersion, with k remaining a good quantum number.

However, as the impurity concentration increases, there is a critical value such that the disorder becomes strong enough to prevent transport. For concentrations above this value, the system is no longer periodic, thus the carriers can no longer be described exactly by Bloch's functions. In this case, the carriers undergo multiple scatterings by the impurities and become trapped to a finite region, preventing the drift and diffusion motions as studied here. This trapping is known as Anderson localization, in honor of Philip W. Anderson, who first proposed this effect. Anderson was awarded the Physics Nobel Prize in 1977 for this and many other contributions to the theory of condensed matter physics.

Problems

5.1 Use a development similar to that of Sect. 4.3 to demonstrate that the effective mass of the holes is given by the expression (5.7).

5.2 Show that in an intrinsic semiconductor, with parabolic bands, the Fermi level is given by Eq. (5.22).

5.3 (a) Show that the effective concentrations of electrons and holes, N_c and N_v, can be calculated numerically with the expression

$$N_{c,v}(T) = 2.54 \left(\frac{m_{c,v}^*}{m_0} \frac{T}{300} \right)^{3/2} \times 10^{19} \, \text{cm}^{-3},$$

where T is the temperature in K. Apply this expression to Si and Ge and compare with the values in Table 5.2.

(b) Calculate the values of n_i at $T = 300$ K for Ge, Si and GaAs, from the data in Table 5.2, and compare with the values in that Table and in Fig. 5.8.

5.4 Calculate the distance between the Fermi level and the middle of the gap in pure Si and in GaAs at 300 K. Explain why E_F is not in the middle of the gap.

5.5 Using the data in Table 5.2, calculate the ionization energy of donor impurities in Si using the hydrogen atom model developed in Sect. 5.3.1.

5.6 Consider a type p semiconductor with acceptor impurity concentration N_a, all ionized, at a temperature such that $n_i \ll N_a$.

(a) Using the law of mass action (5.30) and the charge neutrality Eq. (5.33), obtain the expressions for the concentrations of electrons and holes (5.41) and (5.42).

(b) Using the results of item (a) and Eqs. (5.28) and (5.32), show that the Fermi level is given by (5.43) or (5.44).

(c) Show that E_i given by (5.22) is consistent with the expressions obtained in item c).

5.7 Three silicon wafers are doped with as impurities with concentrations 10^{16}, 10^{17}, and 5×10^{18} atoms/cm^3. Consider $T = 300$ K and assume that all impurities are ionized.

     (a)   Calculate the Fermi level in each wafer.

     (b)   Check if the approximation of Eq. (5.15) for the Fermi–Dirac function is good in the three cases.

     (c)   Calculate the resistivity of each wafer.

5.8   The probability of electron occupation of the discrete energy levels of the impurities is not given simply by the Fermi–Dirac statistics. It can be shown that the concentration of ionized donor impurities is given by (see Ashcroft and Mermin)

$$N_d^+ = \frac{N_d}{1 + (1/2)\, e^{(E_d - E_F)/k_B T}},$$

where E_d is the energy level and N_d the concentration of the impurities.

     (a)   Check if the assumption of complete ionization is good in the three wafers of Problem 5.7.

     (b)   Make a plot of N_d^+/N_d as a function of T for the wafer of Problem 5.7 with the largest concentration, assuming that E_g does not vary in the temperature range of 0 to 400 K.

5.9   A wafer of GaAs is doped with donor impurities with a concentration of 10^{17} atoms/cm^3. Assuming that all impurities are ionized, calculate the resistivity of the wafer and compare with the value obtained in problem 5.7 for Si with the same concentration.

5.10   Calculate the concentrations of donor impurities that make Si and GaAs degenerate ($E_F = E_c$).

5.11   Consider a silicon wafer doped with acceptor impurities with concentration $N_a = 2 \times 10^{14}$ cm^{-3} and assume that they are all ionized.

     (a)   Calculate the concentrations of electrons and holes at $T = 300$ K. In this situation, is the semiconductor considered intrinsic or extrinsic?

     (b)   Calculate the concentrations of electrons and holes at $T = 600$ K, considering that at this temperature the gap decreases to 1.0 eV and that the effectives masses are approximately the same as at room temperature. In this situation, is the semiconductor intrinsic or extrinsic?

5.12   (a)   Explain, qualitatively, using few words and some plots, why the Fermi level in a type n semiconductor is closer to the conduction band than the valence band, and in a type p it is closer to the valence band.

     (b)   Explain, qualitatively, using few words and some plots, how the Fermi level varies with temperature in a type n semiconductor.

5.13   A thermistor is a resistor whose resistance varies with temperature. Consider a thermistor made of intrinsic Silicon, that at $T = 300$ K has a resistance of 500 Ω.

(a) Assuming that the mobility does not vary with temperature, calculate the rate of change of resistance with temperature around 300 K, expressed in $\Omega \ /^{\circ}C$.

(b) What is, approximately, the resistance of the thermistor at $T = 302$ K?

5.14 A Germanium bar has a Length of 1 cm and a square cross-section with side of 1 mm.

(a) Calculate the resistance between the two ends of the bar at $T = 300$ K in the case of the intrinsic semiconductor.

(b) Consider that the bar is doped with a certain concentration of donor impurities N_d. Assuming that the mobility is the same in the pure material, what is the value of N_d for the resistance to be 10 Ω at $T = 300$ K?

5.15 A semiconductor bar with concentration of majority carriers 10^{16} cm^{-3} has width $w = 1$ mm and thickness $d = 0.5$ mm. What is the Hall voltage on the bar when subjected to a magnetic field $B = 0.1$ T (1 kG) and traversed by a current of 100 mA?

5.16 A semi-infinite bar made of a semiconductor material has a stationary hole distribution shown in Fig. 5.22. This distribution is maintained by a certain constant current I, entering at the end of the bar at $x = 0$ through a metallic contact.

(a) Using expression (5.57) for the diffusion current, calculate the current $I = I_p\ (x = 0)$ as a function of L_p, D_p, and the excess concentration δp at $x = 0$.

(b) Show that this current is equal to the total charge at $x > 0$, obtained by integrating the hole distribution $\delta p(x)$, divided by the hole lifetime τ_p. Explain why this calculation leads to the same result as in item (a).

5.17 Consider a semiconductor in the form of a bar like the one in Fig. 5.20, with an electron concentration in excess of equilibrium described by a Gaussian function of the position x, at a time t, given by

$$\delta n(x, t) = \frac{\Delta N_0}{2\sqrt{\pi D_n t}} e^{-x^2/4D_n t},$$

where ΔN_0 is the number of electrons per unit area at $t = 0$, in the region between two sections spaced by a very small distance Δx around $x = 0$.

(a) Show that this Gaussian function is a solution of the diffusion equation for electrons, Eq. (5.65).

(b) Show that at very small t this distribution tends to $A\delta(x)$, where A is a constant and $\delta(x)$ is the Dirac delta function. Calculate the value of A.

(c) Make a qualitative plot of $\delta n(x)$, for a generic instant t_1. At this instant, calculate the width Δx of the distribution, defined as the distance between two points at which the value of the distribution is $\delta n(0)/2$. Obtain the relationship between the diffusion coefficient D_n, the width Δx and the instant t_1. From this result, suggest a method to measure the diffusion coefficient.

Further Reading

N.W. Ashcroft, N.D. Mermin, *Solid State Physics* (Holt, Rinehart and Winston, New York, 2011)

A. Bar-Lev, *Semiconductors and Electronic Devices* (Prentice-Hall, New Jersey, 1993)

W.F. Beadle, J.C.C. Tsai, R.D. Plummer (eds.), *Quick Reference Manual for Silicon Integrated Circuit Technology* (Wiley, New York, 1985)

D.A. Fraser, *The Physics of Semiconductor Devices* (Claredon Press, Oxford, 1986)

R.E. Hummel, *Electronic Properties of Materials* (Springer, Berlin, 2011)

K. Kano, *Semiconductor Devices* (Prentice-Hall, New Jersey, 1998)

C. Kittel, *Introduction to Solid State Physics* (Wiley, New York, 2004)

D. Neamen, *Fundamentals of Semiconductor Physics and Devices* (McGraw-Hill, New York, 2002)

D.J. Roulston, *An Introduction to the Physics of Semiconductor Devices* (Oxford University Press, Oxford, 1999)

B.J. Streetman, S. Banerjee, *Solid State Electronic Devices* (Prentice Hall, New Jersey, 2005)

S.M. Sze, M.K. Lee, *Semiconductor Devices: Physics and Technology* (Wiley, New York, 2012)

F.F.Y. Wang, *Introduction to Solid State Electronics* (North-Holland, Amsterdam, 1989)

Chapter 6
Semiconductor Devices: Diodes

The topics studied in the previous chapter for homogeneous semiconductors can now be used to understand the behavior of devices made of regions with different dopings. In this chapter we consider the p–n junction formed by a semiconductor containing both p- and n-regions. The p–n junction is a device itself, used in rectification, switching, and other operations in electronic circuits. It is also a building block for various other semiconductor-based devices, such as the transistor, that will be considered in the next chapter. Here we study several features and applications of p-n junctions and also of heterojunctions formed by semiconductors and metals.

6.1 The p-n Junction

The fact that several regions of the same semiconductor material can be doped with different impurities enables the fabrication of a wide variety of electronic devices. Almost all semiconductor devices contain at least one p–n junction, that consists of a piece of a semiconductor material with a p-type region next to an n-type region, separated by a thin transition layer. The thickness of the transition layer depends on the manufacturing method, ranging from 0.01 to 1 μm. The behavior of electrons and holes at the junctions of a device determines the current–voltage $(I - V)$ characteristics of its various terminals. For this reason, this Chapter starts with a detailed study of the p–n junction. It will serve as the basis for understanding the operation of many semiconductor devices. In the following section we derive the $I - V$ characteristics of the junction diode, the simplest device of all, consisting of just one p–n junction. In the following sections we shall describe the operation of other types of diodes. Transistors and others active devices will be presented in Chap. 7. More detailed information on the techniques for the fabrication of semiconductor devices and on other specific properties can be found in several books listed in the Further Reading. Semiconductor devices for applications in opto-electronics will be presented in Chap. 8.

© The Author(s), under exclusive license to Springer Nature Switzerland AG 2022   153
S. M. Rezende, *Introduction to Electronic Materials and Devices*,
https://doi.org/10.1007/978-3-030-81772-5_6

6.1.1 Fabrication of p-n Junctions

The technology for preparation of semiconductor junctions has evolved tremendously since the early days of the commercial fabrication of semiconductor devices in the 1950s. The methods most used today are diffusion and ion implantation, mentioned in Sect. 5.3. Figure 6.1 shows the basic steps in the manufacture of a *p–n* junction by diffusion, with the planar technology introduced in the beginning of the 1960s. The first step consists in the preparation of the semiconductor crystal wafer, the **substrate**, shown in Fig. 6.1a. Most semiconductor devices are made with single crystal Si wafers. The wafer, with thickness of some tenths of mm, is obtained by slicing a cylindrical Si ingot, like the one in Fig. 1.11, and polishing one of its surfaces. In general, the Si crystal ingot is grown with high concentration of *n*-type impurities, which is denoted by n^+. The high n concentration enables the formation of ohmic contacts with the metallic layer deposited later (Fig. 6.1f).

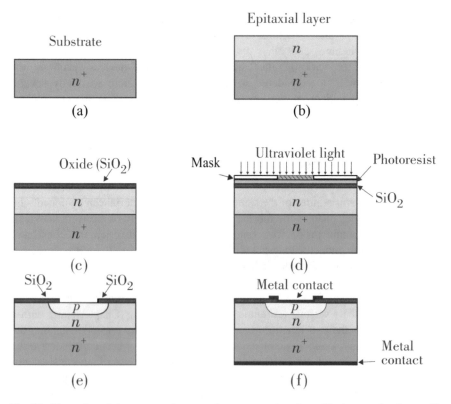

Fig. 6.1 Illustration of the steps used to manufacture a *p–n* junction with planar technology. **a** Si wafer used as a substrate. **b** Substrate with a layer of epitaxial Si doped with *n*-type impurities. **c** Oxide layer grown on the Si layer. **d** Photolithographic process used to produce a pattern on the positive photoresistive film. **e** Diffusion of *p*-type impurities through the window in the oxide layer. **f** Complete structure of the junction with the metal contacts

The next step consists in growing on the n^+ substrate a layer of *n*-type Si with lower impurity concentration, using a technique for epitaxial growth (Fig. 6.1b). The wafer is then taken to an oven with an oxygen atmosphere to form a thin layer of SiO_2 oxide (Fig. 6.1c). Next, a photolithographic process is used to selectively remove the oxide from some regions to allow the diffusion of impurities. The process makes use of the effect of ultraviolet (UV) radiation on a film of a photoresistive liquid, or photoresist, or simply resist, that is spread over the oxide layer in a spin coater. Positive photoresists have a chemical structure that changes after UV exposure, so that they become soluble in a developer solution. On the other hand, negative photoresists become insoluble after UV exposure. Both are also commonly used in photolithography techniques. We consider here the first type. After the spread of the positive resist, the wafer is placed in a baking oven for drying. The formation of the pattern in the Si layer is made by shinning UV light through a photomask containing opaque and transparent areas, as in Fig. 6.1d.

The next step consists in using a solvent to remove the resist from the exposed regions. Then, the wafer in placed in an acid bath, which corrodes the oxide layer in the regions where the resist was removed. This process opens a window in the oxide layer through which diffusion of type *p* impurities is made in an oven at high temperature (on the order of 1000 °C), containing a gas with the impurities (Fig. 6.1e). Finally, the structure is completed with the deposition of metallic films for the external contacts (Fig. 6.1f).

The planar technology is used to manufacture a simple diode junction, or a transistor with multiple junctions, or a complex integrated circuit containing a large number of diodes, transistors and resistors on the same Si chip. A very important component in the processing of the wafer is the mask containing the pattern of the circuit to be produced. Until the 1990s, the original layout was drawn on a paper in large scale to increase its resolution, and later photographically reduced to the full scale of the mask. Currently, the whole process is made on computers using sophisticated softwares. For modern high-integration circuits, in which the lateral dimensions of the structures are much less than 1 μm, the masks are produced on the real scale by electron beams.

6.1.2 The Potential Barrier at the p-n Junction

To treat mathematically the equations that describe the charge and the electric potential in a junction it is necessary to make some approximations to the actual junction. The first is to reduce the problem to one dimension. As we can see in Fig. 6.1f, due to the shape of the junction and contacts, the motion of electrons and holes in most part of the device occurs in the direction normal to the surface that separates the *p* and *n* regions. Therefore, the assumption that quantities vary only in one direction, say *x*, is a good approximation to the actual problem. The second approach refers to the separation between the *p* and *n* regions. In the real junction, the variation in the concentration of impurities near the interface is gradual. The difference in the concentrations of the impurities, $N_a - N_d$, gradually changes from positive in the p region, to negative in the n region, as shown by dashed line of Fig. 6.2a. However, to simplify the problem, we assume that the junction is abrupt, that is, $N_a - N_d$ varies sharply from a positive constant value at $x < 0$ to a negative constant value at $x > 0$, like a step function, as in the full line of Fig. 6.2a. Figure 6.2b shows the one-dimensional model for an abrupt p–n junction, which we consider in this section.

To understand what happens at the junction in equilibrium, let us assume that the p and n regions of the semiconductor are physically separated before the junction is formed. In this situation, the Fermi level is close to the conduction band on the n-side and next to the valence band on the p-side, as illustrated in Fig. 6.3a. Suppose now that the two materials are brought into contact to form the junction. Since electrons are in excess of holes on the n-side, there is electron diffusion from side n to side p. Similarly, holes diffuse from side p to side n. The diffusion of charges from one side to the other produces two layers of charges, illustrated in top of Fig. 6.3b, formed by ionized impurities, donors on the n-side and acceptors on the p-side. These layers of charges create an electric field $\vec{E}$ directed from n to p, which exerts a force on the electrons and the holes opposing the continuation of their diffusion motion. The electric field pushes the holes back to side p and the electrons back to side n, by

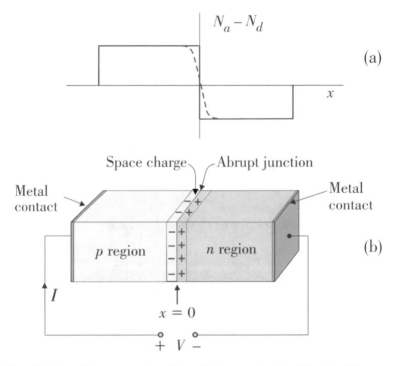

Fig. 6.2 **a** Variation of the concentration of impurities in a p–n junction. The dashed line represents the variation in a real junction, while the solid line represents an ideal abrupt junction. **b** One-dimensional abrupt junction model

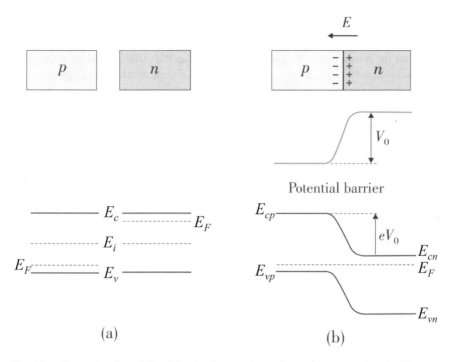

Fig. 6.3 **a** Energy bands and Fermi levels of two regions of a semiconductor doped with *p* and *n* impurities, physically separated. **b** Charge, electric field, electric potential, and energy levels in the space charge region in the *p–n* junction

means of a drift current that opposes the diffusion current. In the equilibrium regime, drift and diffusion currents cancel each other, both for electrons and holes, so that the total current is zero. In this situation, the distribution of charges and the electric field acquire a stationary configuration.

The region near the junction where there are unbalanced charges, shown in Figs. 6.2 and 6.3b, is called **space charge region**. This region is also called **transition layer**, or depletion layer. The electric field created in this region corresponds to a potential difference V_0 between side *n* and side *p*. This potential difference tends to prevent the motion of majority carriers on the *p*-side (holes) to the *n*-side, and majority carriers on the *n*-side (electrons) to the *p*-side. Due to the shape of the potential variation, illustrated in Fig. 6.3b, it is called a **potential barrier**. The formation of the potential barrier is the most important physical phenomenon that occurs at the junction, because it is responsible for its electric characteristics. The shape of the potential barrier also has an important effect on the behavior of the energy levels in the junction. Since the energy of the electron is related to the electrostatic potential ϕ by $E = -e\,\phi$, the difference between the energies of the conduction band E_{cp} in the *p*-side and E_{cn} in the *n*-side is given by

$$E_{cp} - E_{cn} = -e\,(\phi_p - \phi_n) = eV_0. \tag{6.1}$$

Therefore, the difference in energies is, in units of eV, the height V_0 of the potential barrier. This means that when the junction is formed, the references to the energy levels on the p and n sides adjust themselves so that the difference in the energies of the conduction band on the two sides, as well as of the valence band, corresponds to the potential difference due to the electric field created by the charges at the junction. This change in the reference levels is a consequence of the fact that the Fermi level E_F must be the same on both sides of the junction, as shown in Fig. 6.3b. The figure also shows that, since the lowest possible value of E_F on the p-side is E_{vp}, and the highest value of E_F on the n-side is E_{cn}, the limit value of the potential barrier is $V_0 = E_g/e$. The height of potential barrier approaches this limit when the two regions of the junction are heavily doped.

In fact, the explanation for the shape of the potential barrier could have started by analyzing the Fermi level, since it is related to the chemical potential of a thermodynamic system, which must be uniform when the system is in equilibrium. We can draw an analogy between the Fermi level and the level of water in a reservoir, since all water molecules have energy (gravitational) lower than those at the surface. When two reservoirs with different levels are interconnected, part of the water of the tank with higher level goes to the other tank until the levels are the same. What happens when two semiconductors are placed in contact is analogous. The charges flow from one side to the other until the Fermi levels become equal. When this occurs, the system reaches equilibrium. The value of the potential difference V_0 of the barrier at the junction at equilibrium, also called **contact potential**, can be calculated in various ways: the simplest is based on the facts that the Fermi level is uniform at the junction and the intrinsic semiconductor is the same in the two regions. With Eq. (5.32) we can write the relation between the energies and the equilibrium concentrations of holes, p_{p0} on the p-side, and p_{n0} on the n-side, in regions away from the junction

$$p_{p0} = n_i \, e^{(E_{ip}-E_F)/k_B T},$$

$$p_{n0} = n_i \, e^{(E_{in}-E_F)/k_B T}.$$

The ratio between the two concentrations is then

$$\frac{p_{p0}}{p_{n0}} = e^{(E_{ip}-E_{in})/k_B T}. \tag{6.2}$$

Since the intrinsic semiconductor is the same at the p and n regions, we see in Fig. 6.3b that the difference between the intrinsic Fermi levels on both sides is precisely the value of the potential barrier in units of eV, $E_{ip} - E_{in} = eV_0$. Substitution of this expression into Eq. (6.2) gives

$$V_0 = \frac{k_B T}{e} \ln \frac{p_{p0}}{p_{n0}}. \tag{6.3}$$

This same result can also be obtained by integration of Eq. (5.70), that expresses the fact that in a junction in equilibrium the total current of holes is zero (Problem 6.3). We can also relate the contact potential with the concentrations of electrons on both sides of the junction. Starting from Eq. (5.31) we obtain

$$V_0 = \frac{k_B T}{e} \ln \frac{n_{n0}}{n_{p0}}. \tag{6.4}$$

This result can also be obtained from Eq. (6.3) using the law of mass action. Equations (6.3) and (6.4) can be rewritten in the form

$$\frac{p_{p0}}{p_{n0}} = \frac{n_{n0}}{n_{p0}} = e^{eV_0/k_B T}. \tag{6.5}$$

Finally, using the relations obtained in Chap. 5, we can express the contact potential in terms of the concentrations of impurities on both sides of the junction. In the *p*-region, the holes are the majority carriers and their concentration is, by Eq. (5.42), $p_{p0} \approx N_a$. On the other hand, in the *n*-region, according to (5.38), $p_{n0} \approx n_i^2/N_d$. Using these values in Eq. (6.3) we obtain,

$$V_0 \approx \frac{k_B T}{e} \ln \frac{N_a N_d}{n_i^2}, \tag{6.6}$$

and using Eq. (5.25) this result can be written as

$$V_0 \approx \frac{E_g}{e} - \frac{k_B T}{e} \ln \frac{N_c N_v}{N_a N_d}. \tag{6.7}$$

This is a very convenient expression to calculate the contact potential, using only the parameters of the semicontuctors of the junction.

Example 6.1 Consider a *p–n* junction made with Si, having concentrations of impurities $N_d = 10^{16}$ cm^{-3} and $N_a = 10^{18}$ cm^{-3}. Calculate the contact potential of the junction at $T = 300$ K.

Using $k_B T = 0.026$ eV and the values of E_g, N_c and N_v in Table 5.2, we obtain with Eq. (6.7)

$$V_0 = 1.12 - 0.026 \times \ln \frac{2.6 \times 10^{19} \times 1.02 \times 10^{19}}{10^{18} \times 10^{16}}.$$

$$V_0 = 1.12 - 0.026 \times 10.18 = 0.85 \text{ V}.$$

For a junction of Ge with the same concentrations of impurities as in the Example 6.1, it can be shown that (Problem 6.1) $V_0 = 0.45$ V. Notice that as the concentrations of impurities increase, the second term in Eq. (6.7) decreases and V_0 approaches E_g/e. Thus, the maximum value of the contact potential is 0.66 V in a Ge junction and 1.12 V in a Si junction.

6.1.3 Charge and Field at the Junction in Equilibrium

The contact potential calculated in the previous section is the potential difference between a point in the p-side and a point in the n-side, both away from the junction. To calculate the electric field it is necessary to obtain the variation of the potential gradient in the space charge region, which depends on charge distribution in the region. Instead of solving the complete problem self-consistently, we shall approximate the charge distribution by a simple function to calculate the field and the potential from it. To obtain this distribution we consider what happens in the space charge region, illustrated in Fig. 6.4a.

Electrons and holes are in permanent transit, passing from one side of the junction to the other. Some electrons pass from the n-side to the p-side by diffusion, recombine with holes or are "pushed" back to the n-side by the force of the electric field. The same goes for holes on the other side. As a result, there are few electrons and holes in the space charge region because they are swept by the electric field. This exhaustion of mobile charges from the space charge region is the reason for the name **depletion region**. In this way, the charges in the region are due to the ions of uncompensated impurities, donors on the n-side and acceptors on the p-side. Having the donor impurities, with concentration N_d, lost their electrons, their charge is positive. On the other hand, acceptor impurities, with concentration N_a, receive electrons and become negative. Thus, in a first approximation, we can consider that in the n-side, the charge density has value $\rho = +eN_d$ that is uniform in a layer of thickness l_n, and zero outside. On the other hand, in the p-side the density is $\rho = -eN_a$ in a layer of thickness l_p and zero outside, as illustrated in Fig. 6.4b. This is called the depletion approximation. As the total charge must be zero, since the junction is electrically neutral, the moduli of the total charges on the two sides are the same. Since the charge in each side is equal to the product of the charge density by the volume, it is easy see that the thicknesses of the layers are related to the concentrations of impurities by

$$l_n N_d = l_p N_a. \tag{6.8}$$

Since the total thickness of the space charge region is $l = l_p + l_n$, we can express the thicknesses of the two charge layers in terms of the concentrations of impurities in the two sides

$$l_p = \frac{N_d}{N_a + N_d} l, \qquad l_n = \frac{N_a}{N_a + N_d} l. \tag{6.9}$$

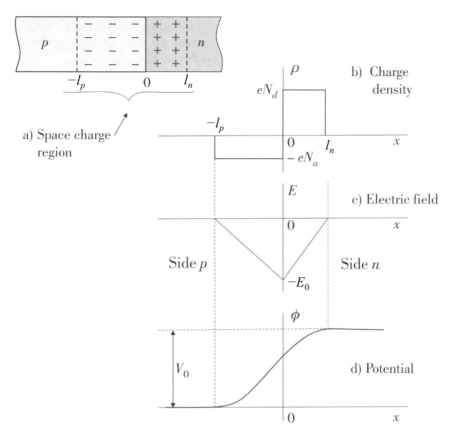

Fig. 6.4 Variation along x of the charge density, electric field, and electrostatic potential in the one-dimensional model of the *p–n* junction

These equations show that the thickness is larger on the side with lower doping. To calculate the electric field from the distribution of charges, we use Gauss's law in the differential form, Eq. (2.1). Considering a spatial variation only in the x direction, using the relation $\vec{D} = \varepsilon \vec{E}$, Eq. (2.1) can be written as

$$\frac{dE}{dx} = \frac{\rho(x)}{\varepsilon},\tag{6.10}$$

where $\rho = -eN_a$, at $-l_p < x < 0$ and $\rho = eN_d$ at $0 < x < l_n$. Integration of (6.10) with these densities results in a linear variation of the electric field on each side, as shown in Fig. 6.4. For $-l_p < x < 0$ we have

$$E(x) = -\frac{e\,N_a}{\varepsilon}x - E_0,\tag{6.11}$$

where E_0 is an integration constant, that corresponds to the value of E at $x = 0$. Since $E\,(x = -l_p) = 0$, because the field is zero outside the space charge region, from (6.11) we have $E_0 = eN_a l_p/\varepsilon$. From (6.10) we also see that, at $0 < x < l_n$

$$E(x) = \frac{e\,N_d}{\varepsilon}\,x - E_0, \tag{6.12}$$

where the integration constant, determined by $E\,(x = l_n) = 0$, must be the same one obtained with (6.11). In fact, using Eq. (6.8), we see that

$$E_0 = \frac{e\,N_a\,l_p}{\varepsilon} = \frac{e\,N_d\,l_n}{\varepsilon}. \tag{6.13}$$

Figure 6.4c shows the variation with x of the electric field given by Eqs. (6.11) and (6.12). Note that the field is different than zero only in the space charge region, and is directed along $-x$, as expected. From the expressions of the electric field, we can obtain the variation of the potential $\phi\,(x)$, using the relation $E\,(x) = -d\phi/dx$. The function whose derivative is Eq. (6.11), with the sign reversed, is

$$\phi(x) = \frac{1}{2}\frac{e\,N_a}{\varepsilon}\,x^2 + E_0\,x + C,$$

where C is a constant with a value that depends on the choice of the reference for the potential. Taking as reference $\phi\,(x = -l_p) = 0$, and replacing the expression for E_0 in (6.13), we obtain $C = e\,N_a\,l_p^2/2$. Thus, the potential at $-l_p \le x \le 0$ is

$$\phi(x) = \frac{e\,N_a}{2\varepsilon}\,(x + l_p)^2. \tag{6.14}$$

To calculate $\phi\,(x)$ at $x \ge 0$, we integrate Eq. (6.12) in an analogous way and determine the integration constant by equating the expressions for the potentials in the two sides at $x = 0$. The result is, for $0 \le x \le l_n$

$$\phi(x) = -\frac{e\,N_d}{\varepsilon}\left(\frac{1}{2}x^2 - l_n x - \frac{1}{2}l_p\,l_n\right). \tag{6.15}$$

The variation of the potential with x given by Eqs. (6.14) and (6.15) is shown in Fig. 6.4d. As expected, it has the shape of the potential barrier in Fig. 6.3b. The value of the contact potential V_0 is the difference between the potentials at points $x = l_n$ and $x = -l_p$, which is simply the value of the potential at $x = l_n$. Using (6.15) and the expression for E_0 in (6.13) we obtain

$$V_0 = \phi(l_n) = \frac{e\,N_d}{2\varepsilon}\,l_n\,l = \frac{1}{2}\,E_0\,l. \tag{6.16}$$

Since the potential difference between two points is the integral of the electric field, the result (6.16) could have been easily obtained by the area of the triangle that represents the variation of $E(x)$ shown in Fig. 6.4c. Starting from Eqs. (6.9) and (6.16), it is possible to relate the thicknesses of the charge layers with the concentrations of the impurities and the contact potential. It is easy to show that

$$V_0 = \frac{e}{2\varepsilon} \frac{N_a N_d}{N_a + N_d} l^2, \tag{6.17}$$

from which we obtain for the thickness of the space charge region

$$l = \left[\frac{2\varepsilon V_0}{e} \left(\frac{1}{N_a} + \frac{1}{N_d} \right) \right]^{1/2}. \tag{6.18}$$

To obtain an expression for the thickness as a function only of the parameters of the semiconductor, we substitute (6.6) into (6.18). The result is

$$l = \left[\frac{2\varepsilon k_B T}{e^2} \left(\frac{1}{N_a} + \frac{1}{N_d} \right) \ln \frac{N_a N_d}{n_i^2} \right]^{1/2}. \tag{6.19}$$

With this expression and Eqs. (6.9) it is possible to calculate the thicknesses l_p and l_n of the charge layers on both sides of the junction. Finally, we note that, as the potential difference between the two sides is produced by two charge layers, the junction has a capacitance C. Denoting by A the area of the cross section of the junction, the total charges in the layers are $+Q$ and $-Q$, where $Q = eN_d l_n A$. In the case the charges are distributed in the two layers, the capacitance is defined by $C = dQ/dV$. From Eqs. (6.9) and (6.18) we then obtain (Problem 6.5) for the junction capacitance

$$C = \frac{\varepsilon A}{l}, \tag{6.20}$$

where l is given by Eq. (6.19). It can be seen that the junction capacitance varies inversely proportional to the thickness l of the space charge region. As we shall see in the next section, l can be changed by applying an external voltage, so that the value of the capacitance can be controlled by the applied voltage.

Example 6.2 Consider a *p–n* junction made of Si, as the one in Example 6.1, having a circular cross section with diameter 200 μm. Calculate: (a) The thickness of the space charge region; (b) The value of the maximum electric field at the junction; (c) The capacitance of the junction.

(a) To calculate the thickness, we use Eq. (6.18), with the absolute value of the electron charge $e = 1.6 \times 10^{-19}$ C and the permissivity of vacuum

$\varepsilon_0 = 8.5 \times 10^{-12}\,\mathrm{Fm^{-1}}$. From Table 5.2 we have for the dielectric constant of Si, $\varepsilon/\varepsilon_0 = 11.8$, and from Example 6.1 we have $V_0 = 0.85\,\mathrm{eV}$. Hence

$$l = \left[\frac{2 \times 11.8 \times 8.85 \times 10^{-12} \times 0.85}{1.6 \times 10^{-19}} \left(\frac{1}{10^{18} \times 10^6} + \frac{1}{10^{16} \times 10^6} \right) \right]^{1/2}\,\mathrm{m.}$$

$$l \approx \left[\frac{2 \times 11.8 \times 8.85 \times 10^{-12} \times 0.85}{1.6 \times 10^{-19} \times 10^{22}} \right]^{1/2} = 3.3 \times 10^{-7}\,\mathrm{m.}$$

or $l = 0.33\,\mu\mathrm{m}$.

(b)   From Eq. (6.16) we have

$$E_0 = \frac{2V_0}{l} = \frac{2 \times 0.85}{3.3 \times 10^{-7}} = 5.2 \times 10^6\,\mathrm{V/m.}$$

(c)   To calculate the capacitance we use Eq. (6.20) with the area $A = \pi R^2$, where $R = 10^{-4}$ m is the radius of the circular section. Thus

$$C = \frac{11.8 \times 8.85 \times 10^{-12} \times 3.14 \times 10^{-8}}{3.3 \times 10^{-7}} = 9.9 \times 10^{-12}\,\mathrm{F} = 9.9\,\mathrm{pF.}$$

6.2   Current in the Biased Junction: *I-V* Characteristics

When a junction is biased, that is, subjected to a voltage from an external circuit, there is a change in the equilibrium conditions so that an electric current flows through, with a direction that depends on the sign of the applied voltage. The essential feature of the *p–n* junction is its asymmetry relative to the direction of the applied voltage. Voltages in opposite directions produce currents with very different intensities. This can be understood by examining the effect of the external voltage on the potential barrier.

When an external voltage V is applied at the junction terminals, it appears almost entirely at the space charge region. This is so because the density of carriers in this region is much smaller than in the neutral regions of the semiconductor, and therefore it has much larger resistance. Thus, the external voltage is added to, or subtracted from, the contact potential V_0 in equilibrium, depending on its direction, as illustrated in Fig. 6.5. When the voltage V is applied from side p to side n, in the so-called **direct polarization,** or **forward bias,** it decreases the height of the potential barrier, that becomes $V_0 - V$. (Fig. 6.5b). On the other hand, if V is directed from n to p, called **reverse bias**, the barrier increases and has a value of $V_0 + V$ (Fig. 6.5c). The result is that the current flowing through the junction when the voltage is applied in the forward direction is larger than in the reverse direction, giving the *p–n* junction an

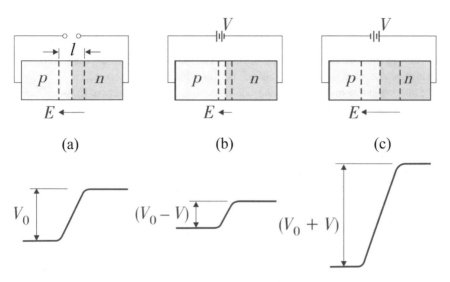

Fig. 6.5 Effect of external voltage on the thickness of the space charge region and on the potential barrier height. **a** Equilibrium situation. **b** Direct, or forward, bias. **c** Reverse bias

asymmetric response to the applied voltage. This is the basis of operation of diodes and junction transistors.

It is easy to verify that the electric field and the thickness of the space charge region also vary with the applied voltage. For a forward bias voltage V the potential difference in the barrier decreases, therefore the field also decreases. As a result, the thickness of the space charge region decreases and can be calculated with Eq. (6.18) with V_0 replaced by $V_0 - V$. On the other hand, when the junction is polarized in the reverse direction, the barrier height, the electric field, and the thickness of the space charge region, all increase.

Let us now consider what happens to the various components of the current in the junction, in order to calculate its *I-V* characteristics. Let us adopt the convention that V is positive if applied in the direction of forward bias, and negative in the reverse direction. When a positive voltage is applied to the terminals of the junction, a current I enters in the metallic contact on the p-side (Fig. 6.2) and leaves by the contact on the n-side. In the two neutral regions of the semiconductor junction, the current is entirely due to drift and dominated by the majority carriers, holes on the p-side and electrons on the n-side. These carriers move towards the space charge region, undergo recombination and also go to the other side, where they move by diffusion. To calculate the value of the current I produced by a voltage V, it is necessary to understand, in detail, the various components of the currents in all regions of the junction.

Consider what happens with the holes that move in the p-side towards the depletion region. Upon reaching the region close to the space charge region, many of them recombine with electrons coming from the n-side. Those that "survive" arrive at the depletion region, where the density of carriers is much smaller and therefore there is little recombination. Upon reaching the border of the depletion region, at the plane $x \approx +l_n$ in Fig. 6.4, the **holes are injected** into the n-side where they become minority carriers. In this side the holes diffuse further into the n-region while recombining with electrons, resulting in a variation of the concentration as that obtained in Sect. 5.4.4. The holes injected into the n-side have a concentration δp in excess of the equilibrium value p_{n0} that decays exponentially with x. Since the diffusion length $L_p = (D_p \tau_p)^{1/2}$ is of the order of 10^{-3} to 10^{-1} cm, it is much larger than the thickness of the space charge layer, $l \sim 1\ \mu\text{m} = 10^{-4}$ cm. Thus, the variation of the hole concentration p_n on the n-side of the junction has the shape shown in Fig. 6.6. Electrons on the n-side, moving in the opposite direction of holes, are injected into the p-side where they are minority carriers, and have a behavior analogous to the one just described for holes.

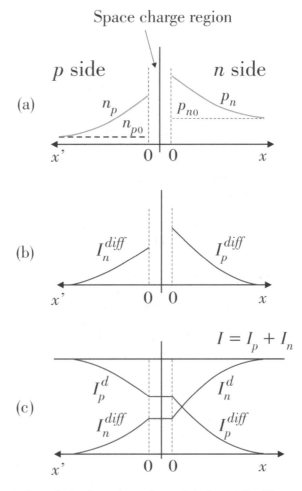

Fig. 6.6 Concentrations of minority carriers and currents in the vicinity of the space charge region in a forward biased p–n junction

The calculation of the total current that flows through the junction can be done entirely based on the diffusion currents of minority carriers on both sides. As described earlier, holes arriving at the junction from the p-side are injected into the n-side and move by diffusion. Similarly, electrons flowing in the opposite direction are injected into the p-side where they diffuse. To calculate the two diffusion currents, we need first to obtain the concentrations of the carriers. For this we use two coordinate systems, shown in Fig. 6.6. On the n-side the coordinate is represented by x, with the origin $x = 0$ at the boundary of the space charge region (plane at $x = + l_n$ in Fig. 6.4). On the p-side the coordinate x' is in the $-x$ direction, where $x' = 0$ is the plane $x = -l_p$ of Fig. 6.4. According to Eq. (6.5), the ratio between the equilibrium concentrations of holes on the two sides of the junction is

$$\frac{p_{p0}}{p_{n0}} = e^{eV_0/k_BT}. \tag{6.21}$$

When an external voltage V is applied at the junction, the height of the potential barrier becomes $V_0 - V$, so that the difference between the intrinsic Fermi levels on both sides is $E_{ip} - E_{in} = e(V_0 - V)$. As shown in Fig. 6.5, this result is consistent with the difference between the Fermi levels on sides p and n of $\Delta E_F = -eV$, because the junction is not in equilibrium. Thus, the ratio between the concentrations of holes at the boundaries of the spatial charge region on sides p and n, obtained by a development analogous to the one that led to Eq. 6.2, is given by

$$\frac{p_p(x' = 0)}{p_n(x = 0)} = e^{e(V_0-V)/k_BT}. \tag{6.22}$$

In case the current at the junction is not very high, the concentrations of the majority carriers almost do not change from the values in equilibrium with the application of the external voltage. Thus, $p_p(x' = 0) \approx p_{p0}$. Substitution of this result in Eq. (6.22) and division by (6.21) gives

$$\frac{p_n(x = 0)}{p_{n0}} = e^{eV/k_BT}. \tag{6.23}$$

This result shows that the concentrations of the minority carriers in the boundaries of the space charge region increase exponentially with the applied voltage, in the case of forward bias. On the other hand, they decrease exponentially with the voltage in the case of reverse bias. From Eq. (6.23) it is simple to obtain the diffusion current of holes on the n side using the results of Sects. 5.4.3 and 5.4.4. The hole concentration in excess of equilibrium at $x = 0$, obtained from Eq. (6.23), is

$$\delta p_n(x = 0) \equiv p_n(x = 0) - p_{n0} = p_{n0}(e^{eV/k_BT} - 1). \tag{6.24}$$

The variation of δp_n along x, obtained using this result in Eq. (5.84), is

$$\delta p_n(x) = p_{n0} \left(e^{eV/k_B T} - 1\right) e^{-x/L_p}.$$ (6.25)

This variation is shown in Fig. 6.6a. From this result we can obtain the diffusion current density of holes in the $+x$ direction on the n-side. Using (6.25) in Eq. (5.57) we have

$$J_p^{diff}(x) = e \frac{D_p}{L_p} p_{n0} \left(e^{eV/k_B T} - 1\right) e^{-x/L_p}.$$ (6.26)

Denoting by A the area of the junction cross-section, the intensity of the diffusion current of holes at $x = 0$ is then

$$I_p^{diff}(0) = e A \frac{D_p}{L_p} p_{n0} \left(e^{eV/k_B T} - 1\right).$$ (6.27)

Assuming that the electron–hole recombination in the space charge region is negligible, the current of holes does not vary in this region, as illustrated in Fig. 6.6c. A similar development to that of Eqs. (6.21)–(6.27), gives for the diffusion current of electrons in the space charge region

$$I_n^{diff}(0) = e A \frac{D_n}{L_n} n_{p0} \left(e^{eV/k_B T} - 1\right).$$ (6.28)

In steady state, the total current must be the same in any section of the junction, and also equal to the current I that flows into and out of the junction through the metal contacts. Thus, we can obtain I by the sum of the drift currents of electrons and holes in the space charge region. As one can see in Fig. 6.6, since the total current I does not vary along x, the electron drift current on the n side and the hole drift current on the p side are given by the differences between I and the diffusion currents of holes and electrons, respectively. In this way, we can calculate the total current without explicitly using the currents of the majority carriers. Adding Eqs. (6.27) and (6.28) we obtain

$$I = I_s \left(e^{eV/k_B T} - 1\right),$$ (6.29)

where

$$I_s = e A \left(\frac{D_p}{L_p} p_{n0} + \frac{D_n}{L_n} n_{p0}\right).$$ (6.30)

Equation (6.29) is called the **diode equation**. It was first derived by William Shockley, one of three physicists who received the Physics Nobel Prize in 1954 for the invention of the transistor. This equation gives the current I in the junction as a function of the applied voltage V. It is important to draw attention to the fact that in the derivation of Eqs. (6.29) and (6.30) it became clear that the current in the p–n junction is **dominated by the minority carriers**. Note that for negative voltages with values much larger than $k_B T/e$ (0.026 V at room temperature), the current tends to $-I_s$. For this reason, I_s is called **reverse saturation current**. Its value can be calculated with Eq. (6.30) using only the parameters of the semiconductor of the junction. Considering that at room temperature the impurities are almost completely ionized, we can use Eqs. (5.38) and (5.41) for the equilibrium concentrations of carriers. Replacing them in Eq. (6.30) we obtain a more convenient expression for the saturation current

$$I_s = e A n_i^2 \left(\frac{D_p}{L_p N_d} + \frac{D_n}{L_n N_a} \right). \tag{6.31}$$

Since n_i varies exponentially with the energy gap E_g, one can see from Eq. (6.31) that the saturation current varies considerably from one semiconductor to another. Clearly, it is also quite sensitive to changes in temperature. Consider a junction made with germanium, with impurity concentrations of $N_a = 10^{18}$ cm^{-3} and $N_d = 10^{15}$ cm^{-3}. A junction like this, with $N_a \gg N_d$, is called $p^+ - n$. In this case, the first term in Eq. (6.31) completely dominates the second. Considering a junction with area $A = 10^{-4}$ cm^2, and a recombination time $\tau_p = 0.1$ μs, with the values of n_i and D_p in Table 5.2, we obtain with Eq. (6.31)

$$I_s = 2.5 \times 10^{-7} \text{A} = 0.25 \, \mu\text{A}.$$

Figure 6.7 shows the I–V curve calculated with Eq. (6.29) with this value of I_s. Figure 6.7b shows an expanded region around the origin. We see that for $V = -0.1$ V the current is already practically equal to that of reverse saturation. With direct bias, $V > 0$, the current increases exponentially with V. Figure 6.7a, made on a current scale 10^5 times larger than in (b), exhibits a more familiar aspect of the I-V characteristics of the junction. The I-V curve is strongly asymmetric relative to the sign of the voltage. With reverse bias the current is negligible compared to the one of direct bias, which is the essential feature of the diode. A striking feature of the curve in Fig. 6.7a is the abrupt increase of the current that occurs at a voltage around 0.3 V. This feature is simply a result of the exponential growth of I with V. The critical value of the voltage for which the current grows abruptly depends essentially on the semiconductor. This can be seen by replacing (6.31) in Eq. (6.29) and using Eq. (5.23) for n_i. Neglecting the unity in the presence of the exponential in Eq. (6.29) we obtain

$$I = e A \left(\frac{D_p}{L_p N_d} + \frac{D_n}{L_n N_a} \right) N_c N_v \, e^{(eV - E_g)/k_B T}. \tag{6.32}$$

We see then that the current grows exponentially with the difference between V and the energy gap in eV. The critical voltage for the abrupt growth of the current is in the range 0.2–0.4 V for germanium junctions and 0.6–0.8 V for silicon junctions.

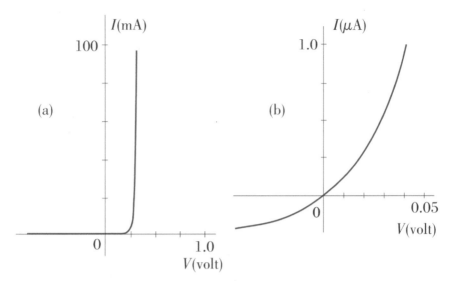

Fig. 6.7 *I–V* characteristic of an ideal *p–n* junction calculated with the diode Eq. (6.29) for $I_s =$ 0.25 μA, a value adequate for a junction made with germanium. The curve in (**b**) is the same as in (**a**), made on an enlarged scale to show the behavior around the origin

Finally, it should be noted that in real junctions the *I–V* response deviates from Eq. (6.29) due to the following factors: The electron–hole recombination in the space charge region is not completely negligible; the concentrations of the majority carriers do not remain in equilibrium when the current increases considerably; the junction is not abrupt, like in the model we considered in this section. These effects are treated in other more specialized books on semiconductor devices.

6.3 Heterojunctions

A junction formed by two intrinsically different materials is called **heterojunction**, in contrast to the one studied in the previous section, which is a **homojunction**. When the materials on the two sides of the junction are different, the energy diagram shows a discontinuity at the interface between them, instead of the continuous behavior of Fig. 6.3. Usually, heterojunctions are those formed by different semiconductors, like GaAs and (GaAl)As, used in semiconductor lasers. However, junctions between metals and semiconductors are also heterojunctions and are used to manufacture some devices. Junctions involving metals have some properties and applications similar to those of *p-n* junctions, but also have special characteristics that are important in some devices. This is the case of metal–semiconductor junctions, which are useful in high frequency devices, and metal–insulator-semiconductor junctions, used in highly integrated digital circuits.

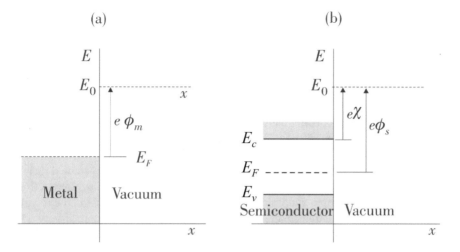

Fig. 6.8 Illustration of the concept of work function in the energy diagrams for two cases: **a** Isolated metal. **b** Isolated semiconductor

The behavior of a material in a heterojunction depends strongly on its **work function** W_0. As studied in Sect. 2.3, the work function is defined as the energy needed to "pull out" an electron from the material and remove it away from its surface. Having studied the quantum properties of electrons in metals and semiconductors, we can now understand better the concept of the work function. In the case of a metal, since the highest energy electrons are at the Fermi level, it is easy to see that the work function is given by $W_0 = E_0 - E_F$, where E_0 is the energy of the electron in vacuum and away from the material, as shown in Fig. 6.8a. In metals it is customary to write $W_0 = e\phi_m$, where ϕ_m is an electric potential with a value typically in the range of 2–6 V. In semiconductors, the definition of work function is also $W_0 = E_0 - E_F$. However, since there are no electrons at the Fermi level, $W_0 = e\phi_s$ is not the minimum energy needed to remove electrons from the semiconductor. The energy necessary to remove electrons in the conduction band from the material is $E_0 - E_c \equiv e\chi$, where $e\chi$ is called **electronic affinity**. Figure 6.8 illustrates schematically the work functions of an isolated metal and an isolated semiconductor in vacuum. Note in the figure that E_0 is the energy level of an electron in vacuum, and it is the same whether it has been removed from a metal or a semiconductor. Hence, when a metal and a semiconductor are separated, their Fermi levels have different relative positions, because they depend exclusively on their respective work functions $e\phi_m$ and $e\phi_s$.

6.3.1 Metal-Semiconductor Junction

When a metal is placed in direct contact with a semiconductor, charges are transferred from one side to the other in order to equalize the two Fermi levels, similarly to what happens in a p–n junction. The direction of motion of the charges then depends on the relative values of the work functions. The difference to the case of junctions of two semiconductors is that here holes cannot go from the semiconductor to the metal, since they are quasi-particles that exist only in semiconductors. This transfer creates layers of charges on both sides of the junction resulting in a potential barrier, called Schottky barrier, in honor of physicist Walter Schottky who studied metal–semiconductor contacts in the 1930s. The shape of the barrier is quite different from the one in a p–n junction, and it depends on the type of semiconductor, the relative values of the work functions in the two materials, and on the electronic affinity. The shapes of the Schottky barrier for two typical cases are shown in Fig. 6.9.

Figure 6.9a corresponds to the junction of a metal with a type n semiconductor with a smaller work function, that is, with $\phi_s < \phi_m$. Since $e\phi_m$ is the energy necessary to pull an electron out of the metal and $-e\chi$ is the energy needed to inject it in the semiconductor, the height of the energy barrier $e\phi_B$ that an electron must overcome to transfer from the metal to the semiconductor is given by $e\phi_B = e(\phi_m - \chi)$. Examination of the relative positions of E_F and E_c in Figs. 6.8 and 6.9a, shows that the energy difference between the peak of the barrier and the minimum of the conduction band E_c is $e(\phi_m - \phi_s)$. This difference characterizes the contact potential between the metal and the semiconductor in equilibrium, $V_0 = \phi_m - \phi_s$, that prevents the transfer of electrons from the semiconductor to the metal. This

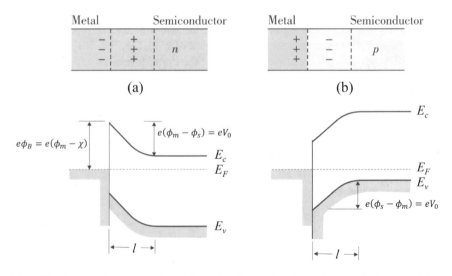

Fig. 6.9 Energy diagrams of metal–semiconductor junctions in equilibrium. **a** Type n semiconductor with $\phi_s < \phi_m$. **b** Type p semiconductor with $\phi_s > \phi_m$

potential can be reduced or increased by applying an external voltage with direct or reverse bias, respectively. For this reason, the metal–semiconductor contact has *I-V* characteristics similar to that of a *p–n* junction.

Figure 6.9b illustrates the Schottky barrier in the case of a type *p* semiconductor with $\phi_s > \phi_m$. In this case, in order to align the Fermi levels, it is necessary to have an accumulation of positive charges on the metal side and negative on the semiconductor side. This occurs with the transfer of electrons from the metal to the semiconductor, where they ionize acceptor impurities in a depletion layer. The charge layers on both sides produce a potential barrier in equilibrium $V_0 = \phi_s - \phi_m$ that prevents the continuation of the charge transfer. As in the previous case, this barrier can be increased or decreased by applying an external voltage.

An important difference between the *p–n* junction and the metal–semiconductor junction is that in the first the current is dominated by **minority carriers**, while in the second it is determined by **majority carriers**. The process by which the majority carriers produce the current in the forward biased metal–semiconductor junction involves the emission of electrons from the metal, similar to the thermionic emission in the hot cathode of a vacuum tube. Its quantitative study can be found in some references cited at the end of this chapter.

Finally, it is important to note that in the case of a metal-type *n* semiconductor contact with $\phi_m < \phi_s$ and a metal-type *p* semiconductor with $\phi_s < \phi_m$, the contact potential is negative and there is no formation of a potential barrier. Contacts of this type are called **ohmic**, because their resistances do not depend on the direction of the current.

6.3.2   Heterojunctions of Semiconductors

In a heterojunction of semiconductors, the different energy gaps of the two sides produce a discontinuity of the energy bands at the interface. A very important heterojunction for applications in optoelectronics is the one formed by GaAs and the alloy $Ga_{1-x}Al_xAs$, also denoted by (GaAl)As. In this alloy a certain fraction *x* of Al atoms replaces the Ga atoms randomly in the crystal lattice. Since GaAs and AlAs crystallize in the same structure (zinc-blende, Fig. 1.8a) and have almost identical lattice parameters, the substitution of Ga by Al does not produce distortions in the lattice. The main effect of Al in the GaAs lattice is to increase the energy gap. As GaAs has $E_g = 1.43$ eV and AlAs has $E_g = 2.16$ eV, the gap of $Ga_{1-x}Al_xAs$ depends on the Al concentration *x*. The gap E_g varies with *x* in an approximate linear fashion between the values in the pure crystals.

Also, due to the fact that the lattices are almost identical, it is possible to grow $Ga_{1-x}Al_xAs$ on the surface of a GaAs crystal, producing an almost perfect crystalline interface. This makes possible to manufacture heterojunctions in which electrons and holes pass from one side to the other without much scattering caused by imperfections at the interface. The growth of (GaAl)As on GaAs was traditionally done with the LPE technique, described in Sect. 1.4. Currently, with the MBE technique and other

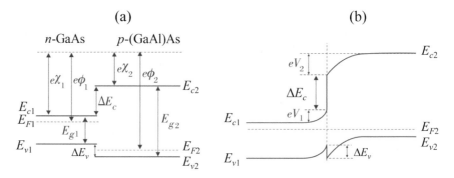

Fig. 6.10 Energy diagrams of a heterojunction of n-GaAs and p-(GaAl)As. **a** Separate materials. **b** Junction in equilibrium

methods of film manufacturing, it is possible to deposit individual atomic layers, one after the other, producing lattices, interfaces, heterojunctions, and almost perfect superlattices.

Figure 6.10 shows the energy diagrams of n-type GaAs and p-type Ga$_{1-x}$Al$_x$As, with a certain Al concentration x, when the two materials are separate. In this case, each material is characterized by different electronic work function and affinity, referred to the vacuum level. When the two materials are brought into contact, electrons and holes pass from one side to the other. Like in the p-n homojunction in equilibrium, the Fermi levels on both sides are equal. However, since the values of E_g are different, in a heterojunction there are discontinuities in the minimum of the conduction band, E_c, and in the maximum of the valence band, E_v, as illustrated in Fig. 6.10b. By examining the energy diagrams we see that

$$\Delta E_c = e\,(\chi_1 - \chi_2), \tag{6.33}$$

$$\Delta E_v = E_{g2} - E_{g1} - \Delta E_c. \tag{6.34}$$

These discontinuities are the same, regardless of the materials being separate or in contact, since they depend only on the electronic affinities and on the band gaps. We see in Eqs. (6.33) and (6.34) that when the values of χ and E_g on the two sides are equal, $\Delta E_c = \Delta E_v = 0$. Since the discontinuities exist in the separate materials, they have nothing to do with the formation of the charge layers in the two sides of the junction, that create the potential barrier V_0. As discussed in the previous section, in the p-n homojunction V_0 is equal to the difference between the energy levels E_c of the conduction band in the two sides. As we can see in Fig. 6.10b, in the heterojunction one has to subtract the discontinuity ΔE_c, so that height of the potential barrier is

$$V_0 = V_1 + V_2, \tag{6.35}$$

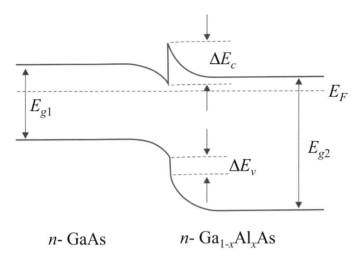

Fig. 6.11 Heterojunction of n-GaAs and n- Ga$_{1-x}$Al$_x$As

where V_1 and V_2 are shown in Fig. 6.10b. The different energy gaps of the semiconductors in a heterojunction makes possible to manufacture a huge variety of potential barrier shapes for electrons in the conduction band and holes in the valence band. This allows the investigation of quantum properties of particles in potentials manufactured with ingenious shapes, as well as the fabrication of sophisticated devices. An important heterojunction for scientific research and for applications is shown in Fig. 6.11. It is formed by two semiconductors doped with type n impurities, n-GaAs and n- Ga$_{1-x}$Al$_x$As. Due to values of the electronic affinities of the two materials, the discontinuities in the bands are $\Delta E_c = 0.85 \, \Delta E_g$ and $\Delta E_v = 0.15 \, \Delta E_g$, where $\Delta E_g = E_{g2} - E_{g1}$. These discontinuities serve to block the diffusion of GaAs carriers into (GaAl)As, which is an important mechanism for the semiconductor lasers that will be studied in Chap. 8.

6.4   The Junction Diode

The diode is a two-terminal electronic device in which the electric current can flow in only one direction. An ideal diode should have zero resistance for the current in one direction, like a short circuit, and infinite resistance, as an open circuit, for the current in the opposite direction. The actual diodes, however, have a small resistance, but not zero, in one direction, and a large resistance, but not infinite, in the other direction. Figure 6.12 shows the circuit symbol of the diode and the I-V characteristic of an ideal diode. The triangular part of the symbol represents the tip of an arrow, indicating the direction of current flow in the diode. The thermionic diode, which existed before the semiconductor era, is made of a vacuum tube inside which there are two elements,

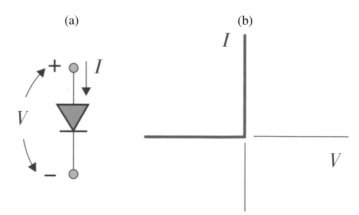

Fig. 6.12 **a** Circuit symbol of a diode. **b** I-V characteristic of the ideal diode

cathode and anode. The cathode is heated by a filament and emits electrons, while the unheated anode only receives electrons from the cathode. When a positive voltage is applied between the anode and the cathode, electrons emitted by the cathode flow towards the anode and produce a current. When the voltage is applied in the opposite direction, the anode cannot emit electrons, because it is not heated, resulting in zero current. Semiconductor diodes are made with p-n junctions or metal–semiconductor contacts. While in the vacuum tube diode the asymmetry is due to the fact that the cathode emits electrons, but the anode does not, in semiconductor diodes the asymmetry is caused by the potential barrier. The junction diode consists of only one p–n junction with two metal contacts for the current input and output. On the p-side the contact between the semiconductor and an aluminum film naturally forms a good ohmic contact, due to the relative values of the work functions. On the n-side the ohmic contact is obtained by means of a stronger doping, producing an intermediate n^+ region as in Fig. 6.1. By analogy with the vacuum tube diode, the p terminal is called anode, while the n terminal is called cathode.

Junction diodes have an I-V characteristic like the one shown in Fig. 6.7. When they are biased in the forward direction, the current reaches substantial intensities when the voltage is close to or larger than a critical value E_0, which depends on the junction semiconductor. In Ge diodes this value is the range of 0.2–0.4 V, while in Si diodes it varies from 0.6 to 0.8 V. Figure 6.13a shows a circuit that is approximately equivalent to the diode junction. For $V < E_0$ the current in the circuit is zero, since the presence of the battery causes the voltage at the terminals of the ideal diode to be negative. For $V > E_0$ the current increases linearly with the difference $V - E_0$, due to resistor R in series with the diode and the battery.

The equivalent circuit of Fig. 6.13 applies to DC voltages. For alternating voltages with relatively high frequencies, it is necessary to consider the effect of the junction capacitance. When the voltage applied to the diode varies rapidly, the charge in the depletion region does not follow immediately. This limits the frequency response of the diode. This effect can be represented by a capacitor in parallel with the circuit

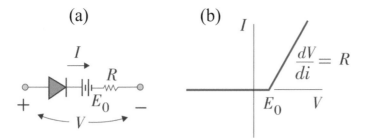

Fig. 6.13 **a** Approximate equivalent circuit of the junction diode. The battery produces the critical voltage E_0 and the resistor determines the finite slope of the I-V characteristic, shown in (**b**)

of Fig. 6.13a, with a capacitance that is, in part, given by Eq. (6.20). The time delay in the diffusion motion of the charges in the vicinity of the junction also contributes to the capacitance. The relative values of the diffusion capacitance and of the spatial charge capacitance depend on the geometry and on the material of the junction.

Diodes are important components in electronic circuits because they perform many functions. Some will be presented in the next section. They are used either as discrete components or in integrated circuits. Figure 6.14 shows three common physical aspects of discrete commercial diodes. Each type of diode is identified by an alphanumeric code (Examples: 1N23 and 1N56. The number 1 before the letter N is used to identify diodes). The identification of the anode and cathode terminals, as well as the I-V characteristic and other diode parameters, are listed in the specifications of each type of diode.

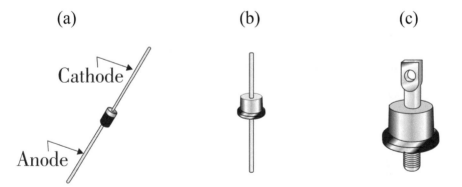

Fig. 6.14 Common physical aspects of discrete commercial diodes. **a** Low current. **b** Intermediate current. **c** High current

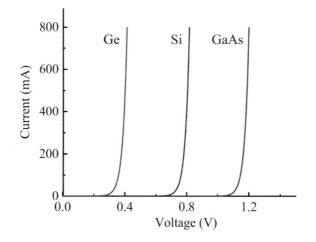

Fig. 6.15 *I-V* curves for *p–n* junction diodes made of Ge, Si, and GaAs, calculated in Example 6.3

Example 6.3 Calculate the saturation currents and make the *I-V* plots of three ideal diodes, at $T = 300$ K, one made with Ge, one with Si, and one with GaAs, considering that they all have the following parameters: $N_a = 10^{17}$ cm^{-3}; $N_d = 10^{15}$ cm^{-3}; $A = 10^{-4}$ cm^2; $\tau_p = \tau_n = 0.5$ μs.

The saturation current is given by Eq. (6.31). Since $N_a \gg N_d$, we can neglect the second term. Using $L_p = (D_p \tau_p)^{1/2}$ we write for I_s

$$I_s \approx \frac{e A n_i^2 D_p^{1/2}}{\tau_p^{1/2} N_d}.$$

Using the diode data and the parameters in Table 5.2 converted to SI units, we have for the Ge diode.

$$I_s = \frac{1.6 \times 10^{-19} \times 10^{-4} \times 10^{-4} \times (2.5 \times 10^{13} \times 10^6)^2 \times (50 \times 10^{-4})^{1/2}}{(0.5 \times 10^{-6})^{1/2} \times 10^{15} \times 10^6} \text{ A}$$

$$I_s = 1.0 \times 10^{-7} \text{ A}.$$

For the Si diode we have

$$I_s = \frac{1.6 \times 10^{-19} \times 10^{-4} \times 10^{-4} \times (1.5 \times 10^{10} \times 10^6)^2 \times (12.5 \times 10^{-4})^{1/2}}{(0.5 \times 10^{-6})^{1/2} \times 10^{15} \times 10^6} \text{ A}$$

$$I_s = 1.8 \times 10^{-14} \text{ A}.$$

For the GaAs diode comes

$$I_s = \frac{1.6 \times 10^{-19} \times 10^{-4} \times 10^{-4} \times (10^7 \times 10^6)^2 \times (10 \times 10^{-4})^{1/2}}{(0.5 \times 10^{-6})^{1/2} \times 10^{15} \times 10^6} A$$

$$I_s = 7.2 \times 10^{-21} \text{ A}.$$

The I-V curves for the three diodes are obtained by calculating I as a function of V, numerically, with the Diode Eq. (6.29), using the three values for I_s above. The curves, shown in Fig. 6.15, clearly demonstrate the effect of the different energy gaps on the critical voltages of the diodes.

6.4.1 Applications of Diodes

Diodes have several applications in electronic circuits. One of the most important is the rectification of alternating voltage in power supply sources, used to produce a DC voltage for the operation of electronic equipment. Figure 6.16a shows a simple half-wave rectifier circuit, consisting of a transformer, a diode, and a capacitor. The transformer has the function of changing the amplitude of the AC voltage of the utility, usually with rms value of 110 or 220 V, to the value appropriate for the power supply. Figure 6.16b shows the time variation of the sinusoidal voltage v (t) in the secondary of the transformer, whose average value is zero, applied at the input of the rectifier. Considering that the critical diode voltage E_0 is much smaller than the

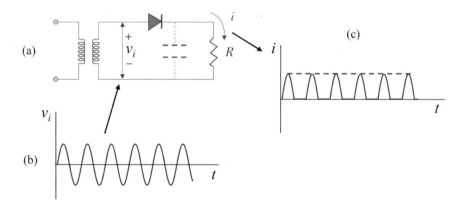

Fig. 6.16 Illustration of the operation of a simple half-wave rectifier circuit. The voltage v (t) in the secondary of the transformer produces a current i (t) in the diode in series with the load. The dashed line represents the waveform obtained with the addition of a capacitor to the circuit

peak value of v_i (t), the diode has negligible resistance in positive half-cycles of v_i (t), and very high resistance in negative half-cycles. As a result, the current i (t) that flows through the diode in series with the load of resistance R, follows v_i (t) on positive half-cycles, but is negligible during negative half-cycles, so that its shape has the form shown in Fig. 6.16c. The rectified current is unidirectional, and has a nonzero average, but not it is constant, as desired for a DC source. However, the simple addition of a capacitor to the circuit in the position shown by the dashed lines, makes the current approach the DC shape. The capacitor charges during the positive half-cycles of the diode current and discharges over R in the negative half-cycles, so that the current takes the approximate form of the dashed line in Fig. 6.16c. Actually, there is a small ripple in the current, but its amplitude that can be made sufficiently small by using a capacitor such that the time constant RC of the circuit is much larger than the period of the AC voltage.

Diodes used in rectifier circuits do not need to have a fast time response, since the AC voltage in these circuits has low frequency, typically 50 or 60 Hz in the distribution grids. But they need to satisfy two basic requirements: the first is that they must have maximum current larger than that required by the load. Due to collisions of electrons and holes with the crystal lattice, the current flow heats the diode, and there is a limit value above which the junction damages due to overheating. Thus, each diode, depending on its physical characteristics, supports a maximum current. The other requirement is that the peak voltage in the negative half-cycle has to be smaller than the breakdown voltage of the diode in the reverse polarization. The origin of this breakdown voltage will be presented in Sect. 6.6.

Another application of the diode, based on its rectification property, is as a peak detector, used in the detection of amplitude-modulated (AM) waves in radio receivers. In AM radio transmission, the high frequency wave, called a carrier, has an amplitude that varies according to the low frequency signal (for example, audio) of the information. A typical AM wave is shown in Fig. 6.17a, in which the high frequency sinusoidal carrier is modulated by a signal, also sinusoidal, of low frequency. If the voltage of this AM wave is at the input of a circuit with diode, capacitor, and load resistance, like the one in Fig. 6.16, without the transformer, the output voltage in the load has the shape shown in Fig. 6.17b. The half-wave

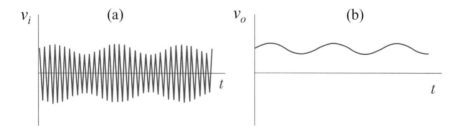

Fig. 6.17 a Input high frequency sinusoidal wave amplitude modulated by a sinusoidal audio signal. The line formed by the peak values corresponds to the audio signal. **b** Output audio signal produced by the diode peak detector and a capacitor

rectifier circuit with capacitor cuts the carrier negative half-cycles, producing a signal that corresponds to the variation in the value of the positive peaks. This signal is approximately equal to the sinusoidal audio signal that modulates the amplitude of the carrier wave. The diodes used in peak detectors work with very small voltages and therefore can have a low maximum current. On the other hand, unlike diodes for rectifying power sources, they must respond to high frequencies.

The two diode applications presented here involve voltages and currents that vary continuously in time, or analog waveforms. Diodes also have many applications in digital circuits, that operate with signals in the form of short pulses. In combination with resistors, diodes are used as clampers, or limiters, and also in logic circuits, such as simple AND and OR gates circuits. When used in combination with transistors, diodes are used in a large variety of circuits for logic operations.

6.5 Schottky Barrier Diode

As mentioned in Sect. 6.3.1, a metal–semiconductor contact with a Schottky barrier has an I-V characteristic similar to that of the p–n junction and is, therefore, also a diode. In fact, historically the first semiconductor device ever built was the metal–semiconductor contact diode. In the early years of electronics, it was used to manufacture signal detector diodes by pressing a metallic needle onto the surface of the semiconductor PbS, called galena, found in nature in crystalline form. The so-called galena radios consisted only of a LC tuning circuit, a galena diode detector, and a headset, and were used to listen to radio signals and in simple communication systems. It was also with metal–semiconductor contacts that Bardeen and Brattain built the first transistor in 1947, which constituted the most important technological breakthrough of the twentieth century. During the 1950s, metal–semiconductor contact diodes and transistors, simply called point-contact diodes, were abandoned due to the difficulty in reproducing its I-V characteristics. They were then replaced by the p–n junction devices that are among the most used to date. However, with the improvement of manufacturing technology and the theoretical understanding of their operation, metal–semiconductor devices became very important in digital circuit applications.

Although the Schottky barrier diode has an I-V characteristic similar to the junction diode, there are important differences between the two types of diodes. They stem fundamentally from the fact that the current in the Schottky barrier is due to majority carriers, while in p–n junction it is due to minority carriers. When the voltage at a junction is abruptly switched from direct to reverse polarity, minority carriers are not removed from the other side instantly due to the recombination time, as discussed in Sect. 5.4.4. This effect limits the frequency response of p–n junction diodes. In Schottky barrier diodes there are no minority carriers to be removed, so that the response time is much shorter. For this reason, they have wide application in high frequency or fast switching detector circuits.

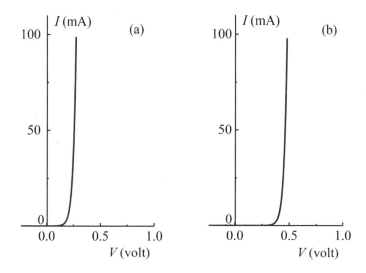

Fig. 6.18 Comparison between the *I-V* characteristics of a Schottky diode (**a**) and a Si *p–n* junction diode (**b**). Both curves were calculated with Eq. (6.29), using in (**a**) $I_s = 2.5 \ \mu A$ and in (**b**) $I_s = 1 \ nA$

Another important difference between the Schottky diode and the *p–n* junction diode is in the value of the critical voltage E_0 of the *I-V* characteristic. As we saw in Sect. 6.1, this voltage results from a combination of the saturation current value I_s and the effect of the exponential growth of I with V, given by Eq. (6.29). In a Schottky diode made with the same semiconductor and with the same area as a *p–n* junction diode, the saturation current is much higher in the first, because it is due to majority carriers. Thus, as shown in Fig. 6.18, Schottky diodes have a critical voltage much smaller than *p–n* junction diodes, so that they are closer to an ideal diode in direct bias.

A final remark. Schottky diodes do not support high currents, because they have quite small contact area, and thus easily the current density becomes very large and damages the contact. For this reason, they are not suitable for rectifier circuits. Their most important application is in detector circuits, which require high frequency response and high sensitivity, expressed by the large inclination of the *I-V* curve close to the origin.

6.6   Breakdown in Reverse Bias: Zener Diode

According to the model presented in Sect. 6.2, in the reverse polarized *p–n* junction there is small current with intensity that tends to a saturation value as the reverse voltage increases. In fact, this only occurs as long as the reverse voltage is smaller, in absolute value, than a certain value V_B, called **breakdown voltage**. If the voltage reaches this critical value, the current increases sharply as a result of a process of electronic disruption of the junction. Since the external circuit limits the current intensity by not letting it exceed a maximum value (which depends on the characteristics of the device), this breakdown process may not be destructive. It is perfectly reproductive and can be repeated infinitely many times. Two different mechanisms may be responsible for the breakdown process in a junction, the so-called Zener effect and the avalanche multiplication. Although the two mechanisms are different, both result from the effect of the electric field in the space charge region of the p–n junction on the charge carriers. In reverse biased junctions, this field increases due to the increase in the height of the potential barrier, as illustrated in Fig. 6.5c. The breakdown process happens when the field reaches a critical value.

The **Zener effect** occurs at relatively small voltages, on the order of a few volts, in highly doped semiconductor junctions. As can be seen in Eq. (6.18), if the concentrations of impurities N_a and N_d on the two sides of the junction increase, the depletion layer thickness decreases. With concentrations of the order of 10^{19}–10^{20} cm^{-3}, for reverse voltages of a few volts the thickness is of the order of 10^{-5} cm, which results in fields of the order of 10^6 V/cm. Electric fields with this intensity break the covalent bonds and ionize atoms of the crystal lattice. The electrons released during ionization are accelerated in the opposite direction to the field, passing to the n-side of the junction and producing a current in the reverse direction, much higher than the reverse saturation current.

In junctions with low concentrations of impurities, the electric field in the depletion region may not be sufficient to produce direct ionization of the semiconductor atoms, therefore, there is no Zener effect. However, there will always be a value of the reverse voltage for which there is a breakdown in the junction through another mechanism, the **avalanche**. As the name suggests, this is a process in which successive events occur, resulting in a multiplication in the number of carriers. The first event results from the acceleration by the electric field of an electron that enters the junction from the p-side. If the electron has enough energy, its collision with the atoms of the crystalline lattice can produce an electron–hole pair, resulting in a multiplication factor of two in the number of carriers. Then, the electron created is accelerated to the n-side, while the hole is accelerated to the p-side. If the reverse voltage is high enough, each of them will produce an electron–hole pair, which in turn will produce other pairs, in a chain reaction process. The value of the reverse voltage for which this avalanche produces a sudden growth in the reverse current is called breakdown voltage V_B. The value of V_B can vary from a few volts to a few thousand volts.

Regardless of the mechanism responsible for the breakdown at the junction, Zener or avalanche, the complete I-V characteristic of the diode has the shape shown in Fig. 6.19a. In the reverse bias, the diode has a large resistance, and the current intensity is small with a value close to I_s, as long as the voltage is less than V_B. For values of the voltage near V_B, any variation in V produces large variations in the reverse current caused by the junction breakdown. Diodes that are manufactured to operate in the breakdown region are called **Zener diodes**. Despite the name, in general the breakdown mechanism of the Zener diodes is the avalanche process.

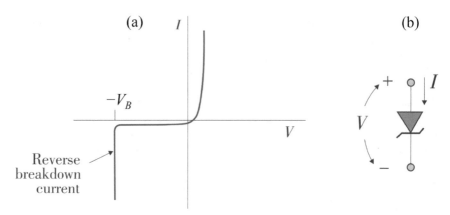

Fig. 6.19 **a** I-V characteristic of the junction diode, showing the sudden increase in reverse current at the breakdown voltage $-V_B$. **b** Circuit symbol of the Zener diode

Zener diodes have circuit symbol shown in Fig. 6.19b, and can be manufactured to have V_B ranging from 1 V to hundreds of volts. In good quality Zener diodes, the breakdown current is represented by an almost vertical line, which means that the voltage at the terminals is kept constant and equal to $-V_B$, regardless of the current value.

A very important application of Zener diodes is as a **voltage regulator** in power supply sources. Figure 6.20 shows a typical simple circuit consisting only of a resistor and a Zener diode connected to an AC rectifier. The input voltage has a small ripple, say around an average value of 12 V, and with the waveform of Fig. 6.16c. The presence of a Zener diode, say with $V_B = 9$ V, clamps the output voltage at 9 V, regardless of the variation of the input. The difference between the input voltage and the output voltage is applied in the resistor, whose role is to "absorb" the input fluctuations.

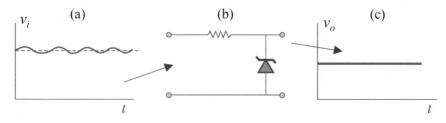

Fig. 6.20 Illustration of the use of a Zener diode in a voltage regulator. **a** Variable voltage at the input. **b** Simple circuit. **c** Constant voltage at the regulator output

6.7 Other Types of Diodes

6.7.1 Varactor

Equations (6.18) and (6.20) show that the p–n junction diode has a capacitance that varies with the voltage of the potential barrier. Considering that in the reverse bias with a voltage V the height of the potential barrier is $V + V_0$, and assuming that $V \gg V_0$, Eq. (6.18) gives for the thickness of the depletion region (Problem 6.4)

$$l \propto V^{1/2}. \tag{6.36}$$

A consequence of this result and of Eq. (6.20) is that the capacitance of the junction varies with the reverse voltage as

$$C \propto V^{-1/2}. \tag{6.37}$$

Thus, a diode with a very small saturation current, subjected to a **reverse bias voltage**, behaves like a capacitor whose capacitance is controlled by the voltage. This is called **varactor**, a term formed by joining parts of the words variable reactor, also known as **varicap** (of variable capacitor), or **VVC** (voltage-variable capacitance). The dependence of the capacitance with the voltage given by Eq. (6.37) is only valid for an abrupt junction, such as the one in Fig. 6.2. If the variation in the impurity concentrations in the junction is gradual, the dependence of C with V will be different. Using techniques of epitaxial growth or ionic implantation, it is possible to manufacture junctions with different profiles of concentrations, chosen in order to obtain $C(V)$ functions suitable for specific varactor applications. Varactors, or varicaps, are used in the LC tuning circuits of radio and TV receivers, in place of the old manually variable plate capacitors. The fact that its capacitance can be controlled by the voltage allows the electronic control the tuning frequency of the receiver circuits. For this application, one uses varactors with a capacitance that varies as $C \propto V^{-2}$. In this case, the tuning frequency of the circuit, $\omega = (LC)^{-1/2}$, is proportional to the voltage applied to the diode. Varactors are also used in active filters, harmonic generators and in microwave circuits.

6.7.2 Tunnel Diode

The **tunnel diode** is made with a p–n junction in which, in a certain range of direct bias voltage, the current is dominated by the tunneling of electrons through the potential barrier at the junction. As shown in Sect. 3.3.3, there is a finite probability for an electron to cross a barrier with a maximum potential larger than its kinetic energy. This is the tunnel effect, that is of entirely quantum nature.

As we saw in Sect. 6.2, the current in a common p–n junction is due to the diffusion motion of minority carriers in both directions. This results in a current that decreases exponentially with the applied voltage, and tends to zero as $V \to 0$. The tunnel diode is made with semiconductors heavily doped on both sides of the junction, favoring the tunneling of electrons from the n-side to the p-side, producing a tunnel current larger than the diffusion current when V is small. For this to happen it is essential, as we shall see below, that the two sides of the junction are heavily doped.

The energy level diagram presented in Sect. 5.3 is valid when the concentration of impurities is relatively small, $N \ll 10^{20}$ cm^{-3}. In this situation, the impurities are very far from each other, so that the interaction between them is negligible. When the concentration of impurities is on the order of 10^{20} cm^{-3} or larger, the interaction between them is no longer negligible. In this case, a phenomenon like the one illustrated in Fig. 4.1 occurs, the impurity energy levels are no longer discrete, they form bands. If the impurities are donors, they form an energy band that overlaps with the conduction, such that the Fermi level is above of the minimum of the conduction band, $E_F > E_c$. As a result, states with energy above E_c and below E_F are filled with electrons, even at $T = 0$. Semiconductors in this situation are called n-type **degenerate**. Analogously, a semiconductor heavily doped with p-type impurities has $E_F < E_v$, so that the states between E_F and E_v are occupied with holes.

The tunnel diode is made of a p–n junction in a semiconductor strongly doped on both sides, that is, with degenerate p and n regions, that has an energy diagram as shown in Fig. 6.21. In the equilibrium condition, with external voltage $V = 0$, the Fermi level E_F is the same in two sides of the junction. As $E_F > E_v$ on the p-side and $E_F < E_c$ on the n-side, there are filled states in the conduction band on the n-side with energy close to that of empty states in the valence band on the p-side. These states are separated spatially by the thickness of the depletion region which, due to the high impurity concentration is quite narrow (see Eq. 6.18). As we saw in the Sect. 3.3.3, filled states separated from empty states by a narrow potential barrier and with finite height, create the conditions favorable for the **tunneling of electrons**. When $V = 0$, as we see in Fig. 6.21a, there are no filled and empty states with exactly the same

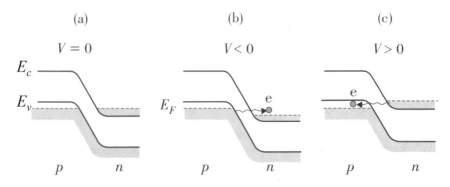

Fig. 6.21 Energy diagrams of a p–n junction in a tunnel diode. **a** $V = 0$, junction in equilibrium. **b** $V < 0$, tunneling current in the reverse direction. **c** $V > 0$, tunneling current in the forward direction

energy. In this situation there is no electron tunneling. However, if the voltage is nonzero, both in the cases of reverse or direct bias, there are states filled on one side with the same energy as empty states on the other side, because the Fermi levels on the two sides are different, resulting in tunneling. If $V < 0$ there is tunneling of electrons from the p-side to the n-side, as shown in Fig. 6.21b, resulting in reverse current. On the other hand, if $V > 0$, we see in Fig. 6.21c that the tunneling is in direction from n to p, producing a direct current.

With direct polarization, $V > 0$, initially the current increases with the voltage, because the number of empty states at the same level as filled states increases with V. However, with the progressive increase in V, above a certain value of V the conduction band on the n-side is above the valence band on the p-side, reducing the tunneling current. In this way, when a direct bias voltage is applied to the diode, the tunneling current initially increases with V, goes through a maximum and then decreases, resulting in a characteristic I-V as shown in Fig. 6.22a. Since the diffusion current increases monotonically with V, the sum of diffusion and tunneling currents results in the I-V characteristic for the tunnel diode that has the curious shape shown in Fig. 6.22b.

An important feature of the I-V curve of the tunnel diode is that in certain voltage range $dI/dV < 0$. This corresponds to a negative resistance for AC signals, whose value can be controlled by the voltage V applied to the diode. When operating in this region of negative resistance, the tunnel diode supplies AC power to the circuit, as opposed to a normal resistance that always absorbs energy. Thus, it finds applications in oscillators and signal amplifiers. As the tunneling mechanism does not present a delay due to the drift and diffusion processes, the tunnel diode also has applications in fast switching circuits.

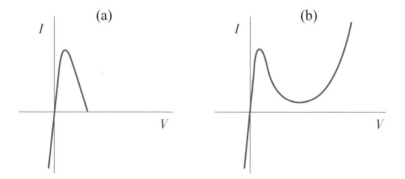

Fig. 6.22 Tunnel diode I-V characteristic. **a** Only the tunneling component of the current. **b** Complete I-V characteristic including the diffusion current

6.7.3 IMPATT Diode

The name IMPATT is made with the letters of the words IMPact Avalanche Transit Time. The IMPATT diode operates with reverse voltage, close to the value for avalanche, having a structure such as to create a field profile that drives an electron packet to travel from one end to the other of the device, producing a high oscillation frequency. The structure of the IMPATT diode is shown in Fig. 6.23a. It consists of a p^+-n junction, in which the n region is long and terminated by a narrow n^+ region with stronger doping. When the diode is reversed polarized, the variation of the potential has the shape shown in Fig. 6.23b. The electric potential has a large variation in the region of the reverse polarized junction, resulting in a peak of the electric field, shown in Fig. 6.23c. In region n the potential varies monotonically, corresponding to an approximately uniform field. The n^+ region has a lower resistivity, resulting in a smaller potential drop and therefore smaller electric field. The IMPATT diode normally operates with an external resonant circuit, in an oscillation regime, so that the variations of the field and the potential of Fig. 6.23 correspond to the average values of those quantities. In the following paragraph we describe qualitatively the mechanism that produces the oscillation on the diode.

When an external voltage is applied to the diode, an electric field is quickly created with variation as shown in Fig. 6.23c, where E_{av} is the field value required to produce avalanche in the p^+-n junction. When the field reaches the value E_{av} at

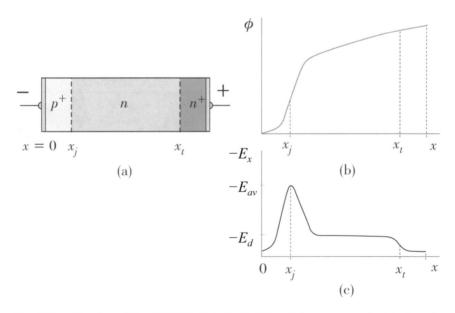

Fig. 6.23 a Structure of the IMPATT diode. **b** Variation of the average potential along the semiconductor with the reverse polarized diode. **c** Variation of the average value of the electric field

the junction, an avalanche produces a large number of electron–hole pairs. The holes created during the avalanche move in the $-x$ direction in the p^+ region and reach the metallic contact, where they recombine with electrons from the external current. On the other hand, the electrons created in the avalanche form a packet that travels through the n-region in a drift motion under the action of the electric field of value E_d. As soon as the electron packet leaves the junction region and enters in n region, it produces a potential drop around itself, which causes a decrease in the electric field at the junction. This makes the field fall below of the value E_{av}, interrupting the avalanche process. The electron packet transits in region n for a certain time, until it reaches region n^+ and passes to the external circuit. When the packet leaves the n region, the electric field at the junction increases again, reaches the value E_{av} causing a new avalanche and restarting the process. If the IMPATT diode is connected to a LC circuit, or a resonant microwave cavity, whose oscillation period is twice the transit time of the electron packet, the oscillation remains indefinitely. During each half cycle of the oscillation, a packet of electrons produces current in the same direction as the half cycle, supplying energy to the circuit and compensating for the losses. IMPATT diodes are used as microwave generators and can produce tens of watts.

6.7.4 Gunn Diode

Another device used as a microwave oscillator is the Gunn diode, invented by J. B. Gunn in 1963. This device is called a diode because it has two terminals. However, unlike all diodes presented previously, instead of being formed by a p–n junction, it is made of only a uniformly doped sample of n-GaAs. The oscillation mechanism of the Gunn diode is based on the **negative resistance** it presents in a certain voltage range, similar to that of the tunnel diode. However, here the negative resistance results from an intrinsic property of GaAs.

Figure 6.24a shows part of the energy bands of GaAs, obtained from the complete band structure in Fig. 5.2. In the semiconductor doped with type n impurities, in equilibrium electrons occupy states close to the minimum of the conduction band at point Γ_1, having effective mass $m_1^* = 0.068\,m_0$ (Table 5.1). When a small electric field E is applied to a n-GaAs sample, electrons with momentum around the point Γ_1 move in the material, with drift velocity proportional to the field. This produces a current density J proportional to E, as in Eq. 5.52, so that the material has a linear $J - E$ curve, as in the initial part of the curve in Fig. 6.24b. When the electric field increases and reaches a certain critical value $E_{cr} \sim 3 \times 10^5$ V/m, electrons gain enough energy to go to the minimum of the conduction band at point X_1, with energy higher by $\Delta E = 0.36$ eV. Note that since $\Delta E \gg k_B T$, the transition to the minimum of X_1 does not occur by thermal excitation, and this is an essential condition for the operation of the device. Since the effective mass at point X_1 is quite larger than at Γ_1, due to the smaller curvature of $E(k)$ at X_1 (Eq. 5.3), the conductivity of the material given by Eq. (5.46) decreases. The negative differential resistance range of Fig. 6.24b corresponds to field values for which a fraction of the electrons in the conduction

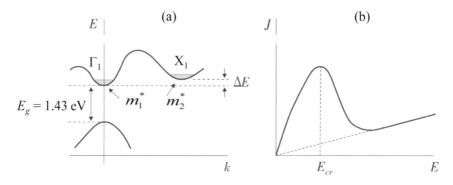

Fig. 6.24 Properties of *n*-GaAs. **a** Detail of the conduction band showing two minima that can be occupied by electrons. In the minimum at point Γ_1, electrons have effective mass $m_1^* = 0.068\,m_0$, where m_0 is the free electron mass. In the minimum at point X_1, the effective mass is $m_2^* = 1.2\,m_0$. **b** Current density–electric field characteristic of the material

band is around point Γ_1, and a fraction is at point X_1. With the progressive increase in E, almost all electrons pass to X_1 and the characteristic J-E is linear, but with a slope much smaller than the initial one.

There are several mechanisms by which the oscillation can occur in the Gunn diode. We will consider here only the dipole layer mode, or domain mode, that occurs in relatively long samples. Figure 6.25a illustrates a sample of *n*-GaAs submitted to an external potential difference between the negative and positive terminals, respectively cathode and anode. The sample has electrons in the conduction band, whose negative charges are compensated by the positive charges of the ionized donor impurities fixed in the crystal. When the external voltage is applied, electrons injected through the metallic contact of the cathode create a layer of negative charge, which together with the impurities forms a layer of electric dipoles, or a **domain**. The dipole layer causes a sharp variation in the potential around it, and consequently a peak in the electric

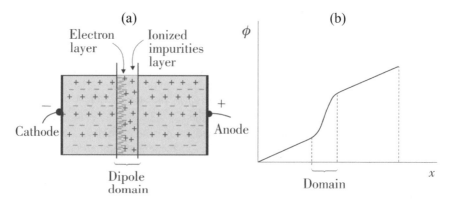

Fig. 6.25 Operation of the Gunn diode. **a** Illustration of the sample of *n*-GaAs with a dipole layer traveling from the cathode to the anode. **b** Variation of the electric potential in the sample

field. The field gradient exerts a force on the dipole layer, which moves towards the anode. Figure 6.25a shows a domain traveling from the cathode to the anode, while Fig. 6.25b illustrates the variation along the sample of the resulting potential. When the domain reaches the anode, a current pulse is produced in the external circuit. If the voltage applied to the diode has an appropriate value, the electric field in the domain will be in the region of negative resistance, resulting in energy supply to the external circuit. If this is a LC circuit, or a resonant cavity, the energy pulse tends to maintain the oscillation, as long as the transit time of the dipole domain is approximately equal to half the period of the oscillation. After its extinction in the anode, another domain forms in the cathode and the cycle is repeated. Gunn diodes are widely used as microwave oscillators. They operate with voltages in the range of 5–20 V, which represents an advantage over IMPATT diodes, which normally require tens of volts. Since the speed of the domain motion increases with increasing applied voltage, the oscillation frequency increases with the voltage. For this reason, the Gunn diode oscillation is easily modulated in frequency, by the superposition of a time-varying signal to the bias voltage.

Problems

6.1   A p–n junction made with Ge has impurities on each side with concentrations $N_d = 10^{16}$ cm^{-3} and $N_a = 10^{18}$ cm^{-3}.

(a)   Calculate the positions of the Fermi level on each side at $T = 300$ K, relative to the conduction and valence bands. .

(b)   Draw the energy diagram of the junction in equilibrium, indicating the values of the relevant energies, and from it determine the contact potential V_0

6.2   Calculate the maximum electric field, the thickness of the depletion region (in μm), and the capacitance of the p–n junction of problem 6.1, considering that it has a circular cross-section of diameter 300 μm.

6.3   In the equilibrium situation of a p–n junction, the diffusion current created by the concentration gradient cancels the drift current due to the potential gradient, both for electrons and holes. This equilibrium condition is expressed by Eq. (5.70) for holes. Calculate the integral of this equation in one dimension and using the Einstein relation obtain Eq. (6.3) for the contact potential.

6.4   A voltage V is applied to polarize an abrupt p–n junction. Considering that for not very high V the condition of equilibrium Eq. (5.70) is not changed much, demonstrate Eq. (6.22) and show that the thickness of the depletion region is given by an equation equal to (6.18), with V_0 replaced by $V_0 - V$.

6.5   Use the result of Problem 6.4 and the expression of the charge in a p–n junction to show that the junction capacitance is given by Eq. (6.20).

6.6   Show that the electron diffusion current in the depletion region of a p–n junction is given by Eq. (6.28).

6.7   Consider two abrupt p-n junctions made with different semiconductors, one with Si and one with Ge. Both have the same concentrations of impurities,

$N_a = 10^{18}$ cm^{-3} and $N_d = 10^{16}$ cm^{-3}, and the same circular cross section of diameter 300 μm. Suppose also that the recombination times are the same, $\tau_p = \tau_n = 1$ μs.

(a) Calculate the saturation currents of the two junctions at $T = 300$ K.
(b) Make I-V plots for the two junctions, preferably with a computer, with V varying in the range -1 to $+1$ V and I limited to 100 mA.

6.8 The breakdown electric field of the Si junction of Problem 6.7 is 10^6 V/cm. Calculate the breakdown voltage of the junction.

6.9 A p^+-n junction has concentration of impurities on the n-side that is negligible compared to the one on the p-side, such that $l_n \gg l_p$. Consider a junction of this type with l_n much smaller than the diffusion length L_p of holes on the n-side. Show that in this directly polarized junction, the electron current is negligible compared to that of holes and that the field on the n-side, outside the depletion region, is given approximately per

$$E(x) = \frac{k_B T}{e} \frac{1}{l_n} \frac{p_n(0)}{N_d + p_n(x)}$$

where x is measured from the boundary of the depletion region.

6.10 A p-n junction diode made with Si directly polarized with a constant current is used as a thermometer. At $T = 27$ °C the voltage at diode is 700 mV.

(a) Calculate the temperature coefficient of the diode in this temperature, that is, the ratio V/T.
(b) What will be the variation in the voltage if the temperature rises to 80 °C? Calculate this variation exactly and compare with the value obtained assuming that it is linear and characterized by the coefficient obtained in item a).

6.11 A p^+-n junction of Si with a cross section area of 10^{-2} cm^2 has impurity concentrations $N_a = 10^{17}$ cm^{-3} and $N_d = 10^{15}$ cm^{-3}. The parameters of Si are given in Table 5.2. Calculate:

(a) The maximum electric field;
(b) The thickness of the depletion region (in μm);
(c) The junction capacitance in the equilibrium situation and also when an external voltage of 0.4 V is applied in the forward direction.

6.12 Equation (6.25) for the concentration of holes on the n-side of a p–n junction is valid for both forward and reverse bias. The same occurs with the analogous equation for the electron concentration on the p-side. For a reverse polarized p–n junction with a voltage much larger than 25 mV:

(a) Give the appropriate expressions and make a qualitative plot for the concentrations of minority carriers, n_p (x') and p_n (x), as functions of the parameters of the junction and of x and x', measured from the boundaries of the depletion region.

(b)   Calculate the variations with x and x' of the minority carriers currents and make the corresponding plots.

(c)   From the results of item (b), calculate the total current at the junction and explain why the majority carriers currents are not necessary for the calculation.

6.13   Consider an abrupt p^+-n junction of Ge, with impurity concentrations $N_a = 5 \times 10^{16}$ cm^{-3} and $N_d = 10^{15}$ cm^{-3} with cross section area 10^{-3} cm^2 and recombination times $\tau_n = \tau_p = 2$ μs. The junction is forward biased with a current of 100 mA.

(a)   Calculate the voltage at the junction.

(b)   Calculate numerically the concentrations of the minority carriers $n_p(x)$ and $p_n(x')$, and plot their variations with x and x', measured from the boundaries of the depletion region.

(c)   Calculate numerically the concentrations of the majority carriers $n_n(x)$ and $p_p(x')$, and make plots showing their variations with x and x'. (They can be calculated using the fact that the charge neutrality outside the depletion region requires that the excess concentrations of electrons and holes in equilibrium are equal at each point, $\delta p(x) = \delta n(x)$.

6.14   A diode made with a Si junction, like the one in Problem 6.7, is placed in the circuit of the figure below. The battery has electromotive force 1.5 V and internal resistance 0.2 V and the resistor has resistance of 20 Ω.

(a)   Using the diode equation, calculate analytically the current and the voltage in the diode.

(b)   Using the I-V curve obtained in Problem 6.7, graphically calculate the current and the voltage in the diode and compare with the values obtained in item (a).

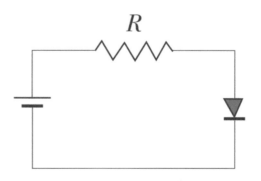

Further Reading

A. Bar-Lev, *Semiconductors and Electronic Devices* (Prentice-Hall, New Jersey, 1993)

D.A. Fraser, *The Physics of Semiconductor Devices* (Claredon Press, Oxford, 1986)

P.E. Gray, C.L. Searle, *Electronic Principles: Physics, Models, and Circuits* (Wiley, New York, 1969)

R.E. Hummel, *Electronic Properties of Materials* (Springer, Berlin, 2011)

K. Kano, *Semiconductor Devices* (Prentice-Hall, New Jersey, 1998)

D. Neamen, *Fundamentals of Semiconductor Physics and Devices* (McGraw-Hill, New York, 2002)

K.K. Ng, *Complete Guide to Semiconductor Devices* (Wiley-IEEE Press, Hobokens, 2002)

D.J. Roulston, *An Introduction to the Physics of Semiconductor Devices* (Oxford University Press, Oxford, 1999)

G.B. Rutkowski, J.E. Oleksy, *Solid State Electronics* (MacMillan- McGraw-Hill, Singapore, 1993)

B.J. Streetman, S. Banerjee, *Solid State Electronic Devices* (Prentice Hall, New Jersey, 2005)

S.M. Sze, M.K. Lee, *Semiconductor Devices: Physics and Technology* (Wiley, New York, 2012)

E.S. Yang, *Microelectronic Devices* (McGraw-Hill, New York, 1988)

F.F.Y. Wang, *Introduction to Solid State Electronics* (North-Holland, Amsterdam, 1989)

Chapter 7
Transistors and Other Semiconductor-Based Devices

This chapter is mostly devoted to the semiconductor device that revolutionized the electronic industry and changed our way of life, the transistor. Initially we treat in detail the bipolar transistor, made of two *p-n* junctions, that for a few decades was the most important device in analog and digital electronic equipment. Then we present field-effect transistors, including the MOSFET, that today is the building block of large-scale integrated circuits used in all digital equipment, such as computers, tablets, and smart phones. Integrated circuits are presented qualitatively at the end of the chapter. We also describe thyristors, that are important power control devices.

7.1 The Transistor

The transistor is a three-terminal device, used to control electric signals. A variable signal applied to the two input terminals electronically controls the signal at the two output terminals, one of which is common with the input. The two most common control functions are **amplification** and **switching**. When used for amplification, the device provides a signal with the same form of time-variation as the input signal, but with larger amplitude. This is illustrated Fig. 7.1 for a sinusoidal variation. The power of the output signal is generally larger than that of the input signal, with the increase in power provided by the DC power supply. In digital applications, a digital signal at the input causes the transistor to switch between two states, one with current and one without current, representing bits 1 and 0. Due to its ability to convert energy from a DC source into energy of a controlled signal, the transistor is called an **active device**.

The invention of the transistor represented one of the most important technological advances of the twentieth century, because it was decisive for the enormous evolution of electronics. Until the mid-1950s, the most widely used electronic control device was the **triode** tube. The triode is formed by a vacuum tube containing a heated cathode that emits electrons and an anode that receives them, but with a third electrode

S. M. Rezende, *Introduction to Electronic Materials and Devices*, https://doi.org/10.1007/978-3-030-81772-5_7

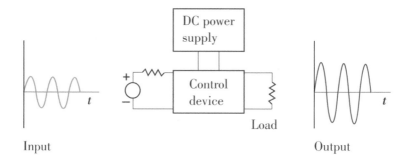

Fig. 7.1 Input and output signals in an amplification device, such as a transistor

made of a wire mesh, called grid. A variable voltage applied between grid and cathode controls the flow of electrons from the cathode to the anode passing through the grid. In this way, a voltage signal between the grid and the cathode controls the output current at the anode, making the triode tube an active control device. After the 1950s, vacuum tubes in electronic equipment gradually gave way to semiconductor transistors and diodes, called **solid state devices**.

The development of the transistor resulted from basic investigations of the properties of semiconductors. In 1947 Walter Brattain and John Bardeen were studying surface properties of germanium with metal rectifier contacts at the Bell Telephone Laboratories. In their studies they observed that the current in the semiconductor diode varied when another current passed through a second metallic contact placed next to the first. In December of that year they announced the discovery of the new amplification device, named by them as **transistor**, meaning an element of variable transconductance. Despite its large potential, the point contact transistor had many problems: it was very fragile; the contact degraded with the air humidity; its internal noise was very large. The next step in the development of the transistor took place in 1948, when William Shockley, also from Bell Laboratories, published a theoretical work proposing the structure of the junction transistor. From then on, many industrial laboratories invested in the study and manufacturing techniques of junction transistors, and within a few years they became commercial devices. In 1956 the Physics Nobel Prize was awarded to Shockley, Bardeen, and Brattain, for their seminal contribution to physics and electronics.

Currently there are two main types of transistors: the **bipolar junction transistor**, usually called simply a junction transistor, and the **field-effect transistor**. The bipolar junction transistor is made of two *p-n* junctions fabricated on the same semiconductor wafer, in which the current in the first junction controls the injection of minority carriers in the second junction. Since minority carriers can be both electrons and holes, this transistor operates with positive and negative charge carriers, hence the name bipolar. The field effect transistor can be made by two junctions, or by metal–oxide–semiconductor contacts. In both types, the input voltage controls the flow of majority carriers that pass from the input to the output of the device. These carriers can be either electrons or holes, depending on the type of impurity of the semiconductor, so that the field-effect transistor is a unipolar device. In the next section we shall present the principles of operation and the modeling of bipolar junction transistors. Field-effect transistors will be studied in Sect. 7.5.

7.2 The Bipolar Transistor

The bipolar junction transistor was the most important semiconductor device until the 1990 decade. With the dissemination of digital electronics, the field-effect transistor became the most used semiconductor device. With the planar technology, described in Sect. 6.1.1, the junction transistor can be made isolated from other devices in the same semiconductor wafer. Thus, it can be manufactured alone, to become a single device with three terminals, or in conjunction with many other diodes, transistors, and passive components, forming an **integrated circuit**. Its basic structure is shown in Fig. 7.2. It consists of three layers with different dopings, made in the same semiconductor, forming two *p-n* junctions with opposite polarities. The three layers are called **emitter, base**, and **collector**, which are connected to the external circuit through metallic contacts to which conducting wires are welded. The structure of Fig. 7.2 is that of a *p-n-p* transistor. If the dopings *p* and *n* are exchanged, a *n-p-n* transistor is obtained. The operations of the two types are entirely analogous, with the roles of electrons and holes interchanged.

Figure 7.3 shows the schematic representation of a *p-n-p* transistor with a simple external circuit for biasing its junctions. We represent by I_E, I_B, and I_C, respectively, the emitter, base and collector currents, that are considered positive when they have the directions indicated in the figure. The *p–n* junction between the emitter and the base is called simply the emitter junction, while the one with the base–collector is called the collector junction. V_{EB} and V_{CB} represent the voltages at the emitter and collector junctions, respectively. The circuit configuration in Fig. 7.3 is called **common base**, since the base terminal is common between the two input and the two output terminals of the device. Although this is not the most widely way to

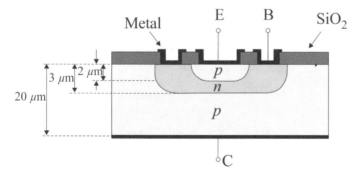

Fig. 7.2 Planar structure of the bipolar junction transistor with some typical dimensions in discrete devices. The letters E, B, and C represent the terminals of the emitter, base, and collector, respectively

Fig. 7.3 Schematic
representation of a
p-n-p transistor with a simple
common base polarization
circuit

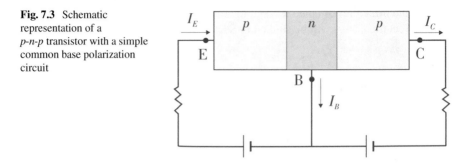

polarize transistors in practical circuits, it is the most convenient for understanding
the operation mechanisms of the transistor.

In the normal operation of the bipolar transistor, the emitter junction is polarized
in the forward direction, while the collector junction is reversed polarized. Thus, the
resistance of the emitter junction is small and the current I_E is relatively high. In
the *p-n-p* transistor, far from the junction, this current consists essentially of holes,
which are the majority carriers in the emitter and collector regions (component 1 in
Fig. 7.4). At the emitter junction, holes in the emitter region are injected into the base,
where they move by diffusion, contributing to part of the emitter current, denoted by
I_{Ep}. On the other hand, electrons go from the base to the emitter and contribute to
another part of the current, I_{En}, illustrated in Fig. 7.4 (component 7). As we saw in
Eqs. (6.29)–(6.31), it is possible to have $I_{Ep} \gg I_{En}$ if the concentration of impurities
is much larger in side *p* than in side *n*. If the base thickness were large, the collector
junction would be isolated from the emitter junction and the system would behave as
two diodes in series with opposite directions. In this case the collector current would

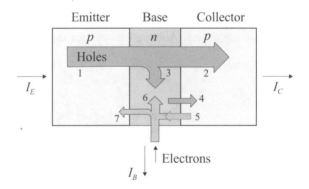

Fig. 7.4 Illustration of the flow of electrons and holes in a *p-n-p* transistor: 1-Holes in drift motion
in the emitter; 2-Holes that reach the collector by diffusion; 3-Holes that disappear in the base
due to recombination; 4 and 5-Holes and electrons thermally generated and that form the reverse
saturation current of the collector junction; 6-Electrons that recombine with holes in component 3;
7-Electrons injected from the base into the emitter forming the current I_{En}

be very small, given by the reverse saturation value I_s of Eq. (6.30) and, therefore, independent of the emitter current.

However, the base thickness is relatively small, less than the diffusion length L_p of holes in the base. Thus, even though they are minority carriers, the holes injected in the base do not have time to completely recombine with electrons, and most of them reach the depletion region of the collector junction, where they are accelerated by the electric field to the other side of the junction. As they reach the p-type collector, the holes are again majority carriers and acquire a drift motion under the action of the external field, forming most of the collector current I_C (component 2 of Fig. 7.4). Thus, the collector current is mainly due to the **holes injected from the emitter**, that is much larger than the reverse saturation current of the collector junction (components 4 and 5 of Fig. 7.4). The sum of components 2 and 4 forms the contribution of the holes to the collector current, which is denoted by I_{Cp}, while the contribution of the electrons, represented by the component 5 in Fig. 7.4, is denoted by I_{Cn}.

To conclude the qualitative explanation of the transistor operation, it is necessary to look closely to the important role of the base current I_B. As we saw before, a large fraction of the emitter current I_E passes to the collector because the thickness of the base is very small. This results in a collector current with a value close to, but smaller than the emitter current. Using the charge conservation equation, $I_E = I_B + I_C$, we see that with I_C a little smaller that I_E the base current I_B is small but nonzero. In fact, I_B results from the flow of electrons from the external circuit to the base through the contact B, and consists of three distinct contributions: the first corresponds to the electrons that recombine with part of the holes injected in the base from the emitter (component 6 in Fig. 7.4). This contribution can be minimized by making the base thickness much less than L_p; the second, denoted by I_{En}, is due to electrons that pass from the base to the emitter (component 7 in Fig. 7.4); from these contributions we have to subtract a third one, I_{Cn}, produced by the flow of electrons generated thermally in the collector and passing to the base through the collector junction (component 5 in Fig. 7.4).

As we shall see later, the basic condition for the transistor to be an amplifier is to have a small base current, typically $I_B \sim 10^{-2} I_E$. Due to the **proportionality between the currents, a small variation in the base current results in a large variation in the emitter current and, therefore, also in the collector current**. Thus, in a good transistor it is necessary to minimize I_B. This is achieved by making the base thickness small, such as to decrease the electron–hole recombination, and with a much lower doping than the emitter, in order to reduce I_{En}. However, since the base thickness cannot be too small, due to the physical limitations, the recombination mechanism is still significant. This fact establishes a minimum limit for I_B. In the next section, we shall present the derivation of the various components of the currents based on microscopic mechanisms. To conclude this section, let us define some relationships between the currents I_E, I_B, and I_C, that are used to characterize important transistor parameters.

As we saw earlier, the collector current consists essentially of holes injected by the emitter that do not disappear in the base by recombination with electrons. In the linear region of the transistor operation, this current is proportional to the I_{Ep} component of the emitter current. Thus, we have

$$I_C = B\, I_{Ep}, \tag{7.1}$$

where B is called the **base transport factor**, which represents the fraction of holes injected by the emitter that can reach the collector. On a p^+-n-p transistor with a very narrow base, $B < 1$. On the other hand, the I_{Ep} component of the current is slightly smaller and also proportional to the emitter current I_E, so we define

$$I_{Ep} = \gamma\, I_E, \tag{7.2}$$

where $\gamma < 1$ is called the **emitter injection efficiency**. If the reverse saturation current I_{Cn} at the collector junction is negligible, we can consider $I_C = I_{Cp} + I_{Cn} \approx I_{Cp}$. In this case, with the relations (7.1) and (7.2) we can express the collector current in terms of the emitter current as

$$I_C = \gamma\, B\, I_E \equiv \alpha\, I_E, \tag{7.3}$$

where $\alpha \equiv \gamma\, B$ is the **current transfer factor**, which is also smaller than 1. From (7.3) and the current continuity equation it is possible obtain the relation between the base and collector currents. Using (7.3) in $I_B = I_E - I_C$ we have

$$I_B = \frac{I_C}{\alpha} - I_C = \frac{1 - \alpha}{\alpha} I_C,$$

or else

$$I_C = \beta\, I_B, \tag{7.4}$$

where

$$\beta = \frac{\alpha}{1 - \alpha}, \tag{7.5}$$

is the **amplification factor** or **current gain**. This factor is a characteristic parameter of each transistor, but it also varies with the polarization voltages at the junctions. In a good transistor, α is smaller but close to 1, so that the factor β is large. Equation (7.4) expresses the basic characteristic of transistors in the linear regime. It shows that by means of a small variation in the base current, it is it is possible to control the variation in the much larger current that flows from the emitter to the collector. The physical explanation of the proportionality between the base and collector currents is the following. The collector current I_C is basically formed by the holes injected into the base by the emitter current, and that reach the collector because they do not have enough time to recombine with electrons in the base, since this it is very narrow (thickness much smaller than the hole diffusion length). Therefore, I_C increases as the emitter current I_E increases. The difference between I_E and I_C is the base current I_B, which is formed mainly by electrons that recombine with the holes injected by the emitter and that do not reach the collector. So, if the base current I_B varies, the number of electrons available for recombination varies, which forces I_C to vary as well, otherwise there would be an accumulation of charges at the base. In this way, a variation in I_B results in a variation in I_C and I_E. In a certain range of variation, the relationship between I_B and I_C is linear, as expressed in Eq. (7.4). In the next section we will obtain the expressions that relate the parameters B, γ, α, and β, with the microscopic quantities of the semiconductor that make up the transistor.

Example 7.1 A p-n-p transistor in the steady-state regime has the following components of the emitter and collector currents: $I_{Ep} = 10$ mA, $I_{En} = 0.1$ mA, $I_{Cp} = 9.98$ mA, and $I_{Cn} = 0.001$ mA. Calculate the transistor parameters B, γ, α, and β, and the base current.

The base transport factor B is given by (7.1)

$$B = \frac{I_{Cp}}{I_{Ep}} = \frac{9.98}{10} = 0.998.$$

The injection efficiency of the emitter is given by (7.2), where $I_E = I_{Ep} + I_{En}$. So

$$\gamma = \frac{I_{Ep}}{I_{Ep} + I_{En}} = \frac{10}{10 + 0.1} = 0.99.$$

The current transfer factor $\alpha = \gamma B$ is then

$$\alpha = 0.99 \times 0.998 = 0.988.$$

Neglecting I_{Cn} in the presence of I_{Cp}, the current gain is calculated with Eq. (7.5),

$$\beta = \frac{\alpha}{1 - \alpha} = \frac{0.988}{1 - 0.988} = 82.33,$$

The base current can be calculated exactly by the difference between the emitter and collector currents,

$$I_B = I_E - I_C = \left(I_{Ep} + I_{En}\right) - \left(I_{Cp} + I_{Cn}\right) = 10.1 - 9.981 = 0.119 \text{ mA}.$$

We can also calculate I_B using the relations (7.4) and (7.5), where (7.5) was obtained neglecting the contribution of the saturation current to I_C. The result is

$$I_B = \frac{I_C}{\beta} = \frac{9.981}{82.33} = 0.121,$$

This value differs from the previous one by 0.002 mA, which corresponds to a difference of only 1.7%.

7.3 Currents in the Bipolar Transistor

Like in the junction diode, the currents in the bipolar junction transistor are determined by the diffusion motion of minority carriers near the interfaces of the junctions. The fundamental difference between the diode and the transistor is that, while in the diode the solution of the diffusion equation is subjected to the boundary conditions at the junction interface, in the transistor it is necessary to consider the two interfaces of the junctions. To calculate the currents we must then solve the diffusion equation for the carriers concentrations in the three regions of the transistor and impose the boundary conditions at the two junction interfaces. After obtaining the variations in the concentrations of the minority carriers, we shall calculate the diffusion currents as we did for the diode in Sect. 6.2.

7.3.1 Calculation of Currents in the One-Dimensional Model

We shall consider here a *p-n-p* transistor with the one-dimensional model illustrated in Fig. 7.5. This model is good for the device in Fig. 7.2 because the lateral dimensions are much larger than the thicknesses of the layers. Let us assume that the thicknesses of the emitter and the collector are very large compared to the diffusion length, while the base has an arbitrary thickness. In the emitter and in the collector the minority carriers are electrons, whose concentrations are, respectively, $n_E(x)$ and $n_C(x)$. The holes injected by the emitter are minority carriers at the base, described by the concentration $p_B(x)$. Since the emitter is long, $n_E(x)$ is described by an exponential that falls away from the emitter junction. The corresponding diffusion current is then given by same expression obtained in Sect. 6.2 for a *p–n* junction. From Eq. (6.28) we can write the contribution of electrons to the emitter current

$$I_{En} = e\, A\, \frac{D_{nE}}{L_{nE}}\, n_E\, (e^{eV_{EB}/k_B T} - 1), \tag{7.6}$$

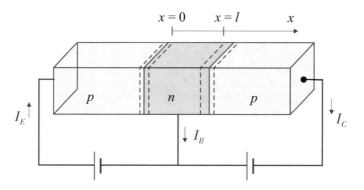

Fig. 7.5 One-dimensional model used to calculate the currents in a *p-n-p* transistor. Note that the $x = 0$ and $x = l$ coordinates are at the ends of the depletion regions of the emitter and collector junctions

where A is the area of the transistor cross-section, n_E is the equilibrium concentration of electrons in the emitter (we have omitted 0 in the subscript to simplify the notation), D_{nE} and L_{nE} are the diffusion coefficient and length, respectively, and V_{EB} is the voltage between emitter and base. In the same way we can write the contribution of electrons to the collector current as

$$I_{Cn} = -e\,A\,\frac{D_{nC}}{L_{nC}}\,n_C\,(e^{eV_{CB}/k_B T} - 1), \tag{7.7}$$

where the notation is analogous to that of Eq. (7.6). Notice that the sign is negative because the positive sign of I_C corresponds to the current flowing from side n to side p of the collector junction, and also that V_{CB} is usually negative. In the case of holes in the base, the solution for the variation of the concentration $p_B(x)$ is more complicated because it is necessary to consider the full solution of the diffusion equation, given by Eq. (5.83)

$$\delta p_B \equiv p_B(x) - p_B = C_1\,e^{-x/L_p} + C_2\,e^{x/L_p}, \tag{7.8}$$

where p_B is the equilibrium concentration of holes in the base and L_p is the diffusion length. Note that we have omitted the subscript 0 in p_B to simplify the notation. To obtain $\delta p_B(x)$ it is necessary to impose the boundary conditions at the emitter and collector junctions, at $x = 0$ and $x = l$, and calculate the constants C_1 and C_2. At the emitter–base junction, neglecting the thickness of the space charge region, we use Eq. (6.24) to obtain an expression for the hole concentration in terms of the polarization voltage

$$\delta p_B(x = 0) \equiv \Delta p_E = p_B\,(e^{eV_{EB}/k_B T} - 1), \tag{7.9}$$

Similarly, at the collector junction we have

$$\delta p_B(x = l) \equiv \Delta p_C = p_B \left(e^{eV_{CB}/k_BT} - 1 \right) \tag{7.10}$$

Note that in a transistor under normal operating conditions, the emitter junction is forward biased ($V_{EB} > 0$), while the collector junction is reverse biased ($V_{CB} < 0$). In this situation, and for $V_{EB} \gg k_BT$ and $|V_{CB}| \gg k_BT$, where $k_BT/e = 0.025$ V at $T = 290$ K, the boundary conditions (7.9) and (7.10) can be written approximately as

$$\Delta p_E \approx p_B \, e^{eV_{EB}/k_BT} \gg p_B, \tag{7.11}$$

$$\Delta p_C \approx -p_B \ll \Delta p_E. \tag{7.12}$$

The result (7.12) is due to the fact that, in the reverse biased junction, the holes in excess of equilibrium in the base are "pulled" quickly to the collector by the strong electric field, so that their concentration is very small. Considering that the solution (7.8) of the diffusion equation at the emitter and collector junctions are

$$\Delta p_E = C_1 + C_2,$$

$$\Delta p_C = C_1 \, e^{-l/L_p} + C_2 \, e^{l/L_p},$$

and using the boundary conditions (7.11) and (7.12) we obtain for the coefficients

$$C_1 = \frac{\Delta p_E \, e^{l/L_p} - \Delta p_C}{2 \sinh(l/L_p)}, \tag{7.13}$$

$$C_2 = \frac{\Delta p_C - \Delta p_E \, e^{-l/L_p}}{2 \sinh(l/L_p)}. \tag{7.14}$$

Before proceeding with the analysis of the currents, let us examine the behavior of the minority carrier concentrations in the three regions of the transistor. Under normal operating conditions, the equilibrium concentration of holes in the n-type base is very small, so that we can consider $\Delta p_C \approx p_B \approx 0$. Substituting Eqs. (7.13) and (7.14) in (7.8), and using this approximation, we obtain for the hole concentration in excess of equilibrium in the base, that is, for $0 < x < l$

$$\delta p_B(x) = \Delta p_E \frac{\sinh[(l - x)/L_p]}{\sinh(l/L_p)}. \tag{7.15}$$

In the case of electrons in the emitter and in the collector, their concentrations are given by simple exponentials, as discussed earlier. Figure 7.6 illustrates the variations

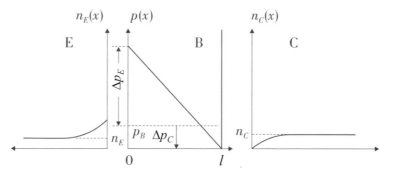

Fig. 7.6 Variations of the concentrations of the minority carriers in a *p-n-p* transistor with forward bias at the emitter junction and reverse bias at the collector junction. The value of $\Delta p_C \approx p_B$ is exaggerated, since $p_B \ll \Delta p_E$

of the minority carriers concentrations in the three regions of a p^+-n-p transistor with forward bias conditions. Note that in general the base is made with small thickness, $l \ll L_p$, so as to minimize the base current. For this reason, the variation in hole concentration is approximately linear. Having obtained the concentration of holes in the base, we can calculate its contributions to the currents. Equation (5.57) gives for the diffusion current of holes in the base

$$I_p(x) = -e\,A\,D_p\,\frac{d\delta p_B}{dx},$$

that applied to Eq. (7.8) gives

$$I_p(x) = e\,A\,\frac{D_p}{L_p}\left(C_1\,e^{-x/L_p} - C_2\,e^{x/L_p}\right). \tag{7.16}$$

The components of the emitter and collector currents due to holes are given by the values of Eq. (7.16) at $x = 0$ and $x = l$, respectively,

$$I_{Ep} = I_p(x = 0) = e\,A\,\frac{D_p}{L_p}(C_1 - C_2), \tag{7.17}$$

$$I_{Cp} = I_p(x = l) = e\,A\,\frac{D_p}{L_p}\left(C_1\,e^{-l/L_p} - C_2\,e^{l/L_p}\right). \tag{7.18}$$

Replacing Eqs. (7.13) and (7.14) in (7.17) and in (7.18) and using the definitions of hyperbolic functions we have

$$I_{Ep} = e\,A\,\frac{D_p}{L_p}\left(\Delta p_E\,\coth\frac{l}{L_p} - \Delta p_C\,\cosh\frac{l}{L_p}\right), \tag{7.19}$$

$$I_{Cp} = e \, A \, \frac{D_p}{L_p} \left(\Delta p_E \cosh \frac{l}{L_p} - \Delta p_C \coth \frac{l}{L_p} \right). \tag{7.20}$$

Adding the contributions of electrons and holes given by Eqs. (7.6), (7.7), (7.19), and (7.20), and using the expressions (9.9) and (7.10), we obtain the dependencies of the emitter and collector currents on the polarization voltages and the material parameters

$$I_E = e \, A \frac{D_p}{L_p} p_B \left[\left(e^{eV_{EB}/k_B T} - 1 \right) \coth \frac{l}{L_p} - \left(e^{eV_{CB}/k_B T} - 1 \right) \cosh \frac{l}{L_p} \right]$$
$$+ e \, A \frac{D_{nE}}{L_{nE}} n_E \left(e^{eV_{EB}/k_B T} - 1 \right), \tag{7.21}$$

$$I_C = e \, A \frac{D_p}{L_p} p_B \left[\left(e^{eV_{EB}/k_B T} - 1 \right) \cosh \frac{l}{L_p} - \left(e^{eV_{CB}/k_B T} - 1 \right) \coth \frac{l}{L_p} \right]$$
$$- e \, A \frac{D_{nC}}{L_{nC}} n_C \left(e^{eV_{CB}/k_B T} - 1 \right). \tag{7.22}$$

With Eqs. (7.21) and (7.22) one can calculate all parameters and characteristic curves of the transistor. The base current can be obtained using these two expressions in the continuity equation, $I_B = I_E - I_C$. The transistor parameters can be calculated by substituting Eqs. (7.19)–(7.22) in the definitions (7.1)–(7.5). Since these equations in the general forms are difficult to interpret, we shall calculate the quantities of interest by making some simplifying approximations that can be seen in the following example.

Example 7.2 A p^+-n-p^+ Si transistor at $T = 300$ K has the following characteristics: Cross-section area $A = 10^{-3}$ cm^2; base thickness $l = 1$ μm; concentrations of impurities, in the emitter $N_{aE} = 10^{17}$ cm^{-3}, in base $N_d = 5 \times 10^{15}$ cm^{-3}, in the collector $N_{aC} = 5 \times 10^{17}$ cm^{-3}; recombination times of the minority carriers, in the emitter and collector, $\tau_n = 0.5$ μs, in the base $\tau_p = 1$ μs. Calculate the emitter and collector currents with the emitter–base junction polarized directly with $V_{EB} = 0.7$ V, and the collector–base junction reversed polarized, with $V_{CB} = -10$ V.

To calculate the currents using Eqs. (7.21) and (7.22) it is necessary, initially, to calculate the concentrations of the minority carriers and the diffusion lengths. The equilibrium concentration of holes in the base is calculated with (5.38), with the value of n_i given in Table 5.2. Using SI units we have,

$$p_B = \frac{n_i^2}{N_d} = \frac{1.5^2 \times 10^{20} \times 10^{12}}{5 \times 10^{15} \times 10^6} = 4.5 \times 10^{10} \, \text{m}^{-3}.$$

The equilibrium concentrations of electrons in the emitter and in the collector are calculated with Eq. (5.41)

$$n_E = \frac{n_i^2}{N_{aE}} = \frac{1.5^2 \times 10^{20} \times 10^{12}}{10^{17} \times 10^6} = 2.2 \times 10^9 \, \text{m}^{-3},$$

$$n_C = \frac{n_i^2}{N_{aC}} = \frac{1.5^2 \times 10^{20} \times 10^{12}}{5 \times 10^{17} \times 10^6} = 4.5 \times 10^8 \, \text{m}^{-3}.$$

The diffusion lengths are calculated through their relationship with the diffusion coefficient D and the recombination time τ, $L = (D\tau)^{1/2}$. Using the value of D for Si in Table 5.2, and the values given for τ, we obtain in the SI

$$L_p = \left(12.5 \times 10^{-4} \times 1 \times 10^{-6}\right)^{1/2} = 3.5 \times 10^{-5} \, \text{m} = 35 \, \mu\text{m},$$

$$L_n = \left(35 \times 10^{-4} \times 0.5 \times 10^{-6}\right)^{1/2} = 4.2 \times 10^{-5} \, \text{m} = 42 \, \mu\text{m}.$$

We see then that $l/L_p \ll 1$, and therefore the hyperbolic functions in Eqs. (7.21) and (7.22) can be replaced by their binomial expansions. With $x = l/L_p$, we have $x = 1/35$, so that

$$\coth x \approx \frac{1}{x} + \frac{x}{3} = \frac{35}{1} + \frac{1}{3 \times 35} = 35.0095,$$

$$\cosh x \approx \frac{1}{x} - \frac{x}{6} = \frac{35}{1} - \frac{1}{6 \times 35} = 34.9952.$$

Finally, to compare the relative values of the different terms in Eqs. (7.21) and (7.22), it is necessary to calculate the values of the exponentials containing the polarization voltages. For $T = 300$ K the thermal energy is $k_B T = 0.026$ eV, so that

$$e^{eV_{EB}/k_B T} = e^{0.7/0.026} = e^{26.92} = 4.9 \times 10^{11},$$

$$e^{eV_{CB}/k_B T} = e^{-10/0.026} = e^{-384.6} \approx 0.$$

We see then that, since $\exp(eV_{EB}/k_B T) \gg 1$, both in (7.21) and (7.22), the terms that do not contain this factor can be neglected. So we can write

$$I_E \approx e \, A \frac{D_p}{L_p} p_B \, e^{eV_{EB}/k_B T} \coth \frac{l}{L_p} + e \, A \frac{D_{nE}}{L_{nE}} n_E \, e^{eV_{EB}/k_B T},$$

$$I_C \approx e \, A \frac{D_p}{L_p} p_B \, e^{eV_{EB}/k_B T} \cosh \frac{l}{L_p}.$$

Using the parameters in Table 5.2, the transistor parameters, and the values obtained previously, we have, in the SI

$$I_E = 1.6 \times 10^{-19} \times 10^{-7} \times \frac{12.5 \times 10^{-4}}{3.5 \times 10^{-5}} \times 4.5 \times 10^{10} \times 4.9 \times 10^{11} \times 35.0095$$

$$+ 1.6 \times 10^{-19} \times 10^{-7} \times \frac{35 \times 10^{-4}}{4.2 \times 10^{-5}} \times 2.2 \times 10^{9} \times 4.9 \times 10^{11}$$

$$I_E = 0.44112 + 0.00144 = 0.44256 \, \text{A}.$$

$$I_C = 1.6 \times 10^{-19} \times 10^{-7} \times \frac{12.5 \times 10^{-4}}{3.5 \times 10^{-5}} \times 4.5 \times 10^{10} \times 4.9 \times 10^{11} \times 35.9952$$

$$= 0.44094 \, \text{A}.$$

Clearly, the values of the emitter and collector currents are very close, as was expected. It is important to note that, if we had used in the expressions for I_E and I_C only the first term of the binomial expansions of the hyperbolic functions, and if the I_{En} contribution were neglected, the two currents would be exactly the same. Therefore, since the difference between the two currents is the base current, it is essential to use the first two terms in the series expansions. We also see that, although the contribution of the thermally generated electrons is small, it should not be neglected, since it is presence in I_E, but not in I_C, has an important contribution for the difference between the two currents.

7.3.2 Base Current and Transistor Parameters

As shown in the calculations made in Example 7.2, in a *p-n-p* transistor with normal polarization, the exponential factor $\exp(V_{EB}/k_BT)$ is very large, while the factor $\exp(V_{CB}/k_BT)$ is negligible. We can then obtain an approximate expression for the base current. Neglecting terms that do not contain the factor $\exp(V_{EB}/k_BT)$ in Eqs. (7.21) and (7.22) we have for the base current

$$I_B = I_E - I_C = e \, A \, e^{eV_{EB}/k_BT} \left[\frac{D_p}{L_p} p_B \left(\coth \frac{l}{L_p} - \cosh \frac{l}{L_p} \right) + \frac{D_{nE}}{L_{nE}} n_E \right].$$

It can be shown (Problem 7.2) that this expression reduces to

$$I_B = e\,A\,e^{eV_{EB}/k_BT}\left(\frac{D_p}{L_p}\,p_B\tanh\frac{l}{2L_p} + \frac{D_{nE}}{L_{nE}}\,n_E\right).\qquad(7.23)$$

This result shows that in a *p-n-p* transistor with normal polarization, the base current is dominated by two contributions. The second term in (23) corresponds to the contribution of the electrons injected from the base to the emitter, represented by component 7 in Fig. 7.4. To interpret the other contribution, we introduce in the first term of (7.23) the excess concentration Δp_E, given by Eq. (7.11). Neglecting the term in n_E we obtain

$$I_B \approx e\,A\,\frac{D_p}{L_p}\,\Delta p_E\tanh\frac{l}{2L_p}.$$

Finally, considering that the base thickness is quite smaller than the diffusion length L_p, we can use the approximation $\tanh x \approx x$ to obtain

$$I_B \approx \frac{e\,A\,l\,\Delta p_E}{2\tau_p},\qquad(7.24)$$

where $\tau_p = L_p^2/D_p$ is the recombination time of holes. This equation has a simple physical interpretation. Since the concentration of holes in excess of equilibrium in the base is Δp_E at $x = 0$ (emitter), and $\Delta p_C = 0$ at $x = l$ (collector), the quantity $e\,\Delta p_E A\,l/2 \equiv Q_p$ is the total charge of the holes that disappear in the base due to recombination. Since recombination occurs over a characteristic period of time τ_p, the current that must be supplied to the base to replace the charge that disappears and maintain the steady-state regime is Q_p/τ_p. This is precisely the base current given by Eq. (7.24). This result confirms the qualitative interpretation of the base current described at the end of Sect. 7.2. Equation (7.24) shows that to have a small current I_B one should make the base very narrow compared to L_p, and with a concentration of impurities relatively low so that the time τ_p is long.

Example 7.3 Calculate the base current in the transistor of Example 7.2 using Eq. (7.23), and compare with the value obtained by the difference between I_E and I_C calculated in Example 7.2.

   Substituting in Eq. (7.23) the values of the quantities in Example 7.2 and using $\tanh(l/2L_p) \approx l/2L_p$ we have

$$I_B = 1.6 \times 10^{-19} \times 10^{-7} \times 4.9 \times 10^{11}$$

$$\times \left[\frac{12.5 \times 10^{-4}}{3.5 \times 10^{-5}} \times 4.5 \times 10^{10} \times \frac{1}{2 \times 35} + \frac{35 \times 10^{-4}}{4.2 \times 10^{-5}} \times 2.2 \times 10^9\right]$$

$$I_B = 1.6 \times 10^{-19} \times 10^{-7} \times 4.9 \times 10^{11} \times \left[2.29 \times 10^{10} + 1.83 \times 10^{11}\right]$$

$$I_B = 1.614 \times 10^{-3}\,\text{A} = 1.614\,\text{mA}.$$

It is interesting to note that, in this case, the contribution of thermal electrons (I_{En}) to the base current, given by the second term of the equation above, is larger than the contribution given by first term. In a p^+-n-p transistor, with a much higher concentration of impurities in the emitter that in the base, n_E is much smaller and the recombination term dominates over I_{En}.

The value of I_B obtained by the difference between the currents calculated in Example 7.2 is

$$I_B = I_E - I_C = 0.44256 - 0.44094 = 0.00162\,\text{A} = 1.62\,\text{mA}$$

which is very close to the value calculated with Eq. (7.23). Evidently, the difference between the two values is due to the approximations made in hyperbolic functions and to the numerical roundings.

To obtain the parameters γ, B, α, and β of the transistor, we shall neglect in Eqs. (7.6)–(7.22) the terms in Δp_C and also the reverse saturation current in the collector. This approximation is valid because when used as an amplifier, the transistor always has the collector junction reverse polarized. With this approximation, using Eqs. (7.6), (7.11), and (7.19) in the definition of the injection efficiency (7.2) we obtain

$$\gamma = \frac{I_{Ep}}{I_E} = \frac{1}{1 + I_E/I_{Ep}} = \frac{1}{1 + (D_{nE} n_E L_p / D_p p_B L_{nE})\tanh(l/L_p)}.$$

Using the relationships (5.38) and (5.41) between the equilibrium concentrations of minority carriers and the concentrations of donor impurities (N_d) in the base and of acceptors (N_a) in the emitter, this expression can be written in the form

$$\gamma = \left(1 + \frac{D_{nE} N_d L_p}{D_p N_a L_{nE}}\tanh\frac{l}{L_p}\right)^{-1}. \tag{7.25}$$

Using only the first term of Eq. (7.22) for I_C and the first term of (7.19) for I_{Ep}, with Δp_E given by (7.9), the base transport factor, defined in (7.1), becomes

$$B = \frac{I_C}{I_{Ep}} = \frac{\cosh(l/L_p)}{\coth(l/L_p)} = \operatorname{sech}\frac{l}{L_p}. \tag{7.26}$$

With Eqs. (7.25) and (7.26), one can show that the current transfer factor α defined in (7.3) is

$$\alpha = \frac{I_C}{I_E} = B\,\gamma = \left(\cosh\frac{l}{L_p} + \frac{D_{nE}N_d L_p}{D_p N_a L_{nE}}\sinh\frac{l}{L_p}\right)^{-1}. \tag{7.27}$$

Finally, using Eqs. (7.27) in (7.5) we obtain for the amplification factor β defined in (7.5)

$$\beta = \frac{I_C}{I_B} = \frac{\alpha}{1-\alpha} \approx \left(\cosh\frac{l}{L_p} + \frac{D_{nE}N_d L_p}{D_p N_a L_{nE}}\sinh\frac{l}{L_p} - 1\right)^{-1}. \tag{7.28}$$

Considering that $l/L_p \ll 1$, we can obtain a simpler expression for β. Using the expansions of the hyperbolic functions,

$$\sinh x \approx x, \qquad \cosh x \approx 1 + \frac{1}{2}x^2,$$

the amplification factor given by Eq. (7.28) becomes, approximately,

$$\beta = \left(\frac{l^2}{2L_p^2} + \frac{D_{nE}N_d\,l}{D_p N_a L_{nE}}\right)^{-1}. \tag{7.29}$$

With Eqs. (7.26)–(7.29) one can calculate all transistor parameters from its fabrication data with good precision.

Example 7.4 Calculate the amplification factor of the silicon *p-n-p* transistor with the same parameters of Example 7.2.

Using in Eq. (7.29) the parameters and quantities calculated in Example 7.2 we have

$$\frac{1}{\beta} = \frac{1}{2 \times 35^2} + \frac{35 \times 5 \times 10^{15} \times 1}{12.5 \times 10^{17} \times 42}$$

$$\frac{1}{\beta} = \frac{1}{2450} + \frac{1}{300} = 0.00374.$$

Therefore, the amplification factor given by Eq. (7.29) is

$$\beta = \frac{1}{0.00374} = 267.3.$$

We can also calculate β directly, through the ratio between I_C, obtained in the Example 7.2, and I_B, calculated in Example 7.3. The result is

$$\beta = \frac{I_C}{I_B} = \frac{0.44094}{0.00162} = 272.2.$$

The difference between the two values, of only 1.8%, is due to the approximations made in the derivation of Eq. (7.29) and also to numerical roundings.

7.3.3 The I-V Characteristic Curves

Equations (7.21) and (7.22) describe the currents in a *p-n-p* transistor very well. To understand qualitatively the behavior of the currents as a function of the polarization voltages, it is better to simplify the notation and write them in the following forms

$$I_E = I_{Es}\left(e^{eV_{EB}/k_BT} - 1\right) - \alpha_I I_{Cs}\left(e^{eV_{CB}/k_BT} - 1\right), \tag{7.30}$$

$$I_C = \alpha_N I_{Es}\left(e^{eV_{EB}/k_BT} - 1\right) - I_{Cs}\left(e^{eV_{CB}/k_BT} - 1\right), \tag{7.31}$$

where the new parameters are defined by

$$I_{Es} = \frac{e\,A\,D_p\,p_B}{L_p}\coth\frac{l}{L_p} + \frac{e\,A\,D_{nE}\,n_E}{L_{nE}}, \tag{7.32}$$

$$I_{Cs} = \frac{e\,A\,D_p\,p_B}{L_p}\coth\frac{l}{L_p} + \frac{e\,A\,D_{nC}\,n_C}{L_{nC}}, \tag{7.33}$$

$$\alpha_N = \frac{e\,A\,D_p\,p_B}{I_{Es}L_p}\cosh\frac{l}{L_p}, \tag{7.34}$$

$$\alpha_I = \frac{e\,A\,D_p\,p_B}{I_{Cs}L_p}\cosh\frac{l}{L_p}. \tag{7.35}$$

The relations (7.30) and (7.31) were originally obtained by J. J. Ebers and J. L. Moll and are therefore called Ebers-Moll equations. They have fairly general validity, even if the transistor cannot be represented by the simple one-dimensional model of Fig. 7.4. In the general case, the parameters of the equations are not given exactly by the expressions (7.32)–(7.35), however it is possible show that they obey the relationship

$$\alpha_N I_{Es} = \alpha_I I_{Cs}, \tag{7.36}$$

which is also satisfied by Eqs. (7.32)–(7.35). Clearly, the Ebers-Moll equations are the sums of two diode equations like (6.29). The emitter current in Eq. (7.30) is given by a term characteristic of the emitter junction diode, added to another term proportional to the current in the collector junction diode. Similarly, the collector current (7.31) is the sum of two terms, one for the emitter diode and the other for the collector diode. These equations show that the transistor can be characterized by only four parameters, related to each other by the expression (7.36). These parameters are not generally supplied by the manufacturer, but can be easily measured in the laboratory. Notice in Eq. (7.30) that if the collector–base junction is short-circuited, that is if $V_{CB} = 0$, the measurements of I_E and I_C as a function of V_{EB} provide the values of I_{Es} and $\alpha_N I_{Es}$, respectively. Similarly, making $V_{EB} = 0$, one can measure I_{Cs} and $\alpha_I I_{Cs}$ and thus have a full characterization of the transistor described by Eqs. (7.30) and (7.31).

The I-V characteristic curves of the transistor are nothing more than the graphical representation of the Ebers-Moll equations. Since in the equations there are two voltages, V_{EB} and V_{CB}, and two currents, I_E and I_C, it is necessary to select some quantities and express them as a function of the others. Multiplying (7.30) by α_N and subtracting from (7.31) one has

$$I_C = \alpha_N I_E - (1 - \alpha_N \alpha_I) I_{Cs}\left(e^{eV_{CB}/k_B T} - 1\right).$$

Similarly, multiplying Eq. (7.31) by α_I and subtracting from (7.30) we obtain

$$I_E = \alpha_I I_C + (1 - \alpha_N \alpha_I) I_{Es}\left(e^{eV_{EB}/k_B T} - 1\right).$$

These equations can be written in the form,

$$I_E = \alpha_I I_C + I_{E0}\left(e^{eV_{EB}/k_B T} - 1\right), \tag{7.37}$$

$$I_C = \alpha_N I_E - I_{C0}\left(e^{eV_{CB}/k_B T} - 1\right), \tag{7.38}$$

where

$$I_{E0} = (1 - \alpha_N \alpha_I) I_{Es},$$
$$I_{C0} = (1 - \alpha_N \alpha_I) I_{Cs},$$

are, respectively, the saturation currents of the emitter junction with the collector junction open ($I_C = 0$), and of the collector junction with the emitter junction open ($I_E = 0$). With Eq. (7.38) we can make a plot of I_C as a function of V_{CB} with I_E as a parameter. For $I_E = 0$, if $V_{CB} < 0$, the curve I_C - V_{CB} is equal to that of a reverse polarized junction, as in Fig. 6.7. Even with small values of V_{CB} the current reaches saturation with value $I_C \approx I_{C0}$. If $I_E \neq 0$, we can make I_C - V_{CB} curves for several values of I_E, resulting in the set of curves shown in Fig. 7.7a. Note that for $I_E = 0$

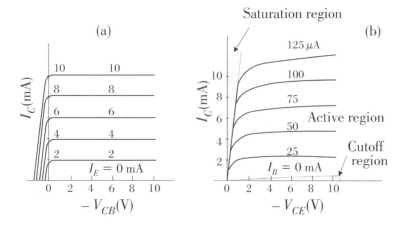

Fig. 7.7 *I-V* characteristic curves of a *p-n-p* transistor: **a** curves with I_E as a parameter used in the common-base configuration; **b** curves with I_B as a parameter for common-emitter configuration

and $V_{CB} < 0$, part of the holes injected into the base by the emitter current reaches the collector junction and produces an additional contribution to the reverse saturation current, $I_C \approx I_{C0} + \alpha_N I_E$. For this reason, the various curves in Fig. 7.7a resemble those of a reverse polarized junction, displaced by $\alpha_N I_E$. The curves in Fig. 7.7a are useful when the transistor is used in the common-base configuration of Fig. 7.3. In this case, if the base current is zero, the collector current is very small. This can be understood by the fact that as $I_C = I_E$, since the collector junction is reversed polarized, both currents must be small. As I_B increases, the difference between I_E and I_C increases and, even though the collector junction is reverse polarized, the injection mechanism makes I_C increase.

When the transistor is used in the common-emitter configuration, it is more convenient to work with the I_C - V_{CE} curves, having the base current I_B as a parameter. Typical curves for a *p-n-p* transistor in the common emitter configuration shown in Fig. 7.7b, clearly show that the current I_C can be controlled by the small current I_B. The various *I-V* curves are characteristic of each type of transistor and are provided by the manufacturer. In fact, they vary a little from one transistor to another, even though they are of the same type, because the curves of the manufacturer represent average data. Since the parameters also vary with temperature, it is common to find curves for a few temperature values.

To conclude this section, we note that in a *n-p-n* transistor the signs of the currents and the voltages are opposite to those of the *p-n-p* transistor. The equations for *n-p-n* transistor have the same form as Eqs. (7.30)–(7.35), with the letters *p* and *n* interchanged, since the roles of electrons and holes are interchanged.

7.4 Applications of Transistors

Bipolar junction transistors have many applications in electronic circuits, the most common are **amplification** and **switching**. Figure 7.8 shows the symbols of the *n-p-n* and *p-n-p* transistors used in circuits, and a typical external view of an encapsulated low power transistor. In the circuit symbols, the only difference between types *n-p-n* and *p-n-p* is in the arrow on the emitter terminal, indicating the direction of the forward current. In the encapsulated transistor there is no visible difference between the two types. Only by looking at the manufacturer's data for the transistor code it is possible to know its type.

To operate in a convenient region of the *I-V* characteristic, the transistor junctions need to be properly biased. Figure 7.9a shows a *n-p-n* transistor in the common-emitter configuration with a simple polarization circuit. Note that the voltages applied to the emitter and collector junctions have opposite directions to the ones in a *p-n-p* transistor. Since the transistor response is highly nonlinear, it is necessary to use graphic methods to determine the so-called **operating point**, whose coordinates are the currents and the voltages in the DC regime. As the resistance of the emitter

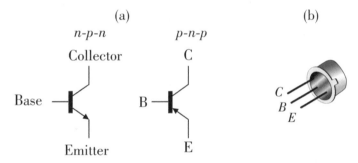

Fig. 7.8 **a** Circuit symbols of *n-p-n* and *p-n-p* transistors. **b** View of an encapsulated low power transistor

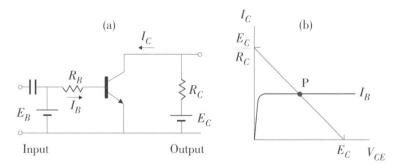

Fig. 7.9 **a** Simple amplifier circuit with a *n-p-n* transistor in the common-emitter configuration. **b** Graphic illustration of the method for determining the operating point P

junction is very small, the base current is given approximately by $I_B \approx E_B/R_B$. To calculate the collector current, we use the curve corresponding to the calculated value of the current I_B in the I-V characteristics for the common-emitter configuration, as shown in Fig. 7.7b. The loop equation for the collector circuit is

$$E_C = R_B\, I_C + V_{CE}(I_C, I_B). \tag{7.39}$$

This equation is represented in the I_C-V_{CE} plane by a line, called the **load line**, that has a position determined by the intersections with the axes I_C and V_{CE}. It is easy to see in Eq. (7.39) that they are given by $I_C = E_C/R_C$ and $V_{CE} = E_C$, as shown in Fig. 7.9b. The intersection between the load line with the I_C-V_{CE} curve of the transistor for the calculated value of I_B, point P of Fig. 7.9a, is the solution of Eq. (7.39) and therefore it is the operating point of the circuit. Depending on the region of the I-V characteristic where the point is located, and also on the shape and amplitude of the input signal, the transistor can perform different functions.

To behave as a good **amplifier**, the operating point should be in the active region of the characteristic curves, shown in Fig. 7.7b. In this region, a variation ΔI_B in the small base current, produced by an AC signal applied to the circuit through the capacitor in Fig. 7.9a, produces a variation ΔI_C in the collector current. As long as the base current does not approach the saturation or cutoff regions, shown in Fig. 7.7b, the variation in the collector current is proportional to that of the base current, $\Delta I_C = \beta\, \Delta I_B$, where β is the amplification factor. We see then that the position of the operating point is essential for the proper operation of the transistor. For this reason, it is customary to use a more complex polarization circuit than that of Fig. 7.9, in which a feedback loop serves to stabilize the operating point.

Another important application of transistors is in switching circuits. Figure 7.10 shows a switching circuit with a p-n-p transistor in the common-emitter configuration, with a simple biasing scheme. The purpose of the circuit is to make the transistor operate in two conducting states, one **on**, with a current, and one **off**, with zero current. In the on state it must behave like a closed switch, that lets a current go through with very low resistance, while in the off state it behaves like an open switch. This control is done on the collector current, by means of a much lower base current. The two states of the transistor can be achieved in the common-emitter configuration, as can

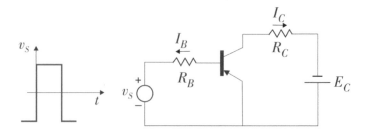

Fig. 7.10 Simple switching circuit using a p-n-p transistor in the common-emitter configuration

be seen in the curves of Figure 7.7b. The load line for the circuit of Fig. 7.10 is obtained in the same way as in Fig. 7.9. However, since there is no battery in the base circuit, the base current is zero in absence of the input signal. In this situation, the collector current is very small and the transistor is cut-off, or in the off state. When a voltage signal v_s with the shape shown in Fig. 7.10 is applied, the circuit operates with the base varying between two values, one that cuts off the collector current and the other that drives the transistor to saturation. The cutoff region, shown in Fig. 7.7b, is reached when the base current is zero or negative. On the other hand, the saturation region is achieved when the base current is positive and sufficiently large. In this situation the collector current is large and the transistor is in the on state. In this way, a small power signal like v_s controls the transistor making it to operate like a switch, that can be open or closed. This switch can control a collector current much larger than the base current, playing a role similar to that of an electromechanical relay, but with many advantages. Since the relay has moving parts and uses mechanical contacts, it is much slower and has much less durability than the transistor.

In an ideal switch, the transition from the off to the on state, or vice versa, should be done almost instantly. Evidently that this does not happen in a real transistor. There is a finite transient time, due to the fact that in the change from the saturation state to the cutoff state, or vice versa, there is removal or introduction of distributed charge in the base. This cannot be done instantly, because it would correspond to an infinite current. The times for decay or growth in the charge of the base are essentially due to the same effects mentioned in the case of the junction diode.

The switching transistor is used in many digital circuit applications, since its two states correspond to bits 0 and 1 of the binary system.

7.5 Field-Effect Transistors

Field-effect transistors (FETs) belong to a family of very important devices for technological applications. Like bipolar transistors, FETs are three terminal devices widely used for amplification and switching. However, from the point of view of the circuit, there is a major difference between the two types of devices. While in the bipolar transistor the output signal is controlled by an input current, in FETs it is controlled by an **input voltage signal**.

The operating mechanisms of field effect transistors are quite different from the ones in bipolar transistors, studied in the previous section. While in bipolar transistors the control of the output signal is done by the diffusion motion of **minority carriers** in the base, in the FETs the control is done by the drift motion of the **majority carriers**. These carriers move from one terminal, called **source**, to another terminal, called **drain**, through a uniform region of the semiconductor, **the channel**. The control of the motion of majority carriers in the channel is done by an electric field created by the voltage applied between a third terminal, called **gate**, and the source. This is the reason for the name **field-effect**.

There are three main types of field-effect transistors: the **junction FET**; the **metal-semiconductor FET**; and the **metal-insulator-semiconductor FET**. In the junction FET, abbreviated by JFET, the voltage applied to the gate varies the thickness of the depletion region of a reverse biased p–n junction, and thus the resistance of the channel. In the metal–semiconductor FET, or MESFET, the gate is formed by a metal–semiconductor junction. The operation of MESFET is very similar to that of JFET, however it has a faster response, and is therefore widely used in high frequencies.

In the metal-insulator-semiconductor FET, the metallic terminal of the gate is isolated from the semiconductor by an insulating layer. In the most common case the insulator is an oxide of the semiconductor itself, such as SiO_2 in the case of silicon. In this case the transistor is called metal–oxide–semiconductor FET, or MOSFET. Due to the presence of the insulating layer, this type is characterized by a high input impedance and low power dissipation. The MOSFETs have huge applications in digital integrated circuits and are essential components in computers and modern information technology equipment.

7.5.1   The Junction Field-Effect Transistor

In the junction field-effect transistor, JFET, a variable voltage applied to the gate controls the effective cross-section of a semiconductor channel through which the majority carriers flow from the source to the drain. Figure 7.11a shows a section of the semiconductor wafer with a JFET of n-type channel, where one can see the n-channel and two p^+ regions of the gates, as well as the metallic contacts for the source (S), the gate (G), and the drain (D). Note that the two p^+ gate regions are electrically

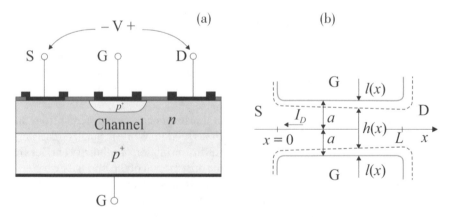

Fig. 7.11 Junction field-effect transistor with n-type channel. **a** Planar structure showing the different regions and terminals of the source (S), gate (G), and drain (D). **b** Symmetrical model for the channel region

interconnected. Due to its symmetry, the structure with two gates is easier to analyze. However, it is also common to manufacture the JFET with only one gate. The p-type channel JFET is entirely analogous to the one with n-type channel, with the regions p and n interchanged relative to those in Fig. 7.11a.

The operation of the JFET is very simple compared to the bipolar junction transistor. Let us consider the case of the n-channel JFET, shown in Fig. 7.11. The voltage V_D applied between the drain and the source produces a current I_D in the channel, formed predominantly by a drift motion of electrons. The electrons move from the source to the drain, while the conventional direction of the current is the opposite. The value of this current is determined by the voltage V_D and by the resistance of the channel, which in turn depends on the concentration of impurities, length, and effective area of the channel cross-section. The area can be controlled by the sizes of the depletion regions of the two p^+-n junctions between the gates and the channel, since there are no conduction electrons in these regions. As we saw in Sect. 6.1.3, the thickness of the depletion region depends on the reverse voltage at the junction, so that the drain current I_D varies with the voltage V_D between the gate and the source. In this way, the variation of the current I_D is controlled by the voltage V_D.

7.5.2   The I-V Characteristics of the JFET

To calculate the I-V characteristics of the JFET, let us consider the symmetrical model for the channel region shown in Fig. 7.11b. The channel has length L, depth D (perpendicular to the plane of the figure), and effective thickness $h(x)$, because there are no conduction electrons in the depletion regions of the two channel-gate junctions. The effective thickness is given by $h = 2(a - l)$, where $2a$ is the distance between the two gate regions and l is the thickness of each depletion region, which depends on the reverse voltage at the junction. This voltage varies with x because the current I_D from the drain to the source produces a potential drop along the channel. Therefore, h also varies with x, so that the area of the effective cross-section varies along the channel. For this reason, the resistance of the channel is not given simply by the usual expression, $\rho L/A$. However, the dependence of the current I_D on the voltages V_D and V_G can be calculated with simple concepts and relationships.

The current density in the channel, given by Eqs. (5.45) and (5.48), can be written in the form

$$J(x) = \sigma \, E(x) = e \, n \, \mu_n \, E(x) = -e \, n \, \mu_n \, \frac{\phi(x)}{dx}, \qquad (7.40)$$

where $\phi(x)$ is the electric potential at the point with coordinate x in the channel, relative to the position of the source ($x = 0$). The current intensity I_D in the channel is given by the product of J with the effective area

$$I_D = 2 \, [a - l(x)] \, D \, J(x), \qquad (7.41)$$

where $l(x)$ is the thickness of the depletion regions of the p^+-n junctions in the section with coordinate x. Considering that at x the reverse voltage is $V(x)$, assuming $N_a \gg N_d$ and negligible contact potential, the thickness $l(x)$ is obtained using Eq. (6.18). It can be shown that (Problem 6.4)

$$l(x) = \left[\frac{2\varepsilon}{e N_d} V(x) \right]^{1/2}, \tag{7.42}$$

where the reverse voltage at the junction $V(x)$ is given by the potential difference between a point at x in the channel axis and the gate, that is $V(x) = \phi(x) - V_G$. Replacing Eqs. (7.40) and (7.42) in (7.41), using this relation for $V(x)$ and making $n = N_d$ we obtain

$$I_D = -2 e N_d \mu_n D \left\{ a - \left[\frac{2\varepsilon}{e N_d} (\phi - V_G) \right]^{1/2} \right\} \frac{d\phi}{dx}.$$

We can now separate the variables ϕ and x and evaluate the integrals on both sides between $x = 0$ and $x = L$. Since the drain voltage relative to the source is $V_D = \phi(L) - \phi(0)$, integration of both sides of the above equation gives

$$\int_0^L I_D \, dx = -2 e N_d \mu_n D \int_0^{V_D} \left\{ a - \left[\frac{2\varepsilon}{e N_d} (\phi - V_G) \right]^{1/2} \right\} d\phi.$$

The integral on the left-hand side is trivial because the current intensity I_D does not vary with x. The integral in ϕ is also simple to perform, leading to

$$I_D = -\frac{2 e N_d \mu_n D a}{L} \left\{ V_D - \frac{2}{3} \left(\frac{2\varepsilon}{e N_d a^2} \right)^{1/2} \left[(V_D - V_G)^{3/2} - (-V_G)^{3/2} \right] \right\}. \tag{7.43}$$

This expression can be simplified using the following considerations. The multiplicative factor to the left of the brackets is the inverse of the channel resistance without the depletion regions, called channel conductance

$$G_0 = \frac{1}{R} = \frac{\sigma \, 2Da}{L} = \frac{2 e N_d \mu_n D a}{L}. \tag{7.44}$$

The effective channel thickness, given by $h(x) = 2[a - l(x)]$, decreases with increasing x, so that the smallest thickness occurs at the drain end, $l = a$, as shown in Fig. 7.12. Since $V(x)$ increases with x, from Eq. (7.42) we see that $l(x)$ also increases with x. Thus, there is a value of the voltage for which the depletion regions of the two gates touch each other at the drain end, causing an obstruction of the channel. This

Fig. 7.12 Variation of the effective channel thickness for two values of the drain voltage. For $V_D = V_p + V_G$, the channel undergoes a constriction at $x = L$

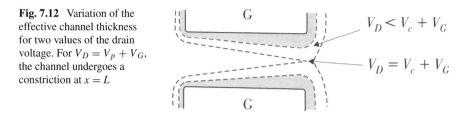

condition is known as **pinchoff**. The value of voltage for which this occurs, given by (7.42) with $l = a$,

$$V_p = \frac{e\,N_d\,a^2}{2\varepsilon},$$

(7.45)

which called the pinchoff voltage. Substituting the definitions (7.44) and (7.45) into (7.43), and noting that the negative sign of (7.43) is due to the fact that the current has the $-x$ direction, we obtain the final expression for the absolute value of the drain current as a function of the drain and gate voltages

$$|I_D| = G_0\,V_p\left[\frac{V_D}{V_p} + \frac{2}{3}\left(-\frac{V_G}{V_p}\right)^{3/2} - \frac{2}{3}\left(\frac{V_D - V_G}{V_p}\right)^{3/2}\right].$$

(7.46)

It is important to note that this expression is only valid if the channel is open at all points, that is, for $V(x) < V_p$. Since the maximum reverse voltage at the junction is $V(L) = V_D - V_G$, Eq. (7.46) is valid only for

$$V_D - V_G \le V_p.$$

(7.47)

For drain voltages larger than the value given by (7.47), the current reaches saturation, with a value obtained from Eq. (7.46) with $V_D - V_G = V_p$. Also note that the gate normally operates at zero or negative voltages relative to the source, so that in all of the above expressions $V_G \le 0$.

Example 7.5 Consider a Si JFET with $N_d = 5 \times 10^{15}$ cm^{-3}, $N_a = 10^{19}$ cm^{-3}, $a = 1$ μm, $L = 15$ μm, and $D = 1$ mm. Calculate parameters G_0 and V_p and plot the $I_D - V_D$ curves for some values of V_G.

Since $N_a \gg N_d$, the thickness of the depletion region can be calculated with Eq. (7.42). So, using in Eq. (7.44) the data in Table 5.2 and the transistor parameters we have

$$G_0 = \frac{2\,e\,N_d\,\mu_n\,D\,a}{L}$$

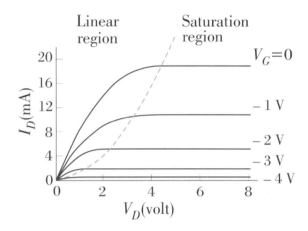

Fig. 7.13 *I-V* characteristics of a junction field-effect transistor, obtained from Eq. (7.46), with $V_p = 3.8$ V and $G_0 = 1.44 \times 10^2$ Ω^{-1}

$$= \frac{2 \times 1.9 \times 10^{-19} \times 5 \times 10^{15} \times 10^6 \times 1350 \times 10^{-4} \times 10^{-3} \times 10^{-6}}{15 \times 10^{-6}}$$

$$G_0 = 1.44 \times 10^{-2} \Omega^{-1}.$$

With Eq. (7.45) we obtain

$$V_p = \frac{e\, N_d\, a^2}{2\varepsilon} = \frac{1.9 \times 10^{-19} \times 5 \times 10^{15} \times 10^6 \times 10^{-12}}{2 \times 11.8 \times 8.85 \times 10^{-12}} = 3.8 \text{ V}.$$

Using these values of G_0 and V_p in Eq. (7.46) we obtain numerically the curves shown in Fig. 7.13.

To understand the behavior of the *I-V* curves in Fig. 7.13, let us consider initially $V_G = 0$. In this situation the current given by Eq. (7.46) is

$$|I_D| = G_0 V_p \left[\frac{V_D}{V_p} - \frac{2}{3}\left(\frac{V_D}{V_p}\right)^{3/2} \right]. \tag{7.48}$$

Note that for small drain voltages, $V_D \ll V_p$, the first term in (7.48) is much larger than the second, so that, $|I_D| \approx G_0 V_D$. This is the linear region of the characteristic curve with $V_G = 0$, indicated in Fig. 7.13. The presence of the term with 3/2 power and negative sign in Eq. (7.48) produces a decrease in the growth rate of $|I_D|$ with increasing V_D. To analyze the behavior of the drain current with $V_G = 0$, we calculate the derivative of $|I_D|$ in (7.48) with respect to V_D

$$\frac{d|I_D|}{dV_D} = G_0 \left[1 - (V_D/V_p)^{1/2} \right]_{V_G=0}. \tag{7.49}$$

We see then that $|I_D|$ reaches a maximum ($dI_D/dV_D = 0$) exactly at $V_D = V_p$ (for $V_G = 0$). At this value of the drain voltage, the current reaches saturation with a value given by Eq. (7.48) with $V_D = V_p$

$$I_{Dsat} = G_0\, V_p/3. \tag{7.50}$$

For $V_D > V_p$ the current maintains this value, which corresponds to the situation of the channel almost completely closed. This is so because if the current decreased, the voltage drop in the channel would also decrease and it would open. This delicate balance keeps the current constant for $V_D > V_p$, with the same value at saturation given by Eq. (7.50).

For nonzero and negative gate voltages V_G, the behavior of the I_D-V_D curves is qualitatively the same as described for $V_G = 0$. The main differences are that the saturation current and the pinchoff drain voltage decrease with the increase of $-V_G$. The dashed line in Fig. 7.13 indicates the geometrical locus of the saturation points for $V_G \neq 0$.

Note that the voltage and current values shown in Fig. 7.13 are typical of a JFET. The transistor works with drain and gate voltages of a few volts and drain current in the mA range. The characteristic curves of the field effect transistor resemble those of the bipolar transistor shown in Fig. 7.7b. The essential difference is that while in the bipolar transistor the control parameter is the base current, in the JFET the control is done by the gate voltage. So, since the gate voltage in the JFET is applied in a reverse biased junction, the input current is very small compared to the base current on the bipolar transistor. In a typical JFET the current at the gate is in the range of 10^{-9} to 10^{-12} A. Since the voltage applied to the gate is of a few volts, the input impedance exceeds 10^8 Ω, so that the power dissipation is quite smaller than in bipolar transistors.

Junction field-effect transistors are used for amplification and switching, in circuits similar to those of Figs. 7.9 and 7.10, in applications that require high input impedance. The circuit symbols of n- and p-channel JFETs are shown in Fig. 7.14.

Fig. 7.14 Circuit symbols of n- and p-channel field-effect transistors (JFET)

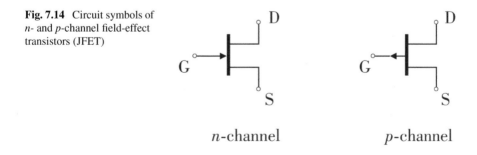

n-channel p-channel

7.5.3   The Metal-Semiconductor Field-Effect Transistor

The operating principle of the metal-semiconductor field-effect transistor, or MESFET, is basically the same as the JFET. The device has three terminals, source, gate, and drain. The majority carriers flow from the source to the drain through a semiconductor channel, type n or type p. The control of this current is done by means of a voltage applied to the gate, which controls the thickness of the channel and therefore its resistance. The difference to the JFET is that in the MESFET the metallic terminal of the gate is in direct contact with the semiconductor of the channel, forming a Schottky barrier junction, instead of a p-n junction as in the JFET. Since in the Schottky potential barrier there is no participation of minority carriers, the response in the change of the channel thickness due to a variation in the gate voltage is faster than in p-n junctions. Therefore, the MESFET is used in high frequency applications. Since GaAs has larger electron mobility than Si, it is the most used semiconductor in the manufacture of high frequency MESFETs. More recently GaN, which is also a direct-gap semiconductor, has become the best material for high-frequency MESFETs for high-power applications, due to its large electron mobility and also high breakdown voltage.

Figure 7.15 shows two common MESFET structures. In both of them the substrate is a high resistivity wafer made with the purest GaAs possible or with a small Cr doping. Since the energy gap in GaAs is large, the Fermi level in the middle of the gap in the intrinsic semiconductor, or with Cr doping, results in resistivities on the order of 10^8 Ωcm. With this value of resistivity, the material is almost insulating, and is called semi-insulating. The channel is formed by a layer of doped GaAs with thickness of the order of 0.1 μm. Since the mobility of electrons in GaAs is 22 times larger than that of holes (see Table 5.2), one uses doping with donor impurities to form n-channel MESFETs for high frequencies applications. Concentrations of impurities of group VI, such as Se, of the order of 10^{17} cm^{-3}, result in conductivities

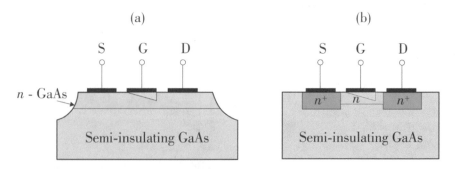

Fig. 7.15 Illustration of GaAs MESFET structures. **a** Simple structure, with metallic terminals of the source, gate and drain, deposited directly on the epitaxial layer that forms the channel. **b** Structure in which the source, channel and drain regions are formed by ionic implantation with type n impurities

adequate to n-channel, in which electrons flow from the source to the drain. The structure of the contacts is made through successive photolithographic processes.

The MESFET structure shown in Fig. 7.15a is quite simple, formed only by a thin layer of n-GaAs grown epitaxially on the substrate and three metallic contacts. The gate contact is made of Al or Ti alloys, W, or Au, which are suitable to form a Schottky barrier in GaAs. The source and drain contacts must be ohmic, so they are made with another metal, generally an alloy of Ge and Au. The fabrication of this structure does not require the use of diffusion processes, it can be done with small and very accurate dimensions. One can then make channels with lengths shorter than 1 μm, which makes possible to minimize the transit time of the electrons and gate capacitance, important requirements for applications in high frequencies. In the structure of Fig. 7.15b the ohmic contacts of the source and drain are made by means of two n^+ regions with impurity concentrations on the order of 10^{18} cm^{-3}. Due to the requirements of precision and good definition of the boundaries between the different regions, the doping that forms the source, channel and drain are made by means of ionic implantation. This type of structure causes an excellent electrical isolation between neighbor transistors manufactured in the same wafer to make an integrated circuit, due to the semi-insulating nature of the substrate. This is not the case with the structure of Fig. 7.15a, because the n-type epitaxial layer on the substrate establishes direct contact between neighboring transistors. To isolate the neighbor transistor elements, each is surrounded by a ditch with depth of about 0.2 μm, which reaches the semi-insulating substrate. The ditch is produced by a corrosion process on a line defined by photolithography.

As mentioned in the beginning of the section, the functioning of the MESFET is basically the same as the JFET. The Schottky junction formed between the gate and the channel is reverse biased, which makes the input impedance of the transistor very high. The voltage applied between the gate and the source determines the thickness of the depletion region, which is given approximately by the same expression (7.42) valid for JFET. Since the potential varies along the channel, the thickness of the depletion region also varies, forming the triangular region indicated by the white area in the structures of Fig. 7.15. The calculation of the current in the channel is done exactly as for the JFET, so that the relationship between the drain current I_D and the gate and drain voltages, V_G and V_D, is given by Eq. (7.46). Thus, the characteristic curves of the MESFET have the same shapes as the curves for the JFET, shown in Fig. 7.13.

GaAs MESFETs can be manufactured on integrated circuits to process analog or digital signals at high frequencies, reaching the microwave range. Currently they have wide application in telephony using frequencies of some GHz. MESFETs are used to make microwave oscillators and amplifiers, which are essential elements for circuits of mobile phones, high-frequency cordless phones, wi-fi routers and other equipment.

The need to increase the frequency and the bandwidth in communication systems has stimulated the development of new MESFET structures. Operation at higher frequencies requires the decrease in transit time for electrons and therefore smaller physical dimensions for the channel. In order to maintain the channel conducting it is necessary to increase its conductivity. This can be done to some extent by increasing the concentration of impurities in the channel. However, excessive concentrations increase the scattering of electrons and compromise the mobility. An ingenious way to increase the concentration of electrons without increasing the concentration of donor impurities is to make the channel with two layers, one of n-(GaAl)As and another of pure GaAs that is grown directly on the substrate. This results in a heterojunction with a band structure similar to that in Fig. 6.11. The composition of the n-(GaAl)As alloy is made in such a way that the Fermi level is above the minimum of the well formed in the discontinuity of the conduction band (in Fig. 6.11 it is a little below). The result is that part of the electrons of the n-(GaAl) layer jump to the GaAs layer, becoming trapped in the interface. Energetically what happens is that electrons occupy the energy states below the Fermi level, staying trapped in the potential well. Since the GaAs layer is intrinsic, the electron scattering is small, resulting in a high mobility channel. The MESFET transistor made with this structure is called HEMT (High Electron Mobility Transistor). HEMTs made of GaAs or GaN are widely employed in mobile telephony operating in the microwave range, with frequencies that can exceed 10 GHz.

7.6 Metal-Oxide-Semiconductor Field-Effect Transistor (MOSFET)

A very important type of field-effect transistor, that has many more technological applications than the JFET, is the metal-insulator-semiconductor field-effect transistor, known as MISFET. It belongs to a broader family of FETs that have insulated gate. In this transistor the control current in the channel is made through the electric field in a capacitor, formed by the metallic contact between the gate and the semiconductor of the channel, separated by an insulating layer. By far the most used MISFET is made with silicon, in which the insulator is silicon dioxide, SiO_2. The device is known as MOSFET, meaning metal-oxide-semiconductor field-effect transistor. The scaling and miniaturization of the MOSFET has been driving the rapid exponential growth of electronic semiconductor technology since the 1960s. The development of high-density MOSFET integrated circuits, such as memory chips and microprocessors, has revolutionized the electronics industry and the world economy, and is central to the digital revolution, silicon age, and information age.

Figure 7.16 shows the planar structure of a n-channel MOSFET. It is formed by two n^+-type regions diffused (or ion implanted) on a p-type semiconductor substrate, one for the source (S) and one for the drain (D). The source and the drain are connected to the circuit by means of aluminum contacts, while the metallic contact of the gate is isolated from the semiconductor by an oxide layer. The conducting channel between the source and the drain is induced in the semiconductor substrate by a voltage applied to the gate, by means of the **inversion** phenomenon that will be explained later.

Fig. 7.16 Illustration of the planar structure of a *n*-channel MOSFET

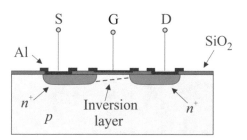

If a voltage is applied between drain and source, in either direction, one of the two *p-n* junctions becomes forward biased, while the other is reverse biased. In this case, if there is no voltage at the gate, there is no channel and therefore the current between source and drain is negligible due to the presence of the reverse polarized junction. When a positive voltage is applied to the gate, a layer of negative charges is induced in the semiconductor, in front of the metallic contact of the gate. This layer of charges provides a conduction channel between source and drain, resulting in a current that varies with the amplitude of the gate voltage. To understand the mechanism of formation of the conducting channel, it is necessary to analyze the behavior of the charges in the capacitor formed by the metal–oxide–semiconductor structure, called **MOS capacitor**, that shall be presented in the next section.

7.6.1 The MOS Capacitor

Figure 7.17 shows the energy diagrams in the three regions of a MOS capacitor with *p*-type semiconductor, for different values of voltage V applied between the metal and the semiconductor. Figure 7.17a shows the energies in the equilibrium situation with $V = 0$, in which the Fermi levels of the metal and the semiconductor are the same. The work functions $e\phi_m$ and $e\phi_s$ of the metal and the semiconductor are indicated in the figure. With the metal and semiconductor in contact with the insulator, $e\phi_m$ and $e\phi_s$ are defined relative to the level of the oxide conduction band, and not to the vacuum level, as was done in the case of Fig. 6.8. For this reason, these quantities are also called modified work functions for the metal-oxide interface. To simplify the analysis of the effect of the applied voltage, in Fig. 7.17 we consider that $\phi_m = \phi_s$. In the general case of $\phi_m \neq \phi_s$, the effect of the different work functions can be easily incorporated in the final result.

Figure 7.17b shows the effect of a voltage $V < 0$ applied between the metal and the semiconductor. In this case, negative charges appear on the metal side and positive charges in the semiconductor, as in a common capacitor. These charges create an electric field E in the direction from the semiconductor to the metal, as indicated in the figure. Since in the *p*-type semiconductor the majority carriers are holes, the appearance of positive charges corresponds to the **accumulation of holes** in the semiconductor-oxide interface. This accumulation of holes is consistent with

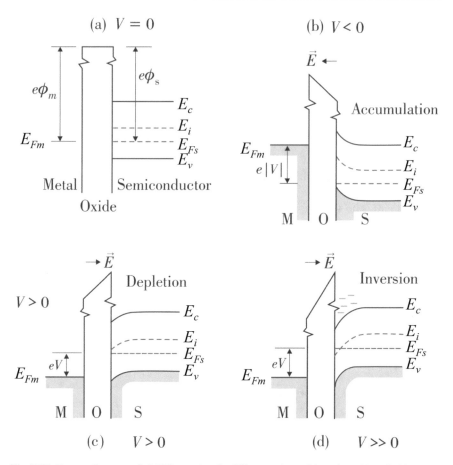

Fig. 7.17 Energy diagrams of a MOS capacitor for different values of the voltage V applied between the metal and the semiconductor (type p)

the behavior of the energies, as we shall see below. When the voltage is applied between the metal and the semiconductor, the electron energies in the metal change by $-eV$ relative to their equilibrium values. Therefore, with $V < 0$ the energies in the metal increase by $e|V|$. As a result, the conduction band in the oxide is bent near the interface and the Fermi level E_{Fm} in the metal rises above the Fermi level E_{Fs} in the semiconductor, so that their difference becomes $E_{Fm} - E_{Fs} = e|V|$. This produces in the vicinity of the interface an upward bending of the energies E_v of the valence band, E_c of the conduction band, and E_i of the intrinsic Fermi level, as shown in Fig. 7.17b. On the other hand, since the oxide layer is insulating, the application of the external voltage does not result in a current in the semiconductor. Thus, the Fermi level E_{Fs} does not vary across the semiconductor, as it does in p-n junctions and metal–semiconductor junctions. In this way, the energy E_i moves away from the level E_{Fs} at the interface. Since the concentration of holes is given by Eq. (5.32),

$$p = n_i \, e^{(E_i - E_{Fs})/k_B T}, \qquad (7.51)$$

one can see that that the hole concentration p increases exponentially with the difference $E_i - E_{Fs}$. Thus, the variation of the energies shown in Fig. 7.17b is consistent with the accumulation of holes at the interface of the semiconductor with the oxide.

The behavior of the energies in the case of a positive electric potential in the metal relative to the semiconductor is illustrated in Fig. 7.17c, d. With $-eV < 0$, the electron energies in the metal decrease with respect to the values in equilibrium, so that the curvatures of E_c, E_v, and E_i near the interface are opposite to those of the diagram in Fig. 7.17b. In this case E_i approaches E_{Fs} at the interface, so that, by Eq. (7.51), the concentration of holes decreases in the oxide. If V is less than a certain threshold value V_T, the difference $E_i - E_{Fs}$ decreases relative to the value in equilibrium but it remains positive in all points, as in the diagram in Fig. 7.17c. In this case the concentration p at the interface is smaller than the value in equilibrium, which leaves a fraction of acceptor impurities not compensated. Thus, the semiconductor is negatively charged when the metal is positively charged, as expected for $V > 0$. The absence of holes in the vicinity of the interface is a phenomenon analogous to the one in the space charge region, or depletion region, of a p–n junction.

If the voltage V exceeds a threshold value V_T, the energy E_i at the interface falls below the level E_{Fs}, as shown in Fig. 7.17d. In this case, as one can see in Eq. (7.51), $p < n_i$, and since $p\,n = n_i^2$, $n > n_i$, so that electrons become the majority carriers. This is a very interesting case in which a p-type semiconductor behaves as a n-type due to an applied voltage and not because of doping. This phenomenon, called **inversion**, is the key to the appearance of the n-channel in the p-type semiconductor of the MOS transistor.

To calculate the voltage applied to the MOSFET above which an inversion layer is induced in the semiconductor, it is necessary first to understand how the electric potential drops in the oxide layer and in the semiconductor. For this we shall initially consider an ideal MOS capacitor with no surface charges and with the same work functions for the metal and the semiconductor, $e\phi_m = e\phi_s$. Later we shall generalize the result for real surfaces.

If a voltage V is applied between the metal and the semiconductor, part of the potential drop occurs in the insulator (V_i) and part in the semiconductor (V_s), so that

$$V = V_i + V_s. \qquad (7.52)$$

This voltage produces charges Q_m on the metal surface and Q_s in the semiconductor, where $Q_m = -Q_s = Q$, as in a capacitor. If $V > 0$, then $Q > 0$. The potential drop in the insulator is related to the charge through the capacitance, as in a capacitor with two metal plates

$$V_i = \frac{Q}{C_i}, \qquad (7.53)$$

Fig. 7.18 Charge distribution in an ideal MOS capacitor with a p-type semiconductor (n-channel), in the depletion approximation. The dashed line indicates the charge created by the inversion when $V > V_T$

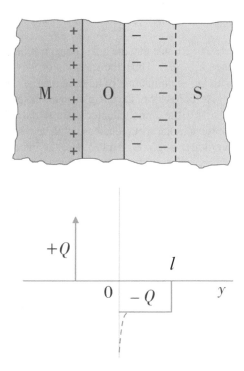

where $C_i = \varepsilon_i A/d$, where ε_i is the permittivity of the insulator, d is its thickness and A the area. To relate the potential drop in the semiconductor with the charge, it is necessary to solve the problem of distributed charge. Since the full problem is very difficult, we shall use an approximation for the charge distribution, as was done for the p–n junction in Sect. 6.1.3. We assume that the charges in the semiconductor are uniformly distributed in a layer of thickness l, as shown in Fig. 7.18. In the depletion approximation we consider that all acceptor impurities in the layer are ionized, so that the total charge in the semiconductor is $Q_s = -Q = -eN_a l\, A$. In this situation, the equation of Gauss's law can be easily integrated to obtain the electric field, and from it the variation of the potential. The relationship between the depletion layer thickness and the potential drop V_s in the semiconductor is the same as in Eq. (6.18), with V_0 replaced by V_s and without the term in Nd. Thus

$$V_s = \frac{e\,N_a}{2\varepsilon_s}\, l^2, \qquad (7.54)$$

where ε_s is the permittivity of the semiconductor. Note that this result was obtained under the assumption that the applied voltage V is sufficient to produce total depletion in the layer of thickness l, but without inversion. With Eqs. (7.52)–(7.54) we can calculate the thickness l as a function of V, as long as V is smaller than the threshold value V_T to produce inversion. Substituting in Eq. (7.52) the expressions for V_i and

V_s in Eqs. (7.53) and (7.54), with $Q = eN_a l A$, we obtain an equation for the voltage in terms of the thickness l

$$V = \frac{e N_a d}{\varepsilon_i} l + \frac{e N_a}{2\varepsilon_s} l^2. \tag{7.55}$$

This result shows that the thickness of the depletion layer increases with increasing voltage V in the capacitor. Actually, this occurs only while V is less than V_T. When V reaches V_T, the inversion produces a thin charge layer at the interface with the oxide, shown by the dashed line in Fig. 7.18. Any additional increase in V above this value results in a growth of the charge in the inversion layer, but not in the thickness of the depletion layer. From Eq. (7.55) we obtain the total capacitance of the MOS capacitor. Using the expression for the charge $Q = eN_a l A$ in Eq. (7.55) and the definition $C = dQ/dV$ we have

$$C = \frac{A}{d/\varepsilon_i + l/\varepsilon_s}. \tag{7.56}$$

This expression can also be obtained by the series association of the capacitors formed by the insulator (C_i) and the semiconductor. Since the thickness l increases with V, the capacitance C decreases with the increase of V in the region $0 \leq V \leq V_T$. For $V \geq V_T$, the value of C stabilizes at C_{min}, as shown in Fig. 7.19. With negative voltages there is an accumulation of holes at the semiconductor surface, so that $l = 0$ and C is due to the capacitor formed only by the dielectric oxide, so that $C = C_i$. Note that when the capacitance is measured with very low frequency, typically less than 100 Hz, the capacitance tends to approach the value C_i, as shown by the dashed lines in Fig. 7.19. The mechanism responsible for this effect is the generation of carriers in the space charge region. When the voltage variation is very slow, the creation of electron-hole pairs in this region masks the capacitance variation. The holes tend to neutralize the acceptor impurities, eliminating the depletion region, while the electrons go to the semiconductor-oxide interface. As a result, $l \rightarrow 0$ and the capacitance tends towards the value C_i.

Fig. 7.19 Variation of the capacitance C with the voltage in the ideal n-channel MOS capacitor. The dashed curves for $V > V_T$ are the results obtained when the measurement of C is made at very low frequencies, typically less than 100 Hz

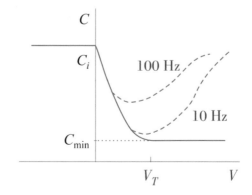

7.6.2 The Threshold Inversion Voltage

To understand the inversion mechanism and the formation of the n-channel in the p-type semiconductor, let us analyze in detail the energy diagram of the semiconductor when the voltage applied to the MOS capacitor is positive. As shown in Fig. 7.20, the conduction and valence bands, as well as the intrinsic Fermi level E_i, bend downwards in the vicinity of the interface. Since the electron energy is related to the electric potential ϕ by $E = -e\phi$, the deviation of the conduction band edge from its equilibrium value E_c is $e\phi$. Since the bending of E_i follows that of E_c, the deviation of E_i at each point y is also $e\phi$, which can be seen in Fig. 7.20. We see then that the potential drop V_s in the semiconductor, due to the applied voltage V, corresponds to the deviation of E_i at the interface with the semiconductor, that is, at $y = 0$. We also see in Fig. 7.20 that if $V_s > \phi_F$, there is a small range of y in which $E_i < E_{Fs}$, and hence the concentration of electrons is larger than of holes. However, it is not enough to have $E_i < E_{Fs}$ for the conduction channel to be significant. The criterion used to characterize a **strong inversion** is that the concentration n of electrons on the surface has to be at least as large as the concentration of holes in the substrate, $p \approx N_a$. From Eq. (7.51) we see that this condition is

$$N_a = n_i \, e^{e\,\phi_F/k_B T}, \tag{7.57}$$

where $e\phi_F$ is the difference between the Fermi levels E_i and E_{Fs} away from the interface. Since $n = n_i^2/p$, we see in Fig. 7.20 that to have $n = N_a$ in $y = 0$ it is necessary that $V_s = 2\,\phi_F$. Using Eq. (7.57) we can write that the condition to have an inversion layer in the semiconductor is

$$V_s \geq V_s^I = 2\phi_F = 2\frac{k_B T}{e} \ln \frac{N_a}{n_i}. \tag{7.58}$$

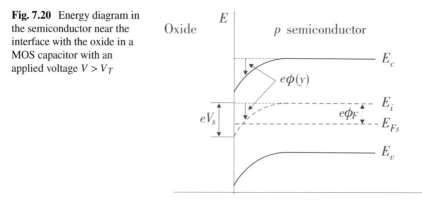

Fig. 7.20 Energy diagram in the semiconductor near the interface with the oxide in a MOS capacitor with an applied voltage $V > V_T$

Substituting this result in Eq. (7.54) we obtain the maximum depletion layer thickness that is achieved under the inversion condition

$$l_{max} = \left(\frac{4\varepsilon_s \phi_F}{e\, N_a}\right)^{1/2} = \left[\frac{4\varepsilon_s k_B T \ln(N_a/n_i)}{e^2\, N_a}\right]^{1/2}. \tag{7.59}$$

This is the situation in which the charge in the depletion region is maximum, and has absolute value given by

$$Q_d = e\, N_a\, l_{max} A = 2(\varepsilon_s e\, N_a \phi_F)^{1/2}\, A. \tag{7.60}$$

Substituting Eqs. (7.53) and (7.58) in (7.52), we obtain the value of the threshold voltage in the MOS capacitor for the creation of the inversion layer

$$V_T = \frac{Q_d}{C_i} + 2\,\phi_F, \tag{7.61}$$

where Q_d is given by Eq. (7.60). This result is valid for an ideal MOS capacitor. In a real capacitor there are two effects that must be considered in the calculation of V_T: (i) the work functions ϕ_m and ϕ_s in general are different; (ii) there are charges inside the oxide and at the semiconductor-oxide interface.

The modified work functions for the metal-SiO$_2$ interface of some metals used in metallic contacts are shown in Table 7.1. In the case of semiconductors, the work function also depends on the concentration of impurities, because it varies with the value of the Fermi level E_{Fs}. Figure 7.21 shows the difference of the modified work functions, $\phi_{ms} = \phi_m - \phi_s$, for the interface between Al and Si, both p-type and n-type, as a function of the concentration of impurities. We see that in this case, ϕ_{ms} is negative regardless of the type of impurity. In this case, the energy diagram in equilibrium ($V = 0$) is similar to that of Fig. 7.17, valid for $\phi_m = \phi_s$ and $V > 0$. This means that even in equilibrium, the metal is positively charged while the semiconductor has negative charges. Thus, to make the bands straight as in Fig. 7.17a, it would be necessary to apply a negative voltage to compensate for the difference in the work functions, with a value exactly equal to ϕ_{ms}.

Another important effect in the MOS capacitors is the presence of charges in the insulator and in the semiconductor-oxide interface. The charges inside the insulator result from the contamination in the manufacturing process, as is often the case with Na^+ ions. These positive charges create an electric field that alters the variation of the potential in the capacitor. The charges at the Si-SiO$_2$ interface result from the existence of surface states created by the symmetry breaking of the crystal lattice

Table 7.1 Modified work functions of some metals in contact with the insulator SiO$_2$

Metal	Al	Ag	Au	Cu
$e\phi_m$ (eV)	4.1	5.1	5.0	4.7

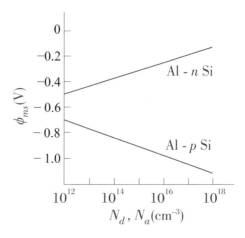

Fig. 7.21 Variation in the difference of the work functions $\phi_{ms} = \phi_m - \phi_s$ at the Al-Si interface as a function of the concentration of impurities

at the surface. In the Si oxidation process for the manufacture of the SiO$_2$ layer, Si atoms are removed from the surface and react with oxygen. When the process is interrupted, some Si ions remain close to the interface, forming a superficial layer of charges. The ensemble of charges in the oxide and at the interface can be represented by an effective charge Q_{ox}. This charge produces an additional potential difference in the capacitor $V_{ox} = Q_{ox}/C_i$, where C_i is the capacitance of the insulator.

The threshold voltage previously calculated is valid for the situation in which, with $V = 0$, the semiconductor bands have no bending. Since the different work functions and the presence of the charge Q_{ox} result in an effective positive voltage $V_{ox} - \phi_{ms}$, the external voltage that must be applied to the capacitor to produce inversion is smaller than V_T obtained for the ideal case, Eq. (7.61). Thus, the value of the threshold voltage in the general case is

$$V_T = \frac{Q_d}{C_i} + 2\,\phi_F + \phi_{ms} - \frac{Q_{ox}}{C_i}, \tag{7.62}$$

This result shows that in order to obtain a small threshold voltage it is necessary to make the capacitance C_i as large as possible. This requires a very small thickness of the insulating oxide, generally on the order of 0.1 μm or less.

Example 7.6 Calculate the threshold inversion voltage V_T for a MOSFET of Al-SiO$_2$-p Si at $T = 295$ K with the following parameters: $d = 0.1$ μm, $N_a = 10^{15}$ cm^{-3}, charge in the oxide per unit of area $Q_{ox}/A = 8 \times 10^{-4}$ C/m^2, oxide dieletric constant 3.9.

Using the value for n_i of Si in Table 5.2 we obtain with Eq. (7.58)

$$2\phi_F = 2\frac{k_B T}{e} \ln \frac{N_a}{n_i} = 2 \times 0.025 \times \ln \frac{10^{15}}{1.5 \times 10^{10}} = 0.56 \,\text{V}.$$

Using Eq. (7.60) we calculate the charge per area unit

$$\frac{Q_d}{A} = 2 \times (11.8 \times 8.85 \times 10^{-12} \times 1.6 \times 10^{-19} \times 10^{15} \times 10^6 \times 0.28)^{1/2}$$

$$\frac{Q_d}{A} = 1.37 \times 10^{-4} \, C/m^2.$$

The capacitance per unit area is determined by the thickness of the oxide layer and its dielectric constant. For SiO_2 $\varepsilon_i = 3.9 \, \varepsilon_0$, thus

$$\frac{C_i}{A} = \frac{\varepsilon_i}{d} = \frac{3.9 \times 8.85 \times 10^{-12}}{10^{-7}} \approx 3.45 \times 10^{-4} \, F/m^2.$$

Using the value $\phi_{ms} = -0.9$ V from Fig. 7.21 and substituting the MOSFET data and the calculated parameters in Eq. (7.62) we have

$$V_T = \frac{1.37 \times 10^{-4}}{3.45 \times 10^{-4}} + 0.56 - 0.9 - \frac{8 \times 10^{-4}}{3.45 \times 10^{-4}}$$
$$V_T = 0.4 + 0.56 - 0.9 - 2.3 = -2.24 \, V.$$

Note that the inversion occurs at low voltage values, that can be supplied by small batteries. This fact is important because it allows the operation of logic circuits fed by small batteries in portable equipment.

7.6.3 The I-V Characteristics of the MOSFET

We are now in a position to understand the mechanism of operation of the metal-oxide-semiconductor field-effect transistor with the structure shown in Fig. 7.16, as well as to calculate the drain current I_D as a function of the drain and gate voltages V_D and V_G. If a positive voltage V_D is applied between drain and source, the p-n junction between substrate and drain becomes reversed polarized. Thus, an electric current flows from the drain to the source (electrons go from the source to the drain) as long as there is an inversion layer over the entire length of the semiconductor-oxide interface. This layer can be induced by a voltage V_G between gate and source larger than a threshold value V_{GT}. This value is different from V_T of Eq. (7.62), because the drain voltage raises the potential of the semiconductor relative to the metallic contact of the gate. Due to the presence of the current I_D, the potential of the semiconductor increases gradually from source to drain. This results in a variation of the threshold voltage across the capacitor and, consequently, in a gradual decrease in the thickness of the inversion layer from source to drain, as shown in Fig. 7.16. Thus, the minimum

Fig. 7.22 Model for the variation of the inversion layer between source and drain in a n-channel MOSFET

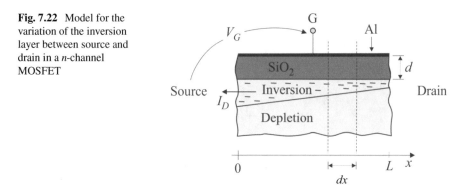

gate voltage V_{GT} that induces a conduction channel across the entire length of the semiconductor is determined by the value of the threshold voltage at the end of the drain

$$V_{GT} = V_T + V_D. \tag{7.63}$$

To calculate the current I_D produced by the voltage V_D it is necessary, initially, to determine the charge in the inversion layer. For this we shall consider the model for the variation of the inversion layer between source and drain in a n-channel MOSFET shown in Fig. 7.22. The MOS capacitor is divided into elementary capacitors of width dx and area $dA = Ddx$, where D is the depth in the direction perpendicular to the plane of the paper. According to Eq. (7.53), the elementary charge in each capacitor is $dQ = V_i(x)\,dC_i$, where $dC_i = C_i dx/L$ and $V_i(x)$ is the voltage drop in the insulator at the point with abscissa x. $V_i(x)$ is determined by the gate voltage V_G, the effective voltage $V_{ox} - \phi_{ms}$, the potential drop V_s in the semiconductor and the potential difference $\phi(x)$ between point x and the source ($x = 0$). Under the condition of inversion, at point x the voltage is $V_s = 2\phi_F$, so that the elementary charge in the capacitor at dx is

$$dQ = \frac{C_i}{L}[V_G + (V_{ox} - \phi_{ms} - 2\phi_F) - \phi(x)]\,dx.$$

Note that in the inversion condition, this charge is the same as the one in the depletion region, whose value in the range dx is $dQ_d = Q_d dx/L$. Any increase in in the voltage results in the appearance of a negative charge in the inversion layer, whose modulus is $dQ_n = dQ - dQ_d$, since the charge in the depletion region does not increase beyond the value given by Eq. (7.60). So we have

$$dQ_n = \frac{C_i}{L}[V_G + (V_{ox} - \phi_{ms} - 2\phi_F) - \phi(x)]\,dx - \frac{Q_d}{L}\,dx.$$

Using Eq. (7.62) in this expression we can write

$$dQ_n = \frac{C_i}{L}[V_G - V_T - \phi(x)]\,dx. \tag{7.64}$$

Under the action of a positive voltage V_D, this (negative) charge moves in the source-drain direction, producing a current I_D in the $-x$ direction. If h is the height of the channel at position x, the volume charge density is $\rho = -dQ_n/Dhdx$. The current density it produces is $J = \rho\mu_n E$, where μ_n is mobility of the charges in the channel and $E = -d\phi/dx$ is the electric field. Thus, the drain current is

$$I_D = J\,D\,h = \mu_n \frac{dQ_n}{dx}\frac{d\phi}{dx}. \tag{7.65}$$

Substituting (7.64) in Eq. (7.65) and passing dx to the left side, we can integrate the two sides separately to have

$$\int_0^L I_D\,dx = \frac{\mu_n C_i}{L}\int_0^{V_D}(V_G - V_T - \phi)\,d\phi.$$

Since I_D does not vary with x, the integral on the left is simply $I_D L$. Performing the integral on the right-hand side and using the expression for V_{GT} in Eq. (7.63) we finally obtain

$$I_D = \frac{\mu_n C_i}{L^2}\left[(V_G - V_T)V_D - \frac{1}{2}V_D^2\right]. \tag{7.66}$$

Note that μ_n is the mobility of electrons near the interface, which in general is smaller than the value inside the p-type substrate. In Eq. (7.66) it is common to use the capacitance per area unit, $c_i = C_i/DL$, instead of C_i. This equation describes the drain current behavior quite well, mainly for low values of V_D. In fact, this result is only approximate because we neglected the variation of Q_d with x. Equation (7.66) shows that for small V_D values, the current I_D grows linearly with V_D, provided that $V_G > V_T$. For larger values of V_D, the term in V_D squared decreases the rate of growth of I_D. Note that the derivative

$$\frac{dI_D}{dV_D} = \frac{\mu_n C_i}{L^2}(V_G - V_T - V_D) \tag{7.67}$$

is zero for $V_D = V_{Ds} \equiv V_G - V_T$. At this voltage value, which is exactly the same as in (7.63), the current is maximum. For drain voltages larger than this value, the conduction channel closes so that the current saturates, in an effect similar to that in JFET. Figure 7.23 shows curves for I_D as a function of V_D for some values of the gate voltage V_G, calculated with Eq. (7.66) for a n-channel MOSFET with $V_T = -2$ V, $C_i/A = 3.45 \times 10^{-4}$ F/m^2, $L = 10$ μm, $D = 300$ μm and $\mu_n = 675$ cm^2/V.s (half the value inside the substrate, given in Table 5.2). These values give $\mu_n C_i/L^2$

Fig. 7.23 *I-V* characteristic
curves of a *n*-channel
MOSFET

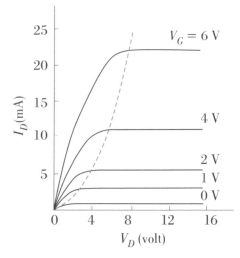

$= \mu_n D C_i / A L = 0.7$ mA/V^2. In Fig. 7.23 the dashed line indicates the points V_D-I_D where the current saturates. Note that the value of the saturation current obtained from Eq. (7.66) with $V_G - V_T = V_D = V_{Ds}$ is

$$I_{Dsat} = \frac{\mu_n C_i}{2L^2} V_{Ds}^2. \tag{7.68}$$

This result shows that the saturation curve is a parabola, represented by the dashed curve in Fig. 7.23.

7.6.4 Applications of MOSFETs

As mentioned in the beginning of this section, high-density integrated circuits with MOSFETs are central to digital electronics. This is so because MOSFETs can be used as building blocks for a large variety of storage and logic devices, that have been manufactured with continuously smaller physical dimensions. The circuit symbols of the *n*-channel and *p*-channel MOSFETs are shown in Fig. 7.24. Note that besides the source, gate, and drain, they have a fourth terminal, corresponding to the contact with the substrate, or with the body of the device, denoted by the symbol B. Since the semiconductor of the substrate forms diode junctions with the source and the drain, it must be maintained in a potential that makes the junctions not conducting. In general, this terminal is connected to the source in the *n*-channel MOSFET and to the drain in the *p*-channel MOSFET.

As we studied in the previous sections, in the operation of the MOSFET the channel is induced in the substrate by the inversion mechanism, produced by a gate voltage. If $V_G < V_{GT}$ the channel is closed and there is no drain current. It is also

Fig. 7.24 Circuit symbols of
n-channel and p-channel
MOSFETs

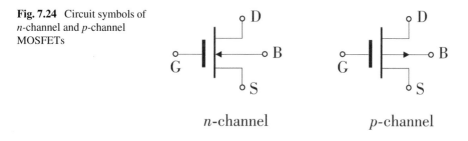

n-channel p-channel

possible to make a MOSFET by doping a n-region between source and drain, so that even without a gate voltage the current I_D can be nonzero. In this type of MOSFET, a negative voltage in the gate repels the electrons of the n-channel and reduces the current, as in a n-JFET. The first type, in which the channel is induced and increases with the voltage V_G, is called of **channel induction** or **channel enhancement**. The second type, in which the channel is depressed with the voltage, is called **depletion**. It is common to use the symbols in Fig. 7.24 to represent the two types of MOSFETs.

A most important feature of MOSFETs is the electric insulation of the gate, that results in an input impedance of the order of 10^{14} Ω, regardless of the direction of the gate voltage. A great advantage of MOSFETs relative to JFETs is in its manufacturing process, which requires a reduced number of steps. This facilitates the manufacture of a large number of transistors with dimensions less than 1 μm, interconnected by means of aluminum contacts on the top surface, constituting **high-density integrated circuits**. The processes of fabrication of these circuits are referred to as **very large scale integration** (VLSI)).

In digital integrated circuits using MOSFETs, it is possible to reduce drastically the power consumption with the use of pairs of transistors interconnected, one n-channel and the other p-channel. This technology, called complementary pair, or **CMOS**, makes possible to manufacture watches, calculators, tablets, mobile phones, computers and other equipment with extremely small power dissipation. As an example of the use of a complementary pair, we show in Fig. 7.25a a CMOS inverter circuit with induction MOSFETs. The two transistors are connected in series and subjected to a voltage $+ V_{DD}$. Figure 7.25b shows the I_D-V_D curves of transistors T1 (n-channel) and T2 (p-channel), for two values of the gate voltages, 0 and $+ V_{DD}$ for T1, and 0 and $- V_{DD}$ for T2. Note that the curves of T2 are placed on the same graph as T1 but with the voltage axis inverted, so that the sum of the two drain voltages is V_{DD}. Thus, the operating point of the circuit is given by the intersection of the curves of T1 and T2, because $V_{D1} + V_{D2} = + V_{DD}$ and $I_{D1} = I_{D2}$.

The circuit of Fig. 7.25a is an inverter logic circuit of type **NO**. Its purpose is to give an output signal $V_o = 0$ (bit 0) when the input signal is $V_i = + V_{DD}$ (bit 1), and output $+ V_{DD}$ (bit 1) when the input is zero (bit 0). This operation can be verified in the plot in Fig. 7.25b. If the input signal is zero, the voltages at the gates of T1 and T2 (relative to the respective sources) are, respectively, $V_{G1} = 0$ and $V_{G2} = V_{DD}$. In this situation, T1 is in the off state and T2 in the on state. The intersection of the two curves is point 1 of the figure, so that the output voltage is close to $+ V_{DD}$. On the

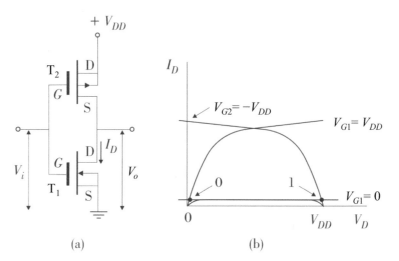

Fig. 7.25 a NO inverter circuit with complementary pair of MOSFETs. **b** Characteristic curves used for the determination of the operating points

other hand, if the input is $+ V_{DD}$, T1 is on and T2 is off, so that the output signal is given by point 0, $V_D \approx 0$. Note that in both situations the current in the circuit is very small. This fact makes possible the fabrication of CMOS circuits with power dissipation below 10 nW.

The peculiar properties of MOS transistors and capacitors are also used for the construction of various types of devices that transfer or store digital information. Among the most important ones are the **semiconductor memories** and **charge-coupled-devices**, or CCD. The CCD is made by an array of MOS capacitors, one next to other, in the same semiconductor substrate. When a voltage pulse is applied to a capacitor, with sufficient amplitude to produce inversion, it creates a charge package that is stored in the capacitor for a certain time. The presence of a charge package in a capacitor represents the binary digit 1, while the absence represents 0. The MOS capacitor is the basic element of semiconductor memories. A set of capacitors and MOS transistors in an integrated circuit, forms a memory that stores the information expressed in binary codes. Currently there is a wide variety of semiconductor memory devices, some of which will be presented in Sect. 7.8.2. When the capacitors are properly interconnected, the application of a voltage pulse to a capacitor produces a potential well in a neighbor capacitor, to which the charge package is transferred through the semiconductor. In this way, it is possible to shift the digit 1 along the series of capacitors, forming a CCD device, used to manufacture shift registers for microprocessors, as well as image sensors, that are presented in Sect. 8.4.4.

7.7 Power Control Devices: Thyristors

Thyristors are devices formed by several *p-n* junctions, that have large application as switches to control high currents. The control is done electronically by means of a relatively small applied to one terminal of the device. The two main members of the thyristor family, that shall be described qualitatively here, are the **semiconductor-controlled rectifier** (SCR), and the bidirectional **triode for alternating current** switching (TRIAC). Most power devices for low frequencies are made with single crystal silicon, because its high thermal conductivity facilitates the flow of the heat generated by the electric current and it can operate at up to 200 °C. For this reason, SCR is often used to denote silicon-controlled rectifier. Until recently, power devices for microwave frequencies were usually made of GaAs, that despite the lower thermal conductivity, has much higher mobility than Si. However, more recently GaN began to be used in most high-power devices because its thermal conductivity is the same as Si but the electron mobility is nearly three times larger.

7.7.1 The Silicon-Controlled Rectifier- SCR

The silicon-controlled rectifier is made by doping a silicon substrate so as to have four layers of impurities, forming a *p-n-p-n* structure, shown in Fig. 7.26a. The device has two terminals at the ends, through which the main current to be controlled circulates, anode (A) in region p_1 and cathode (C) in region n_2. A third terminal in region p_2, called gate (G), serves for the input of the control current. Figure 7.26b shows the model used to represent the four regions of the device, which forms three *p-n* junctions, denoted by J_1, J_2, and J_3.

To understand the operating mechanism of the SCR, let us first analyze what happens in the *p-n-p-n* device without the gate, also known as a Shockley diode.

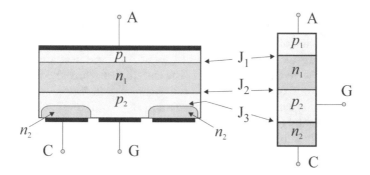

Fig. 7.26 Silicon controlled rectifier. **a** Cross section of the structure in the form of a disc. **b** Model used to describe the *p-n-p-n* device

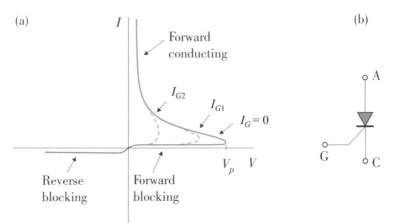

Fig. 7.27 **a** *I-V* curves of the SCR: full line applies to zero gate current; dashed lines apply to two values of the gate current, $I_{G2} > I_{G1}$. **b** Circuit symbol of the SCR

If a positive external voltage is applied between anode and cathode, junctions J_1 and J_3 become directly polarized, while J_2 is reversed polarized. As a result, the resistances of J_1 and J_3 are small, while that of J_2 is very large. Thus, the full external voltage appears in J_2, and if it is less than the breakdown value, the current has the reverse saturation value, which is very small. This is the blocking regime in the direct polarization, indicated by the full line of the *I-V* characteristic shown in Fig. 7.27a. Note that if the external voltage is negative, J_1 and J_3 are reversed polarized, and in this case, they are the ones that limit the current to the saturation value, resulting in the reverse blocking regime, indicated in the *I-V* curve.

The most interesting phenomenon of the *p-n-p-n* device occurs when the positive voltage applied to the gate increases and reaches the breakdown value of junction J_2. In this situation, an avalanche occurs in J_2 and the current tends to increase rapidly, without resistance at junctions J_1 and J_3 that are forward biased. In this case, in region p_1 this current is formed by holes moving from the anode to the cathode, while in n_2 it is formed by electrons going from cathode to anode. Once the conducting process has started, a fraction of the holes of p_1 is injected into region p_2 through n_1, as in a transistor formed by the regions p_1-n_1-p_2. Likewise, electrons in n_2 are injected into n_1, as if n_2-p_2-n_1 were another transistor. The current then starts to be produced by the carrier injection process, ceasing the avalanche process. This results in a rapid decrease of the voltage in J_2 and even a change of its sign, so that the junction becomes forward biased. In the forward conduction regime, shown in Fig. 7.27a, the current can reach high values, being limited only by the resistance the external circuit, or by the rupture of the device. In this regime the three junctions are forward biased. As the potential drop in junction J_2 has the opposite direction to the drops in J_1 and J_3, the total voltage drop in the device corresponds to only one junction, which is about 0.7 V in the case of silicon.

Note that the *I-V* curve in Fig. 7.27a is similar to the one of a thyratron tube, which is a gas-filled tube that passes a current by an arc discharge when triggered by a voltage pulse. The name thrystor was coined because the device is a solid-state analogue of the thyratron. To start the conduction process in the *p-n-p-n* device, it is necessary to increase the external voltage to reach the peak value V_p. However, it is possible to switch from the blocking state to the conduction state by applying a relatively small current to the gate. The dashed lines in Fig. 7.27a show the *I-V* curves for two values of the gate current. The effect of the current entering the gate is to inject holes in region p_2, which being the base of the transistor n_2-p_2-n_1, causes the beginning of the conduction process. This produces injection of electrons from n_2 to n_1, that turns the transistor p_1-n_1-p_2 also conducting. This process **triggers** the SCR, making it switch from the blocking to the conducting regime, without the need to increase the external voltage up to the avalanche breakdown value. The effect of the current in the gate is precisely to reduce the value of the peak voltage, as shown in Fig. 7.27a. Once triggered, the SCR maintains the conduction process, even if the gate current is interrupted. Thus, the SCR can be triggered by a current pulse at the gate. On the other hand, a negative current pulse at the gate can cutoff the device, making it switch from the conducting to the blocking state.

The circuit symbol of the SCR is shown in Fig. 7.27b. The SCR is used in a wide variety of applications in industrial electronics and power control, because it makes possible to control the power delivered to a load by means of low power switching. Figure 7.28 illustrates a simple and very popular application of a SCR used to control the voltage applied to a resistive load from a constant AC line source. The circuit, shown in Fig. 7.28a, consists of a SCR in series with the load (resistance R_L), in which the gate is fed by current pulses with the same period as the line voltage but with a variable delay. The pulses are produced by a simple oscillator synchronous with the line, and with a delay that can be varied by means of a potentiometer. In the negative half-cycles of the line voltage the SCR blocks the current through the

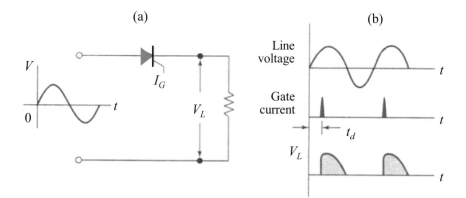

Fig. 7.28 a Schematic circuit for the use of a SCR to control the power delivered to a resistive load. **b** Wave forms of the line voltage, pulsed gate current, and voltage at the load

load, and only when a current pulse is applied to the gate, the SCR switches to the conducting state. This is illustrated in Fig. 7.28b, that shows the wave forms of the line voltage, pulsed gate current, and voltage at the load. By changing the delay τ_d of the gate pulse one can vary the duration of the positive half-cycles in which the SCR conducts, and hence the power delivered to the load. The circuit of Fig. 7.28 is used in lightning dimmers.

7.7.2   The TRIAC

The triode for alternating current switching, or TRIAC, as the name says, is a device to switch AC currents in either direction. It is made of a semiconductor with six doped regions, constituting two SCRs connected in parallel and in opposite directions. The structure of the TRIAC and its circuit symbol are shown in Fig. 7.29. In the structure of Fig. 7.29a one can clearly identify two devices in parallel, one formed by the regions p_1-n_1-p_2-n_2 and the other formed by the regions n_4-p_1-n_1-p_2. Without the gate, they are equivalent two Schockey diodes in parallel and in opposite directions, whose *I-V* characteristic is given by the solid line in Fig. 7.30. This device is called a bidirectional diode, or diode AC switch (DIAC). It can conduct current in any of the two directions, as long as the external voltage reaches the peak value $\pm V_p$.

The gate terminal is used to trigger the TRIAC in any of the two directions by means of current pulses. If the voltage between anode and cathode is positive, a current pulse at the gate triggers the SCR p_1-n_1-p_2-n_2, producing a current from anode to cathode. On the other hand, if the voltage is negative, the current pulse at the gate makes the SCR p_2-n_1-p_1-n_4 conduct from cathode to anode. Note that in the TRIAC the anode and cathode have similar roles and the distinct names is not even justified. They are used only to facilitate the description.

TRIACs are widely used in electronic circuits for AC power control. They can be built in such a way that both in the positive or in the negative cycles, the trigger is

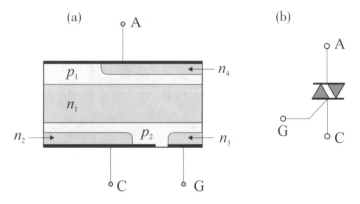

Fig. 7.29   a Cross section of the TRIAC structure. **b** TRIAC circuit symbol

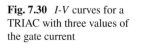

Fig. 7.30 *I-V* curves for a TRIAC with three values of the gate current

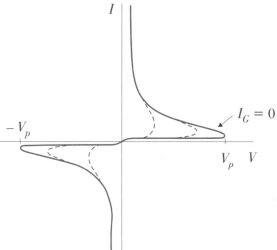

made by positive or negative pulses, or by both. The current in the device is switched off when the voltage between anode and cathode is cutoff. It can then let an alternating current flow if it is triggered twice in each cycle. While the SCR can only be used to control the power delivered to loads that operate with DC voltages, the triac is used in circuits to control AC power.

7.8 Integrated Circuits

Most devices presented in Chaps. 6 and 7 can be encapsulated separately, featuring two or more external terminals, so that they can be connected to other devices, forming an electronic circuit. In this case, they are called **discrete devices**, or components. However, by far most semiconductor devices used in recent times are manufactured in **integrated circuits**. An integrated circuit (IC) is formed by a large number of transistors, diodes, resistors, capacitors and inductors, manufactured in the same semiconductor chip and interconnected with each other through metallic films, composing a complete circuit with microscopic dimensions. The first integrated circuit was produced in 1958, by Jack Kilby, and contained only a few transistors. This achievement was not a scientific breakthrough, it was an innovative way to connect devices whose operating principles were already known. However, the concept of the integrated circuit revolutionized electronics and set the stage for a social revolution and a great advance in scientific equipment and science. For this reason, Kilby was awarded the Physics Nobel Prize in the year 2000.

The first integrated circuits had no more than a few dozen transistors. However, in a short time the integration technology was dominated by several manufacturers and the competition to gain markets led to a race to increase the number of devices in the same circuit. The result was a rapid increase in the integration capacity, with the improvement in performance and decrease in manufacturing costs. At the end of the 1960s, devices in integrated circuits had dimensions of some micrometers, leading to the name **microelectronics** technology. At that time, Gordon Moore, one of founders of Intel, since then one of the largest manufacturers of microprocessors, noted that the number of transistors per circuit doubled every eighteen months, and that the production cost fell by half in the same period. This observation came to be known as **Moore's law**, which has characterized the semiconductor industry for over four decades. Moore's law remains valid until today, and the number of transistors in integrated circuits for microprocessors and other applications has surpassed ten billion. However, the lateral dimensions of the devices have decreased so much that they reach the scale of tens of nanometers, and the thickness of some layers corresponds to a few atoms. This has led to predictions that by 2030 there will be a considerable reduction in the rate of growth of integration, unless new phenomena and new devices are discovered in the coming years. Scientific and technological research of phenomena and material properties on the nanometer scale gave rise to new fields of knowledge, **nanoscience** and **nanotechnology**.

7.8.1 Basic Concepts and Manufacturing Techniques

Integrated circuits (IC) are manufactured by means of several physical and chemical processes, such as those described in Sect. 1.4 for making thin films, and the ones described in Sect. 6.1 for lithography, diffusion, oxidation, etching, and ion implantation. These processes are carried out simultaneously in many small pieces of the same semiconductor wafer, so that the fabrication cost of each individual IC is greatly reduced. Currently the microelectronic industry employs silicon wafers with diameters that vary from 2 in. (51 mm) to 300 mm (11.8 in.), and efforts are under way to introduce 450 mm wafers. ICs are produced in specialized semiconductor fabrication plants, colloquially known as *fabs*. Figure 7.31a shows a photograph of a technician inspecting a 150-mm silicon wafer after its processing for the fabrication of ICs, while Fig. 7.31b shows wafers various diameters. After processing, the wafer is cut with a diamond saw or a laser into small pieces, squares or rectangles, of dimensions that vary from 1 to 20 mm, corresponding to individual ICs. Each of these **chips**, as they are called, is then individually tested. Each approved chip is mounted on a base, interconnected to the external pins through gold or silver leads, and finally encapsulated with an insulating resin (epoxy type). The number of external pins can range from four to a few hundred, depending on the sophistication of the circuit. Figure 7.32 shows the external view of some typical integrated circuits.

Integrated circuits can be classified in several ways. Regarding applications, there are basically two categories, **analog,** also called **linear**, and **digital**. When they are manufactured in the same semiconductor wafer using the same technology, they are called **monolithic**. When the circuit involves different types of technology, for example interconnecting semiconductor devices with magnetic sensors, it is called

(a) (b)

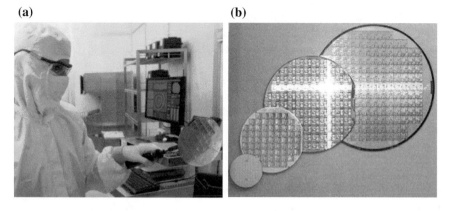

Fig. 7.31 **a** A technician inspects a silicon wafer after processing for IC fabrication (Courtesy of Ceitec Semiconductors). **b** Processed silicon wafers with diameters 2 in. (51 mm), 4 in. (100 mm), 6 in. (150 mm), and 8 in. (200 mm) (Wikipedia)

hybrid. Currently, more than 90% of monolithic circuits are made with single crystal silicon. The other semiconductor most used in the manufacture of ICs is GaAs, which finds an increasing number of applications at high frequencies.

Linear ICs are those that perform analog, or linear, functions. The most common simple linear ICs are the operational amplifiers (set of amplifiers with large gain, high input impedance, and low output impedance), voltage regulators, and switches. In general, these circuits are made with bipolar transistors and are used as discrete components in electronic circuits, or as part of an IC that performs complete functions, such as a radio or TV receiver.

Digital ICs are those that process binary information, in the form of on or off pulses. Digital ICs can be made with bipolar or MOS technology, but the latter is by far the most used currently. As we saw in Sect. 7.6, the use of complementary MOSFETs makes possible to manufacture VLSI circuits with low energy consumption. These

(a)   (b)   (c)

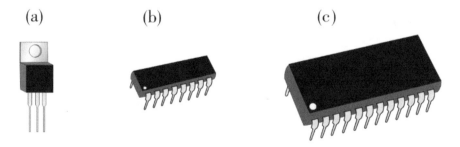

Fig. 7.32 Typical external view of some integrated circuits: **a** Voltage regulator circuit. **b** Common sixteen-pin IC enclosure. **c** Microprocessor for computers and other equipment. The scale of figure (**c**) is different from the other two, since the microprocessor has quite larger physical dimensions

circuits are the basis of microprocessors and high-capacity memories, used in mobile phones, tablets, notebooks, and large computers.

An important issue in integrated circuits is the electrical insulation between neighboring devices, since the semiconductor material of the wafer allows the passage of electric current from one device to another. In the case of MOS and MESFET devices, this is not a problem, since the conduction is restricted to the region of the channel, which makes the operation of each device to be independent of the neighbor. However, in the case of bipolar transistors, precautions must be taken to isolate a transistor from its neighbors. Several techniques are used for isolation. Conceptually, the simplest and most effective is the oxide dielectric insulation used in certain silicon integrated circuits. The manufacturing process starts with the preparation of the Si wafer, with small doping, forming a n-type substrate. Then, the diffusion of donor impurities is made over the entire surface, so as to form a n^+ layer. Then, through a process of photolithography and acid corrosion, a channel is dug in the n^+ layer, until it reaches the n-substrate, surrounding the entire region where the device will be manufactured. The substrate is then placed in an oven with an oxygen atmosphere, producing an insulating oxide layer (SiO_2), which covers the entire surface exposed, including the inner surface of the channel. The next step is the deposition of a layer of polycrystalline Si, which fills the channel, but also covers the entire surface. Finally, the wafer is turned upside down and mechanically polished, in order to remove all layers on the n^+ epitaxial layer. The final result is a set of n^+ regions forming islands, surrounded by insulating channels (as in Fig. 7.35c, which will be explained later). The desired devices are then manufactured on the islands and then interconnected by means of metallic films deposited on the surface. An advantage of the dielectric isolation process is the low parasitic capacitance among the neighboring devices and the elimination of polarization voltages, necessary in the process that will be presented below. The main disadvantage of this method is the large number of processing steps and the need to use a mechanical polishing process.

A very common method of insulation is that of reverse biased junctions. The basic idea of this method consists of forming islands, in which the devices are manufactured, surrounded by reversed biased p-n junctions. Since the current in a reversed biased junction is very small, the islands are effectively insulated electrically from each other. Figure 7.33 illustrates the process for the fabrication of the islands. Initially a n-type epitaxial layer is grown on the p-type silicon wafer (for a p-n-p transistor it would be a layer p on substrate n). The next steps consist in oxidizing the surface and by means of photolithographic processes and corrosion to open windows in the oxide layer in the form of lines that define the islands. Through the windows ones makes diffusion of p-type impurities, producing channels with depth such that they reach the p-substrate. To reverse polarize the p–n junction and produce the insulation of the islands it is necessary to apply a voltage by means of metallic contacts.

An additional issue in integrated circuits with bipolar transistors is the contact with the collector. In the discrete device of Fig. 7.2, the collector contact is located on the opposite side of the emitter and base contacts. This cannot be done in integrated circuits, since the interconnection of the device is made through metallic films in the

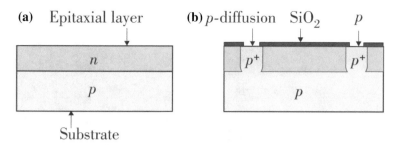

(a) Epitaxial layer **(b)** p-diffusion SiO$_2$ p

n

p

Substrate

Fig. 7.33 Illustration of the insulation method with reverse junctions. **a** p-type substrate with epitaxial n-type layer. **b** p-type insulation channel reaching the substrate

form of lines on the top surface. Figure 7.34 illustrates two methods used to make the collector contact in the same face of the emitter and the base contacts. In (a) the collector is formed by the diffusion of a n^+ region surrounding the base. In this case, the conduction between emitter and collector is made laterally, which results in high collector resistance. In applications that require low collector resistance, the most used structure is that of the buried layer. In this structure the effective contact of the collector is a n^+ layer made by diffusion in the p substrate, before deposition of the n-type epitaxial layer, as in Fig. 7.34b. The connection of the collector with the buried layer is made by means of a n^+ region, produced by diffusion, as illustrated in Fig. 7.34c.

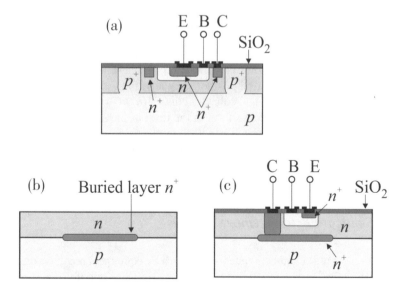

Fig. 7.34 Methods used to make the contacts in bipolar transistors in integrated circuit. **a** Contacts for lateral operation. **b, c** Collector in buried layer

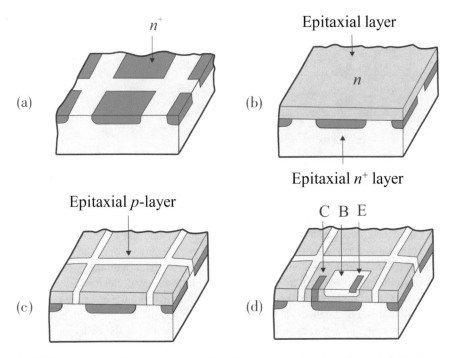

Fig. 7.35 Illustration of the manufacturing steps for a *n-p-n* transistor in an integrated circuit

To conclude this section, we show in Fig. 7.35 the manufacturing steps of *n-p-n* transistor in an integrated circuit. In (a) we see the n^+ layers diffused in the *p*-type substrate. In (b) we see the *n*-type epitaxial layer over the substrate, forming the buried layers. Figure 7.35c shows the insulating channels, that can be made with dielectrics or with reversed biased junctions. Finally, Fig. 7.35d shows the regions of the emitter (*n*), the base (*p*), and collector contact (n^+). To complete the circuit, a metallic film is deposited on the surface, in the form of lines, establishing contacts with the regions of interest through windows in the oxide layer. The terminals of the circuit are then connected to the external pins and the assembly is encapsulated, acquiring the external form such as those shown in Fig. 7.32.

7.8.2 Semiconductor Memory Devices

Among the most important components of electronic equipment are those that store information, called memory devices. These devices were initially developed for computers, for which they are essential components. However, in the last decades, memories have been used in a wide variety of electronic equipment, which incorporate microprocessors in their systems. In these equipments memory devices are essential for storing programs, or codes, or software, that make the system perform its functions, as well as all kinds of information such as texts, images, videos, etc.

There are two major types of memories for computers and other equipment, namely, semiconductor devices and magnetic devices. Magnetic media store information indefinitely, until it is erased or replaced. That is why they are called **non-volatile**. In semiconductor memory devices the storage time depends on its type. Some have storage times of a few milliseconds or less, and are called **volatile**, while others can store the information for many years, and are considered non-volatile. The main advantages of semiconductor memories are the high capacity, high speed of recording and reading, and the fact that they do not need moving parts, like magnetic and optical disks. Magnetic memories, which will be presented in Chap. 9, have larger storage capacity and are non-volatile. Among the removable memory devices, for transport or external storage, the most important currently are semiconductor flash memories, magnetic tapes, disks, and cards, and optical disks (Chap. 8).

Semiconductor memory devices can be manufactured both with bipolar transistor and MOS technologies. However, for over three decades the MOS technology has completely dominated the manufacture of semiconductor memories, due to the larger capacity of integration, lower cost, and lower energy consumption. The basic element of MOS memories is the MOS capacitor, presented in Sect. 6.1. In the capacitor the presence of charge represents bit 1, while no charge represents bit 0.

In general, the basic cells of semiconductor memories are constituted by MOS capacitors and MOSFETs. In each cell, the MOSFET, called pass transistor, serves as a switch to deliver and access charge in the capacitor. Figure 7.36a shows a schematic illustration of one scheme for a simple memory cell formed by a n-channel MOSFET in series with a MOS capacitor. The source (S) and gate (G) terminals of the transistor are used for connections with the addressing electrodes, made by strips of metallic films. The n^+ region of the transistor drain is connected in series with the capacitor formed by the p-type semiconductor of the substrate and the metallic film, separated

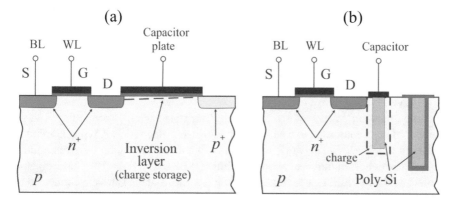

Fig. 7.36 Illustration of two schemes used for a semiconductor memory cell consisting of a MOSFET in series with a MOS capacitor. The insulating layers are shown in brown color. **a** The pass transistor delivers the charge to an adjacent planar MOS capacitor. **b** The capacitor is made in a trench close to the pass transistor. The charge is stored in the inversion region surrounding the trench

by an insulating layer, that can be the oxide SiO_2. It is also common to use a layer of polycrystalline silicon (poly-Si) for the capacitor plate, instead of a metallic film. The capacitor board terminal is in general connected to the ground of the circuit. The p^+ region at the right end of the figure is used to insulate the cell from the neighboring element.

The capacitor charging process, that is, the storage of information corresponding to bit 1, is done by the simultaneous application of two voltage pulses, one between the source of the MOSFET and the ground, and another between the gate and the ground. The values of the voltages peak levels should be sufficient to create an inversion layer between the source and the drain of the MOS transistor and another one under the MOS capacitor terminal. After application of the pulses, the inversion layer in the transistor disappears, but the charge in the inversion layer of the capacitor remains stored. This charge tends to disappear after a certain time due to the thermal generation of carriers, which limits the storage time to a few milliseconds. This time is long enough for the dynamic operation of a memory in equipment with clock period much smaller than 1 μs, as is the case of equipment that operate with clocks in the GHz region as in a dynamic random-access memory (DRAM). For the continuous operation of the equipment, it is then necessary that the memory is periodically "refreshed", that is, that the capacitors of the cells with bit 1 are recharged before the charge disappears. Naturally, the information is lost when the equipment is turned off. Therefore, a memory with cells like the one in Fig. 7.36 is of the volatile type.

The need to have smaller memory cells to increase the packing density in chips has motivated the development of approaches to increase the capacitance while decreasing the capacitor surface area. One of these approaches, illustrated in Fig. 7.36b, employs a trench capacitor instead of the planar capacitor. The trench with straight walls, perpendicular to the plane of the figure, is made by reactive ion etching. Then, by means of chemical vapor deposition (CVD), the trench is filled with polysilicon, that plays the role of one of the capacitor plates, while the p-type substrate is the other plate. The charge is stored in the inversion layer surrounding the trench. Figure 7.36b also illustrates a scheme employed to isolate neighboring cells that requires a smaller area than the one in Fig. 7.36a, that is called trench isolation. In this case, after it is dug with ion etching, the inner walls of the trench are oxidized to form an insulating layer, and then it is filled with polysilicon and covered by an oxide layer.

Semiconductor memories are formed by integrated circuits containing a large number of cells connected to a network of addressing electrodes. A typical circuit is shown in Fig. 7.37. The network has the form of a matrix, in which the interconnections of rows and columns are called word lines (WL) and bit lines (BL). Note that the sources of the MOSFETs are connected to the bit lines and the gates are connected to the word lines, while the capacitors are connected to the ground, allowing to charge the capacitors as explained previously. This matrix arrangement allows to randomly access a cell with any address, both for recording and for reading information. For this reason, this type of memory is called **random access memory** (RAM). The random access enables an operating speed, both for recording and for reading, much faster than in serial access, characteristic of magnetic tapes and disks, in which it is necessary to go through several addresses until reaching a specific desired address.

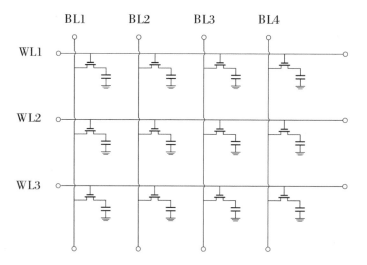

Fig. 7.37 Illustration of a section of the integrated circuit of a random-access memory (RAM), formed by a matrix array of memory cells, each containing a MOSFET in series with a MOS capacitor, interconnected to an addressing network

It is possible to build non-volatile memories with semiconductors using several MOS structures. Figure 7.38 shows a MOSFET structure in which there are two gate electrodes. One of them is metallic, used for external contact, while the other, called floating gate, is generally made of a layer of polycrystalline silicon, also called polysilicon (poly-Si), that properly doped has good conductivity, surrounded by the insulating oxide. In this structure, called floating-gate avalanche-injection MOS (FAMOS), the charge may remain in the floating gate for several years. Charge storage at the floating gate can be done by several processes. A common process consists of applying a voltage to the drain junction, with sufficiently high value to produce a strong reverse bias. This results in avalanche, producing a large acceleration of electrons in the depletion region of the junction. If a relatively high voltage is also applied to the gate terminal, a fraction of the electrons passes to the floating gate by direct injection or by the tunnel effect through the thin oxide layer.

Fig. 7.38 Structure of a
FAMOS device used in
non-volatile semiconductor
memories

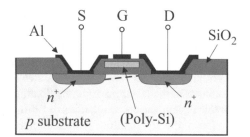

Fig. 7.39 Classification of semiconductor memory devices

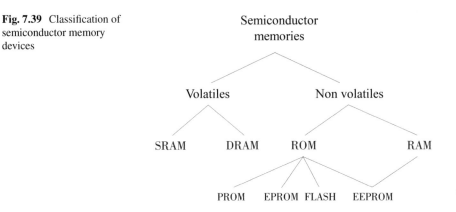

The use of a variety of structures makes possible to build many types of semi-conductor memory devices, volatile or non-volatile, for countless applications. Figure 7.39 shows a classification of memories, represented by their acronyms. Volatile memories are of two types, static random-access memories (SRAM) and dynamic random-access memories (DRAM). Dynamic memories are like those of Figs. 7.36 and 7.37, presented above, that need to be refreshed periodically. The static memories are those that retain the recorded information without the need to be refreshed, as long as the device remains energized. This is done through bi-stable circuits, also known as flip-flops, that can be switched from one stable electric state to another one, representing bit 0 or bit 1.

The variety of non-volatile memories is larger, so they are subdivided in two groups, ROM (Read Only Memory), used only for reading, and RAM, which can be accessed either for writing or for reading. Strictly speaking, ROM memories are also of the RAM type, because, as we saw earlier, this is a name used to designate memories whose addresses can be accessed randomly. However, the name RAM is used for devices with random access, both for recording and for reading information. In memories called EEPROM (Electrically Erasable-Programmable Read Only Memory) the information is used only for reading, but can be recorded or erased electrically. The other main types of ROM memory are: PROM (Programmable Read Only Memory), that is a device in which the information in each cell is recorded permanently by a fuse type process; EPROM (Electrically Programmable Read Only Memory), is a memory that uses FAMOS devices in which the information is recorded electrically. This is the most common memory for storage of the program that starts the operation process of a computer, or other equipment containing microprocessors, when they are turned on. Information in the EPROM memory can be erased globally, this is, at all addresses, by means of ultra-violet radiation or X-rays. The high energy photons take the electrons from the floating gate to the control gate or the substrate. For this reason, the encapsulations of EPROM devices have an opening with an optic window at the top, to allow the radiation to pass through. Finally, the flash memory, an abbreviation for the full name flash EEPROM, is a memory where information is recorded electrically, as in the EPROM, but can be electrically globally erased at once, as in a flash. The flash memories are used in **pendrives** and other devices, which serve to easily store and transport information. Before the development of pendrives, about two decades ago, the main media used to store and transport information outside personal computers were flexible magnetic disks, known as floppy disks. The need to produce memories for several applications, with larger storage capacity and with larger access speed, has boosted research and development of new devices, new materials, and integration structures with increasingly sophisticated technologies. These activities have provided a continuous improvement of memories and a permanent search for new devices, manufactured with semiconductors or other materials, that enable the performance of unusual functions for new equipment or for innovations in the electronic industry.

To conclude this section, we point out that modern IC fabrication techniques employ several thin-film deposition processes of various materials, aiming at increasing the number of devices per chip and improving performance and reliability. Besides the pure and doped semiconductors, simple metals, and silicon dioxide, used in earlier integrated circuits, high-density ICs make use of several other materials, such as silicides, nitrides, and glasses, among others. Here we briefly present some properties and applications of these classes of materials.

The silicides consist of the admixtures of silicon with some metals that form stable metals and semiconductors. They are important in ICs with metallic interconnections of widths smaller than 1 μm because they have resistivity smaller than doped polysilicon. Cobalt silicide ($CoSi_2$), titanium silicide ($TiSi_2$), and nickel silicide ($NiSi$) have lower resistivities and are generally compatible with IC processing. One important application of silicides is as the material for the MOSFET gate electrode, used either alone or with polysilicon above the gate oxide in high-density semiconductor memories.

Silicon nitride (Si_3N_4) is a very hard insulating material with good thermal stability that is used with many purposes in several industrial sectors. In the microelectronic industry it is used as an insulator and chemical barrier in manufacturing integrated circuits, to electrically isolate different structures or as an etch mask in bulk micromachining. As a passivation layer for microchips, it is superior to silicon dioxide, as it is a significantly better diffusion barrier against water molecules and sodium ions, two major sources of corrosion and instability in microelectronics. It is also used as a dielectric between polysilicon layers in capacitors in analog chips. In the fabrication of ICs, silicon nitride films can be deposited by a low-pressure CVD process at intermediate temperatures (~750° C), or by plasma-enhanced CVD process at lower temperatures (~300° C), depending on the properties desired. Silicon nitride deposited by LP-CVD, also called LP-nitride, experiences strong tensile stress, which may crack films thicker than 200 nm. However, it has higher resistivity and dielectric strength than most insulators commonly available in microfabrication. On the other hand, silicon nitride deposited by PE-CVD, called PE-nitride, have much less tensile stress, but smaller resistivity than LP-nitride. Thus, often both are used next to each other to isolate the gate in MOSFETs and provide good mechanical stability.

Besides silicon oxide and silicon nitride, phosphorous-doped silicon dioxide, also called phosphosilicate glass (PSG), and borophosphosilicate glass (BPSG), are often used as insulating materials for maknig thin films in the manufacturing of ICs. Both PSG and BPSG can serve as a passivation layer against moisture penetration and alkali metal contamination. However, the physical mechanism for blocking alkali metal ions is different from that of silicon nitride. PSG and BPSG films can getter alkali ions whereas silicon nitride serves as a diffusion barrier for alkali ions. PSG and BPSG are used mainly as an insulating thin film to make a smooth topography between layers. If the lower metal layer is concave or has a sharp shape, circuit failure may result from an opening that may occur in the upper metal layer during deposition. Since PSG and BPSG deposited at low temperatures become soft and flow upon heating, they provide a smooth surface that effectively isolate adjacent metal layers.

The use of these materials in IC devices with nanometer dimensions is illustrated in Fig. 7.40, showing photographs of cross-sections of a silicon wafer, in the plane perpendicular to the bit-line, of a high-density memory chip, obtained with scanning transmission electron microscopy (STEM). Figure 7.40a shows two DRAM unit cells, such as the one in Fig. 7.36b, each with a trench capacitor used for charge storage and one MOS pass transistor. The isolation between the charge regions of two neighboring cells is assured by the vertical shape of the capacitors and by the triangular prism shaped isolation trench. Note that to save space, the two MOSFETs share the same source, which is in contact with the bit line, while the two capacitors are connected to two different word lines, not shown in the photographs. Figure 7.40b shows a zoom of one MOSFET indicating the materials used for the source, gate, and drain, as well as the PSG layer deposited over them to provide passivation and isolation against moisture penetration and alkali metal contamination.

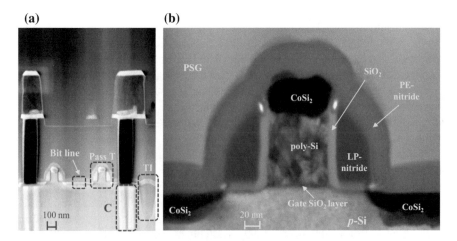

Fig. 7.40 Scanning transmission electron microscopy photographs of a cross-section of a silicon wafer with a high-density memory chip. (a) Two memory cells, each with a n-MOS pass transistor (T) and a trench capacitor (C). The two transistors share the same source in contact with a bit line. The isolation between two memory cells is made by a triangular prism shaped trench (TI). (b) Zoom of one transistor indicating the materials of the various films used in the fabrication of the device (Courtesy of Gunter Fischer, IHP Microelectronics, Germany)

As mentioned in the introduction of this section, the development of materials and processes has made possible a continuous scaling down in size of the components of ICs over the past decades. Typical MOSFET channel lengths were a few decades ago of several micrometers, but modern integrated circuits are incorporating MOSFETs with channel lengths of tens of nanometers, as clearly shown in Fig. 7.40. However, the semiconductor industry maintains a roadmap which sets the pace for further reductions in the size of MOSFETs. Historically, the difficulties with decreasing the size of the MOSFET have been associated with the semiconductor device fabrication process, the need to use very low voltages, and with poorer electrical performance necessitating circuit redesign and innovation. Further reduction in size will depend on the improvement of fabrication technologies in industry, as well as the discovery of new phenomena, materials, and devices in academia in the coming years.

Problems

7.1   The variation in the concentration of holes in excess of equilibrium, $\delta p (x)$, as function of position x at the base of a p-n-p transistor is given by Eq. (7.15). Plot curves of $\delta p_B(x)/\Delta p_E$ (preferably on a computer), for three base thicknesses, $l = 2L_p$, $0.5\, L_p$, and $0.1\, L_p$, where L_p is the diffusion length. Using the plots, explain which thickness is the best for a good transistor.

7.2   Show that in a forward biased p^+-n-p^+ transistor, the base current is given by Eq. (7.23).

7.3   Consider a p^+-n-p^+ symmetric transistor made of silicon, with the following base parameters: $l = 2\, \mu m$, $A = 10^{-3}\, cm^2$, $N_d = 5 \times 10^{15}\, cm^{-3}$, and $\tau_p = 0.5\, \mu s$. Considering that the emitter and the collector have $N_a = 5 \times 10^{17}$ cm^{-3} and $\tau_n = 0.1\, \mu s$, calculate the emitter, base, and collector currents, with the transistor polarized with $V_{EB} = 0.75$ V and $V_{CB} = -10$ V.

7.4   Calculate the parameters α, γ and β of the transistor of Problem 7.3.

7.5   Calculate the parameters I_{Es}, I_{Cs}, α_N, and α_I of the transistor of Problem 7.3 and obtain the emitter and collector currents using Eqs. (7.37) and (7.38). Compare the results with those of Problem 7.3.

7.6   Consider a p^+-n-p^+ symmetric transistor, that is, with the same parameters for the emitter and the collector.

(a)   Write the Ebers-Moll equations for the transistor.

(b)   Obtain the current I as a function of V for the transistor in circuit (a) of the figure below.

(c)   Calculate V_{CB} when the transistor is connected as in circuit (b) in the figure below.

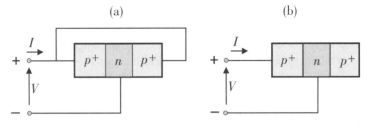

(a)   (b)

7.7   A p^+-n-p^+ symmetric transistor made of Si has the following base parameters: $l = 1$ μm, $A = 10^{-3}$ cm^2, $N_d = 10^{15}$ cm^{-3}, and $\tau_p = 2$ μs. The emitter and the collector have $N_a = 5 \times 10^{16}$ cm^{-3} and a diffusion length half the value at the base. The other Si parameters at room temperature are given in Table 5.2.

(a)   Calculate the saturation currents of the emitter and collector, I_{Es} and I_{Cs}.
(b)   Write the Ebers-Moll equations for the transistor and calculate the numerical values for the parameters I_{Es}, I_{Cs}, α_N, and α_I.

7.8   Consider the p^+-n-p^+ transistor of Problem 7.7. From Eqs. (7.37) and (7.38), obtain an equation for the collector current I_C as a function only of the voltage V_{CE} and the base current I_B. Use a computer and plot the curves for I_C as a function of $-V_{CE}$, for different values of I_B, and compare them with Fig. 7.7b.

7.9   For the transistor of Problem 7.3, calculate the current I when it is connected as in the circuit below, where $V = 500$ mV.

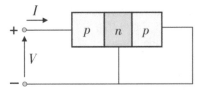

7.10   A p-n-p transistor with the I-V characteristic curves given in Fig. 7.7, is used as an amplifier in a simple polarization circuit, as the one in Fig. 7.9a. If $E_B = E_C = 10$ V and $R_C = 1$ kΩ calculate:

(a)   The value of R_B so that the base current is 50 μA;
(b)   The values of the currents I_C and the voltage V_{CE};
(c)   The current gain of the circuit;
(d)   The maximum voltage of the signal input for the circuit to operate in the linear region.

7.11   Obtain the drain current equation for a JFET such as the one in Fig. 7.11, corresponding to Eq. (7.43), without neglecting the contact potential V_0 of the junctions.

7.12 Obtain the expressions for the conductance and pinchoff voltage for a JFET, neglecting the contact potential of the junctions but without the condition $N_a \gg N_d$ used in the derivation of Eqs. (7.44) and (7.45).

7.13 A junction field effect transistor, like the one in Fig. 7.11, made of silicon, has p^+ regions with doping $N_a = 10^{18}$ cm^{-3} and channel with $N_d = 2 \times 10^{16}$ cm^{-3} and a half-width $a = 0.8$ μm.

    (a) Calculate the pinchoff voltage V_p for channel formation assuming $V_0 = 0$.

    (b) What is the value of V_p if V_0 is not neglected?

    (c) What is the value of V_D at which the current saturates for $V_G = -2$ V?

7.14 What is, approximately, the largest voltage gain of an amplifier circuit made with a JFET with the characteristic curves of Fig. 7.13?

7.15 A GaAs MESFET has a potential barrier $V_0 = 0.8$ V and has a channel with impurity concentration $N_d = 10^{16}$ cm^{-3}, mobility $\mu_n = 7 \times 10^3$ cm^2/V s and dimensions $a = 0.7$ μm, $L = 15$ μm, and $D = 10$ μm. Calculate:

    (a) The channel conductance;

    (b) The threshold voltage;

    (c) The saturation drain current for $V_G = 0$ and $V_G = -1.0$ V.

7.16 A n-channel MOSFET is made with aluminum electrodes and p-type silicon with impurity concentration $N_a = 2 \times 10^{17}$ cm^{-3}. The thickness of the SiO$_2$ layer is 10 nm in the gate region, the charge density at the interface is $Q_i/A = 10^{-4}$ C/m^2, and the other relevant dimensions (Fig. 7.22) are $L = 10$ μm and $D = 300$ μm. Calculate:

    (a) The threshold voltage V_T;

    (b) The saturation current for $V_G = 6$ V;

    (c) The I_d - V_D curve for $V_G = 4$ V.

Further Reading

A. Bar-Lev, *Semiconductors and Electronic Devices* (Prentice-Hall, New Jersey, 1993)

D.A. Fraser, *The Physics of Semiconductor Devices* (Claredon Press, Oxford, 1986)

P.E. Gray, C.L. Searle, *Electronic Principles: Physics, Models, and Circuits* (Wiley, New York, 1969)

R.E. Hummel, *Electronic Properties of Materials* (Springer, Berlin, 2011)

K. Kano, *Semiconductor Devices* (Prentice-Hall, New Jersey, 1998)

D. Neamen, *Fundamentals of Semiconductor Physics and Devices* (McGraw-Hill, New York, 2002)

K.K. Ng, *Complete Guide to Semiconductor Devices* (Wiley-IEEE Press, Hoboken, NJ, 2002)

D.J. Roulston, *An Introduction to the Physics of Semiconductor Devices* (Oxford University Press, Oxford, 1999)

G.B. Rutkowski, J. E. Oleksy, *Solid State Electronics* (MacMillan–McGraw-Hill, Singapore, 1993)

B.J. Streetman, S. Banerjee, *Solid State Electronic Devices* (Prentice Hall, New Jersey, 2005)

S.M. Sze, M.K. Lee, *Semiconductor Devices: Physics and Technology* (Wiley, New York, 2012)
E.S. Yang, *Microelectronic Devices* (McGraw-Hill, New York, 1988)
F.F.Y. Wang, *Introduction to Solid State Electronics* (North-Holland, Amsterdam, 1989)

Chapter 8
Optoelectronic Materials and Devices

The main objective of this chapter is to introduce the physical principles of lasers and photodetectors, as well as some of their applications. Initially we discuss the optical properties of materials, since they are essential to understand the principles of the devices. We present the classical treatment, that is useful for some basic properties, and then the quantum theory of radiation, that is essential to understand the operation of light emitting diodes, photodetectors, and lasers. These devices are described in a manner that can be grasped by undergraduate students. Semiconductor lasers and some of their applications are presented in more detail.

8.1 Optical Properties of Materials

Optical properties are those that characterize how materials respond to an external optical radiation, absorbing, reflecting, emitting, or changing the polarization of light. Some aspects of these properties are undoubtedly among the most easily identifiable in materials. Since immemorial times it has been known that the brightness, color, transparency, and opacity of materials fascinate and intrigue humanity. It is an ancient costume to use metals to manufacture mirrors and to use metals and natural minerals for making jewelry and adornment objects.

Scientific studies of the effect of materials on light gained great impetus with Newton's optical experiments in the seventeenth century. Newton showed that when passing through a glass prism, a beam of sunlight produced a multicolored strip. At one end of the strip formed by the rays that suffer the slightest deviation when passing through the prism is the red color, while in the other end is violet. Nowadays it is known that the color sensation is produced in the brain, and results from the effect of electromagnetic waves in a range of frequencies that interacts with the retina of the human eye. Radiation with wavelengths around 400 nm produces the sensation of the violet color, while at the other end of the spectrum, wavelengths of 700 nm produce the sensation of the red color. Figure 8.1a shows the standard response of the

© The Author(s), under exclusive license to Springer Nature Switzerland AG 2022
S. M. Rezende, *Introduction to Electronic Materials and Devices*,
https://doi.org/10.1007/978-3-030-81772-5_8

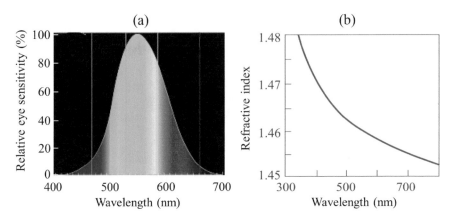

Fig. 8.1 a Relative sensitivity of the human eye as a function of the light wavelength (Wikipedia). **b** Variation of the refractive index of fused quartz with the wavelength

human eye as a function of the wavelength of visible light. The region where the eye is most sensitive is around 555 nm, which corresponds to a yellowish-green color. Figure 8.1b shows the wavelength dependence of the refractive index of fused quartz, which is typical of transparent materials, and explains the separation of the white light beam into various colors. At the violet color (shorter wavelength), the refractive index is larger, resulting in larger deviation when passing through the prism. In the red, the refractive index is smaller and therefore the deviation is smaller. The variation of the refractive index with wavelength is due to the characteristics of the interaction of radiation with matter, which will be studied in Sect. 8.2.

The visible region of the electromagnetic spectrum, with wavelength in the 700–400 nm range, corresponds to a photon energy in the range 1.7–3.1 eV. These values are on the same order of magnitude as the energy gaps in several semiconductors and also the energies of electronic transitions in atoms of several elements. For this reason, in the last decades it was possible to develop semiconductor devices that efficiently convert light into electric current, and vice versa. This gave rise to the field of **optoelectronics**, the branch of science and technology devoted to research and development of materials and devices that employ light and electronic current to process analog and digital signals that form the basis of **optical communications**. A related area, which is also developing rapidly, is **photonics**, devoted to signal processing done entirely in optical devices.

In this chapter we shall study the main phenomena involved in the interaction of electromagnetic radiation with matter, as well as its applications in optoelectronics and semiconductor devices. Other devices based on the interaction of light with different materials will be presented in Chap. 10.

8.1.1 Electromagnetic Waves in Materials

The phenomena of reflection, refraction, and absorption of light in materials, can be macroscopically described by means of Maxwell's equations (2.1)–(2.4). In Chap. 2 these equations were solved for plane waves in a perfectly insulator medium, in which there is no current induced by the fields. In this situation there is no absorption or energy loss, so that the amplitudes of the fields $\vec{E}$ and $\vec{H}$, given by Eqs. (2.7) and (2.8), do not vary during propagation. However, in real materials, metallic, semiconductor, or even insulators, there is always some loss mechanism, so that the wave amplitude decays during propagation. This loss can be described by a current density, related to the electric field through the conductivity σ, $\vec{J} = \sigma \vec{E}$. In this case, the wave equation for propagation in one dimension contains another term that is not present in Eq. (2.6). From Eqs. (2.1)–(2.4) one can easily show that for $\rho = 0$, the electric field varying only in the x-direction is described by (Problem 8.1)

$$\frac{\partial^2 \vec{E}(x,t)}{\partial x^2} = \mu\varepsilon \frac{\partial^2 \vec{E}(x,t)}{\partial t^2} + \mu\sigma \frac{\partial \vec{E}(x,t)}{\partial t}. \tag{8.1}$$

In the visible region of the electromagnetic spectrum, the magnetic effects are negligible, so that we can consider $\mu = \mu_0$. Replacing in Eq. (8.1) the harmonic field solution (2.14), with $\vec{k}$ in the x-direction we obtain

$$k^2 = \mu_0\varepsilon\,\omega^2 + i\omega\mu_0\,\sigma = \frac{\varepsilon_r}{c^2}\omega^2 + i\omega\mu_0\,\sigma, \tag{8.2}$$

where $c = 1/(\mu_0\,\varepsilon_0)^{1/2}$ is the speed of light in vacuum and $\varepsilon_r = \varepsilon/\varepsilon_0$ is the relative permittivity of the material, often called **dielectric constant**. In a medium without losses the ratio between k and ω leads to the definition of the refractive index n, given by Eq. (2.10). This definition can be generalized for a medium with losses, through the **complex refractive index**

$$N(\omega) = \left(\varepsilon_r + i\,\frac{\sigma}{\omega\varepsilon_0}\right)^{1/2}. \tag{8.3}$$

With this definition, Eq. (8.2) takes the same form as (2.10), namely,

$$k = \frac{\omega}{c} N(\omega), \tag{8.4}$$

since $\mu_0 c^2 = 1/\varepsilon_0$. The complex refractive index was represented by $N(\omega)$ not to be confused with the number of particles N and also to make explicit its dependence on ω. This dependency does not arise only from ω that appears in the denominator of the second term in (8.3), but also in the fact that $\varepsilon(\omega)$ and $\sigma(\omega)$ always vary with frequency. In fact, it is precisely the variation of ε and σ with frequency that determines the optical properties of materials, as we shall see later in this chapter.

The advantage of introducing the complex refractive index is that all expressions obtained in Chap. 2 can be used here with the simple substitution of the refractive index n by the complex $N(\omega)$. Let us now write the complex refractive index in terms of the real and imaginary parts

$$N(\omega) = n + i\kappa. \tag{8.5}$$

To relate the two parts of $N(\omega)$ with the parameters ε_r and σ of the medium, we square Eqs. (8.3) and (8.5) and equate them

$$N^2(\omega) = n^2 - \kappa^2 + i2n\kappa = \varepsilon_r + i\frac{\sigma}{\omega\varepsilon_0}.$$

As we shall see later, both ε_r and σ can be complex. Thus, setting $\varepsilon_r = \varepsilon_r' + i\varepsilon_r''$ and $\sigma = \sigma' + i\sigma''$, and replacing the two quantities in the equation above, we obtain for the real and imaginary parts of $N^2(\omega)$

$$n^2 - \kappa^2 = \varepsilon_r' - \frac{\sigma''}{\omega\varepsilon_0}, \tag{8.6}$$

$$2n\kappa = \varepsilon_r'' + \frac{\sigma'}{\omega\varepsilon_0}. \tag{8.7}$$

To understand the meanings of n and κ, we substitute (8.5) in (8.4) and obtain for the modulus of the wave vector

$$k = \frac{\omega n}{c} + i\frac{\omega\kappa}{c} \equiv k' + ik''.$$

Using this expression in Eq. (2.14) we obtain for the electric field of a wave propagating in the x-direction

$$\begin{aligned}
\vec{E}(x,t) &= \text{Re}\left[\vec{E}_0\, e^{i\,(\omega n/c)\,x\, -i\omega t}\right] e^{-(\omega/c)\,\kappa x}\\
&= \vec{E}_0\cos[(\omega n/c)\,x - \omega t]e^{-(\omega/c)\,\kappa x}.
\end{aligned} \tag{8.8}$$

We see then that the field is described by a harmonic function whose amplitude decays exponentially during propagation. Note that n, the real part of $N(\omega)$, is the ratio between the speed of light c and the phase velocity $v_p = \omega/k' = c/n$, and therefore it is the **refractive index** itself. Only in the case of lossless medium ($\kappa = 0$) we have $n = (\varepsilon_r)^{1/2}$, as defined in Eq. (2.10).

Equation (8.8) shows that κ, the imaginary part of $N(\omega)$, produces an exponential decay in the field amplitude. For this reason, it is called **damping coefficient**, or **extinction coefficient**. To understand the meaning of κ, let us study what happens to the wave energy. The quantity that expresses the energy carried by an electromagnetic wave is the Poynting vector, defined by

$$\vec{S} = \vec{E} \times \vec{H}. \tag{8.9}$$

It can be shown that the modulus of $\vec{S}$ is equal to the energy per unit area and per unit time carried by the wave. Using Eqs. (2.11)–(2.13), it is easy to show that for a plane wave in a lossless medium, $\vec{S}$ has the same direction as the propagation vector $\vec{k}$ and has a modulus given by (Problem 8.2),

$$S(\vec{r}, t) = \frac{n}{c\mu_0} E_0^2 \cos^2(\vec{k} \cdot \vec{r} - \omega t + \phi), \tag{8.10}$$

which shows that S varies harmonically in time and space. In energy considerations the most important quantity is its average, so the wave **intensity** is defined as the average value of the modulus of the Poynting vector. Since the average value of cosine squared is equal to 1/2, the wave intensity obtained with Eq. (8.10) is

$$I = \langle S \rangle = \frac{n}{2c\mu_0} E_0^2. \tag{8.11}$$

This relationship shows that in a lossless medium, the intensity of a harmonic plane wave is constant, that is, it does not vary in space or time. It is proportional to the square of the electric field amplitude and it is equal to the average transported energy, per unit area and per unit time. In other words, the intensity is the average power per unit area. In the SI it is expressed in W/m^2.

It is possible to relate the intensity of a light beam with the flow of photons. A beam with intensity I and cross section area A has average power $P = I A$. Since ω is the wave angular frequency, the energy of each photon is $\hbar\omega$. Therefore, the photon flow Φ, defined as the number of photons that go through a cross section of the beam, per unit time and per unit area, is given by

$$\Phi = \frac{I}{\hbar\omega}. \tag{8.12}$$

Note that the number of photons per unit time does not vary along the beam because the amplitude is constant. This result is not valid for a material with losses, in which $N(\omega)$ has real and imaginary parts. In this case, the calculation of the Poynting vector has a complicating factor in the lag between the fields E and H introduced by the imaginary part of $N(\omega)$. However, it is easy to show that the intensity of a wave with the field given by Eq. (8.8) varies in the space as follows (Problem 8.3),

$$I(x) = I(0) e^{-2(\omega/c)\kappa x} \equiv I(0) e^{-\alpha x}. \tag{8.13}$$

This expression shows that κ, the imaginary part of $N(\omega)$, produces along x an exponential decay in the wave amplitude. The rate of decay is characterized by the **absorption coefficient**, defined by

$$\alpha = -\frac{1}{I}\frac{dI}{dx} = 2\frac{\omega}{c}\kappa. \tag{8.14}$$

Note that the absorption coefficient has the dimension of inverse distance. Its inverse, $1/\alpha$, is the characteristic distance of the wave intensity decay. Since the amplitude of the electric field varies with the square root of the intensity, the characteristic length of the field penetration in the material is given by

$$\delta = \frac{2}{\alpha} = \frac{c}{\omega\kappa}. \tag{8.15}$$

Equations (8.6), (8.7) and (8.14) are valid for each value of the frequency ω. As we shall see later, in any material all quantities defined in this section vary with ω. In the visible region of the electromagnetic spectrum, insulators have $\sigma = 0$, and therefore in crystalline form they are transparent ($\alpha = 0$). However, in the ultraviolet region, their conductivity is finite and they strongly absorb radiation. The increase in absorption and in the dielectric constant in the ultraviolet region results in a gradual increase in the refractive index with the frequency in the visible range. This is what makes the refractive index of fused quartz to decrease with the wavelength (which is inversely proportional to ω), as in Fig. 8.1b, and that produces the dispersion of white light. This is also common to many other transparent materials.

8.1.2 Reflectivity of Materials

An electromagnetic wave incident on the surface of any material gives rise to a reflected and a refracted wave. The laws of geometric optics relate the angles of incidence, reflection and refraction, but give no information about the relative wave intensities. To obtain the relationship between intensities, it is necessary to use the boundary conditions at the surface obtained from Maxwell's equations. Let us consider the simple case of a wave with electric field $\vec{E}_1 = E_1\hat{y}$ (in vacuum or in air), incident perpendicularly on the flat surface of a material with complex refractive index $N(\omega)$, as illustrated in Fig. 8.2. The calculation of the reflected $\vec{E}_2 = \hat{y}E_2$ and transmitted (or refracted) $\vec{E}_3 = \hat{y}E_3$ fields is very similar to the

Fig. 8.2 Illustration of incident, reflected and transmitted (refracted) waves on the surface of a material

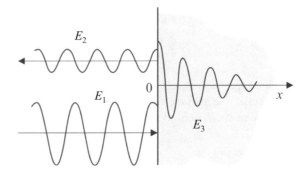

problem of an electron incident on a potential barrier (Sect. 3.3.3 and Problem 3.5). The complex field in air, at $x < 0$, is

$$E_y = E_1 \, e^{i(\omega/c)x - i\omega t} + E_2 \, e^{-i(\omega/c)x - i\omega t},$$

while in the material, supposed to be semi-infinite, it is given by Eq. (8.8)

$$E_y = E_3 \, e^{i(\omega/c) N(\omega)x - i\omega t}.$$

To obtain E_2 and E_3 as a function of E_1, it is necessary to use the boundary conditions at the surface. The continuity of the tangential electric field at $x = 0$ gives

$$E_3 = E_1 + E_2. \tag{8.16}$$

To impose the continuity of the tangential component of the field H, we use Eq. (2.13) with $(\varepsilon_r)^{1/2}$ replaced by $N(\omega)$. The result is,

$$N(\omega) \, E_3 = E_1 - E_2. \tag{8.17}$$

From Eqs. (8.16) and (8.17) we obtain the amplitudes of the fields of the reflected and transmitted waves

$$\frac{E_2}{E_1} = \frac{1 - N(\omega)}{1 + N(\omega)}, \tag{8.18}$$

$$\frac{E_3}{E_1} = \frac{2}{1 + N(\omega)}. \tag{8.19}$$

The **reflectivity** R is defined as the ratio between the intensities of the reflected and incident waves. Using Eq. (8.18) we obtain

$$R = \left| \frac{N(\omega) - 1}{N(\omega) + 1} \right|^2. \tag{8.20}$$

With Eq. (8.5) we can write the reflectivity in terms of the real and imaginary parts of $N(\omega)$

$$R = \frac{(n - 1)^2 + \kappa^2}{(n + 1)^2 + \kappa^2}. \tag{8.21}$$

Note that R is a dimensionless quantity that is often expressed in percentage. For example, ordinary glass has $n = 1.5$ and $\kappa = 0$, which gives $R = 0.04$, or $R = 4\%$. Equation (8.21) shows that the reflectivity depends on both the refractive index and the extinction coefficient. In the case of metals, at frequencies below the visible region, the absorption is very strong, so that $\kappa \gg n$. In this case, the terms with n in Eq. (8.21) can be neglected in the presence of κ^2, so that $R \approx 1$. Thus, the reflectivity approaches 100%, that is precisely one of the features of metals. It is easy to verify that the transmission, defined as the ratio between the intensities of the transmitted and incident waves, is related to R by $T = 1 - R$. This relationship also expresses the conservation of energy in the process whereby a wave is incident on the interface between two media.

Example 8.1 An electromagnetic wave with wavelength of 500 nm strikes the plane surface of an intrinsic CdTe semiconductor sample. Considering that at this wavelength CdTe has negligible conductivity and its dielectric constant has real part $\varepsilon'_r = 8.9$ and imaginary part $\varepsilon''_r = 2.3$ calculate: (a) The phase velocity of the radiation at the given wavelength; (b) The absorption coefficient; (c) The reflectivity; (d) The total transmission of a CdTe plate with parallel faces and thickness 0.2 μm.

(a) To calculate the phase velocity, it is necessary to relate the refractive index n with the real and imaginary parts of the dielectric constant. Making $\sigma = 0$ and replacing (8.7) in (8.6) we obtain the equation for n

$$n^2 - \left(\frac{\varepsilon''_r}{2n}\right)^2 = \varepsilon'_r,$$

which leads to the following biquadratic equation,

$$n^4 - \varepsilon'_r n^2 - \frac{\varepsilon''^2_r}{4} = 0.$$

The positive solution of this equation is

$$n = \frac{1}{\sqrt{2}}\left[\varepsilon'_r + (\varepsilon'^2_r + \varepsilon''^2_r)^{1/2}\right]^{1/2},$$

which gives

$$n = \frac{1}{\sqrt{2}}\left[8.9 + (8.9^2 + 2.3^2)^{1/2}\right]^{1/2} = 3.01.$$

Thus

$$v_p = \frac{c}{n} = \frac{3 \times 10^8}{3.01} = 9.97 \times 10^7 \, \text{m/s}.$$

(b) The extinction coefficient is calculated by a similar procedure,

$$\kappa = \frac{1}{\sqrt{2}}\left[-\varepsilon'_r + (\varepsilon'^2_r + \varepsilon''^2_r)^{1/2}\right]^{1/2}$$

$$\kappa = \frac{1}{\sqrt{2}}[-8.9 + (8.9^2 + 2.3^2)^{1/2}]^{1/2} = 0.38.$$

The absorption coefficient is given by (8.14)

$$\alpha = \frac{2\omega\kappa}{c} = \frac{4\pi\,\nu\,\kappa}{c} = \frac{4\pi\,\kappa}{\lambda}$$

thus

$$\alpha = \frac{4 \times 3.14 \times 0.38}{500 \times 10^{-9}} = 9.55 \times 10^6\,\text{m}^{-1} = 9.55\,\mu\text{m}^{-1}.$$

(c) The reflectivity is given by (8.21)

$$R = \frac{(3.01 - 1)^2 + 0.38^2}{(3.01 + 1)^2 + 0.38^2} = 0.258$$

Thus, the reflectivity of CdTe at 500 nm is 25.8%.

(d) To calculate the total transmission approximately, it is necessary to consider the transmission at the two surfaces and along the thickness d of the plate,

$$T = (1 - R)^2 e^{-\alpha d} = (1 - 0.258)^2 e^{-9.55 \times 0.2} = 0.08.$$

Note that this calculation is only approximate, because rigorously one should solve the full problem of a plate in air, considering the three regions of space. However, since the attenuation of the wave inside the plate is considerable, the approximate calculation is satisfactory.

8.2 Interaction of Radiation with Matter: Classical Model

In this section we study some interaction mechanisms of electromagnetic waves with materials that can be described by classical physics. The objective is to understand some basic phenomena with simple models that allow the calculation of $\varepsilon_r(\omega)$ and $\sigma(\omega)$. As we saw in the previous section, these parameters determine $n(\omega)$ and $\kappa(\omega)$, and therefore the optical properties of materials.

Initially, we discuss the interaction of radiation with free electrons, which plays an essential role in the optical properties of metals. Then we shall study the classical model of the interaction with bound electrons. To understand in detail the optical properties of insulators and semiconductors, it will be necessary to consider the quantum nature of this interaction, that will be studied in the following section.

The interaction of radiation with a material results from the force that the electric field of the wave exerts on the electric charges of electrons and ions. Since the field oscillates at a certain frequency ω, it tends to create a harmonic motion of the charges with the same frequency. However, this motion will only be significant if the charges have a natural vibration mode with frequency close to that of the field. For this reason, in the case of ions, the interaction with the electromagnetic field is important only if it has frequency in the range 1–10 THz (4–40 meV), characteristic of the optical vibration modes of the crystal lattice. For this reason, in insulators and semiconductors, phonons dominate the optical properties in the far infrared. However, in metals, the interaction of the field with free electrons dominates the material response, and makes the reflectivity close to 1, as we shall see next.

In the near-infrared, visible, and ultraviolet regions, the motion of the ions is negligible, so that the optical properties in these ranges are dominated by the interaction of the electric field with the electrons, free or bound to the atoms. We shall study various aspects of this interaction separately in two sub-sections.

8.2.1 Contribution of Free Electrons in Metals

In metals, the behavior of free electrons is determinant for the optical properties in a wide range of frequencies. The motion of electrons under the action of the electric field of a plane wave

$$E = E_0\, e^{-i\omega t}, \tag{8.22}$$

can be calculated with an extension of the concepts presented in Sect. 4.5, for time-varying fields. Since $-eE$ is the force that the field exerts on the electron, the electron equation of motion is

$$m\frac{dv}{dt} + \frac{m}{\tau} v = -e\, E_0\, e^{-i\omega t}, \tag{8.23}$$

where m, v, and τ are, respectively, the mass, velocity, and collision time of the electron. The second term of (8.23) represents the damping in the electron motion due to collisions with the lattice and with impurities. Replacing in (8.23) the solution for the steady-state regime $v = v_0 \exp(-i\omega t)$, we obtain

$$v_0 = \frac{-e\, E_0}{-im\,\omega + m/\tau}. \tag{8.24}$$

Considering N free electrons per unit volume, we obtain for the current density

$$J = -e\, N\, v_0\, e^{-i\omega t}. \tag{8.25}$$

Substituting (8.24) in (8.25) and using the relation $J = \sigma E$, we obtain for the conductivity of the metal at the frequency of the electromagnetic wave

$$\sigma(\omega) = \frac{N e^2 \tau}{m(1 - i \omega\tau)}. \tag{8.26}$$

This result is known as the conductivity of the Drude model. Note that by making $\omega = 0$ in Eq. (8.26), we obtain the DC conductivity given by Eq. (4.30), as expected. Substituting (8.26) in Eq. (8.3) we obtain for the complex refractive index of metals

$$N^2(\omega) = \varepsilon_c + \frac{i N e^2 \tau}{m \omega \varepsilon_0 (1 - i \omega\tau)}, \tag{8.27}$$

where ε_c is the contribution to the dielectric constant of bound electrons. As we will see later, this contribution is more important in the visible and ultraviolet regions of the electromagnetic spectrum, and is approximately constant in the infrared. The fact that $\sigma(\omega)$ is complex makes the expressions for the real and imaginary parts of $N(\omega)$ relatively large and difficult to analyze. For this reason, we will analyze $\sigma(\omega)$ only approximately in two limits, low and high frequencies.

In the low-frequency approximation, we make $\omega\tau << 1$ in Eqs. (8.26) and (8.27). In alkali metals (Li, Na, K, Rb and Cs) and noble metals (Cu, Ag and Au), $\tau \sim 10^{-13}$ s, so that this approximation corresponds to $\omega << 10^{13}$ s^{-1}. Thus, it is valid for the far infrared region. In this region, neglecting $i\omega\tau$ in the presence of 1 in Eq. (8.26), we see that $\sigma(\omega) = N e^2 \tau/m = \sigma_0$ is the DC conductivity given by (4.30). Hence, we can write (8.27) in the form

$$N^2(\omega) \approx \varepsilon_c + i \frac{\sigma_0}{\omega \varepsilon_0}. \tag{8.28}$$

Using the values $\sigma_0 \sim 10^8$ Ω^{-1} m^{-1} (see Fig. 4.17), $(4\pi\varepsilon_0)^{-1} = 9 \times 10^9$ Nm^2/C^2, and $\omega \sim 10^{12}$ s^{-1}, we see that the imaginary part of $N^2(\omega)$ in (8.28) is much larger than the real part, which is $\varepsilon_c \sim 1$. Thus, in the far infrared region metals have complex refractive index $N(\omega)$ given approximately by

$$N(\omega) \approx \left(\frac{\sigma_0}{\omega \varepsilon_0}\right)^{1/2} (i)^{1/2} = (1+i)\left(\frac{\sigma_0}{2\omega \varepsilon_0}\right)^{1/2}$$

Therefore, the real and imaginary parts of the refractive index are equal and much larger than 1,

$$n = \kappa \approx \left(\frac{\sigma_0}{2\omega \varepsilon_0}\right)^{1/2} \gg 1. \tag{8.29}$$

Substitution of this result in Eq. (8.21) shows that the reflectivity is close to unity, $R \approx 1$. This result explains why metals are almost perfect reflectors of electromagnetic waves with frequencies below the infrared region. They are not perfect reflectors because a small fraction of the wave energy penetrates a thin layer in the surface, is absorbed by free electrons and transformed into heat in the collision processes. This is the **skin effect**, characterized by a **penetration length**, or **skin depth**, δ, given by twice the inverse of the absorption coefficient. Replacing (8.29) in Eq. (8.15) and using $c = (\varepsilon_0 \mu_0)^{-1/2}$ we obtain

$$\delta = \frac{c}{\omega \kappa} = \left(\frac{2}{\omega \mu_0 \sigma_0} \right)^{1/2}. \tag{8.30}$$

For copper, at room temperature, with $\sigma_0 = 0.6 \times 10^8 \; \Omega^{-1} \; m^{-1}$, Eq. (8.30) gives $\delta = 0.066$ mm at $\nu = 1$ MHz, and $\delta = 6.6 \; \mu m$ at $\nu = 100$ MHz.

In the high-frequency approximation, $\omega \tau \gg 1$, valid for the near infrared, visible, and ultraviolet regions, we can neglect the unity in the denominator of Eq. (8.27) and write

$$N^2(\omega) = \varepsilon_c \left(1 - \frac{\omega_p^2}{\omega^2} \right), \tag{8.31}$$

where

$$\omega_p^2 = \frac{N e^2}{m \; \varepsilon_0 \; \varepsilon_c}, \tag{8.32}$$

is the square of the **plasma frequency** ω_p of the metal. Its value in common metals is on the order of 10^{15} Hz, corresponding to an energy of about 4 eV, situated at the end of the visible range and the beginning of the ultraviolet. Equation (8.31) shows that for $\omega < \omega_p$, the refractive index $N(\omega)$ is pure imaginary, that is, $n = 0$ and $\kappa = [\varepsilon_c(\omega_p^2/\omega^2 - 1)]^{1/2}$. In this situation, the reflectivity, given by Eq. (8.21), is exactly $R = 1$. For this reason, similarly to the behavior in the infrared, metals are also good reflectors of radiation in the visible region. On the other hand, for $\omega > \omega_p$, $N(\omega)$ is real, and the absorption due to free electrons is null.

Figure 8.3 shows the variation in the reflectivity of silver, in (a) with the incident photon energy, and in (b) with the wavelength on an enlarged scale to highlight details in the visible region. Note that silver has reflectivity almost 100% in the whole energy range from zero to the end of the visible region, so that it is a good reflector over the entire spectrum of white light. The reflectivity drops sharply to near zero in the vicinity of the plasma frequency in the near ultraviolet. For higher energies, R exhibits other variations caused by transitions of bound electrons, that will be studied in the next section. It is interesting to note that in the case of copper, these transitions produce a variation in reflectivity within the visible region. In this case, the reflectivity is high throughout the visible range, but it is higher in the red region than in the blue. For this reason, the copper reflection has an orange color, which contrasts with the "silver color" of silver.

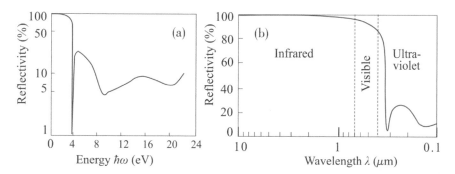

Fig. 8.3 Reflectivity of silver. (a) As a function of the photon energy of the incident electromagnetic wave. Reprinted with permission from H. Ehrenreich et al., Phys. Rev. **128**, 1622 (1962). Copyright (1972) by the American Physical Society. (b) Dependence on the wavelength, on an enlarged scale

Example 8.2 The concentration of free electrons in silver is 5.86×10^{22} cm^{-3} and the collision time is 3.8×10^{-14} s. Calculate: (a) The penetration length in silver of an electromagnetic wave with microwave frequency, $\nu = 1$ GHz; (b) The penetration length of an argon laser beam with wavelength $\lambda = 514.5$ nm; (c) The attenuation suffered by the laser beam passing through a silver film with thickness of 10 nm, not considering the reflections.

(a) Initially, it is necessary to calculate the product $\omega\tau$. With the data given

$$\omega\tau = 2\pi \times 10^9 \times 3.8 \times 10^{-14} = 2.38 \times 10^{-3}.$$

Since $\omega\tau \ll 1$, the penetration length can be calculated with Eq. (8.15). Using the value of the conductivity of silver calculated in Example 4.3, $\sigma_0 = 6.26 \times 10^7$ $(\Omega\text{m})^{-1}$, we have

$$\delta = \left(\frac{2}{\omega\mu_0\sigma_0}\right)^{1/2} = \left(\frac{2}{2\pi \times 10^9 \times 4\pi \times 10^{-7} \times 6.26 \times 10^7}\right)^{1/2}$$
$$\delta = 2.01 \times 10^{-6}\text{m} = 2.01\ \mu\text{m}.$$

(b) The frequency of the argon laser is

$$\omega = 2\pi\nu = 2\pi\frac{c}{\lambda} = \frac{2\pi \times 3 \times 10^8}{514.5 \times 10^{-9}} = 3.66 \times 10^{15}\text{ s},$$

so that

$$\omega\tau = 3.66 \times 10^{15} \times 3.8 \times 10^{-14} = 1.39 \times 10^2.$$

In this case, $\omega\tau \gg 1$ and the complex refractive index is given by Eq. (8.31). Considering $\varepsilon_c = 1$, the plasma frequency of silver, given by Eq. (8.32), is

$$\omega_p^2 = \frac{Ne^2}{m\,\varepsilon_0\varepsilon_c} = \frac{5.86 \times 10^{22} \times 10^6 \times 1.6^2 \times 10^{-38} \times 36\pi \times 10^9}{9.1 \times 10^{-31}}\,\text{s}^{-2},$$

$$\omega_p^2 = 1.86 \times 10^{32}\,\text{s}^{-2}.$$

Thus, $\omega_p = 1.36 \times 10^{16}\,\text{s}^{-1}$ is larger than ω, which makes $N(\omega)^2$ negative and therefore $N(\omega)$ is imaginary. From Eq. (8.31) we have

$$N = i\varepsilon_c^{1/2}\left(\frac{\omega_p^2}{\omega^2} - 1\right)^{1/2} = i\left(\frac{1.86 \times 10^{32}}{3.66^2 \times 10^{30}} - 1\right)^{1/2} = i(13.9 - 1)^{1/2},$$

$$N = i\,3.6.$$

Equation (8.5) shows that the imaginary part of N is the extinction coefficient itself. Thus, $\kappa = 3.6$, and the penetration length calculated with (8.15) becomes

$$\delta = \frac{c}{\omega\kappa} = \frac{3 \times 10^8}{3.66 \times 10^{15} \times 3.6} = 2.27 \times 10^{-8}\,\text{m},$$

$$\delta = 22.7\,\text{nm}.$$

(c) The beam attenuation in a distance $d = 10$ nm is

$$e^{2d/\delta} = e^{20/22.7} = e^{0.88} = 2.41.$$

This means that when passing through the silver film, the intensity of the laser beam decreases by a factor of 2.41. The attenuation can also be expressed in decibel

$$A = 10\log_{10}2.41 = 3.83\,\text{dB}.$$

8.2.2 *Contribution of Bound Electrons*

As mentioned earlier, the optical properties of the materials at energies of the order or larger than 1 eV are mainly due to the transitions of electrons bound to atoms, or valence electrons. The correct treatment of these transitions will be presented in the next section and requires the use of quantum mechanics.

However, it is possible to understand certain aspects of the phenomenon with a simple model due to Lorentz. In this model, based on the classical view of the atom, it is assumed that the application of an electric field results in the displacement of the negative electronic shells relative to the positive nucleus, as illustrated in Fig. 8.4a. This produces an electric dipole moment that contributes to the permittivity of the material. The relative displacement of charges also creates a restoring electrostatic force that influences the motion. In the linear approximation, this force is proportional to the displacement, as in a harmonic oscillator. The simplified model shown in Fig. 8.4b consists of a mass-spring assembly, in which a particle of mass m and charge $-e$, equal to those of the electron, is under the action of a force produced by the electric field of the radiation. For a harmonic time-varying electric field with frequency ω, as in (8.22), the equation of motion of the electron is

$$m\left(\frac{d^2x}{dt^2} + \Gamma\frac{dx}{dt} + \omega_0^2 x\right) = -e\,E_0\,e^{-i\omega t}, \tag{8.33}$$

where x is the displacement of the electron relative to its equilibrium position, ω_0 is the resonance frequency of the oscillator, and Γ is the damping rate of the motion. The first term in Eq. (8.33) is the electron acceleration, which multiplied by the mass is equal to the sum of the forces. The second term is responsible for the damping motion, and corresponds to a counter force proportional to the electron velocity. It is similar to the second term of Eq. (8.23), however it is not the inverse of the collision time because in the present case the electron is bound to the atom. Finally, the third term is a spring restoring force that simulates the binding of the electron to the atom. Denoting by k the spring constant, this force is $-kx$, where $k = \omega_0^2\,m$. The solution of Eq. (8.33) in the steady-state regime is

$$x(t) = \frac{-e\,E_0}{m(\omega_0^2 - \omega^2 - i\omega\Gamma)}\,e^{-i\omega t}. \tag{8.34}$$

The electron displacement, given by (8.34), produces in the atom an instantaneous electric dipole with moment $p = -ex$. If there are N atoms in the material per unit volume, the resulting polarization, which is the electric dipole moment per unit volume, is $P = -Nex$. Recalling the relationship between the displacement vector, the polarization, and the electric field, which defines the permissivity

Fig. 8.4 a Illustration of the classical effect of an external electric field E on the charges in an atom. **b** Simplified model of the atom under the action of the electric field

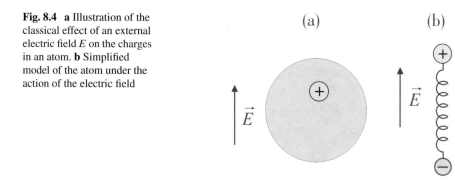

$$\vec{D} = \varepsilon_0 \vec{E} + \vec{P} \equiv \varepsilon_r \varepsilon_0 \vec{E}, \tag{8.35}$$

we obtain for the dielectric constant at a frequency ω

$$\varepsilon_r(\omega) = 1 + \frac{\omega_p^2}{\omega_0^2 - \omega^2 - i\omega\Gamma}, \tag{}$$

where $\omega_p^2 = Ne^2/m\varepsilon_0$. Note that the value of ε_r in this equation tends to 1 when $\omega \to \infty$. In fact, this does not happen, since this result represents only the contribution of one bound electron. The contribution of the other electrons with larger oscillator frequencies causes the real part of ε_r at high frequencies to be larger than 1. Representing this contribution by ε_∞, we can write the dielectric constant at frequencies close to ω_0 as

$$\varepsilon_r(\omega) = \varepsilon_\infty + \frac{\omega_p^2}{\omega_0^2 - \omega^2 - i\omega\Gamma}. \tag{8.36}$$

From Eq. (8.36) we obtain the real and imaginary parts of the dielectric constant

$$\varepsilon_r'(\omega) = \varepsilon_\infty + \frac{\omega_p^2(\omega_0^2 - \omega^2)}{(\omega_0^2 - \omega^2)^2 + \omega^2\Gamma^2}, \tag{8.37}$$

$$\varepsilon_r''(\omega) = \frac{\omega_p^2\,\omega\,\Gamma}{(\omega_0^2 - \omega^2)^2 + \omega^2\Gamma^2}. \tag{8.38}$$

The variations with frequency of the two components are shown in Fig. 8.5 for $\omega_p = 0.7\,\omega_0$, $\Gamma = 0.05\,\omega_0$, and $\varepsilon_\infty = 2.0$. Note that the real part of $\varepsilon_r(\omega)$ is negligible throughout the whole frequency range, except in the vicinity of the resonance frequency ω_0. Since the imaginary part of $\varepsilon_r(\omega)$ is related to the optical

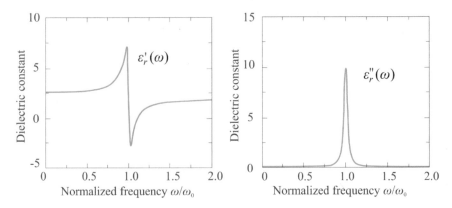

Fig. 8.5 Variation with frequency of the real and imaginary parts of the dielectric constant in the classical model of the radiation-electron interaction, calculated with Eqs. (8.37) and (8.38) using $\omega_p = 0.7\,\omega_0$, $\Gamma = 0.05\,\omega_0$, and $\varepsilon_\infty = 2.0$

absorption in the material, as in Eqs. (8.7) and (8.14), this result means that there is absorption only at $\omega \approx \omega_0$. This same conclusion is reached with the quantum treatment of the interaction of radiation with the material, where $\hbar\omega_0$ is the energy separation between the two electronic quantum levels.

The graphical representation of the function that describes $\varepsilon_r''(\omega)$ is called **absorption lineshape**. Since for $\Gamma \ll \omega_0$, $\varepsilon_r''(\omega)$ is significant only at frequencies $\omega \approx \omega_0$, we can make $\omega_0 + \omega \approx 2\omega_0 \approx 2\omega$ in Eq. (8.38) and rewrite $\varepsilon_r''(\omega)$ in the form

$$\varepsilon_r''(\omega) \approx \frac{\omega_p^2 \, \Gamma/4\omega_0}{(\omega_0 - \omega)^2 + (\Gamma/2)^2} \equiv \frac{\pi \, \omega_p^2}{2\omega_0} f_L(\omega), \tag{8.39}$$

where

$$f_L(\omega) = \frac{\Gamma/2\pi}{(\omega_0 - \omega)^2 + (\Gamma/2)^2}, \tag{8.40}$$

is the **Lorentzian function**. The constant 2π used in the definition makes the area under the curve equal to unity, $\int f_L(\omega)d\omega = 1$. The maximum value of this function occurs at $\omega = \omega_0$, $f_L(\omega_0) = 2/(\pi\Gamma)$, and therefore the peak value is inversely proportional the damping rate Γ. On the other hand, the **linewidth**, defined as the difference between the two frequencies for which $f_L(\omega) = f_L(\omega_0)/2$, is precisely Γ (Problem 8.7). Thus, a small damping rate results in a small linewidth and large peak absorption. This same result will be obtained in the next section by a quantum treatment of the radiation-bound electron interaction.

It is important to note that $\varepsilon_r'(\omega)$ has the form of the derivative of $\varepsilon_r''(\omega)$ with respect to the frequency ω. This is not just a coincidence, it is the consequence of a general result by which $\varepsilon_r''(\omega)$ is given by the integral of a function related to $\varepsilon_r'(\omega)$, and vice versa. These integral equations are called Kramers–Kronig relations that apply to the real and imaginary parts of $\varepsilon_r(\omega)$, and are valid for any mechanism of the radiation-matter interaction. Finally, to conclude this section, let us find the optical constants of a material described by the classical model for bound electrons. Making $\sigma = 0$ and replacing Eq. (8.7) in (8.6), we obtain biquadratic equations for the refractive index n and for the extinction coefficient κ, as shown in Example 8.1,

$$n = \frac{1}{\sqrt{2}} \left[\varepsilon_r' + (\varepsilon_r'^2 + \varepsilon_r''^2)^{1/2} \right]^{1/2}, \tag{8.41}$$

$$\kappa = \frac{1}{\sqrt{2}} \left[-\varepsilon_r' + (\varepsilon_r'^2 + \varepsilon_r''^2)^{1/2} \right]^{1/2}. \tag{8.42}$$

Figure 8.6a shows the functions $n(\omega)$ and $\kappa(\omega)$ obtained with Eqs. (8.37)–(8.42) with the same parameters used in Fig. 8.5. Note that the refractive index is close to 1 in most of the frequency range, except in the vicinity of the resonance frequency ω_0, where it exhibits a pronounced anomaly. On the other hand, the extinction coefficient, responsible for the wave attenuation, is negligible in most of the frequency range but

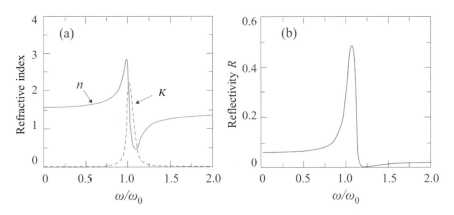

Fig. 8.6 a Variation with frequency of the refractive index n and the extinction coefficient (or damping) κ, calculated with Eqs. (8.37)–(8.42), using the same parameters used in Fig. 8.5. **b** Reflectivity calculated with Eq. 8.21

it peaks at $\omega = \omega_0$. Finally, it is important to note that the refractive index increases with frequency in the region $\omega < \omega_0$, as evidenced in the curve of $n(\lambda)$ for fused quartz, shown in Fig. 8.1b. This is what produces the dispersion phenomenon in transparent materials. This is so because the frequencies of electronic transitions of these materials are above the visible range. From n and κ, we can calculate the reflectivity R of the material using Eq. (8.21). The plot of $R(\omega)$ obtained from $n(\omega)$ and $\kappa(\omega)$ in Fig. 8.6a is shown in Fig. 8.6b. Clearly, the reflectivity has a peak at a frequency slightly above ω_0 and has a lineshape wider than those of $\varepsilon_r''(\omega)$ and $\kappa\ (\omega)$, because there is also an important contribution to the dispersion from $n(\omega)$.

Note that in real materials there are several electronic transitions with different frequencies, and therefore $R(\omega)$ presents several peaks. It is one of these peaks at an energy close to 2 eV that gives copper the orange color.

Example 8.3 A dielectric material has an absorption line due to an infrared phonon with angular frequency $\omega_0 = 2 \times 10^{14}$ s^{-1}, linewidth $\Gamma = 10^{13}$ s^{-1}, and plasma frequency $\omega_p = 0.7\ \omega_0$. Considering that $\varepsilon_\infty = 2.0$, calculate the absorption coefficient and the reflectivity of the material for an infrared beam with frequency equal to that of the absorption peak.

Initially, it is necessary to calculate the real and imaginary components of the dielectric constant. Using Eq. (8.37) with $\omega = \omega_0$ one has $\varepsilon_r' = \varepsilon_\infty = 2.0$. Using (8.38) with $\omega = \omega_0$ we have

$$\varepsilon_r''(\omega_0) = \frac{\omega_p^2}{\omega_0 \Gamma} = \frac{0.7^2 \times 2 \times 10^{14}}{10^{13}} = 9.8.$$

The refractive index n and the extinction coefficient κ are calculated with Eqs. (8.41) and (8.42)

$$n = \frac{1}{\sqrt{2}}\left[2 + \sqrt{4 + 96}\right]^{1/2} = 2.45,$$

$$\kappa = \frac{1}{\sqrt{2}}\left[-2 + \sqrt{4 + 96}\right]^{1/2} = 2.0.$$

The absorption coefficient, given by Eq. (8.14), is

$$\alpha = \frac{2 \times 2 \times 10^{14} \times 2.0}{3 \times 10^8} = 2.67 \times 10^6 \, \text{m}^{-1}.$$

The reflectivity, given by Eq. (8.21), is

$$R = \frac{(2.45 - 1)^2 + 2.0^2}{(2.45 + 1)^2 + 2.0^2} = \frac{6.10}{15.9} = 0.38.$$

Note that the values obtained for ε_r, n, κ, and R, coincide with the values in Figs. 8.5 and 8.6 at $\omega = \omega_0$. Note also in Fig. 8.6 that the peak in R does not occur exactly at ω_0, as pointed out earlier.

8.3 Quantum Theory of the Radiation-Matter Interaction

In the previous section we made the assumption that electrons behave like classical particles, bound to the atom by a harmonic-oscillator type force. This model led to the result that the optical absorption occurs when the frequency of the radiation field is approximately equal to that of the oscillator resonance. We know, however, that in quantum mechanics the electron is described by a wave function, whose modulus squared represents the probability of finding it in a certain position. Now we need to understand how the quantum nature of the electron influences the absorption of light by the material.

As we shall see below, the result of the quantum interaction of light with a two-level system is consistent with that of the classical model. However, there are various aspects of this interaction that are not contained in the classical model and which are essential for the understanding of certain optical properties of materials, such as, for example, the interband transitions and the stimulated emission of light.

8.3.1 Transitions Between Discrete Levels

Let us first consider a system in which an electron can occupy states with discrete energy levels. The system can be an electron in an infinite potential well, or in an atom, for example. As we saw in Chap. 3, if the potential to which the electron is subjected to does not vary in time, the electron can occupy stationary states characterized by a quantum number n, described by wave functions of the type

$$\Psi(\vec{r}, t) = \psi_n(\vec{r})\, e^{-i(E_n/\hbar)\,t}. \tag{8.43}$$

In this case, the expectation value of any operator, given by Eq. (3.14), does not vary in time. Thus, the electron remains in this state indefinitely and its energy E_n is constant. Let us see what happens if the electron is in a state given by the linear combination of two stationary states, for example

$$\Psi(\vec{r}, t) = \psi_1\, e^{-i(E_1/\hbar)\,t} + \psi_2\, e^{-i(E_2/\hbar)\,t}. \tag{8.44}$$

This equation can be rewritten in the form

$$\Psi(\vec{r}, t) = e^{-i(E_1/\hbar)\,t}\big[\psi_1 + \psi_2\, e^{-i\omega_{21}\,t}\big]. \tag{8.45}$$

where $\omega_{21} = (E_2 - E_1)/\hbar$. With the definition (3.14), we can write the expectation value of any operator F in the state (8.45) as

$$\langle F(t)\rangle = \langle F\rangle_1 + \langle F\rangle_2 + \int \psi_1^* F \psi_2 e^{-i\omega_{21}\,t}\, dV + \int \psi_2^* F \psi_1 e^{i\omega_{21}\,t}\, dV$$

where $\langle F\rangle_1$ and $\langle F\rangle_2$ are the expectation values of F in the stationary states 1 and 2, which are constant. If F is a Hermitian operator, that is

$$\int \psi_2^* F \psi_1\, dV = \left(\int \psi_1^* F \psi_2\, dV\right)^* \equiv |F_{12}|\, e^{i\varphi},$$

then we have

$$\langle F(t)\rangle = \langle F\rangle_1 + \langle F\rangle_2 + 2\,|F_{12}|\,\cos(\omega_{21}\,t + \varphi). \tag{8.46}$$

This result shows that if the electron is in a state that is a combination of states with energies E_1 and E_2, the expectation value of an operator varies harmonically in time with angular frequency $\omega_{21} = (E_2 - E_1)/\hbar$. One can show, without difficulty, that the probability of finding the electron in state 1, or in state 2, also varies harmonically in time with frequency ω_{21}. We say that the electron undergoes transitions from state 1 to state 2, and vice versa. For an electron that is initially in state 1, or 2, to undergo transitions between 1 and 2, it is necessary to have an external action that varies in time with frequency ω close to ω_{21}. Using quantum mechanics, we shall see below how to calculate the probability for these transitions to occur.

Consider an electron in an atom, or in any potential well, subjected to an external time-dependent disturbance. This disturbance can be the force of the electric field of an electromagnetic wave, for example. Schrödinger's Eq. (3.24) for the electron is then

$$\left[H_0 + H'(t)\right] \Psi = i\hbar \frac{\partial \Psi}{\partial t}, \tag{8.47}$$

where H_0 is the constant Hamiltonian corresponding to the potential well and $H'(t)$ represents the time-varying interaction due to the external disturbance. Usually the perturbation is small, that is, the external disturbance is much smaller than the interaction that keeps the electron in the well. As we know that the effect of the disturbance is to cause transitions between electronic states, we look for solutions to Eq. (8.47) in the form of an expansion in eigenfunctions ψ_n of the unperturbed Hamiltonian H_0, which we consider known, in the form

$$\Psi(t) = \sum_n a_n(t) \, \psi_n \, e^{-i(E_n/\hbar)t}. \tag{8.48}$$

In order to obtain the coefficients $a_n(t)$ that determine the wave function, we substitute (8.48) into (8.47) and use the equation $H_0\Psi_n = E_n \Psi_n$. Then we have

$$\sum_n a_n(t) \left[E_n + H'(t)\right]\psi_n \, e^{-i(E_n/\hbar)t} = i\hbar \sum_n \left(\frac{da_n}{dt} - i\frac{E_n}{\hbar}a_n\right) \psi_n \, e^{-i(E_n/\hbar)t}.$$

With the cancellation of the terms in E_n on both sides, and switching the left and hand sides, this equation becomes

$$i\hbar \sum_n \frac{da_n}{dt} \, \psi_n \, e^{-i(E_n/\hbar)t} = \sum_n a_n(t) \, H'(t)\psi_n \, e^{-i(E_n/\hbar)t}.$$

Multiplying both sides to the left by $\psi_m^* \, e^{-i(E_m/\hbar)t}$, integrating in the volume, and using Eq. (3.13) that expresses the normalization and orthogonality of the eigenfunctions, we obtain for the coefficients of the expansion in (8.48)

$$\frac{da_m}{dt} = \frac{1}{i\hbar} \sum_n a_n(t) \, e^{i(E_m - E_n)t/\hbar} \int \psi_m^* \, H'(t) \, \psi_n \, dV. \tag{8.49}$$

Since no approximation has been made so far, this equation is entirely equivalent to the time-dependent Schrödinger equation and is exact. It is the basis of the matrix formulation of quantum mechanics. Now let us consider that the disturbance is produced by a time-harmonic field with frequency ω, so that

$$H'(t) = H' e^{-i\omega t}. \tag{8.50}$$

Thus, Eq. (8.49) takes the form

$$\frac{da_m}{dt} = \frac{1}{i\hbar} \sum_n a_n(t) H'_{mn} e^{i(\omega_{mn} - \omega)t}, \tag{8.51}$$

where $\omega_{mn} = (E_m - E_n)/\hbar$ and H'_{mn} is the matrix element of the operator H' between the states m and n, defined by

$$H'_{mn} = \int \psi_m^* H' \psi_n \, dV. \tag{8.52}$$

Equation (8.51) can be used to calculate the time evolution of the state of the system due to the perturbation $H'\exp(-i\omega t)$. Let us assume that an electron is initially in a discrete state n of a Hamiltonian H_0, when a small external perturbation of the type (8.50) is applied. From Eq. (8.51) one can calculate the coefficient $a_m(t)$ corresponding to a state m at an instant t and therefore the probability of finding the electron in this state, given by $|a_m|^2$. It can then be shown (Appendix A) that the **probability per unit time** for the electron to undergo a transition to a set of states m very close to each other is given by

$$W_{mn} = \frac{2\pi}{\hbar} |H'_{mn}|^2 D(E_m = E_n + \hbar\omega). \tag{8.53}$$

where D is the density of states with energy $E_m = E_n + \hbar\omega$. Equation (8.53) is called the **Fermi golden rule** of perturbation theory. In the case that the transition between two energy levels is broadened due to the effect of damping, the density of states is given by a Lorentzian lineshape. In the case of a transition between two electronic energy bands, the density of states to be considered has the shape of the one in Fig. 4.10.

8.3.2 Light Absorption and Luminescence

In this section we use quantum mechanical perturbation theory to calculate the effects of the interaction of an electromagnetic wave with a system of atoms. Initially we consider a set of N independent atoms (per unit volume), with energy levels E_1, E_2, E_3, etc. This is the situation that occurs in gases or in transitions between discrete levels in solids. When the system is in thermal equilibrium at a certain temperature T, electrons undergo transitions from one level to another due to the interactions with thermal phonons, in the case of solids, or atomic collisions, in the case of gases. However, while electrons in a certain number come out of a certain level, other electrons in equal number arrive at that level (in other atoms), in order to maintain the thermal equilibrium distribution of the system. In statistical mechanics it is shown that, in thermal equilibrium, the populations N_i and N_j at the energy levels E_i and E_j (N_i is the number of atoms per unit volume with electrons at level E_i) obey the relation

$$\frac{N_i}{N_j} = \frac{e^{-E_i/k_BT}}{e^{-E_j/k_BT}} = e^{-(E_i-E_j)/k_BT}. \tag{8.54}$$

This is the **Boltzmann distribution**, which applies to a system of distinguishable particles, which in the present case is the collection of atoms. According to Eq. (8.54), the population of a certain level i decreases exponentially with increasing energy E_i, or with decreasing temperature. This is an expected result, because it is the thermal excitation that takes electrons from a certain state to other states with higher energy levels.

The presence of an electromagnetic field in the system tends to produce transitions between energy levels whose separation is close to the energy of the photons $\hbar\omega$. Equation (8.53) shows that the transition probability induced by the field from level m to level n is equal to that from n to m. Thus, the tendency of the radiation is to make the populations N_n and N_m approach each other. However, this equality would only occur if the field intensity were high enough to overcome the role of thermal excitation. This effect is important in the case of lasers. In this section we assume that the field is small and that the thermal equilibrium is not disturbed.

Let us use quantum theory to calculate the optical parameters of a system of atoms, considering, to simplify, that they have only two energy levels, E_1 and E_2 ($E_2 > E_1$), with populations N_1 and N_2. When an electric field

$$\vec{E} = \hat{y}\, E\, e^{-i\omega t}$$

is applied to the system, electrons are subjected to an interaction with energy $-e\phi(y)$ $= eEy$, which results in a perturbation Hamiltonian given by

$$H'(t) = e\,E\,y\,e^{-i\omega t} \equiv -E\,p_y\,e^{-i\omega t},$$

where $p_y = -ey$ is the y component of the electric dipole moment operator. According to the Fermi golden rule (8.53), the transition probabilities, per unit time, for the system to go from level 1 to level 2, or vice versa, are given by

$$W_{12} = W_{21} = \frac{2\pi}{\hbar}\, E^2\, p_{12}^2\, D(E_2 - E_1 = \hbar\omega), \tag{8.55}$$

where p_{12} is the matrix element of the p_y operator between states 1 and 2 (we omit the index y to simplify the notation). If there are N_1 electrons in the lower energy level (per unit volume), the power absorbed from the electromagnetic field by the system is $N_1\,W_{12}\,\hbar\omega$, because $N_1 W_{12}$ is the number of photons absorbed per unit time and per unit volume. On the other hand, $N_2\,W_{12}\,\hbar\omega$ is the power emitted by electrons

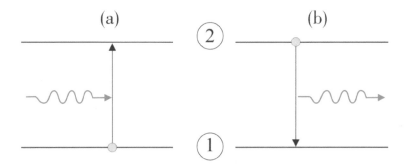

Fig. 8.7 Electronic transitions in a two-level system, by absorption (**a**) and by emission (**b**) of photons

that go from level 2 to level 1 by the emission of photons. The photon absorption and emission processes by electronic transitions in a two-level system are illustrated in Fig. 8.7.

The net power per unit volume absorbed by the system is then

$$P = \frac{2\pi}{\hbar} (N_1 - N_2) E^2 \, p_{12}^2 \, D(\hbar\omega) \, \hbar\omega.$$

Note that the absorbed power per unit volume can be identified as $-dI/dx$, since the intensity I of the wave is the transmitted power per unit area. Thus, using Eqs. (8.11) and (8.14) in the relation $dI/dx = -P$ and noting that $D(\hbar\omega) = D(\omega)/\hbar$, we obtain for the extinction coefficient

$$\kappa = \frac{2\pi}{n \, \hbar \, \varepsilon_0} (N_1 - N_2) \, p_{12}^2 \, D(\omega). \tag{8.56}$$

Considering for $D(\omega)$ a Lorentzian lineshape $f_L(\omega)$ given by (8.40), substitution of Eq. (8.56) into (8.7) provides the imaginary part of the dielectric constant

$$\varepsilon_r'' = \frac{2}{\hbar \varepsilon_0} (N_1 - N_2) \, p_{12}^2 \, \frac{\Gamma}{(\omega_{21} - \omega)^2 + (\Gamma/2)^2}. \tag{8.57}$$

where $\omega_{21} = (E_2 - E_1)/\hbar$. This expression has the same form as Eq. (8.39) with ω_{21} replaced by ω_0, and $\omega_p^2/4\omega_0$ replaced by $2(N_1 - N_2) p_{12}^2/\hbar\varepsilon_0$ (see Problem 8.8). This shows that the classical result is consistent with quantum theory, as anticipated. However, there are important details of the quantum result that do not appear in the classical treatment. From Eq. (8.57) we conclude that to have absorption of energy in electronic transitions between two levels E_1 and E_2, it is necessary that: (1) The frequency of the radiation is $\omega \approx (E_2 - E_1)/\hbar$; (2) The population of the lower level is larger than of the upper level, that is, $N_1 - N_2 > 0$; (3) The matrix element p_{12} of the electric dipole moment operator between the states of the two levels is different from zero. This last condition gives rise to the selection rules for **electric dipole transition** that determine which transitions are possible by absorption or emission of photons. As mentioned in Sect. 3.4, the selection rules for transitions in the hydrogen atom with linearly polarized field are $\Delta l = \pm 1$ and $\Delta m_l = 0$ (Problem 8.10).

The process of absorption of light, that we have just treated, occurs when an electromagnetic radiation interacts with a system of atoms producing transitions from lower energy quantum levels to higher energy levels. Another very important optical process is the **spontaneous emission of radiation**, which occurs when atoms pass from an excited state to another lower energy state, even in the absence of external radiation. The probability per unit of time for spontaneous emission with transition from level 2 to level 1, as in Fig. 8.7b, is also given by Eq. (8.55). However, in this case the field E that appears in (8.55) is that associated with the quantum fluctuations of the ground state of zero photons. If there is an electric dipole moment between states 1 and 2, the transition from 2 to 1 occurs with the emission of a photon, and is called **radiative transition**. The characteristic time of this transition is given by $\tau_R = 1/W_{12}$. If the dipole moment between the two states is zero, the transition from 2 to 1 can also occur, but in this case, instead of emitting only a photon, there is also emission of a phonon or some other elementary excitation, with much lower transition probability. This kind of transition is called **non-radiative**.

The process by which atoms are excited to higher energy states, and subsequently decay through radiative transitions, is called **luminescence**. Among the various mechanisms that produce luminescence in materials, the most important are photoluminescence and electroluminescence. Photoluminescence is one in which the atoms are taken to excited states through the absorption of photons of larger energy. This process is important in solid-state lasers with impurities, which will be presented in Sect. 8.6.2. Electroluminescence is the process by which the light emission is caused by an electric stimulus, like the passage of an electric current, such as the one that occurs in light emitting diodes and diode lasers, or by the incidence of an electron beam, or by the application of an intense electric field.

8.3.3 Absorption and Emission of Light in Insulators and Semiconductors

In the case of crystals, the quantum treatment of the radiation-matter interaction must take into account the fact that electrons are described by wave functions with wave vector $\vec{k}$. In addition, they have energy $E(k)$ in the form of bands and not discrete levels, as in the previous section. The application of the Fermi golden rule (8.53) for crystals with energy bands in the reduced scheme to the first Brillouin zone shows that the electronic transitions between bands must conserve energy and momentum. In the case of transitions involving only photons the conservation equations are

$$E_f - E_i = \pm \hbar \omega, \tag{8.58}$$

$$\vec{k}_f - \vec{k}_i = \pm \vec{k}, \tag{8.59}$$

where E_i and E_f are the electron energies in the initial and final states, respectively, $\vec{k}_i$ and $\vec{k}_f$ are the corresponding wave vectors, ω and $\vec{k}$ are the frequency and wave vector of the photon absorbed ($E_f > E_i$) or emitted ($E_f < E_i$) in the transition. For photons in the visible region, $k \approx 10^5$ cm^{-1}, so that it is negligible compared to the values at the Brillouin zone edges. Thus, the electronic transition between bands occurs with $k_f \approx k_i$. Figure 8.8 shows two absorption transitions between the valence and conduction bands in a direct gap insulator or semiconductor. The transition with minimum energy is the one that occurs at the zone center, $k_f = k_i = 0$, and with photons of energy equal to the gap, $\omega_g = E_g/\hbar$. Photons with energy smaller than E_g go through the material without causing absorption transition between the bands. On the other hand, photons with $\omega > \omega_g$ are easily absorbed because there is a large number of electronic states with $k_f = k_i > 0$. Using Eq. (8.53) to calculate the probability of transition between the valence and conduction bands, with density of states given by Eqs. (5.12) and (5.13), it can be shown that the absorption coefficient in a direct gap semiconductor varies with frequency in form,

$$\alpha(\omega) \propto (\hbar\omega - E_g)^{1/2}/\omega, \tag{8.60}$$

for $\hbar\omega > E_g$, and $\alpha(\omega) = 0$ for $\hbar\omega < E_g$. This expression, illustrated in Fig. 8.9, shows that the absorption coefficient increases rapidly with ω above the critical value $\omega_g = E_g/\hbar$, due to the increase in the density of states.

In indirect gap semiconductors, the transitions are more complicated. As illustrated by the arrow in Fig. 8.10, the transition of an electron from the top of the valence band to the minimum of the conduction band in an indirect gap semiconductor, such as silicon and germanium, requires a major change in the wave vector. This cannot be done only with the absorption of a photon, since it has $k \approx 0$. However, this transition can occur with the absorption of a photon with

Fig. 8.8 Photon absorption in direct gap semiconductors. The photon with minimum energy that is absorbed has frequency $\omega_g = E_g/\hbar$

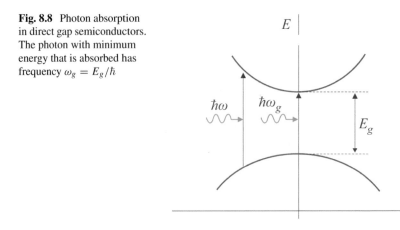

Fig. 8.9 Variation with the photon frequency of the absorption coefficient in a direct gap semiconductor

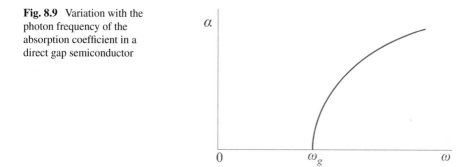

energy $\hbar\omega$ and negligible wave vector ($k \approx 0$), accompanied by the absorption of a phonon with energy $\hbar\Omega$ and wave vector $\vec{k}$, or the emission of a phonon with energy $\hbar\Omega$ and momentum $-\vec{k}$. In this case, the equations for conservation of energy and momentum are

$$E_f - E_i = \hbar\omega \pm \hbar\Omega, \qquad (8.61)$$

$$\vec{k}_f - \vec{k}_i = \pm\vec{k}, \qquad (8.62)$$

where the $+$ and $-$ signs in Eq. (8.61) correspond, respectively, to the processes of absorption and emission of a phonon. Note that in the case of the phonon absorption process, the minimum energy of the photon needed to produce the transition is $E_g - \hbar\Omega$, while in the case of the phonon emission process the minimum energy is $E_g + \hbar\Omega$. However, indirect gap semiconductors can also have a direct gap transition ($k \approx 0$) from the maximum of the valence to a relative minimum of the conduction band, with energy $E_g + E'$, as illustrated in Fig. 8.10. Since the indirect gap transition involves three elementary excitations, its probability is smaller than in the direct

Fig. 8.10 Electronic transitions from the top of the valence band to two minima of the conduction band in an indirect gap semiconductor such as Si. The indirect transition involves phonons and has energy E_g. The direct transition has energy $E_g + E'$

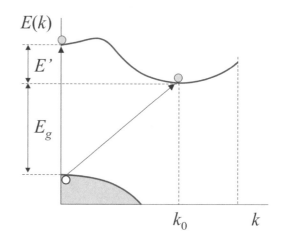

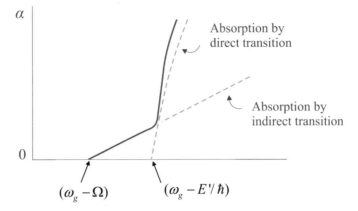

Fig. 8.11 Variation of the absorption coefficient with frequency in an indirect gap semiconductor

process in which phonons do not participate. For this reason, the combination of indirect and direct processes results in an absorption coefficient that varies with frequency as illustrated in Fig. 8.11.

Since the phonon energy (0.05–0.1 eV) is much smaller than typical values of the energy gap ($E_g \sim 1$ eV), in a first approximation the **threshold energy** below which the intrinsic semiconductor (or insulator) does not absorb photons by interband transition is E_g. On the other hand, photons of energy larger than E_g are strongly absorbed resulting in the generation of electron-hole pairs. This process makes possible the use of semiconductors in detectors of electromagnetic radiation. The reverse process by which photons are emitted in the recombination of electron-hole pairs, is called **luminescence**. This process is the basis of the operation of light emitting diodes and semiconductor lasers. Table 8.1 shows the energy gaps and the corresponding wavelengths for several important semiconductors, also indicating the nature of the gap, direct (d) or indirect (i). It can be seen in the table that some direct gap semiconductors are suitable to manufacture light emitting diodes operating at various wavelengths. Also, by combining various compounds in the form of alloys, such as $Ga_xAl_{1-x}As$, it is possible to obtain materials with gaps varying continuously in extensive ranges of the visible and infrared regions.

The processes of interband light absorption and emission in insulators are the same as in semiconductors. However, since the energy gap in insulators is of the order of 10 eV, photons in the visible region do not have enough energy to produce interband transitions. This is the reason why insulating crystals, such as diamond, sapphire, and sodium chloride, for example, are almost entirely transparent to visible light.

To illustrate the most important optical properties of insulators, we show in Fig. 8.12 the transmission spectra of sapphire, which is the crystalline form of Al_2O_3. Figure 8.12a shows the transmission as a function of wavelength, represented on a logarithmic scale to highlight the entire range of transparency. It shows a transmission above 80% in the wavelength between 200 nm (energy of

Table 8.1 Gap energies and corresponding wavelengths at room temperature in important semiconductors

Semiconductor	Gap	E_g(eV)	λ_g (μm)
Si	I	1.12	1.11
Ge	I	0.67	1.88
AlN	I	5.90	0.21
AlAs	I	2.16	0.57
GaN	d	3.40	0.36
GaP	i	2.26	0.55
GaAs	d	1.43	0.86
InP	d	1.35	0.92
InAs	d	0.35	3.54
InSb	d	0.18	6.87
CDS	d	2.53	0.49
CdTe	d	1.50	0.83

6.2 eV) and 2500 nm (0.5 eV), extending from the infrared to the near ultraviolet and covering the entire visible region (400–700 nm). The depression in the transmission in the form of a dip around $\lambda = 3000$ nm and the strong decrease at $\lambda > 6000$ nm, result from the absorption of infrared radiation by optical phonons in sapphire. On the other hand, the drop in transmission in the ultraviolet region, at $\lambda < 200$ nm, is due to the absorption by interband transitions produced by photons with energy above 6 eV.

In the transparency range, there are no electronic transitions to absorb the photons and, therefore, the absorption coefficient is negligible. In this situation the imaginary component of the refractive index is negligible and the real component is $n = (\varepsilon_r)^{1/2}$.

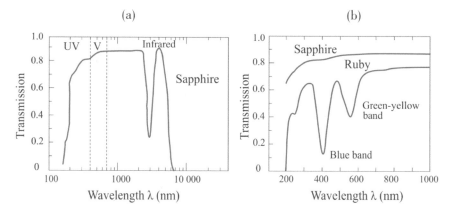

Fig. 8.12 Transmission spectra measured in samples of Al_2O_3. **a** Pure crystal, called sapphire, on a logarithmic wavelength scale, to show the entire range of transparency. **b** Comparison of the spectra of sapphire (pure Al_2O_3) and ruby (Al_2O_3 with 0.05% Cr^{3+})

This is the case of sapphire, that has $n = 1.77$. With this value of n, the reflectivity, given by Eq. (8.21), is $R = 0.077$, while the transmission of one surface is $T = (1 - R) = 0.923$. It turns out that the spectrum shown in Fig. 8.12a was measured in a sample in the form of a thin plate, which reflects the light wave at the two surfaces. For this reason, the transmission is given by $(1 - R)^2 = 0.85$, which is the value observed in Fig. 8.12a. The presence of a small amount of Cr^{3+} impurities in sapphire produces two bands of strong absorption, shown in Fig. 8.12b. The band at higher energy is in the blue region, centered on $\lambda = 400$ nm (3.1 eV), and the one at lower energy is in the green-yellow region, centered on $\lambda = 550$ nm (2.25 eV). The presence of these bands gives the crystal of $Cr^3:Al_2O_3$ a red color, as will be explained in the next section. This crystal is ruby, found in nature as a precious stone. Ruby crystals can also be grown synthetically through the techniques presented in Chap. 1.

Example 8.4 A sample of CdTe in the form of a plate with parallel faces, with thickness 0.3 μm, has anti-reflective layers for visible light on both sides. Calculate the transmission of a light beam from a He–Ne laser, with wavelength 632.8 nm, incident perpendicularly to the plate.

In Example 8.1 we calculated the absorption coefficient of CdTe at the wavelength $\lambda_1 = 500$ nm, and obtained $\alpha(\omega_1) = 9.55 \times 10^6$ m^{-1}. Since CdTe is a direct gap semiconductor, the variation of the absorption coefficient with energy in the vicinity of the band gap is given by Eq. (8.60). Then, the absorption coefficient at the wavelength $\lambda_2 = 632.8$ nm, can be calculated with

$$\frac{\alpha(\omega_2)}{\alpha(\omega_1)} = \frac{(\hbar\omega_2 - E_g)^{1/2}\omega_1}{(\hbar\omega_1 - E_g)^{1/2}\omega_2} = \frac{(\hbar\omega_2 - E_g)^{1/2}\lambda_2}{(\hbar\omega_1 - E_g)^{1/2}\lambda_1}.$$

The energy gap of CdTe, given in Table 8.1, is $E_g = 1.5$ eV. The photon energy at the wavelength λ_2 is

$$\hbar\omega_2 = \frac{hc}{\lambda_2} = \frac{6.63 \times 10^{-34} \times 3 \times 10^8}{632.8 \times 10^{-9}} = 3.14 \times 10^{-19} \text{ J},$$

that gives in eV

$$\hbar\omega_2 = \frac{3.14 \times 10^{-19}}{1.6 \times 10^{-19}} = 1.96 \text{ eV}.$$

Similarly,

$$\hbar\omega_1 = \frac{hc}{\lambda_1} = 1.96 \times \frac{632.8}{500} = 2.48 \text{ eV}.$$

Thus,

$$\alpha(\omega_2) = 9.55 \times 10^6 \times \frac{(1.96 - 1.5)^{1/2} \times 632.8}{(2.48 - 1.5)^{1/2} \times 500} = 8.28 \times 10^6 \, \text{m}^{-1}.$$

Since there are no reflections at the surfaces, the transmission is given by

$$T = e^{-\alpha d} = e^{-8.28 \times 0.3} = 0.084$$

Therefore, the transmission is 8.4%.

8.3.4 Absorption and Emission of Light in Crystals with Impurities

In semiconductor or insulator crystals containing impurities, the presence of discrete energy levels between the valence and conduction bands creates additional sources of photon absorption and emission processes. Figure 8.13 illustrates impurity emission processes in n-type and p-type semiconductors. In (a) an electron goes from the conduction band to an empty level of an acceptor impurity emitting a photon with energy $E_c - E_a$. In (b) an electron at the donor impurity level recombines with a hole in the valence band emitting a photon of energy $E_d - E_v$. Although the number of impurities in a solid is very small compared to the number of crystal ions, the processes of photon emission and absorption involving impurity levels are very important, especially in indirect gap semiconductors. This is due to the fact that the wave function of an electron bound to an impurity has a spatial location with uncertainty Δx of the order of the interatomic distance a. This uncertainty in the electron position results in an uncertainty in its momentum Δp, given by Eq. (2.46), $\Delta x \Delta p \geq \hbar/2$. For $\Delta x \sim a$, the uncertainty in the electron wave vector is $\Delta k \sim (2a)^{-1}$,

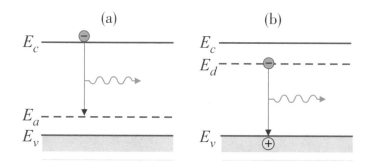

Fig. 8.13 Illustration of two types of electron–hole recombination involving impurity levels with emission of photons. **a** Acceptor impurity in p-type semiconductor. **b** Donor impurity in n-type semiconductor

which covers a wide range of the Brillouin zone. As a result, transitions involving impurities can occur by emission or absorption of photons, without the participation of phonons to conserve momentum. This makes these transitions much more efficient than interband transitions in indirect gap semiconductors. Due to the ease of electrons and holes to recombine through this photon emission process, impurities are called **recombination centers**.

Transitions between discrete levels are also very important in insulators containing impurities of certain elements, especially those of the 3d transition group and rare earths. As we will see in the next chapter, in these elements the formation of impurity ions leaves unfilled inner electronic shells, which often have energy levels within the insulator or semiconductor energy gap. This is the origin of the absorption bands that appear in Al_2O_3 crystals (sapphire) with Cr^{3+} impurities, shown in Fig. 8.12b. The presence of impurities introduces in the gap two sets of levels that form the bands $^4F_1$ and $^4F_2$, in addition to the discrete levels ^2E and $^2F_2$, shown in Fig. 8.14. When the crystal is illuminated with white light, there are transitions from the ground state $^4A_2$ to the bands $^4F_1$ and $^4F_2$ by photon absorption. These transitions are responsible for the absorption bands shown in Fig. 8.12b. Subsequently, electrons in the $^4F_2$ band decay quickly to the ^2E level, by non-radiative transitions, and then return to the ground state $^4A_2$ by radiative transitions with the emission of photons with wavelength 694.3 nm. In this way, the crystal absorbs white light and emits red light, giving ruby its bright red color. This is the photoluminescence process that is the basis of operation of the ruby laser, the first laser ever built. The invention of the ruby laser by Theodore Maiman in 1960 revolutionized the field of optics in science, and set the stage for the development of optoelectronic and photonic technologies, that together with electronics and computer science, revolutionized the costumes of mankind.

Fig. 8.14 Energy levels and important transitions in Cr^{3+}:Al_2O_3, the ruby crystal

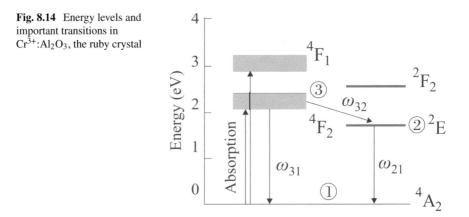

8.4 Photodetectors

Photodetectors are devices that convert light into an electric signal by means of various phenomena. The first technologically important phenomenon was the photoelectric effect, discovered at the end of the nineteenth century, and studied in Chap. 2. It is the basis for the operation of traditional photoelectric cells, made of a vacuum bulb containing a photocathode and an anode, to which an external voltage is applied (positive at the anode). When photons are incident on the photocathode, electrons emitted by means of the photoelectric effect are accelerated to the anode producing an electric current. This constitutes a simple photoelectric cell. With the introduction of electrodes between the photocathode and the anode it is possible to multiply the number of electrons and amplify the current. This is the principle of operation of photomultiplier tubes, which are extremely sensitive devices. Currently there are photomultiplier tubes for scientific applications that detect radiation by counting photons individually, at levels of few counts per second.

Like in electronics, the development of photodetectors and photoemitters manufactured with semiconductors made possible to replace the photo tubes and vacuum light bulbs, providing a huge boost to optoelectronics. The photodetectors most used today in the visible and near infrared regions are semiconductor photodiodes and photoresistors. These devices do not operate in the middle or far infrared regions, since the photons do not have enough energy to produce electron-hole pairs. In these regions, one uses thermal photodetectors, in which the absorption of light produces heating in the sensor element and varies its electrical resistance. In this section we shall study only photoresistors, photodiodes, and CCD image sensors, which are the most important photodetectors for optoelectronics. In these devices, the basic mechanism for converting light into an electric current is the generation of electron-hole pairs by absorption of photons. This process causes a decrease in the intensity of the light as it penetrates into the material, determined by the absorption coefficient α of the material at the light frequency. The variation the light intensity along the propagation direction x is given by Eq. (8.13)

$$I(x) = I_0 \, e^{-\alpha x},$$

where I_0 is the intensity of radiation at $x = 0$. Since the intensity decays exponentially with distance, to ensure that almost all incident photons are absorbed, the thickness of the material should be much larger than α^{-1}. Figure 8.15 shows the variation of the absorption coefficient with wavelength for several important semiconductors. In a photodetector, generally the aim is to work with materials with $\alpha \sim 10^6 \ m^{-1}$ in the operation range of the device. This ensures that almost all photons are absorbed in a distance from the surface of just a few μm. With this condition it is seen that the best materials for photodetection in the visible region are Si and GaAs. Detectors for the wavelengths used in optical communications, $\lambda = 1.3 \ \mu m$ and $1.5 \ \mu m$, are made with $Ga_{0.3}In_{0.7}As_{0.6}P_{0.4}$ and $Ga_{0.5}In_{0.5}As$, respectively.

Fig. 8.15 Variation of the absorption coefficient α with the wavelength for several semiconductors [Wilson and Hawkes]

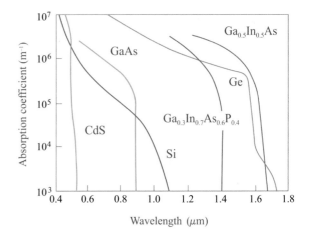

Considering that the semiconductor has a thickness such that all radiation is absorbed, the rate of creation of electron-hole pairs is determined by the initial light intensity I_0. Therefore, the number of photons absorbed per unit time and per unit area is given by Eq. (8.12), $\Phi = I_0 / \hbar\omega$. Actually, there is always some absorption process that does not result in the creation of electron-hole pairs. The **quantum conversion efficiency** η is defined as the ratio between the number of pairs produced and the number of photons absorbed. Thus, the number of pairs created per unit time and area is $\eta I_0 / \hbar\omega$. Therefore, the rate of generation of carriers, defined as the number of pairs created per unit volume and per unit time is

$$g = \frac{\eta I_0}{\hbar\omega\, d}, \tag{8.63}$$

where d is the thickness of the semiconductor. Since electrons and holes are created in pairs, the variation δn in the concentration of electrons due to the radiation is equal to variation in the concentration of holes, $\delta p = \delta n$. The time rates of changes in the carrier concentrations are then

$$\frac{\partial \delta p}{\partial t} = \frac{\partial \delta n}{\partial t} = g. \tag{8.64}$$

This equation shows that if the light intensity incident on the semiconductor is constant and if there are no other mechanisms besides the generation of electron-hole pairs, the number of carriers grows linearly in time, indefinitely. Actually, whenever the concentrations grow above equilibrium, recombination processes tend to restore the equilibrium. The rate at which the pairs are destroyed is determined by the recombination time of the minority carriers, τ_p or τ_n, depending on the semiconductor type. Using τ_r to represent this time, the rate of recombination is given by the ratio between the excess of minority carriers, δp or δn, and the time τ_r. Since $\delta p = \delta n$, the recombination rate is

$$r = \frac{\delta p}{\tau_r} = \frac{\delta n}{\tau_r}. \tag{8.65}$$

In steady state $r = g$, so that the concentrations of electrons and holes generated by light, per unit time, are given by

$$\delta n = \delta p = g\tau_r = \frac{\eta I_0 \tau_r}{\hbar \omega d}, \tag{8.66}$$

This expression determines the number of carriers created in semiconductor photodetectors, which will be presented in the following sections.

8.4.1 Photoresistors

Photoconductivity is the phenomenon by which the conductivity of a material changes with the incident light intensity. Photoconductivity is at the root of operation of the simplest photodetector, the **photoresistor**, that is also called light dependent resistor, or LDR. The simplest structure of a LDR is made of a small slab of an intrinsic semiconductor, or with a very small doping, having at the ends two metallic electrodes for the application of an external voltage, as shown in Fig. 8.16. In the absence of light, the resistance of the LDR is large because the number of carriers is small. When it absorbs light, the number of carriers increases because of the creation of electron-hole pairs. This can produce a sizeable decrease in the resistance relative to its initial value, resulting in an increase in the current between the electrodes. To calculate the effect of light on the current we use Eq. (5.52) for the conductivity,

$$\sigma = n e \mu_n + p e \mu_p$$

The absorbed radiation produces a variation in the concentrations of the carriers, given by Eq. (8.66), resulting in an increase in conductivity given by

Fig. 8.16 Simple structure of a photoresistor, or LDR

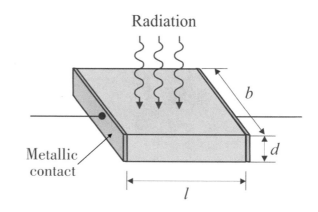

$$\Delta\sigma = g\,\tau_r\,e\,(\mu_n + \mu_p). \tag{8.67}$$

If the voltage applied to the electrodes is V, the variation in the current density is $\Delta J = \Delta\sigma\ V/l$. Therefore, the variation in the current intensity is

$$\Delta I = \frac{b\,d}{l}g\,\tau_r\,e\,(\mu_n + \mu_p)\,V. \tag{8.68}$$

It is common to define the photoconductivity gain as the ratio between the current variation due to the change in conductivity produced by the external voltage, given by Eq. (8.68), and the current due to the electron-hole pairs generated by light. Since this is the total charge of the electrons generated by the radiation per unit time, the gain is

$$G = \frac{\Delta I}{e\,g\,b\,d\,l}.$$

Using Eq. (8.68) in this expression, we obtain for the photoconductivity gain

$$G = \frac{\tau_r\,(\mu_n + \mu_p)\,V}{l^2}. \tag{8.69}$$

This result shows that the gain increases with the value of the applied voltage and with the decrease in the distance between the electrodes. Evidently, the values of μ_n, μ_p, and τ_r depend on the material used.

Figure 8.17 shows the top view of the photoconductive element used in commercial photoresistors, and the view of a typical encapsulated device. The photoconductive element is formed by an insulating substrate with the shape of a disk with diameter ranging from a few mm to several cm. A photosensitive semiconductor, such as CdS, CdSe, PbS, InSb, $Hg_xCd_{1-x}Te$, is deposited over the

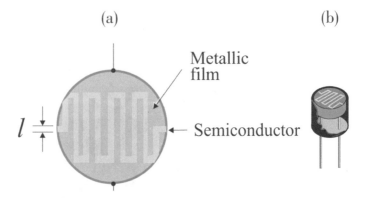

Fig. 8.17 **a** Top view of the photoconductive element with the metal electrode. **b** View of a typical commercial photoresistor

substrate, and on top a metallic film (Al, Ag, or Au) to form the electrodes. The metallic film is evaporated through a mask that leaves the exposed area of the photoconductive material in the form of a zigzag. This results in a large area for illumination of the semiconductor, combined with a small value of the distance between the electrodes, in order to produce a high gain.

The most used materials to manufacture photoresistors to operate in the visible region are CdS and CdSe. In the near infrared, PbS is used, and in the medium infrared, the most common are InSb and $Hg_xCd_{1-x}Te$. These materials have high values for the absorption coefficient in the spectrum range of their operation, and also relatively high mobilities μ_n and μ_p and recombination time τ_r. In addition, these materials are favorable for the formation of **traps** caused by defects in the crystal lattice or impurities. These traps have the role of temporarily trapping electric charge carriers with a certain sign. For example, Mn^{2+} impurities behave as electron traps. Thus, while carriers with a certain charge are trapped, carriers with the opposite charge can move from one electrode to another with less probability of recombination. This results in an effective increase in τ_r, and consequently higher device gain.

An important consideration in any photodetector device regards the noise generated in the absence of radiation. The noise amplitude determines the minimum level of radiation that can be detected. In the case of photoresistors and photodiodes, the main source of noise is thermal generation of electron-hole pairs. Since the probability of thermal generation is proportional to $\exp(-E_g/2k_BT)$ [Eq. (5.25)], the noise depends on the material used and the operating temperature. For this reason, since the materials used in infrared photodetectors have energy gap E_g smaller than those used in the visible region, they have higher noise. To reduce the noise in photodetectors, it is common to cool the photoconductive element. This can be done electrically using compact thermoelectric devices, which easily produce temperatures of the order of $-30\,°C$ ($\sim$240 K). Although this temperature represents a reduction relative to room temperature of only 20%, the effect on the noise is sizeable due to its exponential variation with T^{-1}.

In general, photoresistors are slow devices because they are made with semiconductors that have very long recombination times. For this reason, their use is restricted to applications that need high gain values (10^3–10^4) and that do not require a fast response. For example, photoresistors made with CdS and CdSe, that have response time of the order of 50 ms, are used in light intensity meters of cameras.

To conclude this section, we present in Fig. 8.18 a simple circuit for polarization of a photoresistor. The photoresistor, or LDR, represented in the circuit through its most common symbol, is placed in series with the load resistor R_L. When the incident light intensity varies, the current in the circuit follows the variation of the light. This produces a voltage across R_L, whose variation provides a measure of the light intensity. When only the AC component of the voltage is of interest, a capacitor is used at the output to block the DC part. The value used for R_L depends on the value of the resistance R_D of the LDR, and also on its relative variation with the maximum light intensity.

In case the relative variation of R_D is small (up to 10%), the largest variation in V_L is obtained with $R_L = R_D$. On the other hand, when the variation of R_D is very large,

(a)   (b)

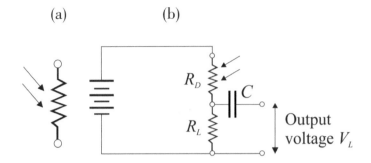

Fig. 8.18 **a** Circuit symbol of a photoresistor, or LDR. **b** Simple circuit used to polarize an LDR

the linearity between the variations of the light intensity and V_L occurs approximately with $R_D \gg R_L$. For this reason, if the output voltage is high, photoresistors must be manufactured with the highest possible value of R_D. This is another reason why the geometry of the photoresistor is made in the form of a long zigzag ribbon, as shown in Fig. 8.17a.

8.4.2 Photodiodes

Photodiodes are light detectors in which the electric signal is produced by the generation of electron–hole pairs caused by absorption of photons in the immediate vicinity of the depletion region of a *p-n* junction. Electrons and holes created in pairs by the electromagnetic radiation are accelerated in opposite directions by the electric field of the junction. Since the field has direction from side *n* to side *p*, holes are accelerated in the $n \rightarrow p$ direction, while electrons move in the $p \rightarrow n$ direction, as illustrated in Fig. 8.19. This produces a current generated by the

Fig. 8.19 Illustration of the electron-hole pair generation by absorption of photons in the depletion region of a *p-n* junction in a photodiode, followed by the acceleration of charges in opposite directions

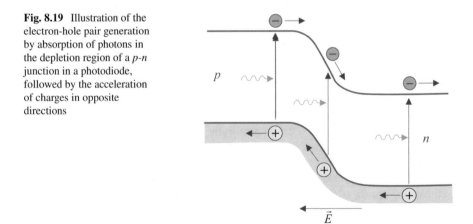

radiation in the direction $n \to p$, which is the direction of the reverse current at the junction. A major advantage of photodiodes relative to photoresistors is that they do not need an external voltage to produce a photocurrent.

The detection of radiation in photodiodes can be done in two different modes of operation: in the **photovoltaic mode** the photodiode operates with open circuit, with no applied voltage. In this case, when the junction is illuminated, a voltage appears between the p and n sides that can be measured externally. In the **photoconductive mode** the device is short-circuited, or operates under an external voltage in the reverse bias direction. In this situation, a current flows in the reverse direction when the junction is illuminated. The choice of the photodiode operation mode depends on its application. Any one of the modes can be used to detect light. The photovoltaic mode is used to convert light energy into electric energy, as in the case of solar cells.

In any mode of operation, the photodiode under radiation behaves as a p-n junction whose current has two components: the first is the one that exists without pair generation by the absorbed photons. It is called **dark current** and is given by Eq. (6.29)

$$I_d = I_s (e^{eV/k_B T} - 1), \tag{8.70}$$

where I_s is the reverse saturation current, given by Eq. (6.30), and V is the voltage at the junction. The other component is the current produced by the electron-hole pairs generated by the photons absorbed in the vicinity of the junction. If I_0 is the intensity of the absorbed radiation and η is the quantum efficiency of the conversion, the number of pairs created per unit volume and per unit time is given by Eq. (8.63), $g = \eta I_0 / \hbar \omega d$. To calculate the total number of pairs created one must consider that minority carriers generated outside the depletion region (thickness l), but within a distance of the order of the diffusion length (L_n and L_p), are able to diffuse to the depletion region and then are accelerated by the field across to the other side of the junction. Since, in general, L_n, $L_p \gg l$, the contribution of these pairs to the current is important, making the effective volume for generation of pairs to be $dA \approx (l + L_n + L_p) A$, where A is the area of illumination of the junction. The current at the junction produced by the light is then

$$I_L = e \, g \, d \, A = \frac{\eta \, e \, I_0 A}{\hbar \omega}.$$

As $P_L = I_0 A$ is the light power incident on the effective area of the junction, using the relation $\hbar \omega = h c / \lambda$, we can write this contribution in the form

$$I_L = \frac{\eta \, e \, P_L \lambda}{h c}. \tag{8.71}$$

The quantum conversion efficiency depends on the material used and also on the wavelength λ of the radiation. Since this current has the direction of the reverse polarization, the total current in the photodiode is given by

$$I = I_s(e^{eV/k_BT} - 1) - I_L.\tag{8.72}$$

Figure 8.20 shows the *I-V* characteristics of a photodiode in the dark regime ($P_L = 0$) and under illumination, for two values of light power P_L. The effect of the radiation contributes a negative portion to the current, regardless of V, which increases proportionally to the light intensity. Equation (8.72) and its graphical representations are used to analyze the two modes of operation of the photodiode.

In the **photoconductive** mode the photodiode operates in short circuit. In this case, $V = 0$ and $I_{sc} = -I_L$. The corresponding operating point is shown in the *I-V* curve of Fig. 8.20 corresponding to the light power P_2. In the **photovoltaic** mode the photodiode operates with open circuit, so $I = 0$. In this case the absorption of light gives rise to a voltage at the diode terminals, whose value is obtained directly from Eq. (8.72),

$$V_{oc} = \frac{k_BT}{e} \ln\left(\frac{I_L}{I_s} + 1\right).\tag{8.73}$$

The operating point $I = 0$, $V = V_{oc}$, is the intersection of the voltage axis with the *I-V* curve, shown in Fig. 8.20. Actually, in all applications the photodiodes do note operate strictly in one of the modes above. As we shall see in the next section, solar cells operate close to the photovoltaic mode, while photodetectors operate near the photoconductive mode.

To make a photodiode act as a photodetector, an external voltage is applied so that the junction operates in the third quadrant of the *I-V* characteristics, such that $I \approx I_s - I_L$. If the thermal pair generation is much less than the absorption of photons, the reverse saturation current can be neglected in comparison to I_L. In this case, the current in the photodiode will be proportional to the radiation power incident at

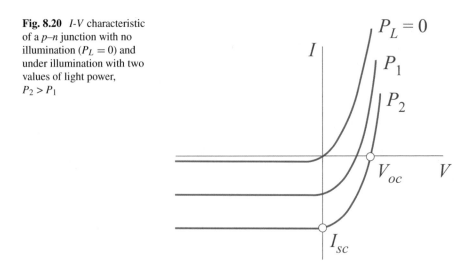

Fig. 8.20 *I-V* characteristic of a *p–n* junction with no illumination ($P_L = 0$) and under illumination with two values of light power, $P_2 > P_1$

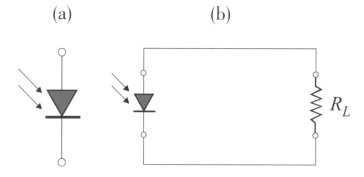

Fig. 8.21 **a** Circuit symbol of a photodiode. **b** Simple circuit for using the photodiode as a radiation detector

the junction. Besides the linear response, the photodiode has other advantages over the photoresistor as a radiation detector. The most important ones are the speed of response, better stability, and larger dynamic range of operation. In applications that do not require a very fast response, it still has the advantage of being able to be used in a very simple circuit, formed only by a small load resistance (of a microameter, or connected to an electronic voltmeter), illustrated in Fig. 8.21.

For small light intensities the photocurrent is small, so that if R_L is small, $V = R_L I \ll V_{oc}$. In this case, the operating point is close to I_{sc}, $V = 0$, so that the current is proportional to the incident power. The advantages of using an additional battery to reverse bias the diode, are the increase in the response speed and also in the dynamic range of operation.

The most used material to manufacture photodiodes for the visible region is silicon. Figure 8.22 shows the **responsivity** of a commercial Si photodiode as a function of wavelength. This quantity, which is often used to characterize the response of photodetectors, is the ratio between the photocurrent and the incident light power, I_L/P_L. The dashed line shown in the figure is the responsivity of an

Fig. 8.22 Responsivity of a Si photodiode (solid line). The dashed line indicates the response of an ideal photodetector, obtained with Eq. (8.71) with $\eta = 1$

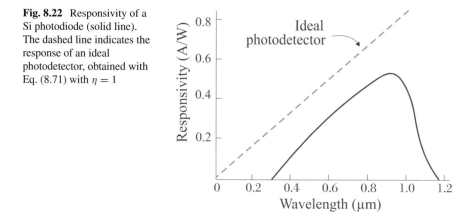

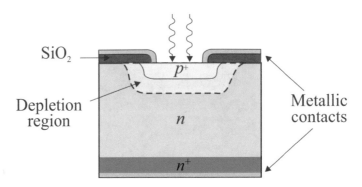

Fig. 8.23 Structure of the p^+-n-n^+ junction of a photodiode

ideal photodetector, calculated with Eq. (8.71) using $\eta = 1$ for any wavelength (Problem 8.11). We see in the figure that the responsivity of silicon approaches that of an ideal photodetector in most of the visible region. Figure 8.23 shows the typical structure of a Si photodiode. It is formed by p^+ and n^+ regions at the ends to facilitate the ohmic contact with the metal films. The main difference for the common diode structure, as the one shown in Fig. 6.1, is the opening in the metal contact for the light entrance. It is also common to deposit on the entrance surface an anti-reflective dielectric layer to increase the conversion efficiency. Since the electron-hole pairs are created in the depletion region or in its vicinity, the thickness of the p^+ side should be as small as possible so that the light is not absorbed before reaching the junction. According to Eq. (6.9), thickness of the depletion region in a p^+-n junction on the n-side is much larger than on the p^+-side. Therefore, the thickness of the region n should be sufficiently large to ensure that all radiation incident on the photodiode is absorbed.

Another structure commonly used in photodiodes is that of the PIN diode, in which an intrinsic semiconductor layer is interposed between the p^+ and the n^+ regions of the p-n junction, as illustrated in Fig. 8.24. The acronym PIN indicates the intrinsic

Fig. 8.24 **a** Model of the PIN photodiode structure. **b** Variation of the electric field along the photodiode

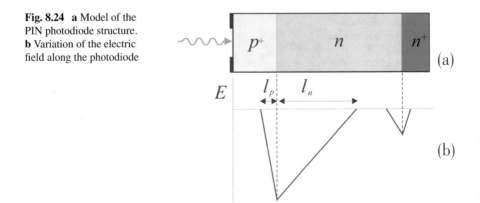

semiconductor between the p and n sides. Actually, the layer is not really intrinsic, but it has a very small concentration of donor impurities ($N_d < 10^{13}$ cm^{-3}), so that it has high resistivity. This results in a depletion region that extends over to the n^+ side, so that the useful thickness of the photodiode is much larger than in the simple p-n structure. This improves the response in the long wavelength region, as it ensures that all radiation is absorbed in this region with the lowest absorption coefficient.

Other widely used photodetectors are the **avalanche photodiode** and the **phototransistor**. The avalanche photodiode operates under a reverse voltage large enough to produce avalanche multiplication, which results in current gain. This allows the device to act with a small load resistance, thus increasing its response speed. On the other hand, the high gain enables the generation of an appreciable voltage in the resistance. The phototransistor is a device in which the emitter-base junction can be illuminated in a manner to generate electron-hole pairs. This results in an emitter current that varies with the light intensity, allowing the detection of light with current gain.

8.4.3 Solar Cells

The solar cell is a photodiode with a large area of exposure to radiation, whose operation aims to supply energy to an external load. For this to occur, it is necessary that the photodiode operates in the fourth quadrant of the I-V characteristics, in such a way that the power absorbed by the device, given by the product VI, is negative. In this situation, the photodiode converts light energy into electric energy. The circuit used for the operation of a solar cell is the same as the one in Fig. 8.21, except that the value of R_L, instead of being small, must be chosen to maximize the delivered power. The operating point of the circuit is determined by the intersection of the load line of the resistor R_L with the I-V curve of the solar cell, as shown in Fig. 8.25. Note that the area of the gray rectangle, shown in the figure, represents the electric power $P_d = VI$ delivered to the load. Thus, the best value of R_L is the one for which P_d is maximum. The values V_m and I_m of the voltage and the current for operation in the condition of maximum P_d, determined by $dP_d/dV = 0$, are given by (Problem 8.14)

$$V_m = \frac{k_B T}{e} \ln\left[\frac{1 + (I_L/I_s)}{1 + (eV_m/k_B T)}\right] \approx V_{oc} - \frac{k_B T}{e} \ln\left(1 + \frac{eV_m}{k_B T}\right), \quad (8.74)$$

$$I_m = I_s \frac{eV_m}{k_B T} e^{eV_m/k_B T} \approx I_L\left(1 - \frac{k_B T}{eV_m}\right). \quad (8.75)$$

Since Eq. (8.74) is a transcendental equation, it is not possible to obtain an analytical expression for V_m, from which the expression for the optimal value of R_L would be obtained. However, using the value of V_{oc} given by the I-V curve, Eq. (8.74) can be solved numerically, leading to the value of V_m. With this value one can obtain the value for I_m with Eq. (8.75) and therefore the resistance

Fig. 8.25 Graphical determination of the operating point of a circuit with a solar cell in series with a resistor R_L

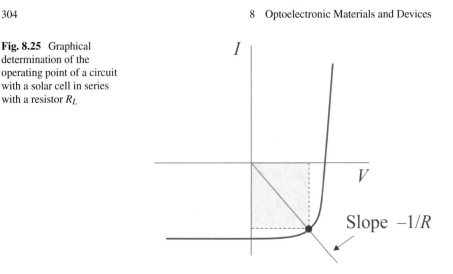

$R_L = V_m/I_m$. The conversion efficiency of the solar cell is the ratio between the maximum electric power and the incident light power P_L. One can see that this efficiency increases with the increase in V_{oc} and in the ratio I_L/P_L.

Currently, the best commercial solar cells are made with crystalline Si, with the structure shown in Fig. 8.26. The junction is formed by a thin n-type layer produced by strong doping ($N_d \sim 10^{18}$ cm^{-3}), in a p-type substrate ($N_a \sim 10^{15}$ cm$^{-3} - 10^{17}$ cm^{-3}). The thickness of the n-region is made small so that most of the incident light goes through with little absorption over a wide frequency range. To increase the exposure area and at the same time keep the contact resistance small, the upper electrode is made in the form of a comb, with fine teeth, as illustrated in the figure. Crystalline Si solar cells in general have circular shape, with diameters of the order of 10 cm or larger, since this is the shape obtained by cutting the Si ingots. Cells made from amorphous or polycrystalline Si have rectangular or square shapes, that have the advantage of occupying the area of a panel when placed next to each other.

Fig. 8.26 Structure of a rectangular Si solar cell. **a** Cross section. **b** View from above

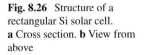

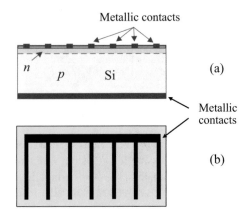

Today good Si solar cells have conversion efficiencies in the range of 15–20%. The solar radiation in the middle of a clear day, at sea level, has intensity in the range 70–80 mW/cm^2. This produces in a cell with area 40 cm^2, an open circuit voltage $V_{oc} \sim 0.7$ V and a short-circuit current $I_{sc} \sim 1$ A. Since the operating values are a little lower than these, it is clear that solar cells must be combined in series and in parallel to produce voltages and currents suitable for different types of loads. In general, the cells are placed on large panels, interconnected, in order to collect solar energy in large areas to produce sufficient electric power for various applications. Until about two decades ago, solar energy was used for specific applications, such as the supply of electricity in remote households. However, the sharp decrease in the manufacturing costs of Si solar cells, combined with the development of power electronics for conversion from DC to AC, have made photovoltaic energy generation competitive with other forms of electric power generation. Today, the energy generated by solar panels on top of house roofs, or in large "solar farms", fed into the electricity distribution grids constitutes a sizeable percentage of the energy generation in some countries. Also, there is intense research activity to produce more efficient solar cells, made with different semiconductor materials with various crystals and physical structures, at costs competitive with Si cells.

8.4.4 CCD Image Sensor

A black and white image in two dimensions, static as in a photograph, is formed by a large number of small area elements, called pixels, each with a tone ranging from white to black, passing through all gradations of gray. Larger number of pixels result in better image resolution. A moving image, as in video, cinema, or television, is formed by a sequence of static images, each one with little difference from the previous. They are shown one after the other, with a small time interval, in such a way that the human perception system has the sensation of a continuous motion. The image on standard television is formed by frames with 525 horizontal lines, with a 60 Hz frame display rate. An **image sensor** is a device that produces an electric signal corresponding to an optical image. It is used in photo or video cameras. The electric signal from the sensor can be stored in analog or digital form, or transmitted through cables or electromagnetic waves. One of the most used image sensors is of the type known as **charge-coupled device**, or **CCD**.

The CCD is part of a class of charge-transfer device structures, developed at the Bell Laboratories in 1969. They are dynamic devices, that move a packet of charge from one unit to another neighboring unit, along a chain, in a sequence determined by the pulses of the system clock. These devices have a variety of applications in electronics, such as in memories, in various logic functions, signal processing, and image sensors. The CCD image sensor is made up of a set of metal-insulating-semiconductor (MIS) capacitors, fabricated in the same semiconductor wafer as in an integrated circuit, forming a network in two dimensions. The semiconductor most used for visible light sensors is Si, while for the infrared the most used ones are InSb

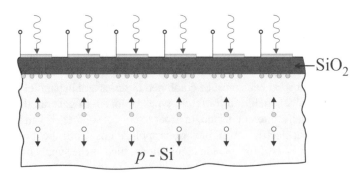

Fig. 8.27 Basic structure of a Si CCD image sensor

and HgCdTe. In the case of silicon, the insulator is SiO$_2$, and the capacitors are of the MOS type, studied in Sect. 7.6. Figure 8.27 illustrates the basic structure of the CCD image sensor, also called MIS- or MOS-photodetector. The metal electrodes of the capacitor gates are thin films, with thickness of the order of 100–300 nm, that let the incident light pass through. Each capacitor has a size of the order of $10 \times 10 \ \mu m^2$ and corresponds to one pixel of the image. The set has a lateral dimensions that can vary from a few mm to several cm. Currently, CCD image sensors have more sophisticated structures, with a polycrystalline silicon gate, instead of metal, and with buried electrodes.

The image is formed in the area of the device by means of the optical system of the camera, causing a certain flow of photons to focus on each pixel. The photons with energy larger than the energy gap create in the region near the semiconductor surface electron-hole pairs, with a rate proportional to the intensity of light in each pixel. A voltage applied between the gate and the electrode on the other side of the wafer (or on the buried electrode, as studied in Sect. 7.8), attracts electrons to the surface and removes the holes, which diffuse into the substrate and are captured in the external circuit, as illustrated in Fig. 8.27. During a time interval characteristic of the device operation (ranges from 100 μs to 100 ms), electron packets form under each capacitor gate, each with total charge proportional to the light intensity integrated in the interval. After this exposure interval, the information stored in each line of capacitors in the form of charge packets, is displaced quickly (in a much shorter time than the exposure interval) to the edge of the line, producing an electric current signal corresponding to the image on the line. The time-varying signal corresponding to a line is followed by the signal of the next line, and so on, in a vertical scan process, from top to bottom. The signal corresponding to the set of lines forms a frame. A static image is formed by only one frame, while an image in motion is formed by a sequence of frames, typically at a rate of 60 Hz.

The transfer of the charge packets from each capacitor to the end of the line is made by the action of a sequence of voltage pulses, applied to the capacitor gates, in a process characteristic of transfer devices, or charge-coupled devices. This is the reason for the name CCD of this type of image sensor. Among the different types of CCD structures, the most used ones are those of two phases and three phases.

Figure 8.28 illustrates the charge transfer in a three-phase structure. Figure 8.28a shows a few capacitors along a line, the external connection scheme for application of voltage pulse sequences, and a packet of charges in capacitor 1, at a certain time t_1. Figure 8.28b illustrates the variation the electric potential and the charge along the capacitor chain, at four time instants. Figure 8.28c shows the variation in time of the electric potential in the three phase lines, ϕ_1, ϕ_2, and ϕ_3. They are periodic functions with two values, one high and one low, with period determined by the system clock. They all have the same shape. However, the phase at ϕ_2 lags the one at ϕ_1 by a time interval corresponding to one-third of the period, while the phase at ϕ_3 lags ϕ_2 also by one-third of the period. Figure 8.28c shows that at time t_1, the potential ϕ_1 is high,

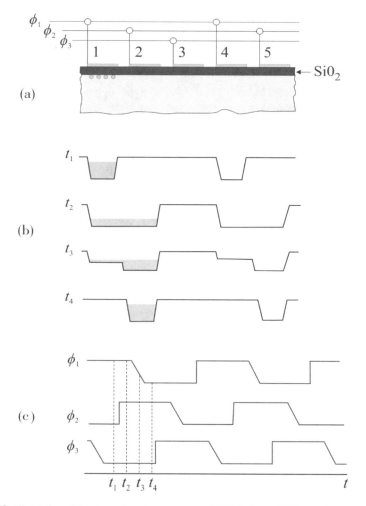

Fig. 8.28 Illustration of the charge transfer process in a CCD device. **a** MOS capacitor connections. **b** Variation in the electric potential of the charge distribution. **c** Variation in time of the potentials of the three lines

while ϕ_2 and ϕ_3 are low. Since the electron charge is negative, the potential energy is in the form of a well in the region of capacitor 1, which maintains the charge packet in that region, as illustrated in the diagram of Fig. 8.28b. At time t_2 the potential ϕ_2 is high, while ϕ_1 remains high, so that the energy well extends to capacitor 2, making the original charge split between capacitors 1 and 2. At the instant t_3 the potential ϕ_1 is smaller than ϕ_2, which remains high, so that most of the charge moves to capacitor 2, a process that is completed when the potential ϕ_1 reaches a low value while ϕ_2 remains high (time t_4). Thus, in each cycle of the potential, the charge moves from a capacitor to the neighboring one until it reaches the end of the line, giving rise to the current signal corresponding to the original pixel of the image at the position of capacitor 1.

8.5 Light Emitting Diodes (LED)

The conversion of an electric signal into a light signal is a very important phenomenon in electronics and other fields of technology. One of its most elementary application is in indicators and light displays used in electric and electronic equipment, appliances, sound and video equipment, scientific and industrial equipment, watches, etc. Another important application is in the generation of images from an electronic signal, as in displays of mobile phones, tablets, computers, television sets, and several other equipment. Starting in the 1980s, this function became more important with the dissemination optical communications, in which an electric signal containing information to be transmitted is encoded into a light signal generated by a light emitting diode or a semiconductor laser. This light signal propagates through an optical fiber to the receiver, where it is decoded and converted again into an electric signal in a photodetector, reproducing the original information.

The simplest and most traditional way of generating light with an electric current is through heating. When an electric current flows through a metallic wire, the atoms of the metal vibrate due to the collisions of electrons in the current. This results in heating of the wire and also in electromagnetic radiation produced by vibrating atomic charges. This radiation occurs in a wide range of the electromagnetic spectrum, which can extend from the infrared to the visible regions, with a peak at an energy that increases with the material temperature. For a wire to be sufficiently heated and emit radiation in the visible region, it must be made of a material with high melting point and placed in vacuum or in an inert atmosphere, to avoid combustion. The incandescent lamps are made with tungsten threads, heated to temperatures of about 6200 °C. At this temperature the peak of the radiation spectrum occurs in the visible region. However, most of the energy in the electric current is converted into heat or infrared radiation, making the efficiency of conversion into visible light very low. In typical incandescent lamps, only 13% of the electric energy are converted into light energy. Besides the inefficiency, these lamps generate a lot of heat and have an extremely slow response. For a long time they were used in indicators and device displays in electronic devices, but since the 1970s they have been replaced by light emitting diodes and other solid-state devices, such as liquid crystals displays.

The emission of light in an incandescent lamp occurs due to heating, a classical physical mechanism. The operation of modern optoelectronic devices is based on quantum processes of radiation emission, by means of **luminescence** processes.

The operation of a light emitting diode, or LED, is based on a special form of electroluminescence, produced by injection of charge carriers in a *p-n* junction. As we saw in Sect. 6.2, when a *p-n* junction is biased in the forward direction, holes on the *p*-side and electrons on the *n*-side move in opposite directions towards the depletion region. Holes injected into the *n*-side recombine with electrons that are arriving in the depletion region, while electrons injected into the *p*-side recombine with holes. In this way, all electrons and holes that participate in the current recombine in the vicinity of the depletion region, in a layer of thickness L_p on the *p*-side and L_n on the *n*-side. If the semiconductor of the junction has an indirect gap, such as Si or Ge, the recombination produces phonons, in addition to photons and, therefore heat. This makes light emission very inefficient in *p-n* junctions made with indirect gap semiconductors. However, if the semiconductor has a direct gap, the recombination of each electron-hole pair results in the emission of one photon. Figure 8.29 illustrates the process of minority carrier injection on both sides of a *p–n* junction, producing pair recombination and emission of photons by inter-band transitions. In diodes made with direct gap semiconductors this process is extremely efficient in converting electric energy into light. If the electrons in the conduction band have minimum energy E_c, the photons emitted in the interband transition have energy equal to the semiconductor gap, E_g. In general, due to the thermal excitation energy, the average energy of the electrons is about $E_c + k_B T/2$. This causes the energy of the photons emitted in the transition region to be a little higher that E_g. In addition to the interband transition, shown in Fig. 8.29, it is possible to have at the *p-n* junction transitions involving impurity levels, as illustrated in Fig. 8.13.

Most materials used in LEDs are ternary alloys, such as $Ga_x Al_{1-x} As$ and $GaAs_{1-x} P_x$. GaAs is a direct gap semiconductor with small resistivity, which can be easily doped with *n*- or *p*-type impurities, for making *p-n* junctions with high luminescence efficiency in interband transitions, with wavelength of about $0.87\ \mu m$. This value corresponds to the radiation in the near infrared. Since GaP

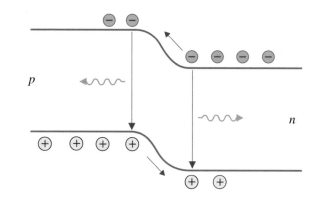

Fig. 8.29 Recombination of electron–hole pairs with emission of photons in interband transitions, due to the injection of minority carriers in a forward biased *p-n* junction

has a larger energy gap, the alloys formed by GaAs and GaP have interband transitions with wavelengths shorter than in GaAs. It is interesting to note that unlike GaAs, GaP has an indirect gap. With this, the gap in the $GaAs_{1-x}P_x$ alloy is direct for $x < 0.45$, as GaAs, but it becomes indirect for $x > 0.45$. The alloy composition $GaAs_{0.6}P_{0.4}$, with direct gap, is widely used to manufacture LEDs that produce red light in interband transitions with $\lambda = 0.65$ μm.

The $Ga_xAl_{1-x}As$ alloy is also widely used to manufacture LEDs with high efficiency. It is common to find devices made with heterojunctions of n-type $Ga_{0.3}Al_{0.7}As$ and p-type $Ga_{0.6}Al_{0.4}As$. In this system, electrons in the n-type are injected into the p-side, where they produce transitions to the acceptor impurity levels, as in Fig. 8.13a, with emission of photons with wavelength 0.65 μm (red). The radiation produced on the p-side passes through the n-side without absorption, since it has a larger energy gap, making the efficiency of these LEDs close to 100%. During the 1990s, the technology for manufacturing efficient LEDs with GaN and its alloys was developed by Shuji Nakamura, Isamu Akasaki, and Hiroshi Amano. As shown in Table 8.1, the energy gap of GaN is 3.4 eV, corresponding to a wavelength of 360 nm, in the near ultra-violet region. The development of the blue LED made possible the manufacture of panels containing clusters of LEDs with the three basic colors of the visible spectrum, simulating a white light source. Also, the use of UV LED inside glass bulbs internally covered with a phosphor material, made possible the fabrication of efficient LED bulbs for lighting. Nakamura, Akasaki, and Amano were awarded the Physics Nobel Prize in 2014 *"for the invention of efficient blue light-emitting diodes which has enabled bright and energy-saving white light sources"*.

Figure 8.30 shows a typical structure of a Ga(AsP) LED that operates in the red region. Similarly to photodiodes, the metallic contact on the top side has an opening that serves for the window to allow the transmission of radiation. Usually, the p-side is a thin layer on the top, made with doping much smaller than on the n-side. This makes the radiation to be produced on the p-side, close to the exit window, by electrons injected from the n-side, which minimizes the absorption of radiation emitted at the junction. The several layers of the LED structure are made by epitaxial growth over a GaAs substrate. Since $GaAs_{0.6}P_{0.4}$ has a lattice parameter very different from GaAs, it is not grown directly on the substrate, to avoid the formation of crystalline defects that act as nonradiative recombination centers. This is the reason for having the

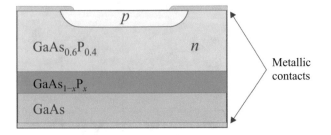

Fig. 8.30 Typical structure of a light emitting diode (LED)

Fig. 8.31 Typical structure
of LED lamps used in
indicators of electric and
electronic equipment

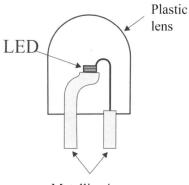

intermediate layer of $GaAs_{1-x}P_x$, shown in Fig. 8.30. It is made with a concentration
x that varies gradually from 0 to 0.4, so as to produce a matching between the crystal
lattice parameters of $GaAs_{0.6}P_{0.4}$ and GaAs.

LEDs that operate in the visible are widely used to make lamp indicators for
panels of electro-electronic equipment. These lamps are made with a wide variety of
shapes, sizes, and colors. Figure 8.31 shows a typical structure of a LED lamp. The
LED chip is mounted on one of the metallic pins used as an external terminal. The
contact with the other terminal is made by a wire welded on the metallic film on the
side of the LED window. The set is encapsulated in a colored plastic, with an upper
part in the form of a lens, to partially collimate the radiation.

The most important applications of infrared LEDs are in optical communication
systems. They are made mainly with the quaternary alloy $Ga_xIn_{1-x}As_yP_{1-y}$.
Depending on the concentrations, the LED made with this alloy can emit in any
wavelength in the range 1.1–1.6 μm, used in optical communications. As will be
shown in Sect. 8.8, these systems are based on the transmission of information by
means of an infrared light beam, which propagates confined in an optical fiber with
a diameter of some μm. The LEDs used for this purpose are made with the
structure shown in Fig. 8.32, known as Burrus type, invented by Charles A. Burrus.

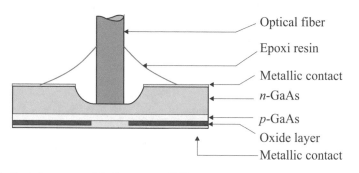

Fig. 8.32 Typical structure of the Burrus type LED

(a)   (b)

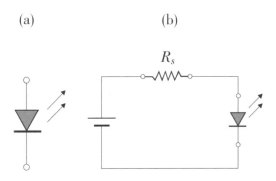

Fig. 8.33 **a** Circuit symbol of a LED. **b** Simple LED supply circuit

In this structure, the metallic contact with the semiconductor is confined to a region with diameter similar to that of the optical fiber. This makes the active light emission region small, resulting in an efficient coupling with the optical fiber. The fiber is rigidly mounted to the structure and secured by epoxy resin, as shown in Fig. 8.32.

The supply circuits for LEDs are quite simple. To have light emission with constant intensity one simply applies a DC current in the forward diode direction. In optical communication systems it is necessary to incorporate a current modulation circuit to produce the corresponding variations in the light intensity. Figure 8.33 shows the circuit symbol of the LED and a simple power supply circuit. The series resistor R_s is necessary to limit the current that passes through the LED, because as it operates with forward bias, its resistance is very small.

LEDs that operate in the visible are also widely used today to make alphanumeric light displays. Figure 8.34 shows two types of very common displays. The 7-segment system shown in Fig. 8.34a is used to indicate the digits from 0 to 9. Each segment is

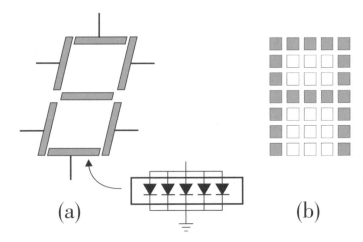

Fig. 8.34 Illustration of two types of LED displays. **a** 7-segment numeric display. **b** Alphanumeric 7×5 matrix display

formed by a set of LEDs, connected in parallel and encapsulated in a same piece, in order to produce uniform illumination throughout its length. Figure 8.34b shows the 7×5 matrix of individual LEDs, which allows to display digits and letters, forming an alphanumeric display.

8.6 Stimulated Emission and Lasers

The radiation produced by a traditional light source, such as incandescent and fluorescent lamps, or by a LED, consists of photons emitted spontaneously by independent atoms or molecules. The **spontaneous emission** process occurs when a quantum system changes from one state with a certain energy level to another one of lower energy due to random fluctuations. As a result, the phase of the resulting field varies randomly in space and time, so that the radiation is **incoherent**. Another type of radiation is the one produced by a **laser**, a device known by the acronym for **L**ight **A**mplification by **S**timulated **E**mission of **R**adiation. The radiation of a laser results from the emission induced in atoms or molecules, or **stimulated** by a macroscopic electromagnetic field. In this process, the phases of the fields of the emitted photons are correlated, and as a result the radiation is **coherent**. In addition to coherence, the radiation from a laser is highly monochromatic, that is, it has frequencies in a narrow spectrum range. The intensity depends on the type of laser and the magnitude of the excitation, and can vary over a wide range of values.

The development of the laser has a long history. In 1917, Albert Einstein established the theoretical foundations for the laser by proposing the processes of absorption, spontaneous, and stimulated emission of electromagnetic radiation. After the development of quantum mechanics, there were several proposals for related phenomena. Then, in 1950, Alfred Kastler proposed the method of optical pumping, which was experimentally confirmed two years later by Brossel, Kastler, and Winter. Kastler received the Physics Nobel Prize in 1966 for his contributions to the development of the laser.

In 1953, Charles H. Townes and his graduate students J. P. Gordon and H. J. Zeiger produced the first maser, a device operating on similar principles to the laser, but amplifying microwave radiation. In the following years, Charles H. Townes and Arthur L. Schawlow, and independently Nikolay Basov and Aleksandr Prokhorov, proposed detailed theories for the operation of lasers in the infrared and visible regions. The Physics Nobel Prize was awarded to Townes, Basov, and Prokhorov in 1964, and to Schawlow in 1981. As previously mentioned, the first laser operating in the visible region was the ruby laser, invented in 1960 by Maiman. Several types of lasers developed in the following years, such as other solid-state lasers and gas lasers, have many applications in industry, in medicine, and in science. The semiconductor lasers operating at room temperature, developed at the end of the 1960s, set the stage for a revolution in optical communications, that made possible high-speed Internet and video communication.

The main components of a laser are: the active medium containing the atoms or molecules that produce the radiation emission; the resonator or optical cavity; and the pumping mechanism. The cavity is formed by two mirrors, one perfect and one partial, that reflect the radiation back and forth through the active medium between them. The radiation is emitted through the partial mirror. The structure resonates at certain wavelengths, resulting in a macroscopic electromagnetic field that produces the stimulated emission in the atoms or molecules in the active medium. This emission amplifies the field in the cavity and maintains the laser radiation. The main features of the laser are determined by the nature of the active medium. We begin this section by studying the mechanism of stimulated emission, that is essential to understand the role of the active medium.

8.6.1   The Mechanism of Amplification by Stimulated Emission

As we saw in Sect. 8.3, a quantum system with two energy levels, $E_2 > E_1$, with populations N_2 and N_1, has an absorption coefficient given by Eqs. (8.14) and (8.56) that can be written as

$$\alpha = 2\frac{\omega}{c}\kappa = \frac{4\pi\omega}{n\,c\,\varepsilon_0\,\hbar}\,(N_1 - N_2)\,p_{12}^2\,D(\omega), \qquad (8.76)$$

where p_{12} is the matrix element of the electric dipole moment between the two levels and $D(\omega)$ represents the spectral line shape of the transition between the two levels. When a radiation with frequency ω crosses the medium, its intensity varies in space according to Eq. (8.13)

$$I(x) = I(0)\,e^{-\alpha x}$$

In thermal equilibrium, the population N_1 in the lower energy level is larger than in the higher energy level, N_2, so that $\alpha > 0$. In this situation, the radiation is absorbed by the transitions from E_1 to E_2, so that its intensity decreases as it propagates in the medium. However, if there is an external mechanism for **population inversion**, making $N_2 > N_1$, we have $\alpha < 0$, so that the radiation is **amplified by the stimulated emission**. Thus, for $N_2 > N_1$, we define the radiation gain of the medium as $\gamma(\omega) = -\alpha(\omega)$. Naturally, the system has losses, mainly caused by the radiation that comes out of the resonant cavity. In this way, it is necessary that the pumping process causes the population inversion to exceed a threshold value $(N_2 - N_1)_T$ such that the total gain is larger than the total losses. In this situation the system generates radiation by stimulated emission.

The threshold value of the population difference is determined by the condition for which the intensity gain over the length of the active medium is equal to the losses. Losses have two origins, the attenuation along the beam, caused by diffraction and by interactions with other excitations, and radiation loss out of the optical cavity. The first is large for frequencies different from the cavity resonance frequencies. For this reason, the laser operates only at the wavelengths corresponding to the resonances of the cavity, given by

$$\lambda' = \frac{2L}{m},$$
(8.77)

where λ' is the wavelength in the active medium, L is the distance between the cavity mirrors, and m a positive integer number. Considering that λ' is related to the wavelength in vacuum (practically equal to air) by $\lambda = n\,\lambda'$, where n is the refractive index of the medium, we obtain for the laser operating frequencies,

$$\nu = \frac{c}{\lambda} = m\,\frac{c}{2nL}.$$
(8.78)

Lasers operate at one or more frequencies given by this expression that are in the range of the gain curve of the active medium. In general, lasers operate simultaneously in various cavity modes, called longitudinal modes, each with a line width of few MHz, which is much less than the width of the gain curve. For example, the gain curve of the He–Ne gas laser has a width of about 1.5 GHz, which supports 10 longitudinal modes spaced by 150 MHz, which is the value obtained from Eq. (8.78) for an optical cavity with length of 1.0 m and $n = 1$. Figure 8.35 shows a typical spectrum of a He–Ne laser.

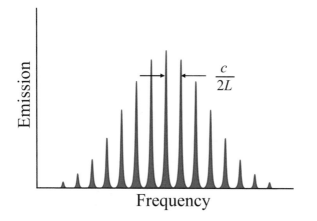

Fig. 8.35 Light emission spectrum of a He–Ne laser showing the longitudinal modes

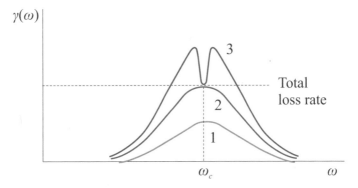

Fig. 8.36 Illustration of the gain curves of a laser for three values of the population difference $\Delta N = N_2 - N_1$, with $\Delta N_3 > \Delta N_2 > \Delta N_1$

Figure 8.36 illustrates the behavior of the gain curve γ (ω) of a laser, for three values of the population difference $\Delta N = N_2 - N_1$. In curve 1, ΔN is such that the gain is less than the total loss rate at any frequency value. In this situation, the laser does not emit radiation. Curve 2 corresponds to the rate of threshold pumping, for which ΔN is such that the maximum of γ (ω) is equal to the loss rate. With a higher pumping rate, the gain is larger than the loss in a certain frequency range. In this situation, the system maintains a radiation with frequency ω_c determined by the resonant cavity. The dip that appears in curve 3 results from the fact that the intense radiation created in the cavity increases the transitions from level 2 to level 1, so that the two populations tend to become equal. The steady-state operating regime is reached when the gain rate is equal to the loss rate. The essential requirement for stimulated emission of radiation is to have a net gain produced by the population inversion in the active medium. There are several methods for inverting the populations in two-level systems, the most important of which are:

- Optical pumping or photon excitation;
- Electronic excitation;
- Inelastic collision between atoms;
- Injection of carriers in semiconductors.

The population inversion between two levels involved in the stimulated emission in homogeneous systems requires the existence of at least another quantum energy level. Figure 8.37 shows two modes of operation in a three-level system. In (a) the atom is excited from the ground state E_1 to a state with energy E_3 by means of some efficient pumping process. Then the system relaxes from E_3 to E_2 by fast and not radiative transitions, causing an accumulation of population at level E_2 and population inversion relative to the level E_1. The laser radiation then occurs by transitions from the level E_2 to E_1. Figure 8.37b illustrates another possible mode of operation in a three-level system, in which radiation occurs in transitions from the higher level to another intermediate level, which in turn relaxes to the ground state.

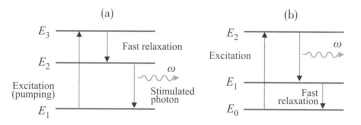

Fig. 8.37 Stimulated emission processes in 3-level systems

Optical pumping processes use an external light source, which can be a high-power flash lamp, or another type of laser, to increase the population of a band above the two levels of interest. This method of excitation is used in lasers with solid materials, called **solid-state lasers**, either in crystalline form, such as sapphire and yttrium aluminum garnet (YAG), or glasses, doped with appropriate impurities. **Electronic excitation** processes are generally used in **gas lasers**. In an electric discharge in a gas, electrons of the current are accelerated by the applied voltage and collide with ions of the gas. In this collision they transfer energy to the ions, producing electronic transitions from the ground state to excited states. This is the process used in argon lasers.

Another important pumping process in gases is the atom-atom collision, in which during an electric discharge a certain type of atom collides with another type, leaving the latter in an excited state to radiate. This process is important in lasers with gas mixtures, such as Helium-Neon. Finally, another important method is that of injection of carriers in a semiconductor junction. As we saw in Sect. 8.5, the electric current in a p^+-n junction causes holes on the p^+-side to diffuse into the n-side, resulting in an excess of holes relative to electrons. Thus, in the junction region there is a population inversion, in the sense that there are more minority carriers than in thermal equilibrium. This results in the recombination of electron-hole pairs and in the generation of photons, by spontaneous emission, as in a LED, or by stimulated radiation, as in the semiconductor laser. In the following subsections we shall present some details of several commercially important lasers. Due to its major importance in optoelectronics, the semiconductor diode laser will be presented in more detail in the next section.

8.6.2 Solid-State Lasers

The active medium in solid-state lasers is a piece of transparent crystalline material, or glass, in the form of a cylinder (rod), slab, or block, doped with impurity ions that have energy levels suitable for stimulated emission. The optical cavity is, in general, formed by two external mirrors, one of which is fully reflective while the other transmits a small fraction of the incident radiation. It is through the partial mirror that part of the radiation energy stored in the cavity is transmitted outwards, producing the laser beam. The excited states of the impurities are populated by optical pumping, produced by flash lamps or by another laser.

Figure 8.38 shows the arrangement used in the original ruby laser. Ruby is a sapphire crystal, Al_2O_3, containing Cr^{3+} impurities in small concentrations, from 0.01 to 0.1%. The energy levels of the Cr^{3+} ions in Al_2O_3 and the three transitions involved in the laser action are shown in Fig. 8.14. The optical pumping takes electrons from the ground state, level 1, to a relatively broad energy band 3. Then, the electrons decay in a short time, of the order of 10^{-8} s, to level 2. Since the decay time from 2 to 1 is relatively long ($\sim 10^{-3}$ s), there is an accumulation of electrons at level 2, and hence inversion of population relative to level 1. The stimulated emission from 2 to 1 generates the red light of the ruby laser, with wavelength 694.3 nm. Since the flash lamp is activated by the discharge of a capacitor, indicated in Fig. 8.38, the pumping light is in the form of pulses with duration of few ms. For this reason, instead of generating continuous radiation, the laser emits light pulses with a repetition rate determined by the circuit of discharge-flash lamp. The choice of the repetition rate depends on the ability to cool the ruby rod. This cooling can be done by circulating water in contact with the rod, as shown in Fig. 8.38. In current solid-state lasers with impurities that use a flash lamp, this is not wrapped around of the rod, as in the original arrangement of Fig. 8.38. The lamp is a cylindrical tube, placed parallel to the solid rod, inside a metallic cavity, with elliptical section, internally polished. The lamp is placed in one of the foci of the ellipse and the rod in the other, so that the radiation from the flash is focused on the rod.

One of the most important solid-state lasers employs for active medium a crystal of yttrium aluminum garnet (YAG), that has chemical formula $Y_3Al_5O_{12}$, with neodymium impurities. The so-called Nd-YAG laser action occurs in the energy levels of the Nd^{3+} impurities. Figure 8.39 illustrates the energy levels and the important transitions of the Nd-YAG laser. The optical pumping, produced by the radiation of a flash lamp, or a semiconductor diode laser, takes the electrons from the ground state to a broad energy band of excited states. From there they fall to the

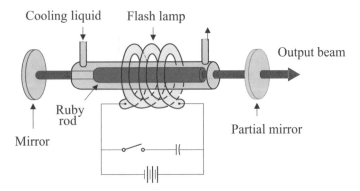

Fig. 8.38 Arrangement used in the ruby laser pumped by a flash lamp

Fig. 8.39 Energy scheme and electronic transitions responsible for the radiation with wavelength 1064 nm in the Nd-YAG laser

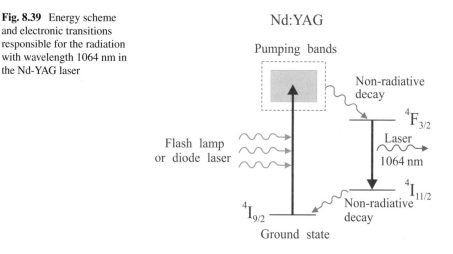

Nd:YAG

Pumping bands

Non-radiative decay

$^4F_{3/2}$

Flash lamp or diode laser

Laser 1064 nm

$^4I_{11/2}$

Non-radiative decay

$^4I_{9/2}$

Ground state

$^4F_{3/2}$ state by means of non-radiative transitions. The transition from this state to $^4I_{11/2}$ produces laser radiation at a wavelength $\lambda = 1064$ nm, situated in the near infrared. The gain of the Nd-YAG laser is about 75 times larger than in ruby. For this reason, it can be pumped with continuous light from a diode laser, as shown in Fig. 8.40. The beam of the diode laser passes through two lenses and is focused on the axis of the Nd-YAG rod. The entrance surface of the rod is spherical and covered by dielectric layers that transmit the 809 nm radiation and reflect the 1064 nm radiation. The flat surface at the other end has a reflective layer, forming the optical cavity. When operated with flash lamps, it reaches quite high peak powers. Despite operating in the infrared, the Nd-YAG laser is widely used for applications in the visible region. This is achieved by passing the pulsed beam through a second-harmonic generator crystal, that converts most of the radiation into green light, with wavelength 532 nm (See Sect. 10.2.2).

Solid-state lasers pumped by a light beam from a diode laser, as in Fig. 8.40, produce continuous wave (CW) radiation. Lasers pumped by flash lamps produce light in the form of pulses emitted periodically. The pulses are long, lasting a few ms, and the repetition rate is low, with few shots per second, because these are

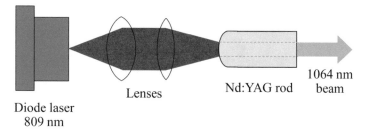

Diode laser 809 nm

Lenses

Nd:YAG rod

1064 nm beam

Fig. 8.40 Nd-YAG laser pumping scheme with continuous radiation from a semiconductor diode laser

characteristic of the electric discharge in flash lamps. Pulsed lasers are important in applications that require high power light pulses, since the energy accumulated in the period is emitted in a much shorter time span. There are other methods for producing light pulses in solid-state and other types of lasers, such as gas and liquid dye lasers, with continuous pumping. Two methods that enable to obtain very short pulses are Q-switching and mode-locking. In both methods, the cavity mirrors are outside the active medium, since the mechanism requires the insertion of a device in the beam path within the cavity.

The Q-switching method consists of deteriorating the Q-factor of the optical cavity during a certain time, preventing the laser action. Since the pumping is continuous, during the time the Q is low and there is no stimulated radiation, the population of the excited states increases and exceeds the threshold for light emission with the normal Q. Then, periodically the Q is restored, providing the emission of high-power short pulses. One of the mechanisms used to vary the Q is the modulation of the light polarization by means of an electro-optic modulator (Sect. 10.2) placed inside the cavity. In this mechanism, the light polarization is changed in one beam pass, so that the beam reflected in a mirror has a polarization different from the incident one and does not produce constructive interference necessary for the resonance.

The method of mode-locking also uses an internal modulation in the cavity, but the mechanism is based on the existence of a large number of longitudinal modes, as in Fig. 8.35. It can be shown that the amplitude modulation with frequency equal to the separation of the modes produces a locking of the mode phases. Since they have different frequencies, periodically the phases of all modes coincide, resulting in a train of radiation pulses. The period of pulse emission is the inverse of the frequency spacing between the modes.

Another solid-state laser widely used today due to its versatility is the titanium-sapphire laser ($Ti^{3+}:Al_2O_3$). This laser can operate in CW or pulsed regimes (Q-switching or mode-locking). Ti^{3+} impurities have an absorption band with a peak around 500 nm, as shown in Fig. 8.41, and a broad emission band, so that the laser can operate in the entire 660 nm-1180 nm range, with proper use of filters and

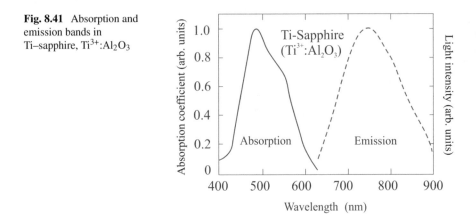

Fig. 8.41 Absorption and emission bands in Ti–sapphire, $Ti^{3+}:Al_2O_3$

mirrors in the optical cavity. Its conversion efficiency, defined as the ratio of the emitted optical power by the electrical input power, is approximately 0.01%. This value is low compared to the Nd:YAG laser (0.5%), but it is of the same order as in gas lasers. The Ti-sapphire laser is commonly pumped by an argon laser (514 nm), or by the second harmonic of a Nd:YAG laser (532 nm). For optical pumping powers of the order of 10 W, about 1.5 W can be emitted in the Ti–sapphire laser in CW regime. It also operates in a mode-locking regime, emitting ultrashort pulses. Since the width of the Ti-sapphire transition line is 100 THz, pulses of duration as short as 10 fs (1 fs $= 10^{-15}$ s) are produced by Ti–sapphire lasers.

8.6.3 Gas Lasers

In gas lasers, the stimulated emission occurs between quantum states of atoms or molecules, which are usually excited by collisions in an electric discharge through the gas. Figure 8.42 shows the basic components of a gas laser. The high voltage applied to the electrodes at the ends of the tube maintains an electric discharge in the gas, which may be confined or circulating. When the optical cavity is formed by external mirrors, the ends of the tube are made with transparent parallel plates, with an inclination at Brewster's angle to minimize the reflection losses. In small gas lasers, the mirrors are made inside, at the tube ends.

The helium–neon (He–Ne) laser was the first gas laser invented, and it is still used today in simple low power applications. In the discharge through the mixture of the two gases, the He atoms are easily excited by collisions. The excitations of these atoms are transferred to the 2S and 3S states of Ne, which coincidentally have almost the same energies of He. Transitions with stimulated emission occur in the Ne atoms between the levels illustrated in Fig. 8.43, at the wavelengths indicated. The 3S-3p and 2S-2p transitions occur in the infrared, while the 3S-2p transition has $\lambda = 632.8$ nm, located in the red region of the spectrum. The He–Ne laser is simple to manufacture and operates continuously with low current, which is why it is used in a large variety of low power applications (few mW).

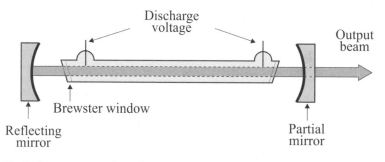

Fig. 8.42 Basic components of a gas laser

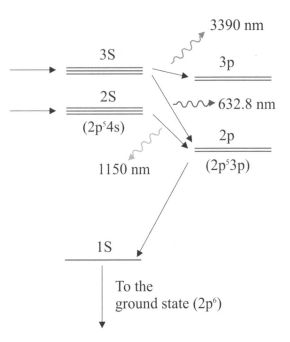

Fig. 8.43 Energy levels and laser transitions in Ne atoms

Another important gas laser with visible light radiation is the argon laser. It operates with electronic transitions in the Ar^+ ions, producing radiation in several lines of the visible spectrum. The most intense one occurs at $\lambda = 488$ nm (blue) and 514.5 nm (green). Generally, argon lasers operate continuously with powers ranging from hundreds of mW to dozens of watts, and have countless medical, industrial, and scientific applications. In the category of molecular gas lasers, the most important one is that of carbon dioxide, CO_2. In this system, the quantum levels involved in the laser transitions are associated with the vibrations of the CO_2 molecule. The stimulated emission has a wavelength around 10 μm, corresponding to mid-infrared radiation. The CO_2 laser has an easy and robust construction and produces CW power with tens to hundreds of watts, and is also widely used in industry and medicine.

8.7 Semiconductor Lasers

The semiconductor laser, also known as diode laser, or semiconductor diode laser, is by far the most important one for optoelectronics. While all lasers presented in the previous section are physically big, expensive, and require significant electric power to operate, the semiconductor lasers have submillimetric dimensions, low cost and require low power supply. The semiconductor diode laser was invented in 1962, but it took many years of research and development to have efficient semiconductor lasers such as the ones in current use. The first lasers were formed by simple GaAs junction diodes operating only at the liquid helium temperature (4.2 K) with relatively high currents. At the end of 1960s some laboratories managed to materialize the theoretical proposals of Zhores Alferov and Herbert Kroemer, who showed that the laser gain could increase with the confinement of electron and holes in heterojunctions.

Currently, semiconductor lasers are made with multiple heterojunctions of direct gap semiconductor alloys, operate at room temperature, with low currents, and produce light powers that vary from a few mW, comparable to those of the He–Ne laser, to tens of watts. The semiconductor laser has become an essential component for optical communication systems and for a wide variety of applications in electronic equipment for household, industrial, medical, scientific, and other uses. Due to the success of the heterojunction structure for the fabrication of efficient semiconductor lasers, Alferov and Kroemer were awarded the Physics Nobel Prize in 2000, together with Kilby, the inventor of the integrated circuits.

8.7.1   The p-n Junction Diode Laser

One of the basic mechanisms for the operation of a laser, the population inversion, occurs naturally in a forward biased *p-n* junction made with a direct gap semiconductor. This is because the electrons on the *n*-side that move towards the junction region and are injected into the *p*-side, have in the conduction band on the *p*-side concentration larger than the one in thermal equilibrium. A similar situation happens with the holes in the valence band injected into the *n*-side. The recombination of electron–hole pairs, that occurs to make the concentrations reach equilibrium on both sides, produces spontaneous emission characteristic of the LEDs. However, when the injection is strong enough, the threshold condition for laser operation can be achieved and the diode emits stimulated radiation.

To achieve the laser threshold condition, the *p-n* junction must have large doping on both sides, that is, it must be formed by degenerate semiconductors. At this junction, the Fermi level E_{Fn} on the *n*-side is above the minimum of the conduction band, E_{cn}, while on the *p*-side the E_{Fp} level is below the maximum of the valence band, E_{vp}. Figure 8.44 illustrates the energy bands in a junction of this type. In (a) there is no applied voltage, so that the Fermi level E_F is the same on both sides. In (b) the junction is forward polarized with an external voltage V, so that the energies on the *p*-side decrease relative to the energies on the *n*-side, and the difference between the Fermi energies on the two sides is equal to eV. Finally, Fig. 8.44c shows what happens with a larger voltage: in the transition region of the junction, the conduction band is filled with electrons from the *n*-side, while the valence band receives holes from the *p*-side. This produces population inversion in this region, which results in high recombination rates accompanied by the spontaneous emission of light. The photons created in this process that are confined to the junction region, make the recombination rate to increase even more through the stimulated emission process. The laser action occurs when the current in the diode exceeds a certain threshold value for which the optical gain equals the losses in the system.

Fig. 8.44 Energy diagrams in a *p-n* junction formed by degenerate semiconductors. **a** No applied voltage. **b** Forward biased with a voltage *V*. **c** With a voltage sufficiently large to produce population inversion in the transition region

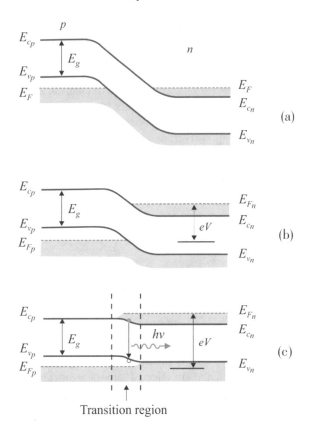

The minimum frequency ν of the photons emitted in the band-to-band transition is given by $\nu = E_g/h$. On the other hand, due to the condition of population inversion illustrated in Fig. 8.44c, we see that the maximum value of ν is given by $E_{Fn} - E_{Fp} \geq h\nu$. Therefore, the operating condition of a *p-n* junction laser is

$$E_{Fn} - E_{Fp} \geq h\nu \geq E_g. \tag{8.79}$$

To increase the gain and decrease the losses, and also to make the radiation go out in only one direction, it is necessary to build an optical cavity at the junction. The two flat and parallel surfaces that form the mirrors of the cavity are made through the cleavage of the junction chip in the crystalline planes, as illustrated in Fig. 8.45. Since the refractive index of GaAs is $n = 3.6$, the reflectivity of the surface, given by Eq. (8.21), is $R = 0.32$. This value is sufficient to create an optical cavity between the two cleavage planes. However, to increase the gain and make the radiation come out in only one direction, one of the sides is covered with a metallic film. In addition, to prevent radiation from coming out laterally, an abrasive is used to roughen the two side surfaces. This eliminates the effect of the resonant cavity in the lateral direction, so that the radiation beam leaves only from the front surface.

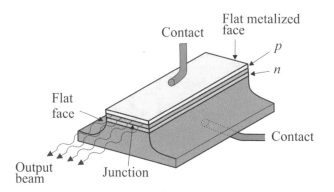

Fig. 8.45 Illustration of a simple *p-n* junction diode laser

Figure 8.46 shows the behavior of the laser power and the radiation spectrum with the current in the junction. Figure 8.46a shows that if the current is less than a threshold value I_T, the intensity of the radiation is small. In this situation the radiation is produced by the spontaneous emission that occurs in the vicinity of the junction, as in a LED. In this case, the radiation spectrum is wide, as illustrated in Fig. 8.46b. However, if $I > I_T$, the radiation becomes much more intense and with a spectrum confined to a narrow band of frequencies. These two characteristics are the main differences between the LED and the junction laser: the laser emits stimulated radiation with a narrow spectrum, while the LED emits spontaneous radiation with a wide spectrum; the laser only operates with a current above a threshold value, whereas the LED operates with any current. Actually, the diode laser operates with various longitudinal modes, with frequencies within the range of the gain, as in Fig. 8.35

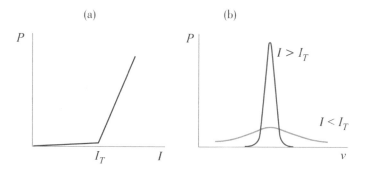

Fig. 8.46 Behavior of the light power emitted by a semiconductor laser. If the current I is larger than a threshold value I_T, the power increases sharply, as in (**a**), and its spectrum becomes narrow, as in (**b**)

Example 8.5 Calculate the spacing between the longitudinal modes of a GaAs diode laser, with an optical cavity of length 1 mm.

Considering λ' the wavelength of light in GaAs, the resonance condition in the optical cavity is

$$L = \frac{m\,\lambda'}{2} = \frac{m\,\lambda}{2\,n} = \frac{m\,c}{2\,n\,v},$$

where the n is the refractive index, λ is the wavelength in vacuum, and v is the frequency. Thus, the difference between the frequencies of two neighboring modes (m and $m \pm 1$) is $\Delta v = c/2nL$. Considering the refractive index of GaAs $n = 3.6$, we have.

$$\Delta v = \frac{c}{2\,n\,L} = \frac{3 \times 10^8}{2 \times 3.6 \times 10^{-3}} = 4.17 \times 10^{10}\,\text{Hz} = 41.7\,\text{GHz}.$$

The semiconductor laser formed by only one *p-n* junction, also called a homojunction laser, was the first to be developed. This type of laser has several problems, such as: its threshold current is very high; to avoid overheating it must be placed at quite low temperatures or operate in pulsed mode; the spectral width of the radiation is large compared to other types of lasers; the light power is small compared with other types of laser; since the radiation is emitted in a region of thickness ($< \mu$m) smaller than the wavelength, the diffraction is large, and the output beam is not collimated. Several of these problems are circumvented in heterojunctions lasers, presented in the next section.

8.7.2 Heterojunctions Lasers

In the homojunction laser, the population inversion that produces the recombination of electron-hole pairs with photon emission occurs only in the space charge, or transition, region, as shown in Fig. 8.44c. However, not all electrons and holes that arrive at the junction participate in this process. Many of them are injected into the other side as minority carriers and diffuse in a region of thickness in the range of 1–$10\,\mu$m. Thus, for the photon emission rate to be larger than the optical losses and to produce the laser operation, it is necessary to have large current densities. This fact is the main reason for the high threshold current of the homojunction laser, which is of the order of 40–$100\,\text{kA/cm}^2$ in GaAs junctions at room temperature. Another effect that contributes to the high value of the threshold current in the homojunction laser is the strong light diffraction. This makes a large fraction of emitted photons leave the junction region, failing to contribute for the stimulated emission.

In heterojunction lasers, these two effects are much smaller, so that the threshold currents are reduced by several orders of magnitude relative to the homojunction laser, and are in the range of 100–500 A/cm². As we saw in Sect. 6.3.2, in a heterojunction there is a potential barrier due to the difference between the energy gaps on the two sides. This allows one to build heterojunction structures with potential barriers that produce confinement of electrons and holes in a thin layer, with thickness of the order of 0.1–0.5 μm. At the same time, since the refractive indices on both sides of the heterojunctions are different, due also to the difference in the energy gaps, there is a confinement of the emitted photons. The increase in the concentration of electron-hole pairs and photons in the same spatial region, results in a larger recombination rate and therefore lower threshold currents.

Figure 8.47 shows three possible structures for heterojunction lasers made with GaAs and (GaAl)As. The structures are made by depositing layers with appropriate thicknesses and compositions on a single crystal n-GaAs substrate. The deposition can be made by the simple liquid phase epitaxy technique (LPE), or the sophisticated molecular beam epitaxy (MBE) technique, described in Chap. 1. The concentration x of Al in the alloy $Ga_{1-x}Al_xAs$ determines the value of the energy gap E_g, that varies from 1.43 eV ($x = 0$) to 2.16 eV ($x = 1$). The p-type semiconductor is made with diffusion of group II atoms, such as Zn, forming acceptors impurities. For n-type doping, elements of group IV are used, such as Sn. The atoms of these elements donate electrons to the atoms of Ga or Al, which are in group III, forming donor impurities.

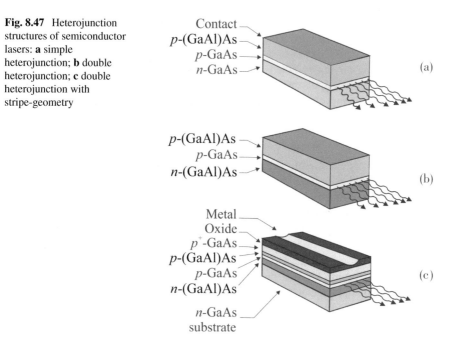

Fig. 8.47 Heterojunction structures of semiconductor lasers: **a** simple heterojunction; **b** double heterojunction; **c** double heterojunction with stripe-geometry

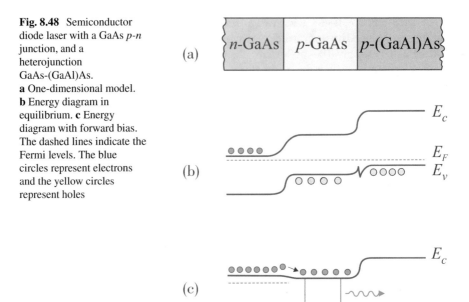

Fig. 8.48 Semiconductor diode laser with a GaAs *p-n* junction, and a heterojunction GaAs-(GaAl)As.
a One-dimensional model.
b Energy diagram in equilibrium. **c** Energy diagram with forward bias. The dashed lines indicate the Fermi levels. The blue circles represent electrons and the yellow circles represent holes

Figure 8.48a shows the one-dimensional model of a laser with a simple *p-n* junction of GaAs and a heterojunction of *p*-GaAs - *p*-Ga$_{1-x}$Al$_x$As. Figure 8.48b shows the energy diagram with no applied voltage and Fig. 8.48c shows the diagram with forward bias. The central region of the structure is made of *p*-type because the injection of electrons to the *p*-side is more efficient than injecting holes in the opposite direction. When a voltage is applied to polarize the *p-n* junction in the forward direction, electrons from the *n*-side are injected into the *p*-type central region. The potential barrier created in the heterojunction prevents the electrons from passing to the *p*-(GaAl)As region. Since the thickness of the central region is much smaller than the diffusion length, the electron-hole pairs are confined in this region and are uniformly distributed.

The difference between the refractive indexes of GaAs and GaAlAs causes photons emitted by electron-hole recombination to be reflected at the interface between the two materials, increasing the stimulated emission rate. Laser operation occurs when the current exceeds a certain threshold value, with emission of photons with energy approximately equal to the GaAs gap, $E_g = 1.43$ eV. This corresponds to radiation in the near infrared region, with wavelength $\lambda = 860$ nm.

The development at the end of the 1960s of the heterojunction laser operating at room temperature, boosted research activities in these lasers, mainly because of their potential in optoelectronics and optical communications. During the 1970s and 1980s, laboratories around the world competed to develop heterojunction laser structures with smaller threshold currents, better collimation of the radiation beam, and better operation stability at several wavelengths.

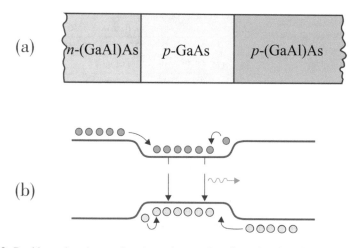

Fig. 8.49 Dual heterojunction semiconductor laser. **a** One-dimensional model. **b** Energy diagram with the polarized *p-n* junction, illustrating the motion of electrons (blue circles) and holes (yellow circles)

The structure of Fig. 8.47b, called a double heterojunction, represents an advance relative to the simple heterojunction because it increases the confinement of carriers and photons in the central region. The structure and scheme of energy bands of the double heterojunction are illustrated in Fig. 8.49. The fact that the GaAs layer is located between two layers of GaAlAs, that have larger energy gap, makes the potential well more pronounced for electrons in the conduction band and for the holes in the valence band. A further improvement for the reduction in the threshold current and laser beam width is achieved using a metal contact in the form of a narrow strip, of width on the order of 20 μm, with the stripe-geometry shown in Fig. 8.47c. The development of this double heterojunction structure represented a major advance for the practical use of semiconductor lasers in optical communications and other applications.

Currently, semiconductor lasers use structures with several layers of alloys with different concentrations of constituents and impurities. The objectives of the sophistication of the structures are: reduction of the threshold current; improvement of collimation and spectral distribution of radiation; greater stability of operation; ease of modulation; and lower manufacturing costs within the quality standards for the desired application. The semiconductor lasers are produced by various techniques of epitaxial growth, such as LPE, MBE and MOCVD.

The semiconductor materials used in heterojunction lasers depend mainly on the desired radiation wavelength. Table 8.2 shows the energy ranges covered by some of the most important semiconductor alloys. The variation of the constituent concentrations makes possible to tune the wavelength of the laser. (InGa)(AsP) lasers are used in optical communications. One of the most important applications of (GaAl)As in infrared lasers was the improvement in reading optical discs, or compact discs (CD and DVD), that became very popular for sound and video

Alloy	λ (μm)	Region of spectrum
$Pb_xSn_{1-x}Te$	7–30	Infrared (ir)
$In_{1-x}Ga_xAs_yP_{1-y}$	1.1–1.6	Near infrared
$Ga_xAl_{1-x}As$	0.7–0.9	Near Infrared
$Ga_xAl_{1-x}In_{1-y}P_y$	0.6–0.8	Red and near ir
$In_{1-x}Ga_xN$	0.4–0.5	Violet-blue

reproduction. On the other hand, the (GaAl)InP red lasers replaced the He–Ne gas lasers in several applications, with the enormous advantage of being powered by small batteries in portable structures. The (InGa)N lasers, with emission in the violet-blue region of the visible spectrum, developed in the late 1990s, were very successful in making possible high-definition digital video discs and, as mentioned earlier, LED lamps for lightning.

8.7.3 Quantum-Well and Quantum-Cascade Lasers

The energy diagram of the double heterostructure laser in Fig. 8.49b represents the variation in one dimension of the energy levels E_v and E_c, corresponding, respectively, to the maximum of the valence band and the minimum of the conduction band. Actually, in each section of the one-dimensional model in Fig. 8.49a, electrons can occupy states with energy above E_c and holes can occupy states with energy below E_v. As illustrated in the energy bands of Fig. 5.7, the occupation of the excited states is due to the thermal energy. Thermally excited electrons have energy in a range k_BT above E_c and holes have energy in a range k_BT below E_v. For this reason, the radiation from the heterostructure laser has a broad line, which at room temperature is on the order of $k_BT/h = 6$ THz.

However, if the GaAs layer is very narrow, the effects of quantum confinement become important. In this case, the states of electrons and holes cannot be described by waves propagating in the longitudinal direction. They are described by stationary waves, similar to a particle in a potential well, as studied in Sect. 3.3.2.

Figure 8.50 illustrates the energy diagram of a heterojunction of (GaAl)As/GaAs/(GaAl)As, exhibiting the potential wells for electrons in the conduction band and for holes in the valence band. In the thin GaAs layer, the minimum energy of electrons is E_1, not E_c, while the maximum energy of holes is E_1', and not E_v. The potential wells created by the difference between the energy gaps of GaAs and (GaAl)As, have depths that depend of the concentrations of Ga and Al. As shown in Table 8.1, for high concentrations of Al, the energy gap approaches 2.16 eV, while the gap in GaAs is 1.43 eV. Thus, in heterojunctions of alloys with high concentrations of Al, the depth of the well in the conduction band is of the order of 0.5 eV. Since this value is much larger than the thermal energy of electrons, $k_BT = 0.026$ eV at room temperature, the effect of quantum confinement

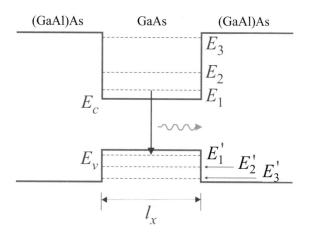

Fig. 8.50 Energy diagram of a quantum-well laser

in thin layers of GaAs is quite pronounced. The exact calculation of energy levels for the well in Fig. 8.50 is more complex than the one for the infinite potential well, presented in Sect. 3.3.2. However, if the thickness l_x of the GaAs layer is not too small, the level E_1 close to E_c can be calculated approximately as if the depth were infinite. Thus, we use Eq. (3.44) with $n = 1$ to calculate, in a first approximation, the lowest energy level of electrons in the well

$$E_1 = E_c + \frac{\hbar^2 \pi^2}{2m_e^* l_x^2}. \tag{8.80}$$

Likewise, the highest energy level of holes is

$$E_1' = E_v - \frac{\hbar^2 \pi^2}{2m_h^* l_x^2}. \tag{8.81}$$

We see that in the double heterostructure laser with a thick GaAs layer, E_1 approaches E_c and E_1' approaches E_v. In this case, the effects of the confinement are small. The laser radiation has a frequency given by Eq. (8.79) and a linewidth of $k_B T/h$. On the other hand, if the thickness l_x is small, such that $E_1 - E_c$ given by Eq. (8.80) is comparable or larger than the thermal energy, the radiation spectrum has a narrow line and the frequency is given by the difference between (8.80) and (8.81), that is

$$h\nu = E_g + \frac{\hbar^2 \pi^2}{2 l_x^2} \left(\frac{1}{m_e^*} + \frac{1}{m_h^*} \right). \tag{8.82}$$

The double heterostructure laser with small l_x is called **quantum well laser**, or QW laser. Quantum well lasers have a structure like the one in Fig. 8.47c, where the p-GaAs layers have a thickness of the order of 10 nm. The main advantages of the quantum well laser are: the linewidth of the radiation spectrum is narrower; the laser frequency can be tuned by choosing the appropriate thickness l_x, and can be larger than the frequency of the GaAs emission.

One disadvantage of the simple quantum-well laser is the small area for recombination of the electron-hole pairs, which compromises the intensity of the radiation. This problem is solved with a **multiple quantum-well** (MQW) laser, also called **quantum-cascade laser** (QCL), formed by dozens of alternating layers with the periodic repetition of the basic unit of Fig. 8.50. The first (GaAl)As layer is deposited on the n-GaAs substrate and doped with donor impurities, to make a n-(GaAl)As layer. All other layers, made alternately of GaAs (thickness of 10 nm or less) and (GaAl)As (thickness of order of 10 nm or more), are doped with acceptor impurities. Since the diffusion length is of order or larger than 10 μm, the electrons injected by the set n-GaAs/n-(GaAl)As reach all layers of the set $(p$-GaAs/p-GaAlAs$)^m$, even if the number m of repetitions is a few dozens. It is also important to note that since the distance between two neighboring wells is smaller than the radiation wavelength, the stimulated emissions of neighboring wells are synchronized, so that the total radiation is coherent.

Quantum-cascade lasers are manufactured by the same production techniques as the heterostructure lasers, that had a major evolution in the 1990s, enabling growth of thin layers of semiconductors with great precision and reliability. One advantage they have relative to diode lasers is that the radiation wavelength is determined by the structure of the layers rather than the lasing material. Thus, device fabricators can tailor the wavelength in a way that cannot be achieved with diode lasers. While diode lasers have wavelength limited to ~2.5 μm, QCLs operate at much longer wavelengths: mid- and long-wave infrared production devices up to 11 μm are available, and some 25 μm emitters have been made on an experimental basis. They are finding new applications in precision sensing, spectroscopy, medical, scientific, and military applications.

Example 8.6 Calculate for a quantum-well laser of (GaAl)As/GaAs(l_x)/ (GaAl)As, the thickness l_x for which: (a) The energy $E_1 - E_c$ is equal to the thermal energy of the electrons at a temperature $T = 300$ K; (b) The laser emission has a wavelength of 820 nm.

(a) The thickness l_x that satisfies the given condition is obtained by equating the energy difference in Eq. (8.80) with the thermal energy. Thus,

$$E_1 - E_c = \frac{\hbar^2 \pi^2}{2m_e^* l_x^2} = k_B T,$$

so that

$$l_x = \frac{\hbar \pi}{(2m_e^* k_B T)^{1/2}}.$$

Using for the effective mass of electrons $m_e^* = 0.068\,m_0$ (Table 5.1) we have

$$l_x = \frac{1.05 \times 10^{-34} \times 3.14}{(2 \times 0.068 \times 9.1 \times 10^{-31} \times 1.38 \times 10^{-23} \times 300)^{1/2}}$$
$$= 1.46 \times 10^{-8}\,\text{m} = 14.6\,\text{nm}.$$

Note that thinner GaAs layers result in an energy spacing $E_1 - E_c$ larger than the thermal energy, and therefore exhibit quantum confinement effects.

(b) The thickness l_x which results in laser radiation with photon energy $h\nu$ is obtained from Eq. (8.82)

$$l_x = \frac{\hbar\pi}{2^{1/2}(h\nu - E_g)^{1/2}}\left(\frac{1}{m_e^*} + \frac{1}{m_h^*}\right)^{1/2}.$$

The wavelength 820 nm corresponds to a photon energy

$$h\nu = \frac{hc}{\lambda} = \frac{6.62 \times 10^{-34} \times 3 \times 10^8}{820 \times 10^{-9} \times 1.6 \times 10^{-19}} = 1.51\,\text{eV}.$$

Using $E_g = 1.43$ eV for GaAs, we have

$$h\nu - E_g = 0.08\,\text{eV} = 1.28 \times 10^{-20}\,\text{J}.$$

Using this value and $m_h^* = 0.5\,m_0$ (Table 5.1) in the expression for l_x we have

$$l_x = \frac{1.05 \times 10^{-34} \times 3.14}{(2 \times 1.28 \times 10^{-20} \times 9.1 \times 10^{-31})^{1/2}}\left(\frac{1}{0.068} + \frac{1}{0.5}\right)^{1/2} = 8.86 \times 10^{-9}\,\text{m}.$$
$$l_x = 8.86\,\text{nm}.$$

Note that, since the GaAs lattice parameter is 0.565 nm, this GaAs layer thickness contains about 16 basic units of quantum wells.

8.8 Some Applications of Semiconductor Lasers and Other Types of Lasers

Semiconductor lasers have become essential components in a large number of equipment and systems developed in recent times. Their applications range from very simple equipment, such as the laser pointer, to sophisticated high-speed optical communication equipment that connect the entire Globe. Many systems had their performance improved and the cost reduced with the replacement of the light sources by semiconductor lasers, such as the optical bar-code scanners used in stores, and a large variety of scientific, medical, and industrial equipment, for example. Other equipments have become possible with the development of the semiconductor lasers, such as optical compact discs used in CD and DVD players. Each application requires a laser with radiation of specific wavelength and other characteristics, and therefore employs specific materials and structures. In general, semiconductor lasers are made with double heterostructures, but certain applications require structures with multiple quantum wells. The development of laser materials, structures, and manufacturing processes is an area of research activities in many academic and industrial laboratories around the world. In this section we shall address only two of the most important applications of semiconductor lasers, optical communications and compact disc players.

8.8.1 Optical Communications

The advent of optical communications was made possible not only by the development of semiconductor lasers and photodiodes, but also optical fibers. The optical fiber is a thin wire with a circular cross section, made of a transparent material, in general glass or plastic. Optical fibers are relatively flexible and are used to guide a light beam through winding paths without interference from the outside.

The basic idea of using an optical fiber as a light guide is very old. A light beam in a transparent material with refractive index n_1, incident on the interface with another material with refractive index $n_2 < n_1$, undergoes total reflection if the incidence angle θ_1 (relative to the normal) is larger than a critical value given by $\theta_c = \sin^{-1}(n_2/n_1)$. Thus, a light beam can propagate within a solid cylinder, undergoing successive reflections on the internal surface so as to be guided along the cylinder. Optically transparent glass or plastic fibers are used as light waveguides for simple applications since the 1930s. However, only after the 1970s optical fibers made of silica (SiO_2) with low losses were developed, making possible to guide light beams for very large distances. The major contribution of Charles K. Kao for the development of low-loss silica optical fibers earned him the Physics Nobel Prize in 2009.

The simplest optical fiber is made of homogeneous material with a certain index of refraction $n > 1$. This fiber is not very useful because its contact with any external material, such as dirt on the surface, can result in scattering and loss of the guided light energy. For this reason, optical fibers are made with two regions, a central core

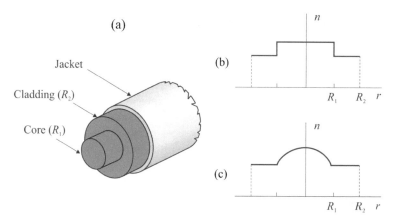

Fig. 8.51 **a** Section of an optical fiber showing the core, the cladding, and the jacket. **b** Step refractive index profile. **c** Graded index profile

with radius R_1 and refractive index n_1, and a cladding with external radius $R_2 > R_1$ and refractive index $n_2 < n_1$. In this way, the total internal reflection occurs at the interface between the core and the cladding, so that the propagation is not disturbed by external interferences.

Figure 8.51 shows the section of an optical fiber and the two common refractive index profiles. In the **step-index fiber**, both the core and the cladding are homogeneous, so that n_1 and n_2 do not vary with the radius. Its main applications are in lighting and image systems. In the **graded-index fiber**, the refractive index of the core varies with the radius, $n_1(r)$, while the cladding is homogeneous. A very common profile form is parabolic, in which n_1 decreases from the axis with the square of the radius. The most used material for the manufacture of optical fibers is fused silica. The variation in the index of refraction is obtained in the manufacturing process through suitable doping with various materials, such as GeO_2, P_2O_5, B_2O_3, etc. For specific applications some types of glass are also used. The protection of the optical fiber is made by a plastic jacket, similar to the ones used in various types of wires.

The propagation of light in an optical fiber is illustrated in Fig. 8.52. In a step-index fiber, the light behaves as if formed by rays that propagate in a straight line, undergoing successive reflections at the internal surface of the cladding. In the

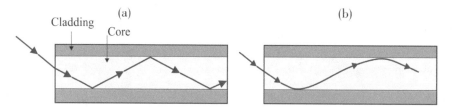

Fig. 8.52 Illustration of the light propagation in optical fibers. **a** Step-index fiber. **b** Graded-index fiber

graded-index fiber, the trajectories of the rays are curved because they undergo continuous refraction due to the variation of the refractive index n_1.

Actually, the view of the wave formed by rays is a simplification of the phenomenon. The propagation of the wave guided by the fiber is described mathematically by the solutions of Maxwell's equations in the cylindrical geometry of the fiber. These solutions correspond to discrete modes of propagation, which can be seen as propagating waves along the fiber axis and stationary wave modes in the cross section. Stationary-wave modes are similar to the wave functions of a particle in a potential well, as in Fig. 3.3, with a certain number of maxima and zeroes of the field along the diameter. Larger diameter/wavelength ratios give larger numbers of maxima. In the simplified view of geometric optics, each stationary-wave mode corresponds to a different angle for the propagation of rays. Thus, fibers with larger diameters can support larger number of propagating modes. A step-index fiber, with core diameter of 125 µm, can propagate thousands of modes with wavelength 0.85, and is called **multimode fiber**. A fiber that allows propagation of only one mode is called **single-mode fiber**. The transverse variation of the electromagnetic field in the cross-section of a single-mode fiber is similar to electron wave function of the mode with $n = 1$, shown in Fig. 3.3. Single-mode fibers have a core with a diameter of about 5–10 µm and cladding with diameter of 125 µm.

A very important characteristic of optical fibers is the dependence of the light intensity attenuation with the wavelength. This is shown in Fig. 8.53 for silica fibers, with the vertical scale expressed in dB/km. The decibel value of an attenuation A is given by $A(dB) = 10 \log 10\, A$. Thus, an attenuation by a factor 10 corresponds at 10 dB, 100 corresponds to 20 dB, 1000 to 30 dB, etc. This notation is convenient to express multiplicative quantities, because the values in dB add up. For example, since the total attenuation in two consecutive stretches of fiber is the product of the attenuations in each one, the value in dB is given by sum of the individual values in dB.

Fig. 8.53 Light attenuation in silica optical fibers as a function of wavelength

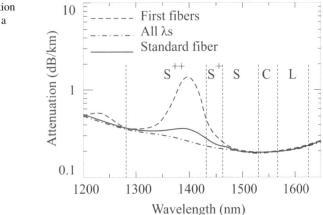

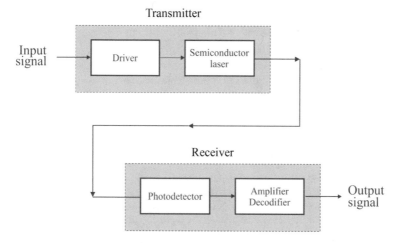

Fig. 8.54 Basic components of optical communication systems

Figure 8.53 shows three attenuation curves. The dashed curve corresponds to fibers used until the mid-1990s. The peaks in attenuation at 1240 nm and 1390 nm are due to vibrational modes of impurity ions OH⁻ caused by the presence of water dissolved in the glass. The solid curve corresponds to commercial fibers manufactured today by processes that reduce the presence of impurities. The dash-dot line corresponds to fibers produced in research laboratories and allows the use in a continuous range of wavelengths.

Figure 8.54 shows the basic components of an optical communication system. The **transmitter** is composed of a light source and a driver circuit. The light source is a semiconductor laser. The driver circuit serves to polarize the laser and also to modulate the light with the input electric signal. The light signal generated by the transmitter is guided by the optical fiber to the **receiver**, where it is converted into an electric signal in the photodetector, amplified and then processed to restore the original signal. In communications over long distances, repeaters are used to amplify the signal between the transmitter and the receiver. The major advantage of optical communication relative to microwave-based systems is the wide spectral width, that allows transmission over a single fiber of thousands of voice and video channels. With the increasing digitalization of information, the capacity of the communication systems started to be expressed in terms of the pulse transmission speed, in bits per second (bit/s).

The first commercial optical communication system was implemented in the second half of the 1980s, operating with GaAs semiconductor lasers, with wavelength around 800 nm and repetition rate of 45 Mbit/s. Since the attenuation of fibers in this region is high, about 2 dB/km, there was a need to have repeaters at every 10 km. The repeaters used at that time were based on electronic amplifiers devices, so that they had to detect the optical signal, amplify the electric signal, and then modulate a new laser beam. In the 1980s, systems operating around 1300 nm were developed, where the fibers have attenuation of about 0.6 dB/km. They employed semiconductor lasers of InGaAsP, able to operate with rates up to 2 Gbit/s, and with a distance of 44 km between repeaters, which used optoelectronic amplifiers. Later, systems were developed for 1550 nm, in which the fibers have attenuation of 0.2 dB/km, being able to transmit at distances of 70 km without the need of repeaters. These systems use InGaAsP lasers with a higher In concentration, operating in the range of 10–40 Gbit/s.

What made possible a major increase in the transmission speed was the development of the optical amplifier at the end of the 1980s. An optical amplifier consists basically of an optical fiber with rare earth impurities. In the range around 1550 nm, silica fibers are used with Er doping, while at 1300 nm, doped fluorate glasses are used with Pr. These impurities have energy levels that absorb radiation in the visible and emit in the infrared. When pumped by a semiconductor laser, this system amplifies an optical signal that propagates in the fiber by means of stimulated emission. Optical amplifiers enabled the advent of a new area of technology, in which signal processing is entirely optical, called **photonics**. In Sect. 10.2 we present some optical devices based on dielectric materials that are used in the processing of optical signals.

The signal amplification and processing by optical means made possible to increase the distance between repeaters. The first totally optical commercial systems appeared in 1990, operating at 10 Gbit/s, and required distances of 60–80 km between repeaters. The scientific advances in electronics and in photonics made possible the development of a new technology of transmission of several channels on a single fiber, namely **wavelength division multiplexing** (WDM), which enabled a large increase in the bit rate. Currently, communication systems operate with transmission rates of a few 1Tbit/s. The wavelength bands used in optical communication are indicated in Fig. 8.54 and are denoted by letters established by the International Telecommunication Union (ITU). The S band corresponds to the wavelength range from 1460 to 1530 nm, the C band is between 1530 and 1565 nm, and the L band is between 1565 and 1625 nm.

8.8.2   Recording and Playback on Compact Discs

An application of semiconductor lasers that became very important at the end of the 90s was recording, initially in **compact discs** (CD) for sound reproduction, and then in **digital video discs** (DVD) for video playback. They were also widely used for recording, storing, and reading digital information in computers. The main advantages of these devices over other recording systems existing at the time, mainly magnetic, are in the large storage capacity of optical discs, easy of transport, and the fact that the reading process does not require physical contact with the disc. Figure 8.55 illustrates the basic elements for storage and reading on optical discs.

In the optical discs for permanent storage, the digital information is recorded in the form of small pits, shown in Figs. 8.55b, c. The pits have long oval shape with dimensions of the order or less than 1 μm, arranged along a spiralling track. The

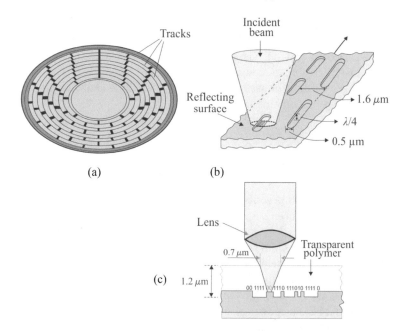

Fig. 8.55 Basic elements of a CD system. **a** View of the optical disc. **b** Illustration of the pits in the disc tracks and the focusing of the reading laser. **c** Lateral view of the pits obtained by cutting the disc along a track. The light reflected from the bottom of the pit with depth $\lambda/4$, lags the light reflected from the top by 180°, producing destructive interference

presence of the pit in a certain position represents bit 1, and the absence represents bit 0. The disk is made of a plastic material, and its preparation can be made by means of an injection process on a flat metallic matrix, containing bumps in the positions that will produce the pits. After the injection, the surface containing the pits in the spiral track is metallized with an aluminum film to reflect light. Finally, the disc is covered with a transparent polymer film to protect the pits and form a smooth final surface, to avoid the accumulation of dirt.

Thermo-optic CD recording systems were developed in late 90s for permanent or rewritable storage, compatible with the earlier systems. In this case, the blank CD consists of a uniform polymer layer, over a flat metallic film. The process of recording a bit 1 in a small region of the layer is made by heating, produced by pulses from a semiconductor laser, focused by a lens, that produces a change in the refractive index of the material after it cools. The depth of the recast region is such that in the reading process, the part of the light that passes through it and is reflected by the metal film has a lag of 180° with respect to the other part. This produces destructive interference in the light indicating the presence of bit 1, as in the traditional CD. This is also the reading process in discs with pits, as shown in Fig. 8.55.

The dimensions shown in Fig. 8.55b are typical of CDs for sound and for digital storage in computers. The reading lasers, in this case, are GaAs double heterostructure or QCL, operating in the infrared with a wavelength of 780 nm. Video or DVD discs require larger storage capacity. For this reason, the dimensions of the pits and the distance between tracks are about 50% smaller, and the reading lasers operate in the red, with wavelength 650 nm. The development of semiconductor lasers operating in the blue made possible to reduce the dimensions of the pits and tracks resulting in larger storage capacity, for use in for high-definition DVD and other applications, with the technology called blue-ray.

8.8.3 Other Types of Lasers

Besides the several lasers described in this chapter, other types of lasers with more unsusal properties have been developed more recently. To conclude this chapter, let us briefly present three types with interesting characteristics and applications, fiber lasers, nanolasers, and random lasers.

In **fiber lasers** the gain medium is an optical fiber doped with impuritites of rare-earth elements that provide the electronic energy levels for the stimulated emission of radiation. This type of laser was first demonstrated in the 1960s, but only in the last two decades it became practical and began to be fabricated commercially. In the first fiber lasers the cavity was is formed by two mirrors placed externally. In modern fiber lasers, the cavity mirrors are directly inscribed in the gain fiber core by means of a Bragg grating, made by means interferometric methods using ultraviolet light. Single-mode or multimode fibers can be employed, and the rare earth ions used for the gain are Nd, Er, Yb, Pr, etc., depending on the desired wavelength and power. The pump mechanism is usually optical, either longitudinally or transversely. A fiber laser can be a very compact device, optically pumped by a semiconductor laser, and can operate in continuous mode, Q-switched, or mode-locked. An advantage of fiber lasers over other types of lasers is that the laser light is both generated and delivered by an inherently flexible medium, which allows easier delivery to the focusing location and target. Another advantage is the high output power that can reach 1 kW. The mechanical flexibility and the high power make fiber lasers an important industrial tool for laser cutting, welding, and folding of metals and polymers.

As the prefix *nano* implies, **nanolasers** are coherent-light emitting devices with nanoscale dimensions. In these lasers the light is generated from nanowires or other nanostructures that encompass the gain medium and resonator. Early studies of ultrasmall lasers started in the 1990s, but the development of nanolasers boosted in the early 2000s, aiming at ultracompact and very low power consumption coherent optical sources. But soon it was realized that many other features would be important as the device dimensions became of the order of the wavelength of the emitted light. For example, the light-matter interaction in nanostructures is different than in macro media. Currently, nanolasers are fabricated with several different structures, such as metallic nanodisks, semiconductor nanowires, plasmonic nanowires, among others. These structures have dimensions that can range from 20 nm to a few micrometers. Nanolaser arrays have also being demonstrated. The applications of nanolasers include their ability to be fast-modulated, on-chip optical computing, ultra-dense data storage, nanolithography, super-resolution imaging, among others.

As we have shown in this chapter, the common lasers are formed by three basic elements: the pump source, the gain medium, and a set of two static mirrors to form the cavity and provide optical feedback. However, in late 1960s, it was proposed that the mirrors could be replaced by a scattering medium, which in turn would provide the optical feedback. After some attempts to build such a laser with rare earth doped powders in the 1970s and 1980s, in 1994 a laser using a Rh6G dye as the gain medium, the second harmonic of a Nd-YAG laser (532 nm) as the pump source, and 250 nm diameter TiO_2 nanoparticles as scatterers to provide optical feedback, the so-called **random laser** was unambiguously demonstrated. Since then, the field of random lasers has grown to become intensively studied, since it comprises a complex open system with gain and disorder, and can be used as platforms for a diversity of applications. Figure 8.56 shows a comparison between a regular laser cavity and a random laser, that has no cavity. The randomness arises in the path of the photons before amplification, which can occur in a diffusive or optically localized regime, and the influence of random cavities in the medium.

Random lasers have been demonstrated using as gain media solid state nanomaterials, liquid suspensions, and semiconductors, that can be pumped optically or electrically, as well as polymers and biomaterials. The scatterers can be passive micron or nanosize powders (TiO_2, Al_2O_3), metallic nanomaterials (Au, Ag) that have plasmonic properties, semiconductor nanoparticles (ZnO), 2D materials ($ZrTe_2$), rare earth ions (Er, Nd), or quantum dots. Random lasers are essentially multidirectional, multimode, low coherence lasers. Directionality can be achieved by managing the pump source or employing a random fiber laser.

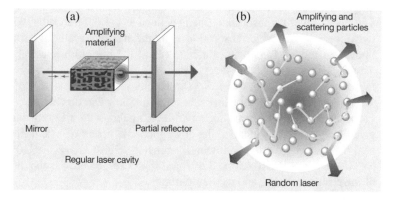

Fig. 8.56 Comparison between a regular laser and a 'random laser'. (a) In a regular laser the light bounces back and forth between two mirrors that form a cavity. After several passes through the amplifying material in the cavity, the gain amplification can be large enough to produce laser light. (b) In a random laser the cavity is absent but multiple scattering between particles in the disordered material keeps the light trapped long enough for the amplification to become efficient, and for laser light to emerge in random directions. Reproduced with permission from D. Wiersma, Nature **406**, 132 (2000)

Applications of random lasers include the use as speckle free sources for optical imaging, optical diagnostic of some types of skin cancer, and optical sensors.

Problems

8.1 Show, using Maxwell's equations (2.1)–(2.4), that in a medium of conductivity σ and without electric charge ($\rho = 0$), the electric field varying only in the x direction is described by the wave equation Eq. (8.1).

8.2 (a) Show that in a sinusoidal plane wave with an electric field of amplitude E_0 propagating in a dielectric material with refractive index n, the Poynting vector is given by Eq. (8.11). (b) Calculate the light intensity in a laser beam with diameter 1 mm and average power of 10 W. (c) Calculate the value of E_0 in the laser beam of item b.

8.3 Show that in a sinusoidal plane wave with an electric field given by Eq. (8.8), the variation of the intensity in space is given by Eq. (8.13).

8.4 (a) Check that the penetration length δ, given by Eq. (8.30), has the unit of meter, in the International System. (b) Calculate the value of δ for a frequency $\nu = 10$ THz, in the far-infrared region. (c) Check that the plasma frequency, given by Eq. (8.32), has the unit of rad/s. (d) Calculate the plasma frequency for copper in Hz and in eV.

8.5 The complex refractive index of germanium for a light beam with wavelength 400 nm is given by $N = 4.14 + i\,2.221$. Calculate:

   (a) The phase velocity of light;
   (b) The absorption coefficient;
   (c) The reflectivity.

8.6 Calculate the total transmission of a germanium plate with parallel faces and thickness 50 nm for a light beam of wavelength 400 nm.

8.7 The linewidth of an absorption line is defined as the difference between the two frequencies for which absorption is half the maximum value at resonance. Show that for the Lorentzian function (8.40) the linewidth is equal to the damping rate Γ.

8.8 In comparing Eqs. (8.57) and (8.38), one can see that $(N_1 - N_2)\,p_{12}^2 / \hbar\varepsilon_0$ must have has the same dimension as ω_p^2 / ω_0. Show that this is true.

8.9 From the dielectric constant for a two-level system with a Lorentzian line, given by Eq. (8.57), calculate the absorption coefficient α, in cm^{-1}, at the peak of the line, for the following parameter values: $\nu_0 = 3 \times 10^{14}$ Hz; $N_2 = 0$; $N_1 = 10^{18}$ cm^{-3}; $\Gamma = 3 \times 10^3$ s^{-1}; $p = ea_0$, where e is the modulus of the electron charge, and a_0 the Bohr radius.

8.10 A hydrogen atom is in an electromagnetic field linearly polarized with photons of energy equal to the separation between the levels $n = 1$ and $n = 2$. Using the eigenfunctions in Table 3.1, show that there is only one electric dipole transition from state $n, l, m = 1, 0, 0$ to state $1, 1, 0$.

8.11 Consider an infrared photo-resistor with dimensions $30 \times 1 \times 0.1$ mm^3, made of intrinsic Ge, with recombination time $\tau_r = 10^{-6}$ s, subjected to a voltage of 10 V.

(a) Calculate the current variation at the terminals, produced by a variation of 1 μW in the intensity of an infrared beam with wavelength 1.1 μm, evenly distributed on the surface, knowing that the quantum efficiency of Ge in this wavelength is 80%.

(b) Calculate the voltage variation in a load resistance $R_L = R_D$ in the conditions of item a.

8.12 A solar cell with area 10 cm^2 is made of Si with impurities concentrations $N_a = 2 \times 10^{16}$ cm^{-3} and $N_d = 5 \times 10^{19}$ cm^{-3}, for which $\tau_n = 10$ μs, $\tau_p = 0.5$ μs, $D_n = 9.3$ cm^2/s, and $D_p = 2.5$ cm^2/s. For normal radiation conditions this cell has $I_L = 500$ mA. Calculate the open circuit voltage and the maximum electric power provided by the solar cell.

8.13 The responsivity of a photodetector is defined as the ratio between the current generated by the photons and the incident light power. Calculate the responsivity of an ideal photodetector as a function of the radiation wavelength and compare its value with any point on the dashed line in Fig. 8.22.

8.14 Show that the values of the current and voltage in a solar cell in the condition of maximum power, are given by Eqs. (8.74) and (8.75).

8.15 The width of the gain curve of a He–Ne laser is 1.5 GHz. Calculate the number of longitudinal modes of an optical cavity of length 0.5 m that can be emitted by the laser.

8.16 Calculate the angle of the Brewster window of a gas laser made with a glass of refractive index $n = 1.46$.

8.17 The core and cladding of a step-index optical fiber have indexes of refraction of 1.48 and 1.46, respectively. Calculate the critical angle of propagation.

Further Reading

G.P. Agrawal, *Fiber-Optic Communication Systems* (Wiley, New York, 2012)

L. Dong, B. Samson, *Fiber Lasers: Basics, Technology, and Applications* (CRC Press, Boca Raton, 2016)

M. Fox, *Optical Properties of Solids* (Oxford University Press, Oxford, 2010)

Q. Gu, Y. Fainman, *Semiconductor Nanolasers* (Cambridge University Press, Cambridge, 2017)

R.E. Hummel, *Electronic Properties of Materials* (Springer, Berlin, 2011)

W.B. Jones Jr., *Introduction to Optical Fiber Communication Systems* (Holt, Rinehart and Winston, New York, 1995)

M. Noginov, *Solid-State Random Lasers* (Springer, Heidelberg, 2005)

S.M. Riad, I.M. Salama, *Electromagnetic Fields and Waves: Fundamentals of Engineering* (McGraw-Hill Book Co., New York, 2019)

A.E. Siegman, *Lasers* (University Science Books, Sausalito, CA, 1986)

S.D. Smith, *Optoelectronic Devices* (Prentice Hall, New York, 1995)

L. Solymar, D. Walsh, *Lectures on the Electrical Properties of Materials* (Oxford University Press, Oxford, 2018)

B.J. Streetman, S. Banerjee, *Solid State Electronic Devices* (Prentice Hall, New Jersey, 2005)

S.M. Sze, M.K. Lee, *Semiconductor Devices: Physics and Technology* (Wiley, New York, 2012)

J. Wilson, J.E.B. Hawkes, *Optoelectronics: An Introduction* (Prentice Hall, New York, 1998)

D. Wood, *Optoelectronic Semiconductor Devices* (Prentice Hall, New York, 1994)

A. Yariv, P. Yeh, *Photonics: Optical Electronics in Modern Communications* (Oxford University Press, Oxford, 2007)

Chapter 9
Magnetism, Magnetic Materials, and Devices

It has been known since Antiquity that iron fragments are attracted by the lodestone, the name given to the natural magnet, now known as magnetite, with chemical composition Fe_3O_4. The word **magnetism** comes from Magnesia, a city in ancient Turkey that was rich in iron ore. In contrast to most electronic materials studied in this book, magnetic materials have had applications in electrical equipment for almost two centuries. Remarkably, they continue to reveal new properties, phenomena, and applications. This chapter begins with the quantum mechanical explanation of the magnetic properties of materials. The conventional applications are briefly presented. A longer section is devoted to magnetic recording, an application that continues to evolve with recently developed spintronics technologies. In one section we describe microwave properties of magnetic ferrites and their device applications, and finally we present the basic concept of magnonics and perspectives for applications.

9.1 Magnetism and Magnetic Materials

Magnetic phenomena were among the first to arouse the curiosity of mankind about the interior of materials. The first reports of experiments with the "mysterious strength" of the lodestone are attributed to the Greeks and date back to 800 BC. The first practical use of magnetism was in the compass, invented by the Chinese in Antiquity. Based on the property of a magnetized needle to orient itself in the direction of the Earth's magnetic field, the compass was an important instrument for navigation in the beginning of the early modern period.

Magnetic phenomena took on a much larger dimension in the nineteenth century, with the discovery of its correlation with electricity. In 1820 Hans Christian Oersted discovered that an electric current in a wire produces a magnetic effect, such as changing the orientation of a nearby compass needle. Later André-Marie Ampère formulated the law relating the current intensity in the wire and the magnetic field it creates. The effect by which a wire with current suffers the action of a force produced

© The Author(s), under exclusive license to Springer Nature Switzerland AG 2022
S. M. Rezende, *Introduction to Electronic Materials and Devices*,
https://doi.org/10.1007/978-3-030-81772-5_9

by the field created by a permanent magnet was observed shortly thereafter. In 1831, Michael Faraday and Joseph Henry discovered the reciprocal effect, magnetic induction. They found that a time-varying magnetic field could induce an electric current in a circuit. At the end of the nineteenth century these three phenomena were perfectly understood and already had countless technological applications, among them the electric motor and the electric generator. The invention of the incandescent lamp, associated with the development of electric generators, made possible a revolution in the customs of humanity with the advent of electric lighting. At the same time, the introduction of electric motors in industry and machine shops revolutionized industrial, manufacturing, and service activities.

Since the end of the nineteenth century, magnetic materials have played a very important role in the operation of many devices and equipment of daily use. In traditional applications, magnetic materials are classified in three categories that will be studied in this chapter. **Permanent magnets**, that have the property to create a constant magnetic field, are used in electric motors, generators, loudspeakers, and a variety of devices. **Soft magnetic materials**, that are used to generate a field proportional to the current in the winding of a coil much larger than the field that would be created only by the current, are used in transformers, relays, electric motors and generators, as well as in many devices. Materials with magnetic properties intermediate between permanent magnets and soft magnets are used to store information in **magnetic recording**. Magnetic recording is the best technology in electronics for the storage of non-volatile and rewritable analog and digital information.

Many of the current applications of magnetic materials have resulted from scientific and technological advances obtained in the past decades in academic and industrial laboratories in many countries. Those advances were only possible thanks to the understanding of the atomic properties of materials, based on quantum mechanics developed in the 1920s and 1930s. Many fundamental contributions to magnetism were made in the following decades. The Physics Nobel Prize was awarded to Louis Néel in 1970, and to J. H. van Vleck and P. W. Anderson in 1977, for their outstanding contributions to magnetism, that has been one of the most fertile and active fields in Condensed Matter Physics. The knowledge accumulated in this field, together with the progress in materials science and engineering, have enabled the discovery of new phenomena and the development of new magnetic materials for applications in electronics. In the last three decades, the development of techniques for the fabrication of very thin magnetic films and nanometric structures led to the discovery of the giant magnetoresistance by Albert Fert and Peter Grünberg. This discovery gave origin to a new field of physics and technology, **Spintronics**, that studies phenomena in which the electron transport is controlled by its spin. Fert and Grünberg received the Physics Nobel Prize in 2007 for their seminal discovery.

The behavior of materials under an external magnetic field is determined by the origin of their **magnetic dipoles**, or **magnetic moments**, and the nature of the interactions between them. Magnetic dipoles originate from the angular momentum of electrons in the ions or atoms in the material. This momentum has a quantum nature, as will be shown in Sect. 9.2.1. However, we shall use in this chapter a combination of quantum and semi-classical treatments aiming to obtain the important results as straightforwardly as possible. Macroscopically, the quantity that represents the magnetic state of a material is the **magnetization vector** $\vec{M}$, defined as the magnetic dipole moment per unit volume

$$\vec{M} = \frac{1}{V} \sum_i \vec{\mu}_i, \qquad (9.1)$$

where the sum runs over all points i at which there are dipole moments $\vec{\mu}_i$ inside a volume V. This volume is chosen large enough so as to have a good macroscopic average, but small relative to the sample size so that the magnetization represents a local magnetic property.

The magnetic field can be expressed by two quantities, either $\vec{B}$, the magnetic induction vector, or $\vec{H}$, the magnetic field intensity vector. While $\vec{H}$ is related to the current that creates the field, $\vec{B}$ depends as much on the current as on the magnetization of the medium. The vector $\vec{B}$ determines the magnetic flux Φ across a surface S, defined by

$$\Phi = \int_S \vec{B} \cdot d\vec{a}, \qquad (9.2)$$

where $d\vec{a}$ is a vector normal to the surface S at each point. In the semi-classical macroscopic theory, the magnetization enters in Maxwell's equations carrying information on the magnetic properties of the material, through the relationship between $\vec{B}$ and $\vec{H}$. In the SI this is

$$\vec{B} = \mu_0(\vec{H} + \vec{M}), \qquad (9.3)$$

where $\mu_0 = 4\pi \times 10^{-7}$ N/A^2 is the magnetic permeability of vacuum. In the CGS the relationship between the fields takes the form,

$$\vec{B} = \vec{H} + 4\pi \vec{M}. \qquad (9.4)$$

We see that in the CGS, in a vacuum, $\vec{B} = \vec{H}$ and $\mu_0 = 1$. The material response to an applied field $\vec{H}$, characterized by the behavior of $\vec{M}$, in many cases can be represented by the magnetic susceptibility χ. In the simplest case, the magnetization is induced in the same direction as the applied field so that χ is a scalar defined by,

$$\chi = \frac{M}{H}. \qquad (9.5)$$

Note that, since M and H have the same dimension, the susceptibility is a dimensionless quantity. The magnetic permeability μ is defined through the relation between $\vec{B}$ and $\vec{H}$

$$\vec{B} = \mu \vec{H}. \tag{9.6}$$

The relationship between μ and χ, obtained from Eqs. (9.3)–(9.6), is, in the two systems of units

$$\mu = \mu_0(1 + \chi), \quad \text{(SI)} \quad \mu = 1 + 4\pi\chi. \quad \text{(CGS)} \tag{9.7}$$

The reason for presenting the relationship between the magnetic quantities in the two systems of units is the fact that both are widely used mainly in science, but also in engineering. For this reason, we present in Table 9.1 the units of the magnetic quantities in the two systems. Note that the unit of M in the CGS is emu/cm^3, where emu (electromagnetic units) is the unit of magnetic moment, and that emu/cm^3 is formally equivalent to gauss (G). However, since gauss is the unit of B, and in the CGS the relationship between B and M is given by Eq. (9.4), emu/cm^3 is used for the unit of M and gauss is used for $4\pi M$. In the SI, on the other hand, the unit of M is A/m, the same as H, while the unit of $\mu_0 M$ and of $\mu_0 H$ is tesla (T). Another important relationship is that of the energy of a magnetic dipole with moment $\vec{\mu}_i$ in a magnetic field $\vec{B}_i$ at point i

$$U_z = -\vec{\mu}_i \cdot \vec{B}_i. \tag{9.8}$$

This equation shows that the energy is minimum when $\vec{\mu}_i$ has the same direction as the field $\vec{B}_i$. Inside a solid, $\vec{B}_i$ is the sum of the external field with the fields created by the ions around point i. This internal field is responsible for the differentiation of the magnetic properties of materials.

The susceptibility value ranges from 10^{-5} on very weak magnetic materials up to 10^6 in strongly magnetic materials. In some cases the susceptibility is small and negative. In other cases, the relationship between M and H is not linear, so the susceptibility varies with the intensity of the magnetic field. Depending on the microscopic origin of the magnetization and the internal interactions, materials are commonly classified in one of the following main categories:

- Diamagnetic

Table 9.1 Units of magnetic quantities in the International (SI) and Gaussian (CGS) Systems

Quantity	SI	CGS	Relation
Φ	Weber (Wb)	Maxwell	1 Wb = 10^8 Maxwells
B	Tesla (T) = Wb/m^2	Gauss (G)	1 T = 10^4 G
H	A/m	Oersted (Oe)	1 A/m = $4\pi \times 10^{-3}$ Oe = (1/79.58) Oe
M	A/m	emu/cm^3	1 A/m = 10^{-3} emu/cm^3
μ	N/A^2	Dimensionless	
χ	Dimensionless	Dimensionless	

- Paramagnetic
- Ferromagnetic
- Ferrimagnetic
- Antiferromagnetic.

Diamagnetism is the weakest type of magnetic response in a system and is characterized by a negative susceptibility of the order of 10^{-5}. The origin of diamagnetism lies in the variation of the orbital angular momentum of the electron induced by the application of the external field. The classical explanation of this phenomenon is based on Lenz's law, by which a variation in the magnetic field generates an induced electric current that tends to oppose this variation, that is, by creating a field opposite to the applied one. This phenomenon occurs in any atom. But since it is very weak, it is important only in materials with no atomic magnetic dipoles that have much more pronounced effects. Diamagnetic materials are those that do not have atomic magnetic dipoles, that is, that have atoms or ions with complete electronic shells. This is the case of the noble gases, He, Ne, Ar, Kr, Xe. It is also the case of solids with ionic bonding, whose atoms exchange electrons to keep their last shells filled, such as NaCl, KBr, LiF, and CaF_2. Since diamagnetism is a very weak property of materials, its detailed properties will not be studied here.

Materials that have permanent atomic magnetic moments are classified in one of the other categories above, or else have a more complex magnetic structure, as is the case of the so-called spin glasses. However, to have practical use in the conventional applications, it is necessary to have large macroscopic magnetization, which occurs only in ferro- and ferrimagnetic materials. These are the materials used in the three applications previously mentioned: permanent magnets; soft magnetic materials; and magnetic recording media.

9.2 Magnetic Properties of Materials

The magnetization of a material originates from the magnetic moment associated with the electron angular momentum. Thus, in order to understand the origin of the moment magnetic, it is necessary to review some properties of angular momentum.

9.2.1 Origin of the Magnetic Moment of Electrons

Classically, the angular momentum $\vec{L}$ of a particle is related to its linear momentum $\vec{p}$ and position vector $\vec{r}$ by the expression $\vec{L} = \vec{r} \times \vec{p}$. Since the linear momentum operator $\vec{p}_{op}$ is given by Eq. (3.6), the angular momentum operator is

$$\vec{L}_{op} = -i\hbar \vec{r} \times \nabla. \tag{9.9}$$

From this result, it can be shown (Problem 9.1) that in a hydrogen atom, in which there is only one electron outside the last filled shell, the electronic orbital is an eigenstate of the operators L_{op}^2 and L_{zop}. Actually, this is true only if the electron spin is ignored. The eigenvalue equations for the two operators are

$$L_{op}^2 \Psi_{nlm_l} = \hbar^2 l(l+1)\Psi_{nlm_l},\tag{9.10}$$

$$L_{z\,op} \Psi_{nlm_l} = \hbar\, m_l\, \Psi_{nlm_l},\tag{9.11}$$

where Ψ_{nlm_l} is the electronic wave function with quantum numbers n, l, m_l. In addition to the orbital angular momentum, the electron has spin angular momentum, which is represented by the operator $\vec{S}_{op}$. If the electron were a classical particle of mass m, the spin could be interpreted as resulting from a rotation of the electron around itself, and whose value would depend on the angular speed of rotation. Actually, the electron is not a classical particle and its spin is an inherently quantum property. Due to the presence of the spin, the complete electronic wave function has to be characterized by the orbital part and also a part that represents the spin state. This part is the eigenfunction of the operators S_{op}^2 and S_{zop}, having eigenvalues respectively $\hbar^2 s(s+1)$ and $\hbar\, m_s$. For the electron $s = 1/2$, so that the quantum number m_s can take only the values $+1/2$ and $-1/2$, representing the spin up or down, relative to a quantization axis. This axis is determined, for example, by the direction of an external magnetic field.

The orbital angular momentum and the spin angular momentum of the electron give rise to the magnetic dipole moment of the atom. Figure 9.1 illustrates the orbital magnetic moment in the Bohr model of the atom. The classical angular momentum of the electron with an orbit of radius r and angular velocity ω is $L = I\omega = mr^2\omega$. Since the electron has charge $-e$, its motion corresponds to a circular loop with current $i = e\omega/2\pi$. This loop creates a magnetic dipole with moment $\mu = iA = i\pi r^2$. Note that since the electron charge is negative, the magnetic moment has a direction opposite to the angular momentum. This magnetic moment makes the atom behave like a tiny magnetic needle, with north (N) and south (S) poles as indicated in Fig. 9.1. From these expressions we can obtain the relationship between the magnetic moment and the momentum angular. In the SI we have

Fig. 9.1 Electron orbit in the Bohr model for the atom illustrating the orbital angular momentum and the orbital magnetic moment

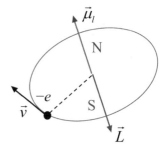

$$\vec{\mu}_l = -g_l \frac{e}{2m} \vec{L}, \tag{9.12}$$

where $g_l = 1$ is the orbital g-factor. If the electron spin were a classical quantity, the relationship between the magnetic moment of the spin $\vec{\mu}_s$ and $\vec{S}$ would be the same as (9.12). However, due to the quantum nature of $\vec{S}$, the relation is

$$\vec{\mu}_s = -g_s \frac{e}{2m} \vec{S}, \tag{9.13}$$

where $g_s \approx 2$. The difference in the orbital and spin g-factors is one of the manifestations of the non-classical origin of the spin.

One consequence of the electron motion around the nucleus is that the electrostatic field in the reference frame of the nucleus produces a magnetic field in the electron reference frame resulting from the relativistic Lorentz transformation. The interaction of the electron spin magnetic moment with this magnetic field gives rise to the **spin–orbit interaction**. Due to this interaction the electronic wave function is not an eigenfunction of L_{zop} and S_{zop} separately. It is an eigenfunction of J_{zop}, the z-component of the total angular momentum operator

$$\vec{J}_{op} = \vec{L}_{op} + \vec{S}_{op} \tag{9.14}$$

In this case, m_l and m_s are no longer "good" quantum numbers, but they are still useful in determining the new quantum number $m_j = m_l + m_s$.

9.2.2 Magnetic Moment of Atoms and Ions

When the atom or ion has several electrons outside the last filled shell, its magnetic behavior is determined by the properties of these electrons. This is so because in a filled shell, electrons occupy orbitals with all possible m_l values, positive and negative, as well as all possible m_s values. In this way, the total angular momentum of the closed shell is zero, and therefore its magnetic moment is also zero. The manner by which the external electrons occupy the orbitals to form the ground state is determined by the conditions of minimum energy. These conditions are given by **Hund's rules**, stated as follows:

1. Electrons occupy states so as to maximize the total spin, $S = \sum m_s$, without violating Pauli principle.
2. Electrons occupy orbitals that result in the maximum value of $L = \sum m_l$, consistent with rule 1 and the Pauli principle.
3. The quantum number of the total angular momentum is $J = |L - S|$ when the shell is less than half full, and $J = |L + S|$ when the shell is more than half full.

Most elements of the periodic table form a solid with their atoms gaining or losing electrons from or to neighboring atoms, so that their last shells are completely filled and form diamagnetic ions. This is the case, for example, of sodium and chlorine in sodium chloride. The sodium atom has 11 electrons, with configuration $1s^2 2s^2 2p^6 3s^1$, while chlorine has configuration $1s^2 2s^2 2p^6 3s^2 3p^5$. The sodium atom loses its 3s electron, which is the only unpaired one, to form the Na^+ ion. This electron goes to chlorine, completing the 3p layer of the Cl^- ion in NaCl. However, this does not happen with the ions of the transition iron group elements, Ti, V, Cr, Mn, Fe, Co, and Ni. The atoms of these elements have an incomplete 3d shell, even though they have electrons in the 4s shell. They are the 4s electrons that are lost in the chemical bond, leaving the 3d shell unfilled and forming an ion with a non-zero total magnetic moment. A similar phenomenon occurs with atoms of rare earths elements of the lanthanide group (Nd, Pm, Sm, Eu, Gd, Tb, Dy, etc.), that lose 6s electrons and stay with the unfilled 4f shell, and also with some elements of the actinide group. These are the elements whose atoms or ions have permanent magnetic moments. For this reason, magnetic materials necessarily contain one or more elements of the transition iron or rare earth groups.

To calculate the magnetic moment of a certain isolated atom or ion, it is necessary to apply Hund's rules to determine the configuration of the states and the corresponding S, L and J values. This is presented schematically, below, for some important magnetic ions.

Fe^{2+}—configuration: $(1s^2 2s^2 2p^6 3s^2 3p^6) 3d^6$.

(Argon atom)

The six 3d electrons are distributed as follows:

Rule 1	m_s	1/2	1/2	1/2	1/2	1/2	$-1/2$	$S = 2$
Rule 2	m_l	2	1	0	-1	-2	2	$L = 2$
Rule 3	$J = L + S = 4$							

The ground state of Fe^{2+} is represented by $^5D_4$, where the capital letter designates the value of L (S for $L = 0$, P for $L = 1$, D for $L = 2$, F, G, H, I, etc.). The superscript is the multiplicity $2S + 1$, and the subscript is the value of J.

Mn^{2+}, Fe^{3+}—configuration: (Argon atom) $3d^5$.

Note that the Mn^{2+} ion is formed by the loss of two 4s electrons, while the Fe^{3+} ion is formed by the loss of two 4s electrons and one 3d electron. The distribution of the five 3d electrons is determined by Hund's rule as follows:

Rule 1	m_s	1/2	1/2	1/2	1/2	1/2	$S = 5/2$
Rule 2	m_l	2	1	0	-1	-2	$L = 0$
Rule 3	$J = L + S = 5/2$						

The ground state of these ions is then $^6S_{5/2}$. To conclude the examples of the use of Hund's rules, consider the case of a rare earth element, Sm^{3+}. Its ion is formed by the loss of two 6s electrons and one 4f electron, so that five electrons remain in the 4f shell. Thus we have:

Sm³⁺—configuration of the last shells: $4f^5 5s^2 5p^6$.

Rule 1	m_s	1/2	1/2	1/2	1/2	1/2	$S = 5/2$
Rule 2	m_l	3	2	1	0	-1	$L = 5$
Rule 3	$J = L - S = 5/2$						

Therefore, the ground state of Sm^{3+} is $^6H_{5/2}$.

The total angular momentum has the same properties as in Eqs. (9.10) and (9.11). Thus, $\hbar^2 J(J+1)$ is the eigenvalue of the operator J_{op}^2 and m_J is the eigenvalue of the operator J_{zop}, where m_J ranges from $-J$ to $+J$. The determination of J for each ion using Hund's rules allows to calculate the magnetic properties of materials containing that ion. Using the relations (9.12)–(9.14), it can be shown that the z-component of the total magnetic moment of a free magnetic ion is, approximately

$$\mu_z = -g \, \mu_B \, m_J, \qquad (9.15)$$

where μ_B is the so-called **Bohr magneton**, given by

$$\text{(CGS)} \quad \mu_B = \frac{e\hbar}{2mc} = 9.27 \times 10^{-21} \, \text{G cm}^3$$

$$\text{(SI)} \quad \mu_B = \frac{e\hbar}{2m} = 9.27 \times 10^{-24} \, \text{Am}^2, \qquad (9.16)$$

where g is the Landé g-factor, given by

$$g = 1 + \frac{J(J+1) + S(S+1) - L(L+1)}{2 \, J(J+1)}. \qquad (9.17)$$

Note that this equation gives $g = 1$ for $S = 0$, and $g = 2$ for $L = 0$, as expected, because $g_l = 1$ and $g_s = 2$. Regarding the units of the Bohr magneton, note that, in the CGS, G cm³ is the same as erg/G, and in the SI, A m² is the same as J/T.

Hund's rules are valid exactly for electrons in isolated atoms or ions, in which the electric field seen by the electrons has spherical symmetry. However, when an ion of an element in the 3d group is in a crystal, the electrons of the 3d layer are also influenced by the crystalline electric field produced by neighboring ions. As a result, the atomic orbitals of the type in Table 3.1 are not eigenstates of the crystalline Hamiltonian. It can be shown that the eigenstates are formed approximately by linear combinations of atomic orbitals with quantum numbers $+m_L$ and $-m_L$. Thus, the effective orbital angular momentum of **ions of elements in group 3d in solids** is $L \approx 0$. This effect, called suppression of orbital angular momentum, makes the magnetic moment of materials containing 3d ions to be almost entirely due to the electron spin. In this case, the moment is calculated with Eq. (9.15) with $g \approx 2$.

In the case of rare earth ions, the orbitals responsible for the magnetic moment correspond to the inner electronic shells. In this case, the crystalline electric field has a negligible effect on the electrons of these orbitals, because of the shielding by the outer electrons. This makes the angular momentum of the ion in the solid equal to that of the free ion, and therefore given by Hund's rules. Consequently, rare earth elements have, in general, larger magnetic moment than elements of the 3d group. In addition, they also have strong interaction between the spin and the crystalline field, through the spin–orbit coupling. Due to the strong chemical reactivity of rare earth elements, the technology for using them in materials took some time to be developed. However, in the past decades they have acquired great importance in the industry of magnetic materials.

9.2.3 Paramagnetism

Paramagnetism is the magnetic property of materials that have permanent atomic magnetic moments but with negligible interaction with each other. In the absence of external fields, paramagnetic materials have zero magnetization. The application of an external field produces a small magnetization in the direction of the field. For this reason, paramagnetic materials have positive magnetic susceptibility, with values in the range $\chi \sim 10^{-5}$–10^{-3}.

The main paramagnetic materials are metals of non-magnetic elements and insulating materials that contain noninteracting atoms or ions of elements of the iron transition group, rare earths, and actinide elements. Metals are paramagnetic, because an applied magnetic field separates the conduction energy band in two bands, one with $+1/2$ spin electrons and one with $-1/2$ spin electrons. This is due to the fact that the energies of the magnetic moments of the spins $+1/2$ and $-1/2$ in the field are different. Thus, the lower energy band has a larger number of electrons than the higher energy band. Since the band with lower energy has a magnetic moment in the direction of the field, the magnetization induced in the material has the direction of the field. Thus, the susceptibility χ is positive and the metal is paramagnetic. This type of magnetism is called Pauli paramagnetism.

Figure 9.2 illustrates some characteristics of paramagnetic materials. The basic characteristic of these materials is the fact that their atomic magnetic dipoles can

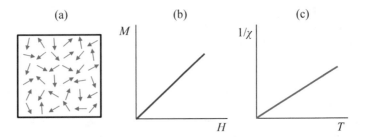

Fig. 9.2 Characteristics of paramagnetic materials. **a** Behavior of magnetic moments in the absence of an external magnetic field. **b** Variation of M with H (the slope of the curve is the susceptibility). **c** Variation of the inverse of susceptibility with temperature

change their orientation freely, without the influence of neighboring dipoles. At a finite temperature, in the absence of an external field, the magnetic moments occupy random directions due to thermal agitation, as illustrated in Fig. 9.2a. With the application of an external field, the average orientation of the dipoles produces a resulting magnetization in the direction of the field. As the field increases, the interaction energy of the dipoles with the field increases relative to the thermal energy, resulting in increasing order in the system. At certain temperature ranges, M is proportional to H, as in Fig. 9.2b. On the other hand, if the field is kept fixed and the temperature increases, thermal agitation increases, resulting in a smaller susceptibility. Experiments carried out by Pierre Curie in the nineteenth century showed that the susceptibility varies with the inverse of the temperature, as shown in Fig. 9.2c. This temperature dependence of the susceptibility is called Curie law. Pierre Curie was awarded the Physics Nobel Prize in 1903 not for his work in magnetism, but for his contributions to phenomena in radioactivity.

The quantum treatment of the basic properties of paramagnetic materials is based on results presented earlier. For a magnetic field B applied in the z-direction, the energy levels of a system of magnetic moments obtained from Eqs. (9.8) and (9.15) are

$$E_m = m\, g\, \mu_B\, B, \tag{9.18}$$

If the system has N independent magnetic moments, as in a paramagnetic material, the ratio between the number N_{m+1} of moments in the level $(m + 1)$ and N_m in the level m, at a certain temperature T, given by Boltzmann statistics, Eq. (8.54), is

$$\frac{N_{m+1}}{N_m} = e^{-g\,\mu_B\,B/k_B T}, \tag{9.19}$$

because $g\mu_B B$ is the energy difference between the two levels. Figure 9.3 illustrates the variation of the population of each level. Equation (9.19) shows that the levels with larger negative values of m have lower energy, and hence they have larger population. As a result, the number of magnetic moments in the direction of B is larger, so that

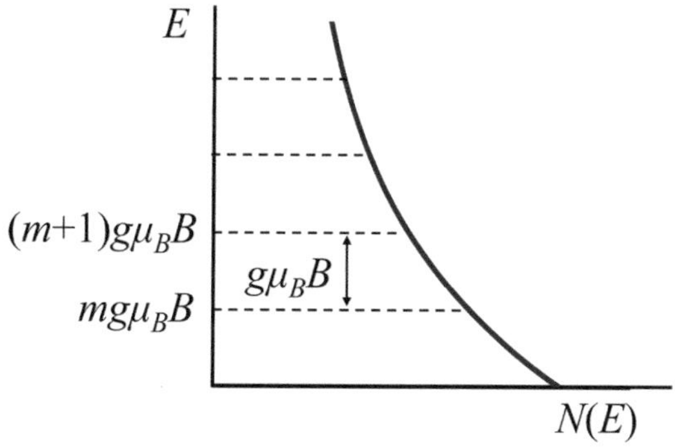

Fig. 9.3 Variation with energy of the population of independent magnetic moments in thermal equilibrium

M is nonzero and points in the direction of the field. Using Eqs. (9.1), (9.5) and (9.19) one can calculate the temperature dependence of the susceptibility $\chi(T)$ and obtain the Curie law. This derivation was one of the first successes of quantum theory of materials. Let us calculate the susceptibility for the simple case of $S = 1/2$ and $L = 0$, in which there are only two energy levels. In this case $J = 1/2$, $m_J = \pm 1/2$, and $g = 2$, so that the magnetization in the direction (z) of the field is

$$M = (N_1 - N_2)\,\mu_B, \tag{9.20}$$

where N_1 is the number of magnetic moments per unit volume in the direction of the field, and N_2 the number in the opposite direction. Substituting Eq. (9.19) in (9.20), and introducing the dimensionless quantity $x \equiv \mu_B B/k_B T$ we obtain

$$M = N\,\mu_B\,\frac{1 - e^{-x}}{1 + e^{-x}} = N\,\mu_B\,\tanh x, \tag{9.21}$$

where $N = N_1 + N_2$ is the total number of magnetic dipoles per unit of volume.

Figure 9.2 shows qualitatively the variation of the magnetization in a paramagnetic material with field and with temperature, given by Eq. (9.21). For $x \ll 1$, that is, for low field values and/or high temperatures, Eq. (9.21) shows that M varies linearly with x,

$$M \approx N\,\mu_B\,x = \frac{N\,\mu_B^2}{k_B\,T}\,B,$$

that gives for the susceptibility

$$\chi = \frac{M}{H} = \frac{N\,\mu_B^2\,\mu_0}{k_B\,T}, \tag{9.22}$$

which is Curie's law. On the other hand, for $x \gg 1$, corresponding to high field values and/or low temperatures, $M \rightarrow N\mu_B$. In this situation, all dipoles are aligned along the field and therefore the **magnetization is saturated**. In quantum theory, this state consists of all moments at the lowest energy level E_1, that is, $N_2 = 0$ and $N_1 = N$. In the general case in which the quantum number J has any value, the calculation of M is a little more complex but can be done analytically. One can show that the value of the saturation magnetization at high fields and/or low temperatures is

$$M_s = N\,g\,J\,\mu_B. \tag{9.23}$$

On the other hand, at low fields and/or high temperatures, the susceptibility is

$$\chi = \frac{M}{H} \approx \mu_0\,\frac{C}{T}, \tag{9.24}$$

where

$$C = \frac{N \, J(J + 1) \, g^2 \mu_B^2}{3 \, k_B} \tag{9.25}$$

is called Curie constant. Note that in the Gaussian system $\mu_0 = 1$ and C has the dimension of temperature, since χ is dimensionless. Evidently, the expressions (9.24) and (9.25) give the same result as Eq. (9.22) for the case $J = 1/2$, $g = 2$. With Eq. (9.23) it is possible to calculate the value of the saturation magnetization in paramagnetic materials.

Example 9.1 Consider a material with a simple cubic lattice, with lattice parameter $a = 2.5$ Å, having $J = 1$ and magnetic moment $2\mu_B$ per unit cell. Calculate: (a) The saturation magnetization; (b) The susceptibility at $T = 300$ K.

(a)   To calculate M_s with Eq. (9.23), note that $N = 1/a^3$. Using $\mu_B = 9.27 \times 10^{-21}$ Gcm^3, we have in the Gaussian system

$$M_s = \frac{2 \times 9.27 \times 10^{-21}}{2.5^3 \times 10^{-24}} = 1.19 \times 10^3 \text{G}.$$

This is the typical order of magnitude of the saturation magnetization observed both in paramagnetic and ferromagnetic materials.

(b)   The susceptibility obtained from Eqs. (9.24) and (9.25) with these same data, at $T = 300$ K, in the Gaussian system ($\mu_0 = 1$) is

$$\chi = \frac{2 \times 2^2 \times 9.27^2 \times 10^{-42}}{2.5^3 \times 10^{-24} \times 300 \times 3 \times 1.38 \times 10^{-16}} = 3.54 \times 10^{-4}$$

The susceptibility calculated in Example 9.1, of the order of 10^{-4}, observed in paramagnetic materials, is several orders of magnitude smaller than in ferromagnetic materials. The fact that both classes of materials have similar magnetizations when all moments are aligned, which occurs at $T = 0$, indicates that the origin of the magnetic moments is the same in both classes. However, the fact that at room temperature many ferromagnetic materials have magnetization of the same order as at $T = 0$ indicates that there is an interaction between their moments that tends to keep them aligned. Materials with a strong interaction between the magnetic moments are presented in the next section.

9.3 Magnetic Materials

Various metals with elements of the iron transition group, such as iron, nickel, and cobalt, pure or in alloys with other elements, exhibit a large magnetization at room temperature when subjected to a small external magnetic field. These materials are said to be **ferromagnetic**. They have the property of being attracted by magnetite, as known since ancient times. However, only at the end of the nineteenth century quantitative measurements of the magnetic properties of these materials were carried out. In the middle of the twentieth century, it was discovered that various materials that were supposed to be ferromagnetic are, actually, **ferrimagnetic**. These two categories of materials have similar magnetic properties and find various applications in electronics.

9.3.1 Spontaneous Magnetization and Curie Temperature

At the end of the nineteenth century, Pierre Curie found that the magnetization of ferromagnetic materials under a small applied field decreases with increasing temperature and vanishes above a certain critical temperature T_c, called Curie temperature. Actually, it is now well established that locally, in small regions called magnetic domains, ferromagnetic materials have finite magnetization even in the absence of an external field. This is called **spontaneous magnetization** and it results from a strong interaction between neighboring atomic magnetic moments that acts to keep them aligned. The qualitative form of the temperature dependence of the spontaneous magnetization M is shown in Fig. 9.4. At $T = 0$, M has a value equal to the saturation magnetization M_s, because all moments are aligned in the same direction. As the temperature increases, M gradually decreases due to the thermal agitation of the moments. At $T > T_c$, the thermal energy is larger than the ordering

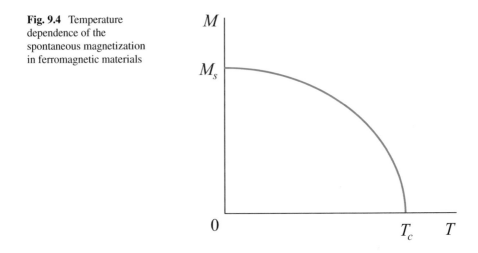

Fig. 9.4 Temperature dependence of the spontaneous magnetization in ferromagnetic materials

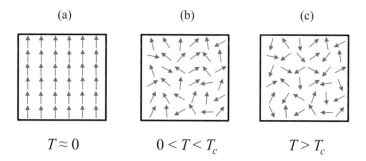

Fig. 9.5 Classical view of the magnetic moments in a ferromagnetic material in three temperature ranges

Table 9.2 Data for some ferromagnetic materials in the CGS system. To obtain the value of $\mu_0 M$ in the SI (tesla) just divide $4\pi M$ by 10

Material	T_c (K)	$4\pi M$ (0) (kG)	$4\pi M$ (300 K) (kG)
Fe	1043	22.016	21.450
Co	1394	18.171	17.593
Ni	631	6.409	6.095
Gd	293	24.881	0
CrBr$_3$	37	3.393	0
EuO	77	24.002	0
EuS	16.5	14.878	0

energy, so that the material has a paramagnetic behavior, with $M = 0$. Figure 9.5 illustrates the classical view of the magnetic moments in these three temperature ranges.

Table 9.2 presents the values of the Curie temperature T_c and the spontaneous magnetization at $T = 0$ and 300 K in some simple ferromagnets. In the CGS system the magnetization is multiplied by 4π, because the value that contributes to the field B is $4\pi M$. Note that various materials have $T_c < 300$ K, and therefore do not have spontaneous magnetization at room temperature. Another interesting observation is that the materials that have larger magnetization, do not necessarily have larger T_c. The reason for this is that the value of M depends on the atomic magnetic moment, whereas T_c depends on the interaction between the moments, as we shall see in the next subsection.

9.3.2 The Molecular Field Model

In the beginning of the twentieth century, when the origin of the atomic magnetic moment was still not understood, Pierre Weiss proposed a theoretical model for ferromagnetism, that even today is still useful. In this model, each atomic magnetic moment is under the action of an effective magnetic field created by its neighbors, which tends to make them align in the same direction. This effective field $\vec{B}_E$, postulated empirically by Weiss, called Weiss molecular field, is proportional to the local magnetization

$$\vec{B}_E = \lambda \, \vec{M}, \tag{9.26}$$

where λ is a parameter characteristic of each material. Thus, each dipole tends to align along $\vec{B}_E$ and therefore with $\vec{M}$, whose direction is given by the average of all neighboring dipoles. This model can be used to calculate the local magnetization as a function of temperature and the applied field H_0. Let us carry out this calculation using the magnetization in Eq. (9.21), valid only for $S = 1/2$ and $g = 2$, with the parameter x determined by the total local field

$$x = \frac{\mu_B \, B}{k_B \, T} = \frac{\mu_B \, (\mu_0 H_0 + \lambda \, M)}{k_B \, T}. \tag{9.27}$$

Thus, the spontaneous magnetization given by Eqs. (9.21) and (9.27) with $H_0 = 0$ is

$$M = N \, \mu_B \, \tanh\!\left(\frac{\mu_B \lambda \, M}{k_B \, T}\right). \tag{9.28}$$

This is a transcendental equation, which does not have analytical solution but can be solved numerically or graphically. Figure 9.6 illustrates the solutions of (9.28) for the spontaneous magnetization at four temperature values. The solid curve represents the function $M(x)$ given by Eq. (9.21), while the dashed lines represent the function $M = k_B T x / \mu_B \lambda$ at different temperatures.

For $T = T_1 \ll T_c$, the solution of Eq. (9.28) is given by point 1, the intersection of the curve with the line corresponding to T_1, which has a small slope. At this point the spontaneous magnetization has a value close to the saturation value M_s. It is easy to see that as T increases, the slope of the line increases and therefore the value of M

Fig. 9.6 Graphic solution of Eq. (9.28) for the spontaneous magnetization in a ferromagnetic material at four temperature values

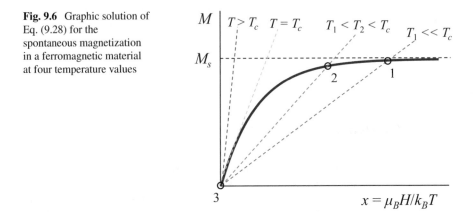

decreases, as in point 2 of Fig. 9.6. This behavior is consistent with the experimental result shown in Fig. 9.4. The Curie temperature T_c is the one for which the line is tangential to the $M(x)$ curve, since it is the smallest value of T for which $M = 0$ (point 3 of the figure). Evidently, for $T > T_c$, the solution remains at point 3 and therefore $M = 0$.

We can obtain an expression relating T_c with the material parameters using the condition that the line is tangential to the $M(x)$ curve at $x = 0$ in Fig. 9.6. The slope of the line is $k_B T/\mu_B \lambda$. For $x \ll 1$, $\tanh x \approx x$, so that the slope of the tangent to the curve for $M(x)$ at $x = 0$ calculated with Eq. (9.21) is $N\mu_B$. Equating the two slopes we obtain

$$T_c = \frac{N\mu_B^2}{k_B} \lambda.$$

Note that the expression that multiplies λ in this result is exactly the Curie constant C, valid for $J = 1/2$, $g = 2$, in Eq. (9.25). For the general case of any J, it can be shown that

$$T_c = C \lambda. \tag{9.29}$$

This result can be obtained in another interesting way. In the ferromagnetic phase, $T < T_c$, it makes no sense to define a local susceptibility, since $M = 0$ with $H_0 = 0$. However, in the paramagnetic phase, $T > T_c$, the local susceptibility is given by Eq. (9.24), $\chi = \mu_0 C/T$. Since the local field is the sum of the external field and the molecular field, we can write

$$M = \frac{\chi_l}{\mu_0} (\mu_0 H_0 + B_E) = \frac{C}{T} (\mu_0 H_0 + \lambda M).$$

With this expression we obtain the susceptibility of a ferromagnetic material in the paramagnetic phase,

$$\chi = \frac{M}{H_0} = \frac{\mu_0 C}{T - T_c}, \tag{9.30}$$

where $T_c = \lambda C$, as in Eq. (9.29). This result, known as the Curie–Weiss law, shows that χ diverges when $T \rightarrow T_c$. This is consistent with the fact that at $T \leq T_c$, M is finite even with $H_0 = 0$, as expected for the ferromagnetic phase.

From the result (9.29) we can estimate the value of the molecular field in ferromagnetic materials. For metals of the iron transition group, from Table 9.2 we have $M \sim 10^3$ G and $T_c \sim 10^3$ K, and with Eq. (9.25) we obtain in the Gaussian system $C \sim 1$ K. Therefore, $\lambda = T_c/C \sim 10^3$, so that the molecular field $B_E = \lambda M$ is on the order of 10^6 G. This value is very high compared to typical magnetic fields produced in laboratories, which are at most on the order of 10^5 G. It is also much larger than the magnetic field that an atomic magnetic dipole creates on its neighbors, which is on the order of $\mu_B/a^3 \sim 10^{-20}/10^{-23} = 10^3$ G.

9.3.3 The Exchange Interaction

The origin of the Weiss molecular field was explained many years after its proposal by means of quantum mechanics. It is associated with the so-called Heisenberg exchange interaction energy, that has its origin in the electrostatic interaction but has a quantum nature, with no classical analogy. This energy results from the difference between the electrostatic energies of two interacting electrons in two situations, one with parallel spins and the other with antiparallel spins. We can understand this interaction with a simple model.

Consider two electrons of neighboring ions, whose spins are $\vec{S}_1$ and $\vec{S}_2$. The Pauli exclusion principle requires the total wave function of the two electrons to be antisymmetric. The total wave function is the product of the spatial wave function and the one that describes the spin state. When the spatial wave function is symmetric, the spins must be antiparallel, so that the total wave function is antisymmetric. On the other hand, when the spatial function is antisymmetric, the spins must be parallel. Since the total electrostatic energy of the set depends on the spatial distribution of the electric charge, it is different in the two cases illustrated in Fig. 9.7. The difference between the values of the electrostatic energy in the two situations is called **exchange energy** between the two spins. This energy depends essentially on the spin states, and it can be shown that it has the form

$$U_{12} = -2J_{12}\,\vec{S}_1 \cdot \vec{S}_2, \tag{9.31}$$

where J_{12} is the Heisenberg integral, also called **exchange constant**. Its value depends on the electronic distributions of the atoms and the distance between them. Since the electrostatic interaction decreases with increasing distance, J_{12} is small for atoms far apart. We see from Eq. (9.31) that when J_{12} is positive, the lowest energy state corresponds to the two spins parallel, which gives rise to a ferromagnetic ordering.

In the case of substances that contain only one element with atomic magnetic moment, J_{12} is, in general, positive. However, when the substance contains elements that mediate the chemical bond between the atoms with magnetic moments, as in the case of O, F and C, for example, J_{12} tends to be negative. In this case, the state

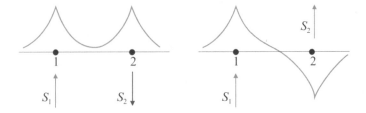

Fig. 9.7 Illustration of the origin of the exchange interaction. The directions of the spins depend of the spatial charge distribution (spatial wave function) of the electrons of the neighboring ions

of minimum exchange energy has the spins antiparallel. This gives rise to antiferro-magnetism and ferrimagnetism. For this reason, it is rare to find ferromagnetism in oxides, fluorides, or chlorides, and most magnetic compounds are ferrimagnetic or antiferromagnetic. Evidently, when $J_{12} = 0$, the material is paramagnetic. It is easy to relate the exchange constant with the molecular field. Consider two neighboring atoms with magnetic moments $\vec{\mu}_1$ and $\vec{\mu}_2$. Assuming that the magnetic moment of the ion is due only to the spin, $\vec{\mu} = -g\mu_B \vec{S}$, Eq. (9.31) can be rewritten in the form

$$ U_{12} = \frac{2J_{12}}{g\mu_B} \vec{\mu}_1 \cdot \vec{S}_2. \tag{9.32} $$

Comparing Eq. (9.32) with (9.8), we see that it is possible to represent the action of the spin 2 on spin 1 in the form of an effective magnetic field

$$ \vec{B}_{12} = -\frac{2J_{12}}{g\mu_B} \vec{S}_2. $$

Generally, the exchange constant is the same between nearest magnetic neighbors, that is represented by J_1, and it is negligible for more distant neighbors. If a spin has z_1 nearest neighbors, the modulus of the average effective field acting on it is then

$$ B_E = \frac{2 S z_1 J_1}{g\mu_B}. \tag{9.33} $$

This is precisely the Weiss molecular field. Comparing Eqs. (9.26) and (9.33) and using the value of the magnetization at $T = 0$, $M_s = g\mu_B S N$, where N is the number of spins per unit volume, we see that the parameter introduced in Eq. (9.26) is

$$ \lambda = \frac{2 z_1 J_1}{N(g\mu_B)^2}. \tag{9.34} $$

Using Eqs. (9.25) and (9.29), with $J = S$, we obtain the relationship between the Curie temperature and the exchange constant,

$$ T_c = \frac{2 S(S + 1) z_1 J_1}{3k_B}. \tag{9.35} $$

This result shows that T_c increases with increasing exchange interaction, because a higher temperature is needed to destroy the magnetic order. This explains the wide variation in the values of T_c of the materials in Table 9.2.

To conclude this section, we shall address the important question of the magnetism in metallic materials. So far, all magnetic properties were treated as if the magnetic moments were associated with localized ions, fixed in the crystal structure. This is valid for insulators, but not for metals. In metals it is necessary to consider the fact that electrons occupy states in energy bands, and not discrete levels as in localized ions. In the case of metals of elements from the iron transition group, the important

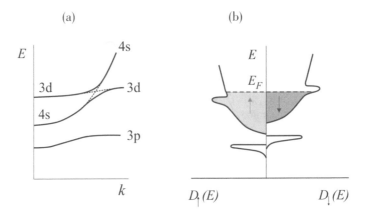

Fig. 9.8 Illustration of the upper energy bands in metals of the iron transition group. **a** $E(k)$ curves for $T > T_c$. **b** Occupation of states at $T < T_c$

bands are those associated with levels 3d and 4 s. The 4 s band is the one of almost free electrons, responsible for most of the conductivity. The 3d band is that of magnetism, for the same reasons discussed in Sect. 9.2.1. In fact, the $E(k)$ curves corresponding to the 3d and 4s bands overlap, as shown in Fig. 9.8a. As a result, there is a mixture of 3d and 4s states and the curves for the density of states have the shape shown in Fig. 9.8b for the up and down spins. Due to the exchange interaction between electronic spins, at temperatures below T_c the energy of an electron in state k with up spin is smaller than the energy of an electron in the same state k but with down spin. As a result, the density of states splits in two, one for up spin with lower energy than the other for down spin, as indicated in Fig. 9.8b. Since the states are occupied up to the Fermi level, the band with lower energy has more occupied states than the other. As a result, some metals of the 3d group have non-zero total magnetic moment, which gives rise to spontaneous magnetization and ferromagnetic behavior. This is the case of Fe, Co and Ni, which have magnetic moments at $T = 0$ with values 2.22 μ_B, 1.72 μ_B and 0.16 μ_B per atom, respectively. Despite the fact that the origins of the magnetic moments in metals and insulators are different, the magnetic macroscopic properties can be treated in a very similar manner in both types of materials.

9.3.4   Ferrimagnetic Materials and Ferrites

When the exchange interaction between two neighboring ions is negative, their spins tend to align in opposite directions. This gives rise to more complex magnetic orderings than ferromagnetic. Figure 9.9 illustrates two types of simple ordering that occur with $J_{12} < 0$, antiferromagnetism and ferrimagnetism. In antiferromagnetism the antiparallel moments are equal, so that the net macroscopic magnetization is zero. For this reason, although antiferromagnetic materials have a strong interaction

Fig. 9.9 Illustration of simple antiferromagnetic and ferrimagnetic orderings

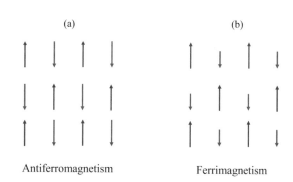

(a)   (b)

Antiferromagnetism   Ferrimagnetism

between magnetic moments, and thus are of great scientific interest, until about the year 2000 they did not have any technological application. Currently they are used in read-heads of magnetic recording based on the giant magnetoresistance, presented in Sect. 9.5.4. Ferrimagnetic materials are also characterized by a negative exchange interaction. However, they have different neighboring magnetic moments, so that the resulting net magnetization is different from zero. Actually, macroscopically, many properties of ferrimagnetic materials are similar to those of ferromagnets.

A class of ferrimagnetic materials very important for electronics consists of **ferrites**. They are ferrimagnetic oxides with crystalline structure similar to the natural spinel $MgAl_2O_4$. Their magnetic properties result from the presence of magnetic ions, such as Fe, Ni, Co, Mn, or rare-earths elements, in place of Mg or Al. They have a crystal structure in which the spins tend to align opposite to each other, but several of their magnetic properties are similar to those of ferromagnets. Two important properties of some ferrites give them considerable technological importance. These are the speed of the magnetization response and the **high resistivity**. The last one makes them suitable to be used in high frequency applications, including in the microwave range, because they do not exhibit eddy currents, which are responsible for heating and energy loss in ferromagnetic metals.

Figure 9.10 shows the crystal structure of the classical spinel $MgAl_2O_4$, which is a well-known mineral found in nature. The spinel crystal structure is preserved when

Fig. 9.10 Crystal structure of the natural spinel $MgAl_2O_4$. The Mg^{2+} ions occupy tetrahedron positions and the neighboring oxygens form a tetrahedron. The Al^{3+} ions occupy octahedral positions (only two octants of the unit cell are shown with all ions, to facilitate the view)

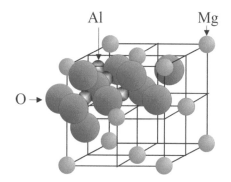

Al   Mg

O →

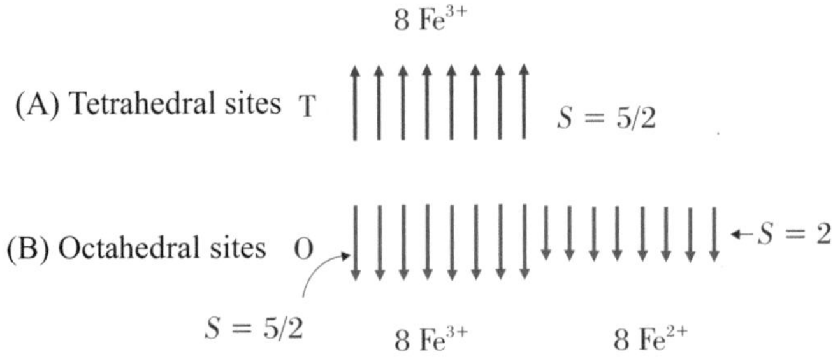

$$8 \, Fe^{3+}$$

(A) Tetrahedral sites T $\uparrow\uparrow\uparrow\uparrow\uparrow\uparrow\uparrow\uparrow$ $S = 5/2$

(B) Octahedral sites O $\downarrow\downarrow\downarrow\downarrow\downarrow\downarrow\downarrow\downarrow\downarrow\downarrow\downarrow\downarrow\downarrow\downarrow\downarrow\downarrow$ $\leftarrow S = 2$

$S = 5/2$ $8 \, Fe^{3+}$ $8 \, Fe^{2+}$

Fig. 9.11 Ordering of the spins in the unit cell of magnetite, $Fe^{2+}O.Fe_2^{3+}O_3$. The spins of the Fe^{3+} ions at positions A and B cancel each other out

Al^{3+} is replaced by Fe^{3+} to produce $Mg^{2+}O.Fe_2^{3+}O_3$, which is called magnesium ferrite. Further, if Mg^{2+} is replaced by a divalent metal M, one obtains the ferrospinel $MO.Fe_2O_3$, also generally called **ferrite**. Nearly any divalent metal such as Ni, Co, Mn, Zn, Cu, Ba, or Cd may be used to produce a ferrite. If the divalent metal is Fe^{2+}, we then have iron ferrite, or magnetite, which is the magnet found in nature, Fe_3O_4 (or $FeO.Fe_2O_3$).

The metal ions in the spinel structure occupy two different symmetrical positions. Position A has tetrahedron symmetry, that is, the oxygen that surrounds the metal ion forms a tetrahedron. Position B is that of octahedral symmetry. In the crystal unit cell there are 8 positions A and 16 positions B. In ferrites, the interaction between the spins of the magnetic ions in position A and a spin in B is negative, so that they order in an antiparallel or ferrimagnetic arrangement. Figure 9.11 shows how the Fe^{3+} ions ($S = 5/2$) and the Fe^{2+} ions ($S = 2$) are distributed in magnetite, $Fe^{2+}O.Fe_2^{3+}O_3$. Notice that the spins of the eight Fe^{3+} ions in position A cancel the other eight of Fe^{3+} in position B. As a result, the resulting magnetic moment is due exclusively to the Fe^{2+} ions, which have spin $S = 2$. The magnetic moment measured experimentally at $T \approx 0$ K is $8 \times 4.07 \, \mu_B$ per unit cell, which is a value close to the one obtained in Fig. 9.11, namely $8 \times g\mu_B S = 8 \times 4 \, \mu_B$. The difference is due to the small contribution of the orbital magnetic moment.

It is possible to obtain a wide variety of ferrites with different magnetizations, suitable for each application, by the appropriate replacement of the metal ions. Pure magnesium ferrite $Mg^{2+}Fe_2^{3+}O_3$, for example, has almost zero magnetization, since the Mg^{2+} ion is not magnetic and the spins of the Fe^{3+} ions at positions A and B cancel out. We show below the chemical formulas of some ferrites in which the metal ions are fractionally substituted, indicating their magnetic moments per unit cell.

Nickel Ferrite $(Fe_{1.0}^{3+})^A(Ni_{1.0}^{2+}Fe_{1.0}^{3+})^B O_4$

Magnetic Moment $= 8 \times 2\mu_B \times (\downarrow 5/2 \uparrow 1 \uparrow 5/2) = 16\mu_B$

Nickel Ferrite with Aluminum $(Fe_{1.0}^{3+})^A(Ni_{1.0}^{2+}Al_{0.4}Fe_{0.6}^{3+})^B O_4$

Magnetic Moment $= 8 \times 2\mu_B \times (\downarrow 5/2 \uparrow 1 \uparrow 0.6 \times 5/2) = 0$

Note that with the replacement of a small fraction of iron by aluminum, the magnetization is canceled out. With the substitution of a smaller fraction we can obtain a magnetization with any value between 0 and 16 μ_B per unit cell, producing a variety of ferrites for different applications.

Ferrites are ceramics, that have great hardness and high melting point. They are generally used in polycrystalline form. The preparation of ferrites starts with the mixture of fine particles of the various metal oxides which enter in their composition, in the desired proportion in the final form of the material. This mixture is heated to temperatures around 1000 °C with the purpose of homogenizing the oxides. It is then grounded again and the resulting powder is pressed to obtain the desired shape. Finally, it is heated to a temperature just below the melting point (1200–1500 °C), acquiring the dense polycrystalline shape. These steps have durations that vary from material to material and their details constitute industrial secrets of the manufacturers. The development of these processes was done during decades of research work in university and industrial laboratories, with the participation of physicists, chemists, electrical and materials engineers.

A very important ferrimagnetic material is **yttrium iron garnet**, well-known by the acronym **YIG**, whose chemical formula is $Y_3Fe_5O_{12}$. Since the Y^{3+} ion is diamagnetic, the magnetic properties of YIG are due to Fe^{3+} ions, three of which have spins opposite to the other two in a ferrimagnetic arrangement. The Fe^{3+} ions have spin $S = 5/2$ and orbital angular momentum $L = 0$, so that the magnetic moment per unit formula is 5 μ_B. YIG has a complex cubic crystal structure, with the unit cell containing eight formula units of $Y_3Fe_5O_{12}$, so that the magnetic moment per unit cell is 40 μ_B. The lattice parameter at 0 K is $a = 12.376$ Å, resulting in a magnetization at $T = 0$ K of $M_0 = 40 \ \mu_B/a^3 = 194$ G. At room temperature $M = 140$ G and the Curie temperature is 559 K. YIG is a good insulator and since Fe^{3+} has $L = 0$, the spins have very small coupling with the lattice, so that its magnetic losses are very low at microwave frequencies. For this reason, YIG plays an important role in the investigations of dynamic magnetic phenomena at high frequencies and has several technological applications, such as the YIG filter presented in Sect. 9.6.4. YIG does not exist in natural minerals. Single-crystal bulk YIG can be grown from the melt by several methods, while films are grown by liquid-phase epitaxy, pulsed laser and sputter depositions.

9.3.5 Magnetization Curve: Magnetic Domains

The net magnetization of a piece of a ferro- or ferrimagnetic material in the absence of an applied magnetic field is generally much smaller than the spontaneous magnetization. This is due to the formation of **magnetic domains**. In this section we shall discuss the origin of these domains and how they influence the magnetization of a magnetic material.

In a ferromagnetic material at a temperature $T \ll T_c$, the magnetic moments tend to align in the same direction due to the exchange interaction, even in the absence of

(a)   (b)   (c)

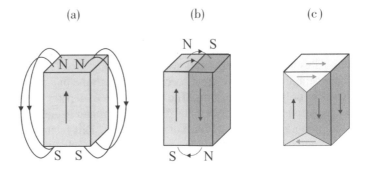

Fig. 9.12 Illustration of the decrease in magnetostatic energy due to the formation of magnetic domains

an external field. If this alignment occurs throughout the material, the magnetization will be uniform, as in Fig. 9.12a. In this case, the magnetic poles generated at the ends create an external magnetic field, as shown in the figure. The energy associated with this field, called magnetostatic energy, given by $(\mu_0/2) \int H^2 dV$, is relatively high, so that this configuration does not remain in equilibrium. If half of the sample has magnetization in one direction and half in the other, as in Fig. 9.12b, the external field is smaller and the magnetostatic energy is reduced approximately to one-half the value in (a). Figure 9.12c shows a situation with even less energy, with the field lines closing internally in the material, so that the external field is negligible. The four regions shown in Fig. 9.12c have, internally, saturated magnetization, but the total average magnetization is negligible. These regions are called magnetic domains, and they form spontaneously to minimize the energy of the system. The main contributions to the energy are: magnetostatic energy; Zeeman energy, due to the interaction of the moments with the externally applied field; exchange energy that increases with increasing angle between neighboring moments; and the crystalline anisotropy energy. This last one is the contribution due to the interaction between the orbital angular momentum and the electric field of the crystal lattice, which tends to make the moments align themselves along some of the crystalline axes of the material.

The shape and size of the domains are determined by the minimization of the total energy. One important feature of the domain arrangement is that the boundary between two domains is not abrupt, since this would result in a high exchange energy. The exchange energy of the boundary is lowered with the formation of a layer where the orientation of the moments varies gradually from one domain to the other. This layer is the **domain wall**, also called Bloch wall. Figure 9.13 illustrates a 180° wall, separating two domains whose magnetizations have opposite directions. Since the orientation of the moments can vary easily, domain walls are highly mobile. These walls have typically thickness of around 100 to 1000 nm. The domain widths range from a few μm to several mm or cm, depending on the characteristics of the material and the applied external field. Figure 9.14 illustrates the behavior of domains in an idealized situation. When the applied field is zero, domains are formed, as in (a), which result in total magnetization null. When a small field is applied along the bar, there is a displacement of the domain walls to decrease the Zeeman energy. The sizes

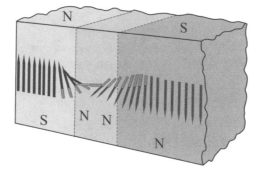

Fig. 9.13 Illustration of a 180° domain wall

of the magnetized domains in the direction of the field increase, while those in the opposite direction become smaller. As a result, the bar has a net magnetization along the field, as shown in (b). An increase in the field produces a larger displacement of the walls and also rotation of the domains, as in (c). Finally, with a much larger field, all domain walls disappear and the bar is magnetized to saturation, as shown in (d).

The shape of the magnetization curve as a function of the applied field, shown in Fig. 9.15, is determined by the behavior of the domains. The curve in (a) corresponds to an initially demagnetized material. For small field values, the initial increase in the magnetization is due to reversible displacements of the domain walls. If the field is removed, the domains return to the initial configuration. With a larger field the magnetization increases due to the displacements of the walls, but these displacements become irreversible due to imperfections in the material. Finally, with higher field values, there is a domain rotation until the magnetization is saturated over the whole material. Figure 9.15b shows the behavior of the magnetization M with the variation of the field H after the material has been saturated. When H decreases, M does not return by the same curve as with increasing field, because of the effects of irreversible domains rotations and displacements. Consequently, even with $H = 0$, there is a finite value of $M = M_r$, called **remanent magnetization**, or remanence, that results from the imprisonment of certain walls that make domains favorably oriented with respect to the field to prevail over unfavorably ones. If H increases in

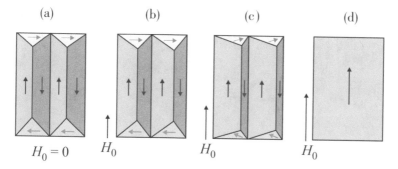

Fig. 9.14 Behavior of magnetic domains in a ferromagnetic bar subject to an external applied field

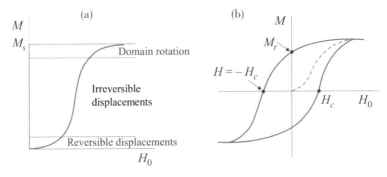

Fig. 9.15 Variation of the magnetization of a ferro- or ferrimagnetic material under an applied external field. **a** Sample initially demagnetized. **b** Hysteresis curve

the opposite direction, M gradually decreases and only with a value $H = -H_c$, called **coercive field**, or coercitivity, the magnetization is canceled. The curve in Fig. 9.15b, called **hysteresis cycle**, or **hysteresis curve**, shows the variation of M in a complete cycle of variation of H. The shape of the hysteresis cycle is determinant in the type of application of a magnetic material, as we shall see in the next section.

To conclude this section, let us see what happens when a material is subjected to an alternating magnetic field H_{ac}, whose amplitude gradually decreases in time. This is used to demagnetize a ferromagnetic material that has a remanent magnetization, as illustrated in Fig. 9.16. The application of an AC field with amplitude that decreases in time, as in (a), results in a trajectory of the magnetization in the M - H plane shown in (b). If the variation of H is periodic, M describes the hysteresis cycle, reaching saturation at the positive and negative ends. However, if the amplitude of H decreases gradually in time, in each consecutive extreme the maximum value of M is smaller than in the previous cycle. As a result, as the amplitude of H tends to zero, the material gradually becomes demagnetized.

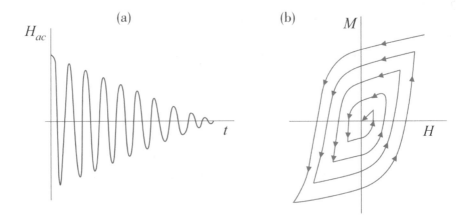

Fig. 9.16 AC demagnetization process of a ferromagnetic material. **a** Alternating field H_{ac} with decreasing amplitude, **b** trajectory of M in the M - H plane

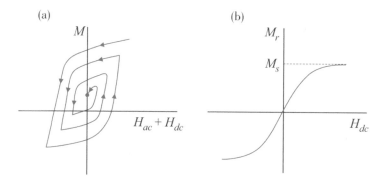

Fig. 9.17 Process of DC magnetization with AC polarization. **a** Trajectory of the magnetization in the plane M - H, where $H = H_{ac} + H_{dc}$; **b** remanent magnetization curve as a function of H_{dc}, in which there is no hysteretic behavior

An important process for the magnetic recording of audio signals is the DC magnetization with AC polarization, illustrated in Fig. 9.17. In this process the material is subject simultaneously to two H fields in the same direction, one constant H_{dc} and the other alternating H_{ac}. The H_{ac} field initially has amplitude larger than H_{dc}, and decreases in time as in Fig. 9.16a. In this case, as H_{ac} decreases, the presence of the H_{dc} field prevents the value of M to vanish, as in the case of AC demagnetization. When H_{ac} becomes zero, the material retains a finite magnetization M_r, whose value depends on H_{dc}, as shown in Fig. 9.17a. Of course, $M_r = 0$ if $H_{dc} = 0$, and $M_r = M_s$ when the value of H_{dc} is high. As a result of this process, the $M_r - H_{dc}$ curve has the shape shown in (b), in which there is no cycle hysteresis. This process allows to magnetize a ferromagnetic material with a magnetization proportional to the DC field, in a certain field range around the origin, to record some information without the deleterious effect of the hysteretic response of the magnetic medium.

9.4 Materials for Conventional Applications

Magnetic materials play an important role in technology as they find applications in a large number of products and processes of the most varied sectors. These applications range from devices with very simple functions, such as the small permanent magnets used to hold pieces of paper on refrigerator doors and door locks for furniture, to numerous sophisticated components used in the electro-electronic and computer industry. One of the most important is magnetic recording, whose market expanded in the second half of the twentieth century. In the electro-electronic sector, the fabrication of magnetic materials is supplanted in volume only by semiconductors, because they are essential elements for many devices and equipment.

From the point of view of the basic magnetic properties, magnetic materials are classified in three broad classes:

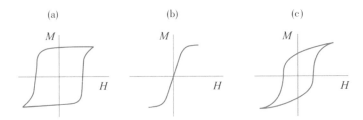

Fig. 9.18 Hysteresis cycles of magnetic materials: **a** Hard magnetic materials, or permanent magnets; **b** soft magnetic materials; **c** materials for magnetic recording

- Hard magnetic materials, or permanent magnets;
- Soft magnetic materials, also called high-permeability materials;
- Magnetic recording media.

The hysteresis cycles characteristic of these materials are shown in Fig. 9.18. Hard magnetic materials, used to make permanent magnets, have high values of remanent magnetization M_r and coercive field H_c, and therefore have a rectangle-like hysteresis cycle, as in (a). Soft materials are those easily magnetizable by applying an external field, and easily demagnetizable with the removal of the field. They have very small coercivity and very narrow hysteresis cycle, as in (b). Finally, magnetic recording media must have a hysteresis cycle intermediate between the previous two, as in (c). They have M_r and H_c large enough to store the information contained in the recording field but quite smaller than in permanent magnets, to allow information to be erased. In the following sections we present some details on the applications of these classes of materials.

9.4.1 Permanent Magnets

Permanent magnets constitute the most notable class of magnetic materials. Their function is to create a static magnetic field in a certain region of space, without the need of an electric current. Permanent magnets, also simply called magnets, are made of hard magnetic materials, so that their magnetization is not easily altered by external fields. They are essential components in a variety of electromagnetic devices, such as generators and motors used to convert mechanical energy into electric energy, and vice-versa. Motors are employed in industry, household appliances, automobiles, airplanes, watches, computers, etc.). They are also essential components in electro-acoustic devices (loudspeakers, headphones, microphones, magnetic needles of recording players, etc.), measuring instruments (galvanometers and scales), torque devices (ultracentrifuges, electric power gauges, etc.), medical equipment, ferrite devices for microwaves, and various instruments and scientific equipment.

The magnetic field created by a magnet has an intensity and spatial variation that depend on its physical shape and the magnetic properties of its material. To

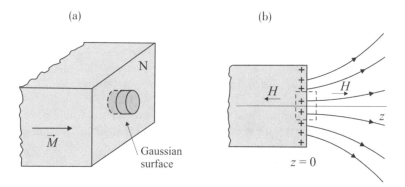

Fig. 9.19 **a** Flat end of a permanent magnet bar with magnetization perpendicular to the surface. **b** Profile view

understand some important issues, let us consider a simple geometry. Figure 9.19 shows a flat end of a magnetized material with magnetization M perpendicular to the surface. The field created by the magnetization can be obtained by replacing relations (9.3) and (9.4) in Maxwell's Eq. (2.2), so that

$$\nabla \cdot \vec{H} = -\nabla \cdot 4\pi \vec{M} \quad (\text{CGS}), \qquad \nabla \cdot \vec{H} = -\nabla \cdot \vec{M} \quad (\text{SI}). \tag{9.36}$$

To integrate these equations, we use Gauss's theorem, in a manner analogous to the calculation of the electric field with Eq. (2.1). The analogy with the electric field suggests the introduction of a fictitious volume density defined by

$$\rho_m = -\nabla \cdot 4\pi \vec{M} \quad (\text{CGS}), \qquad \rho_m = -\nabla \cdot \vec{M} \quad (\text{SI}) \tag{9.37}$$

where ρ_m plays the role of a density of magnetic monopoles. Thus, using Eq. (9.37) we have for the volume integral of Eq. (9.36) inside the Gaussian surface

$$\int \nabla \cdot \vec{H} \, dV = \int \rho_m \, dV = q_m$$

With the application of the Gauss theorem we obtain

$$\oint \vec{H} \cdot d\vec{a} = q_m, \tag{9.38}$$

where, in the SI,

$$q_m = \int \nabla \cdot \vec{M} dV = \oint \vec{M} \cdot d\vec{a} \tag{9.39}$$

represents the non-compensated magnetic dipoles inside the surface S. In the CGS the expression of q_m is the same as (9.39) with the factor 4π, as in (9.37).

Note that although magnetic monopoles, or charges, do not exist, q_m is mathematically equivalent to a magnetic charge, since the magnetic dipoles can be seen as pairs of magnetic charges with opposite signs. Thus, Eq. (9.38) is analogous to Gauss's law of electrostatics. In the geometry of Fig. 9.19, the surface used for the application of Eq. (9.38) is that of a cylinder with bases parallel to the flat surface of the material. Since the magnetization is M at $z < 0$ and zero at $z > 0$, from Eq. (9.39) we obtain $q_m = M A$, where A is the area of cylinder base. The monopoles are distributed on the surface with magnetic charge surface density $\sigma_m = q_m/A = M$. The interpretation of this result is that the discontinuity of M at the surface produces uncompensated magnetic dipoles, or magnetic monopoles. The surface towards which M is directed has positive monopoles, and plays the role of the north pole (N) of the magnet. Assuming that the other end of the magnet (South pole) is very distant, the H field created by the charge density σ_m is analogous to the electric field created by a plane of electric charges. The application of Eq. (9.39) to the cylinder of Fig. 9.19 then gives, in the SI,

$$H_z = \frac{\sigma_m}{2} = \frac{M}{2} \ (z > 0), \qquad H_z = -\frac{M}{2} \ (z < 0). \qquad (9.40)$$

To obtain the result in the CGS, just multiply the right-hand side by 4π. One can see that the field H inside the magnet has the direction opposite to M, and therefore it tends to demagnetize the material. This requires the coercive field to be large enough to avoid demagnetizing the material. With Eqs. (9.4) and (9.40) we obtain the magnetic induction

$$B_z = \frac{\mu_0 M}{2} \ (z > 0), \qquad B_z = \frac{\mu_0 M}{2} \ (z < 0). \qquad (9.41)$$

We then see that the field B has the same value at $z > 0$ and $z < 0$, which is an expected result because of the boundary condition that establishes that the normal component of B is continuous at the surface.

Another simple shape of a permanent magnet for calculating the field is that of a closed horseshoe, illustrated in Fig. 9.20. In this case, the field in the air gap (the space between the two poles) is the sum of the fields created by the two planes of monopoles, the north and south poles of the magnet. Then, in the central region of the

Fig. 9.20 Permanent magnet with the shape of a closed horseshoe

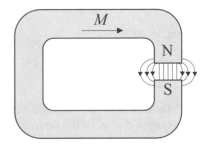

air gap, the magnetic induction vector and the field intensity vector are approximately uniform, with modulus $B = \mu_0 H = \mu_0 M$ (in the SI). On the other hand, inside the magnet and near the surfaces, $B = \mu_0 M$ and $H = 0$.

These two simple examples show that the fields B and H, both inside and outside, depend on the shape of the magnet. They also show that with no external applied field, the value of M is not in general the remanent magnetization M_r, because H is not necessarily zero inside the magnet. Due to the demagnetization effect, inside the magnet $\vec{H}$ has the direction opposite to $\vec{M}$, and therefore the region of the hysteresis curve relevant for a permanent magnet is the second quadrant of the M versus H curve. Since Maxwell's equations involve the fields B and H, it is common to represent the hysteresis cycle with B as a function of H. Figure 9.21 shows the second quadrant of the $B(H)$ cycle of a permanent magnet. Note that for $H = 0$, $B = \mu_0 M_r$. The operating point of the magnet is determined by the intersection of the hysteresis curve with the curve that represents the equation that relates B and H, obtained from Maxwell's equations and the boundary conditions. In the case of the closed horseshoe, the operating point of the inner central region of one of the poles is point 1 in Fig. 9.21. In the case of a magnet with the shape of a thin disc, the operating point is point 2 (Problem 9.8). Generally, one seeks to have the operating point in a situation intermediate between 1 and 2, because the energy stored in the magnet is proportional to the volume integral of the BH product. For this reason, a quantity that indicates the quality of a magnet is the maximum value of the BH product, which corresponds to a certain point P_m of hysteresis curve. A good permanent magnet has a high value of $(BH)_{max}$. This requires high values of M_r and H_c.

Table 9.3 shows the main properties of some materials used for permanent magnets. The first magnets were made of magnetite, Fe_3O_4, the natural magnet. The first artificial hard magnetic materials developed in the beginning of the twentieth century were several types of steel, Fe–C alloys, hardened by special heat treatments. In the 1930s, Japanese laboratories discovered that the coercive field in Fe alloys could be increased by mixing Al, Ni, and Co, giving rise to the Alnico alloys. Table 9.3 shows the composition and the parameters of Alnico 5, a low-cost alloy that is still widely used today. The discovery of Alnico resulted in a major improvement compared to steel, that has (BH) product of only 1MG.Oe (10^6G.Oe). The magnetic properties of these alloys were improved in the 1940s, when Neél

Fig. 9.21 Second quadrant of the B versus H curve of a material used in a permanent magnet

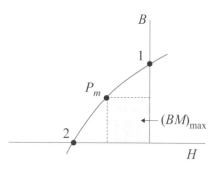

Table 9.3 Main properties of hard magnetic materials, used as permanent magnets, at room temperature

Material	Composition	$4\pi M_s$ (kG)	H_c (kOe)	$(BH)_{max}$ (MG.Oe)
Alnico 5	$Fe_{.51}Al_{.08}Ni_{.14}Co_{.24}Cu_{.03}$	12.5	0.72	5.0
Barium ferrite	$BaFe_{12}O_{19}$	3.95	2.4	3.5
Samarium-Cobalt	$SmCo_5$	9.0	8.7	20.0
Neodymium-Iron-Boron	$Nd_2Fe_{14}B$	13.0	14.0	40.0

and Kittel introduced the concept of single domain particle. The idea is to make the material as an aggregate of particles so small that energetically they do not allow the formation of domain walls. These particles are made with elongated shapes, like needles, oriented in the same direction during the preparation process.

In the 1950s, the Philips Company Research Laboratories developed the barium ferrite, whose remanent magnetization is smaller than in Alnico 5, but the coercive field is much larger. Due to the fact that it has high H_c, barium ferrite is used to make magnets with any shape, that are used for many simple household applications. Rare earth magnets started to be developed in the 1970s. Initially Co-Sm, and then Fe-Nd-B, represented a huge advance in quality of the magnets, as can be seen in Table 9.3. The great increase in the $(BH)_{max}$ product of these materials made possible to manufacture smaller devices with much better performance than with Alnico. The high power generators of wind turbines and the compact engines of electric vehicles only became possible with the development of the Nd-Fe-B magnets.

9.4.2 High Permeability Materials

High permeability materials, also called soft magnetic materials, are used to produce a high magnetic flux by an electric current in a coil, or to produce a large magnetic induction due to a variable external field. These properties must be achieved with some requirements of variation in time and space and with minimum energy dissipation. High permeability materials should then have a narrow hysteresis cycle (very small H_c) and a large slope of the B-H curve near the origin.

Table 9.4 presents the main properties of some high permeability materials used today. In low frequency devices (motors, generators, transformers, reactors, among others) the most common are: pure iron; so-called electric steel, made with steel blades with small concentrations of carbon or silicon; alloys of iron and nickel, or iron and cobalt, in the form of the raw material or amorphous alloy prepared by fast cooling on a cold metal plate. In devices for frequencies above 10 kHz, eddy current losses do not allow the use of iron, steel or metal alloys. Several ferrites are used at higher frequencies, such as the hexagonal ferrites (structure of $BaFe_{12}O_{19}$), spinels (MFe_2O_4), and garnets ($Y_3Fe_5O_{12}$), that have quite high resistivity. The main

Table 9.4 Main properties of some high-permeability materials. μ_{max} is the maximum value of the permeability of the B-H curve. ρ is the resistivity

Material	Composition	μ_{max}/μ_0	$4\pi M_s$ (kG)	H_c (Oe)	ρ ($\mu\Omega$.cm)
Iron	Fe	5×10^3	21.5	1.0	10
Carbon steel	Fe–C(0.05)	5×10^3	21.5	1.0	10
Silicon steel	Fe-Si(3)-C(0.05)	7×10^3	19.7	0.5	60
Permalloy	$Ni_{78}Fe_{22}$	10^5	10.8	0.05	16
Sendust	$Fe_{85}Si_{10}Al_5$	10^4	10.5	–	80
Mumetal	$Ni_{77}Fe_{16}Cu_5Cr_2$	10^5	6.5	0.05	62
Mn-Zn ferrite	$Mn_{50}Zn_{50}$	2×10^3	2.5	0.1	10^8

applications of these materials are in high frequency transformers and inductors used in electronic equipment, microwave devices used in telecommunications, in radar and other applications, as well as in magnetic recording heads.

Figure 9.22 shows a device used to generate a magnetic field proportional to an electric current, that has a variety of applications. It consists of a core of soft magnetic material with permeability μ, around which there is a wire coil with N turns. Let us calculate the magnetic field B in the air gap of the device, created by the current i in the coil. The relationship between the field and the current originates from Ampère's law, obtained from Maxwell's Eq. (2.3). Taking into account that the current i traverses N times the closed path C, shown in the figure, we have, in the SI,

$$\oint \vec{H} \cdot d\vec{l} = Ni. \tag{9.42}$$

Considering $\mu \gg \mu_0$, the magnetic flux produced by the current is entirely confined in the magnetic circuit. This makes the field strength approximately uniform in the cross section of the magnetic core, with value H_i. Assuming that the air-gap spacing d is small, the field is approximately uniform in the region between the poles, having intensity H_e. Equation (9.42) then gives

Fig. 9.22 Magnetic circuit used to generate a magnetic field in the air gap proportional the current i

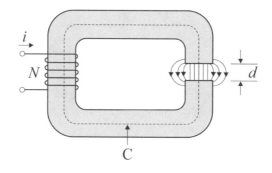
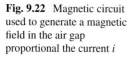

$$H_i l + H_e d = N i, \tag{9.43}$$

where l is the length of curve C inside the core. To obtain the field B in the air gap, we use the relationships between B and H in the core, $B = \mu H_i$, and in air, $B = \mu_0 H e$, and the fact that the normal component of B is continuous at the boundary of two media. Since B only has a normal component at the surfaces of the poles, $B = \mu_0 H_i = \mu_0 H_e$. Substituting this relation in (9.43) we obtain.

$$B = \frac{N i}{l/\mu + d/\mu_0}. \tag{9.44}$$

Denoting by A the area of the cross section, the magnetic flux in the air gap, $\Phi = BA$, can be written in the form

$$\Phi = \frac{Ni}{R}, \tag{9.45}$$

where R is the reluctance of the magnetic circuit, given by

$$R = \frac{l}{\mu A} + \frac{d}{\mu_0 A} = R_c + R_g, \tag{9.46}$$

where R_c and R_g are, respectively, the reluctances of the core and the gap. Equation (9.45) is analogous to the relationship between current and voltage in an electric circuit with resistors in series, $I = E/R$. In the magnetic circuit the flux Φ is analogous to the current I, Ni is analogous to the electromotive force E, and is called magnetomotive force. Of course, R is analogous to resistance, and is given by the sum of the reluctances of the core and the air gap, as in Eq. (9.46). The recording and reading heads used in magnetic recording have magnetic circuits like the one in Fig. 9.22. In these heads it is important to have the highest possible magnetic flux in the air gap, for a certain magnetomotive force Ni. From Eqs. (9.44)–(9.46) we see that this is achieved with a minimum core reluctance R_c. It is customary to define the efficiency η of a recording head by the ratio between the flux in the air gap and its maximum possible value, that would be obtained with $R_c = 0$. We see then that the efficiency is given by

$$\eta = \frac{R_g}{R_g + R_c}. \tag{9.47}$$

Thus, to make η close to 1, one must use materials with very high values of μ. In addition, it is customary to make the cross section of the air gap much smaller than that of the nucleus, in order to make $R_c << R_g$.

Example 9.2 Consider an alnico permanent magnet and an electromagnet with an iron core, both in the form of Figs. 9.20 and 9.22, with circular cross section of 10 cm diameter, total average length of 100 cm and air gap spacing of 2 cm. Considering that the winding in the electromagnet has 800 turns, calculate the current that must pass in the winding so that the magnetic field in the air gap, at a point close to the center of the gap, has the same value as in the alnico magnet.

The field in the air gap of the magnet is calculated using Eq. (9.40), considering that the surface of the north pole is equivalent to a disk with positive magnetic charge density σ_m, while the south pole has a magnetic charge $-\sigma_m$. In this case the field in the air gap is approximately uniform and has value $H = \sigma_m$. The problem is entirely analogous to that of the electric field between the plates of a capacitor with circular section. Using the value for $4\pi M$ of alnico given in Table 9.3, we see that the field in the air gap of the magnet is

$$H = 12.5\,\text{kOe} = 12.5 \times 10^3 \times 79.58\,\text{A/m} = 9.95 \times 10^5 \text{A/m}.$$

The H field in the air gap of the electromagnet, calculated with Eq. (9.44), is.

$$H = \frac{B}{\mu_0} = \frac{N i}{l\mu_0/\mu + d}.$$

Using for μ the maximum permeability of iron, given in Table 9.4, we obtain the current that produces the same field as in a permanent magnet, in the SI,

$$i = \frac{H(l\mu_0/\mu + d)}{N} = \frac{9.95 \times 10^5 \times (1/5000 + 0.02)}{800} = 25.12\,\text{A}.$$

Note that the value of $l\mu_0/\mu$ is much smaller than the air-gap space d. Therefore, it is the space that limits the value of the field.

9.5 Magnetic Recording

The recording of an electric signal, for storage and later reproduction, is one of the most important functions in information processing. The possibility of using a magnetic material for recording information was first demonstrated at the end of the nineteenth century by the Danish engineer Valdemar Poulsen. He invented an instrument that recorded voice signals in a magnetic steel wire. However, as the reproduced signal was very weak and distorted, for many years the invention was nothing more than a technological curiosity. Only in the 1940s, magnetic tapes gained commercial importance in the USA in electronic equipment for audio recording. The recording of video signals was demonstrated for the first time in 1951, also in the USA, gaining commercial importance in the 1960s. Thanks to the development of microelectronics and the evolution of magnetic tapes, audio and video recording and reproduction became very popular in the 1970s. Starting in the 1990s, audio and video magnetic tape recording and playback systems gradually gave way to optical compact disks, described in Sect. 8.8.2, and to semiconductor flash memories, described in Sect. 7.8.2.

Magnetic recording has always played an important role in digital equipment. The first commercial electronic computers produced in the 1950s used disks or cylinders covered with a magnetic layer to store information. They had a storage capacity of 10^3–10^4 bits/inch^2 and very slow access, with times on the order of milliseconds. In the 1960s they were no longer used as the main memory, giving place to the ferrite cores, which allowed faster access. Starting in the 70s, the fast random-access memories (RAMs) started to be made with MOS semiconductor devices. However, the tapes and then the magnetic disks established themselves as the best means of non-volatile and rewritable storage of large amounts of data.

It was in the magnetic recording area that magnetic materials gained importance and had a huge market expansion at the end of the twentieth century, and therefore attracted attention for research and development in company laboratories and universities. As a result, there has been a continuing increase in storage capacity and speed of recording and reading magnetic information. Disk storage density, for example, has been increasing continuously for several decades, now exceeding 10^{11} bits/inch^2. In addition to the use in hard disks, magnetic media became popular in flexible disks, or floppy disks, for external storage and transport of information. Later, the floppy disks gave way to the semiconductor flash memories, presented in Sect. 7.8.2. However, magnetic recording continues to be a success technology, due to several factors: the variety of media formats (tapes, cards, sheets, hard or floppy disks, etc.); low cost; non-volatility; high density; and almost unlimited ability to record, erase and re-record information.

9.5.1 Basic Concepts

Figure 9.23 shows the basic elements of a conventional system for recording and playback with a magnetic tape or disk. Typical tapes are made of a plastic material, like polyethylene, with thickness of the order of 25 μm, covered with a fine emulsion layer of magnetic particles. The tape moves with constant speed, maintaining tribological contact with a recording head and a reading head. Each head is made of a core of high-permeability magnetic material, with a narrow air gap, having a winding for the electric current signal. It was also common to use a single head, both for recording and for reading.

In the recording process, the time-varying current, corresponding to the signal to be recorded, produces a variable magnetic field at the edge of the air gap of

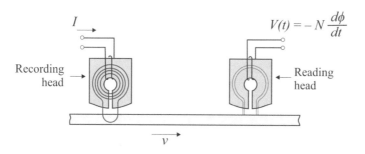

Fig. 9.23 Basic elements of a traditional recording and reproduction system with magnetic tape

the recording head. As the tape is moving, the field creates a magnetization that varies along the tape, portraying the input signal. In the process of reproduction, or playback, or reading, the magnetization of the tape creates a magnetic field that produces a variable magnetic flux in the reading head. This variable flux induces an electric current in the winding by the Faraday effect, which reproduces the original recording signal. In the system shown in Fig. 9.23, the magnetization of the tape has the longitudinal direction. Assuming that the input current varies sinusoidally in time, with angular frequency ω, that is

$$I = I_0 \sin \omega t, \tag{9.48}$$

it can be shown that the magnetization of the tape is, in a first approximation

$$M = M_0 \sin kx, \tag{9.49}$$

where x is the coordinate along the tape (or other medium) and $k = \omega/v$, v being the tape speed. Based on this relationship, the wavelength of the spatial variation is defined as

$$\lambda = \frac{2\pi}{k} = 2\pi \frac{v}{\omega} = \frac{v}{f}, \tag{9.50}$$

where f is the signal frequency. The quantitative analysis of the recording and reproduction processes is done based on the variations given by Eqs. (9.48) and (9.49), since a signal with any time variation can be decomposed in sinusoidal functions by Fourier transformation. Actually, the process of magnetizing a tape with a current in the recording head is reasonably complex, and its analysis requires numerical methods or approximate models to obtain the spatial variation of M. However, Eq. (9.49) is a good approximation for the variation of M. This variation produces an internal field on the tape that tends to demagnetize it, hence the need for the material to have a reasonably high coercive field. As will be shown in the next section, the demagnetization field increases with decreasing wavelength.

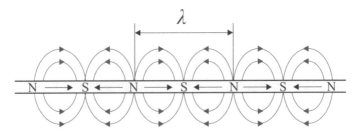

Fig. 9.24 Illustration of the magnetic field created by a magnetization that varies sinusoidally along the tape

The variation of M also creates a field outside the tape. This field, which will be calculated in the next section, has induction lines as shown in Fig. 9.24. When the tape passes over the edges of the air gap of the reading head, the external field generates a magnetic flux that varies sinusoidally in time. This variable flow induces an electromotive force on the reading-head turns, which is approximately proportional to the current of the recording signal. This process is used in the three basic types of magnetic recording: analog audio signal; analog video signal; and digital signals.

In **audio recording**, the frequency of the signal is in the range 50 Hz-20 kHz. For a tape speed in the range $v = 5$–20 cm/s, the recording wavelength is in the range 2.5 μm–4 mm. Since the air gap thickness of the recording and reading heads is of the order of a few μm, it is perfectly possible to record and detect variations of the magnetization in this wavelength range. The audio recording is done by superimposing the audio signal with an AC polarization signal of frequency 100 kHz, to obtain a linearity in the M-H relation, by means of the process described in Sect. 9.3.5. Actually, it was the discovery of this process in the 1920s that enabled audio recording without the strong distortion that occurs in direct recording.

In **video recording**, the signal spectrum covers the range 30 Hz–2 MHz, which causes two problems: the high value of the ratio between the maximum and minimum of this range, around 7×10^4, makes it difficult to operate the recording and reading circuits; the wavelength corresponding to the maximum frequency for a tape speed of 75 cm/s, $\lambda = 0.15$ μm, is too small to be recorded in the usual magnetic media. Two techniques are used to bypass these problems. The first consists of using frequency modulation (FM) of a carrier wave with the video signal. The modulated FM signal is then used directly in the recording. The carrier frequency used is 3.9 MHz and the bandwidth has a total width of 5.6 MHz. This greatly reduces the ratio between the extremes of the band and makes the system little sensitive to fluctuations in the amplitude of the reading signal. The other important technique is the use of a rotary recording head. The tape slides at low speed (2 cm/s) over the head in the form of a rotating cylinder, with high surface speed (5.6 m/s). This results in a high relative speed between the tape and the head, and consequently in a larger wavelength.

Digital recording is, conceptually, very simple, because the signal is a sequence of pulses with only two values, corresponding to the digits 0 and 1. These digits can be stored in a small area of a magnetic medium magnetized in two opposite

directions. Thus, the recording process is direct, without the need for the sophistication used in audio or video recording. The recording can be done on tape or disk,
either in the longitudinal (in-plane) or perpendicular directions. Until the 1990s, the
reading process of magnetic digital information was based on the Faraday induction effect, as will be presented in the next section. However, since the induction
signal is proportional to the magnetic flux, and thus to the area of the storage bit,
this process limits the storage capacity. With the need to increase the data density,
read-heads using magnetoresistance came into use. This is the effect by which the
electric resistance of a magnetic metal changes with the applied magnetic field. Since
the effect depends on the field strength, not the flux, it makes possible to decrease
the bit area. Later developments introduced a read-head made of a magnetic multilayer that has a magnetoresistive effect much greater than in single films, called giant
magnetoresistance (GMR). Current hard-disk drives employ write and read heads in
close proximity, mounted on an actuator arm that can access any position on the disk.
Magnetic digital recording is widely used in computers, and hard disks and magnetic
tapes constitute the main non-volatile process for storing large amounts of data.

9.5.2 Quantitative Analysis of Magnetic Recording

In this section we shall analyze in detail the process of reproducing a signal recorded
on a magnetic medium. To simplify the problem, we consider a magnetic layer
of thickness δ, infinite in the x and z directions. The coordinate system is shown in
Fig. 9.25, where x is the longitudinal direction of the layer motion. Since the magnetic
layer, in a tape or disk, has a finite width in the z direction, the result obtained for
the infinite layer will be approximately valid for the central region and for small
distances from the layer. The purpose of the analysis is to obtain the fields created
by the magnetized layer, and with them calculate the voltage induced in the reading
head. For this we consider that the layer has longitudinal magnetization, varying
sinusoidally in the x direction as

$$\vec{M} = \hat{x} M(x) = \hat{x} M_0 \sin kx. \tag{9.51}$$

Fig. 9.25 Geometry used to
calculate the fields created
by a magnetic tape or disk

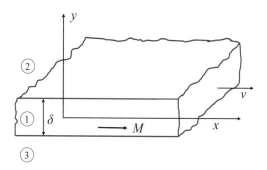

Since M does not vary in the z direction, the geometry of the problem is reduced to two dimensions, x and y. To facilitate the solution of the problem, we introduce the scalar magnetic potential $\psi(x, y)$, defined by the relation

$$\vec{H} = -\nabla\psi. \tag{9.52}$$

This is possible because the curl of the gradient of any scalar function is zero, and in a magnetostatic system ($\partial / \partial t = 0$) without electric current, Eq. (2.4) becomes $\nabla \times \vec{H} = 0$. The equation for ψ, obtained by substituting Eqs. (9.52) and (9.37) into (9.36), is, in the SI,

$$\nabla^2\psi = -\rho_m. \tag{9.53}$$

This equation must be solved for the three regions of Fig. 9.25, and the final solution determined by the boundary conditions on the surfaces at $y = \pm\delta/2$. In region 1 the magnetic density ρ_m is given by Eq. (9.37) applied to (9.51), that is

$$\rho_m = -M_0 \cos kx. \tag{9.54}$$

Equation (9.53) with $\rho_m = 0$ is called Poisson equation. Outside the layer, where $\rho_m = 0$, it reduces the Laplace equation,

$$\nabla^2\psi = 0. \tag{9.55}$$

In two dimensions this can be written in the form

$$\frac{\partial^2\psi}{\partial x^2} + \frac{\partial^2\psi}{\partial y^2} = 0. \tag{9.56}$$

The solutions of this equation for regions 2 and 3 are

$$\psi_{2,3}(x, y) \propto (\sin kx, \cos kx)e^{\pm ky}. \tag{9.57}$$

Note that although ρ_m does not have a y-dependence, the potential within the layer must vary with y, otherwise it is not possible to satisfy the boundary conditions at the surfaces. The solution of Eq. (9.53) with ρ_m given by (9.54) is

$$\psi_1(x, y) = \cos kx\,(A\,e^{ky} + B\,e^{-ky} + C). \tag{9.58}$$

The final solution in the three regions is determined by the boundary conditions for the fields $\vec{H}$ and $\vec{B}$ at the two surfaces. The continuity of the tangential component of $\vec{H}$ at the interface between two media implies that

$$\psi_1(x, \delta/2) = \psi_2(x, \delta/2), \qquad \psi_1(x, -\delta/2) = \psi_3(x, -\delta/2). \tag{9.59}$$

Since $\vec{M}$ has no normal component to the surface, the continuity of the normal component of $\vec{B}$ implies that H_y is continuous, that is,

$$\partial\psi_1/\partial y|_{\delta/2} = \partial\psi_2/\partial y|_{\delta/2}, \qquad \partial\psi_1/\partial y|_{-\delta/2} = \partial\psi_3/\partial y|_{-\delta/2} \qquad (9.60)$$

In regions 2 and 3 the solutions (9.57) contain only the term $\cos kx$, because the term $\sin kx$ cannot satisfy the boundary conditions with the solution (9.58) in region 1. In addition, ψ_2 cannot contain the term $\exp(ky)$ because it diverges at $y \to +\infty$, while ψ_3 cannot have the term $\exp(-ky)$ that diverges at $y \to -\infty$. Hence, the potentials in regions 2 and 3 are given by

$$\psi_2(x, y) = D \cos kx e^{-ky}, \qquad (9.61)$$

$$\psi_3(x, y) = E \cos kx e^{ky}. \qquad (9.62)$$

Application of the boundary conditions (9.59) and (9.60) to the functions (9.58), (9.61) and (9.62) at $y = \pm\delta/2$, together with Eqs. (9.53) and (9.54) form a system of five equations that allow to determine the five unknown coefficients. One can show that the potentials in the three regions are (Problem 9.10)

$$-\frac{\delta}{2} \le y \le \frac{\delta}{2} \quad \psi_1(x, y) = -\frac{M_0}{2k}\cos kx\left[2 - e^{-k(y+\delta/2)} - e^{k(y-\delta/2)}\right], \qquad (9.63)$$

$$y \ge \frac{\delta}{2} \quad \psi_2(x, y) = -\frac{M_0}{2k}\left(1 - e^{-k\delta}\right)\cos kx e^{-k(y-\delta/2)}, \qquad (9.64)$$

$$y \le -\frac{\delta}{2} \quad \psi_3(x, y) = -\frac{M_0}{2k}\left(1 - e^{-k\delta}\right)\cos kx e^{k(y+\delta/2)}. \qquad (9.65)$$

The solution (9.63) provides the magnetic field inside the layer. Its components are

$$H_x = -\frac{\partial\psi_1}{\partial x} = -\frac{M_0}{2}\sin kx\left[2 - e^{-k(y+\delta/2)} - e^{k(y-\delta/2)}\right], \qquad (9.66)$$

$$H_y = -\frac{\partial\psi_1}{\partial y} = \frac{M_0}{2}\cos kx\left[e^{-k(y+\delta/2)} - e^{k(y-\delta/2)}\right]. \qquad (9.67)$$

The field created by the magnetization in the longitudinal direction, Eq. (9.66), has a direction opposite to $\vec{M}$, and for this reason it is called **demagnetization field**. For the magnetic layer to retain the magnetization produced by the recording process, it is necessary that the material has a coercive field larger than the demagnetization field at all points. We see in Eq. (9.66) that the field is zero in the limit $\omega = k = 0$ and increases with increasing frequency. The maximum field value in the limit $\omega, k \to \infty$ is $M_0/2$. This result shows that video tapes should be made with materials of larger coercive fields than audio tapes.

The field outside the tape is obtained from Eqs. (9.64) and (9.65). It is easy show that

$$H_x = -\frac{M_0}{2}\left(1 - e^{-k\delta}\right)\sin kx \, e^{-k(\pm y - \delta/2)}, \tag{9.68}$$

$$H_y = \mp\frac{M_0}{2}\left(1 - e^{-k\delta}\right)\cos kx \, e^{-k(\pm y - \delta/2)}, \tag{9.69}$$

where the upper signs are valid for the region above the tape ($y > \delta/2$) and the lower one for the region below the tape ($y < -\delta/2$). This result indicates that the longitudinal and normal components of the field are 90° out of phase. The factor $\exp(-ky)$ has major importance in the reading signal, since it introduces an exponential decay in the field with the distance to the tape. For example, the field at a distance $d = \lambda$ from the tape is reduced to $\exp(-2\pi) \approx 0.002$ of the value at the surface. Because of this result, it is important to make the tape tensioned to slide in close contact with the reading head. In the case of video recording, since λ is very small, the noise caused by amplitude fluctuations due to the exponential factor is avoided by means of the frequency modulation.

To calculate the signal induced in the reading head, we consider the geometry shown in Fig. 9.26. Only the H_x component contributes to the magnetic flux through the winding in the core. The fact that the core is magnetic results in an increase of the field. Using the method of images, it can be shown that in a core with permeability $\mu \gg \mu_0$, the field B is twice as large as the field in the air. Thus, the magnetic flux that passes through the turns is, approximately

$$\Phi = \eta L \int_{\delta/2+d}^{\infty} B_x dy \approx -\eta\mu_0 M_0 L\left(1 - e^{-k\delta}\right)\frac{e^{-kd}}{k}\sin kx, \tag{9.70}$$

where η is the efficiency of the reading head, L is the width of the tape or the recording track, and d is the distance from the read-head core to the tape. The voltage induced in the N turns of the head is obtained with Faraday's law. With $x = vt$ we have

Fig. 9.26 Illustration of the flux created by the magnetic tape on the reading head

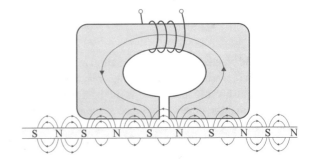

$$V(t) = -N\frac{d\Phi}{dt} = -N\,v\frac{d\Phi}{dx}. \tag{9.71}$$

Substituting Eq. (9.70) into (9.71) we obtain

$$V(t) = N\eta\mu_0 M_0 L v\left(1 - e^{-k\delta}\right)e^{-kd}\cos\omega t. \tag{9.72}$$

This result shows that the electric voltage produced in the reading head by the magnetization of the tape is an alternating signal, lagging by 90° the sine wave of the recording current. The amplitude of the output voltage depends on the signal frequency, the tape speed, the remanent magnetization, and the width of the recording track.

An important requirement for digital recording in computers is the storage capacity, expressed in bits/cm². The increase of this capacity requires the decrease of the area occupied by a bit, and therefore of the track thickness L. However, Eq. (9.72) shows that the voltage read signal decreases as L decreases, showing that the reading process based on the induction effect limits the data storage capacity. For this reason, as mentioned in the previous section, the read-heads of current disk drives are based on sensors that make use of the giant magnetoresistance effect, which will be presented in Sect. 9.5.4.

The need to increase data storage capacity in computers has always been a strong driving force for the development of new processes for magnetic recording and reading processes. One technology that led to commercial devices that was in the market for several years in late 1990s and early 2000s is that of the magneto-optical memory used in removable disks. In this technology the information bits are recorded on a magnetic disk in motion by a thermomagnetic process. The film is previously magnetized in the perpendicular direction, upwards, corresponding to bit 0. In the recording process a semiconductor laser beam modulated by the electric signal containing the information to be recorded (0 or 1), is focused by a lens on the magnetic layer, where there is a magnetic field created by a permanent magnet, directed downwards. This field has a value less than the coercive field of the film at room temperature, so that it does not change the magnetization of the film. To record a bit 1 in a small region of the film, the laser beam heats this region and produces a rapid decrease of H_c, such that the field of the magnet reverses the direction of magnetization. The reading process is based on the magneto-optical Kerr effect (MOKE), by which a beam of polarized light incident on a film is reflected with a polarization that has information about the direction of the magnetization. The main advantages of magneto-optical memories over the conventional floppy disks are the shortest access time and larger memory capacity. The main disadvantage is the higher cost of the recording device and the disk itself, compared to those of floppy disks and compact optical disks.

Example 9.3 Calculate the amplitude of the output signal from an induction read-head on a magnetic tape with audio frequency 1 kHz, having the following parameters: $N = 20$, $\eta = 1$, $\mu_0 M = 0.5$ T, $v = 0.1$ m/s, $L = 1$ mm, $\delta = 10\,\mu$m and $d = 0$.

The wave number of a 1 kHz signal recorded on a tape with speed $v = 0.1$ m/s, given by Eq. (9.50), is

$$k = \frac{2\pi f}{v} = \frac{2\pi \times 10^3}{0.1} = 6.28 \times 10^4 \, \text{m}^{-1}.$$

The signal amplitude is calculated with Eq. (9.72),

$$V = N\eta\mu_0 M_0 \, Lv \left(1 - e^{-k\delta}\right) = 20 \times 1 \times 0.5 \times 10^{-3} \times 0.1 \times \left(1 - e^{-6.28 \times 10^4 \times 10^{-5}}\right)$$

$$V = 4.66 \times 10^{-4} \, \text{V} = 0.466 \, \text{mV}.$$

This value is easily processed for the final reproduction of the recorded signal.

9.5.3   Materials for Recording

Two types of magnetic materials are used in magnetic recording equipment: high permeability materials, which form the core of write and read heads; materials with intermediate hysteresis cycle, which are used in the layers of magnetic storage media.

The main materials used in the head cores are metals alloys, permalloy, sendust, and the oxide ferrites of MnZn and NiZn. The main advantages of permalloy and sendust, whose parameters are in the Table 9.4, are the large value of the saturation magnetization and the high permeability. In addition, they have great resistance to mechanical wear caused by the contact with the moving tape or disk. However, since they have low electric resistivity, the eddy current effect allows their use only at low frequencies. This is why they were used in the heads of audio recorders. The MnZn and NiZn ferrites have smaller magnetization, but have resistivity 10^5 times larger than metallic alloys. For this reason, they are used in heads for recording and reading of video and digital signals.

Regarding the recording media, there are two types of intermediate materials used to make the magnetic layer: **particulate media**, which consist of microscopic particles of oxides or magnetic metals in suspension in a polymeric layer; and **thin films** of ferromagnetic metals or metal alloys.

Particulate media are used to cover audio and video tapes, plastic or paper cards used in a number of applications, and computer floppy disks. They were also used on hard-disk drives, however, they were gradually replaced by thin metallic films. The particulate media are prepared by processes similar to the manufacturing process of paints used to paint walls, wood, artistic canvas, etc. Any paint is formed by diluted solid particles in suspension in a liquid medium, consisting of solvents, diluents and dryers added to a binder. The solid particles are the colored pigments that give the color to the paint, while the binder can be natural, artificial, or organic oil resin. After that the paint is spread on the surface to be painted, the drying process takes place, in which some components of the liquid evaporate and others react chemically in the binder. After drying, the colored pigments are fixed in the binder layer that covers the surface. In the case of a magnetic medium, the binder is a polymer and the particles are made of magnetic oxides or metals, forming what is called magnetic paint. The particles have elongated shape, with length of the order of 1 μm and transverse dimension of the order of 0.1 μm. Due to these reduced dimensions, they can only accommodate a magnetic domain. The magnetic paint is spread over the surface of the base material, which can be a sheet of polyethylene, in the case of tapes, or plastic or cardboard, in the case of cards. During the drying process, it is subjected to a magnetic field of the order of 1 kOe, which serves to align the particles in the direction that will be used to hold the magnetization. After drying, the particles are aligned and separated from each other in the magnetic layer.

Various compounds are used to make the magnetic particles. The oldest one and still used in audio tapes is ferric oxide, with chemical formula γ-Fe_2O_3. It has saturation magnetization $4\pi M_s = 4.65$ kG (or $\mu_0 M_s = 0.465$ T in the SI) and coercive field $H_c = 300$ Oe. However, as the particles are diluted in the magnetic layer, the value of $4\pi M_s$ is reduced in the same proportion as the dilution, generally in the range 30–50%. The ferric oxide was also used on floppy disks, and is currently employed in plastic cards and cardboard tickets. However, its application is restricted to low frequencies and low recording densities due to the small value of H_c. As the wavelength, and therefore the size of the recording bit decreases, the demagnetization field increases, requiring higher values of H_c.

In the 1970s, the Japanese company TDK discovered that the impregnation of a thin cobalt layer on the surface of ferric oxide increases the value of H_c to about 700–800 Oe. This became the standard process used in high quality audio tapes and high-capacity floppy disks. Another substance used to make particles of magnetic inks is chromium dioxide, CrO_2, that has $4\pi M_s = 6.16$ kG and $H_c = 450$ Oe, that are higher than in pure ferric oxide. CrO_2 was widely used in tapes for audio, video, and digital recording, before the discovery of the process of modification of ferric oxide with cobalt. Since then, several substances were used in the magnetic media for tapes that provided higher storage capacities. Currently, manufacturers employ very fine particles of ferrite oxide materials that are chemically stable, to make magnetic tapes for long-term high density digital data storage. Figure 9.27a shows details of a magnetic tape developed jointly by Fujii film and IBM, in which the magnetic layer is made of strontium ferrite (SrFe) or barium ferrite (BaFe), used in a cartridge that can store the staggering amount of 580 terabytes of data. Figure 9.27b shows images

(a)   (b)

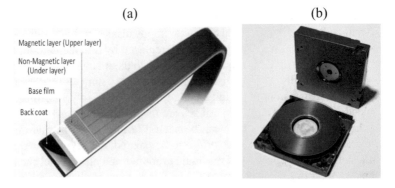

Magnetic layer (Upper layer)

Non-Magnetic layer (Under layer)

Base film

Back coat

Fig. 9.27 a Illustration of the structure of a magnetic tape used for long-term high density digital data storage (Courtesy of Fujifilm Co.). **b** Images of a modern data storage tape cartridge consisting of a single reel (Courtesy of Mark Lantz, IBM Corp.)

of a modern data storage tape cartridge consisting of a single reel. After the cartridge is inserted in the storage equipments, the tape is fed automatically to a reel built into the drive mechanism.

In regard to magnetic hard-disks for computers, during the 1990s, thin metallic or multilayer films of elements of the iron transition group, rare earths, and their alloys, started to be used in the manufacturing technology. A great advantage of films is the high value of M_s. For example, pure Fe and Co films have magnetization an order of magnitude larger than the magnetic oxide particles diluted in magnetic paints. Magnetic thin films are prepared by high vacuum deposition processes, described in Sect. 1.4.5. Some of the most used alloys are CoNiPt, CoCrTa and CoCrPt, which have coercive fields in the range 750–1500 Oe. Films are deposited on an aluminum disk, having a thickness of tens of nm, and covered with a carbon layer to provide resistance to corrosion and for lubrication in the contact with the write-read head. These materials are used for both in-plane and perpendicular magnetic recording.

Magnetic tapes and hard disks are both used for digital data storage in the wide variety of computing equipment manufactured today, from notebooks and personal computers to supercomputers. With the exception of capacity, the performance characteristics of tape and hard drives are very different. The long length of the tape held in a cartridge, normally hundreds of meters, results in average data-access times of 50 to 60 s compared with just 5 to 10 ms for hard drives. But the rate at which data can be written to tape is, surprisingly enough, more than twice the rate of writing to disk.

9.5.4   Spintronic Technologies for Magnetic Memories

The scientific investigations in magnetic thin films, multilayers, and structures on the nanometric scale, that started to attract worldwide attention in the 1980s, have led to several breakthroughs. The most notable one, the **giant magnetoresistance** (GMR), discovered independently by Albert Fert and Peter Grünberg, gave birth to the field of **Spintronics**. This is the field of science and technology devoted to the investigation of phenomena and their applications in which the electron transport is controlled by an action on its spin. Fert and Grünberg received the Physics Nobel Prize in 2007 for their seminal discovery that triggered research in this new field of science and technology. In this section we shall describe two spintronic devices, the GMR reading head, that made possible a huge increase in the hard-disk storage capacity, and the magnetic random-access memory (MRAM), that has an advantage over semiconductor RAM because of its non-volatility. First, we present a qualitative explanation of the GMR, that is the essential for the operation of both devices.

9.5.4.1 The GMR Reading Head

The magnetoresistance (MR) is the phenomenon by which the resistivity ρ of a metal or semiconductor varies with the application of a magnetic field H. The rate of change $\Delta\rho/\Delta H$ depends on the material and also on the value of H. The MR is much higher in ferromagnetic metals than in semiconductors and in non-magnetic metals. In a thick ferromagnetic wire or film, the magnetoresistance is smaller than 1%, but in a magnetic multilayer it can be larger than 100%. The increase in the magnetoresistance in a multilayer of thin films is due to the spin-dependence of the electron transport. As discussed in Sect. 4.5, the electron mean free path in several simple metals is of the order of 100 nm. Therefore, when traversing a bulk material or a thick film, with thickness on the order of 1 µm or larger, the electron undergoes countless collisions and loses the orientation of its spin. For this reason, the electron spin does not have any role in conventional electronics. However, when crossing a thin film with thickness up to tens of nm, the spin orientation is preserved, since the electron does not suffer collisions on the way. Thus, when electrons traverse a structure with various thin metallic layers, collisions occur mainly at the interfaces.

Figure 9.28 shows a trilayer made of two ferromagnetic (FM) metallic films separated by a thin nonmagnetic metal (NM) layer. All layers are made with thickness much less than the electron mean-free path, typically in the range of few nm

Fig. 9.28 Schematic illustration of the mechanism underlying the giant magnetoresistance in a magnetic trilayer with antiferromagnetic coupling between the two magnetic layers. **a** $H = 0$. **b** $H > 0$

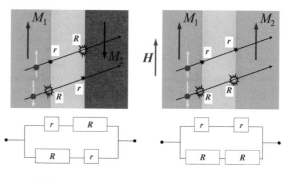

(a) High resistance   (b) Low resistance

to about 30 nm. Two different mechanisms are responsible for the GMR of the trilayer, the coupling between the magnetizations in the two FM layers, and the spin-dependent interfacial electron scattering. The coupling between the two magnetizations, discovered in the 1980s, has a quantum origin, mediated by the conduction electrons in the NM spacer, that can be explained qualitatively as follows. The total electron wave function, consisting of spatial and spin components, has to be anti-symmetric. The spatial wave function in the NM layer is a stationary wave with a number of half-cycles that depend on the spacer layer thickness. For a thickness such that the spatial wave function is symmetric, the spins are antiparallel, while for a thickness such that the spatial wave function is anti-symmetric, the spins are parallel. The spins of the conduction electrons interact with the spins of the FM layers by means of the exchange interaction and induce a coupling between the two magne-tizations, with sign and amplitude that vary with the spacer thickness. In Fig. 9.28 the thickness is such that with no applied external field, the two magnetizations are antiparallel, that is, the coupling is antiferromagnetic, as in (a). If a sufficiently strong field is applied in the plane, the two magnetizations align with the field, as in Fig. 9.28b.

The second mechanism underlying the GMR is the spin-dependent electron scattering at the interfaces of the NM layer with each FM film. The interface scattering is the main source of the electric resistance, since the layers have thickness much smaller than the electron mean-free path. The origin of the spin-dependent resistance lies in the fact that the scattering strength for electrons with spin parallel to the magne-tization is smaller than for electrons with spin antiparallel to the magnetization. Thus, one can represent the electric current produced by electrons flowing from one FM layer to the other by a circuit with two parallel branches, one branch for up-spin elec-trons and the other for down-spin electrons. In each branch there are two resistors, representing the electron scatterings at the two interfaces. The equivalent circuits for the two magnetic configurations are shown in the lower panels of Fig. 9.28. If the two magnetizations are in opposite directions, as in (a), the equivalent circuit has in both parallel branches a high resistance in series with a small resistance, so that the total resistance is high. However, with the two magnetizations in the same direction due to the applied field, one branch has two high resistances in series, while the other has two low resistances is series, as in (b). In this case, the total resistance is low. The large change in resistance with the application of the magnetic field is called giant magnetoresistance, and is characterized by the GMR parameter

$$\rho_{GMR} = \frac{R_{AP} - R_P}{R_P}, \qquad (9.73)$$

where R_P and R_{AP} are, respectively, the resistances with the two magnetizations parallel and antiparallel to each other. The value of the GMR parameter depends on the layer materials, thickness and temperature, and ranges from a few % to over 100%.

An important application of the GMR effect is the **spin valve sensor**, also called **GMR sensor**, a device whose electric resistance is controlled by an external magnetic

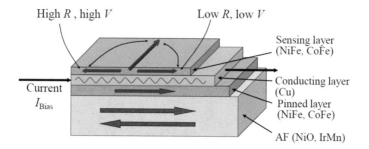

Fig. 9.29 Schematic illustration of the structure of a spin valve sensor, or GMR sensor

field that acts on the magnetization of a sensor layer. The basic structure of a spin valve is shown in Fig. 9.29. It consists basically of four thin layers, sequentially deposited on an appropriate substrate, which can be glass, an insulating crystal (such as MgO), or a semiconductor (such as Si or GaAs). The top layer is the free layer, made of a soft magnetic metallic film, with low magnetic anisotropy, such as $Ni_{0.81}Fe_{0.19}$ (permalloy) or $Co_{0.9}Fe_{0.1}$, with thickness on the order of 10 nm. This is the sensor layer, because the direction of its magnetization, determined by an external field, controls the resistance of the device. The other magnetic layer, called reference or pinned layer, has a magnetization fixed by means of a mechanism that will be explained latter. Between the two magnetic layers there is a thin layer of a non-magnetic metal, usually copper, with thickness of about 5 nm, where most of the sensor current I passes. The resistance of the copper layer depends on the magnetization direction of the sensor layer relative to that of the pinned layer, because of the spin-dependent interface scattering. When the magnetization of the sensor layer is parallel to that of the reference layer, the resistance R of the spin valve is low, and when it is antiparallel the resistance is high. Thus, by passing a constant current, the value of the sensor voltage indicates the direction of the magnetic field acting on the sensor.

The mechanism that holds the reference layer magnetization in a given direction was first observed in the 1950s, but only in late 90s it was investigated in detail in thin films and was fully understood. It is called **exchange bias**, and its origin lies in the exchange interaction between atoms at the interface of a ferromagnetic layer (FM) with an antiferromagnetic layer (AF). As mentioned earlier, AF materials have a negative exchange interaction between neighboring spins. When cooled below a critical temperature, called Néel temperature, T_N, they exhibit an ordered magnetic arrangement, with spins oriented in two or more sublattice directions, such that the total macroscopic magnetization is zero. Thus, the spin arrangement in an AF material is insensitive to external fields. When the FM/AF bilayer is cooled in the presence of an external field, from a temperature above to one below T_N, the sublattice spins at the interface on the AF side become mostly parallel to the magnetization of the FM layer by means of the interfacial exchange interaction. Since the AF spin arrangement is not affected by external fields, the magnetization of the FM layer stays **pinned** in the direction of the field applied during cooling. The AF layer is much thicker than

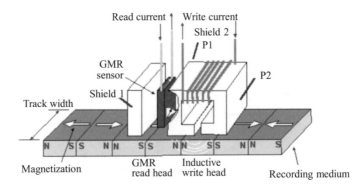

Fig. 9.30 Schematic illustration of the inductive recording head and the GMR reading head used in magnetic hard-disk drives in computers

the other layers, on the order of tens of nm, and it can be made from a variety of materials with T_N above room temperature, such as NiO and IrMn.

An important application of spin valves is in the reading heads of magnetic recording on computer hard disks. Figure 9.30 illustrates the inductive recording head and the GMR reading head used to write and read the magnetic information on the disk. Actually, the figure represents only a schematic illustration to facilitate the view of the components, since all components are manufactured in the form of multilayer of thin films, forming an integrated set. In the element shown, the recording is done with the conventional device, in which the recording current creates a magnetic field in the air gap of the core that magnetizes the magnetic layer of the disk. The reading of the recorded information is done by a GMR sensor. A current pulse passes through the sensor when it is over the magnetized region to be detected. Since the resistance of the sensor varies with the field created by the magnetization, the resulting voltage value indicates the information of the recorded bit. The main advantage of the GMR reading head is that it can detect magnetization in smaller regions than the conventional inductive heads, since it is sensitive to the field created by the magnetization, while the inductive head is sensitive to magnetic flux, which depends on the area occupied by the bit. The introduction of the GMR sensor in commercial hard-disk recording systems started in 1998, and, together with the improvement in the materials used in the magnetic media, has enabled the continuous increase in memory capacity. Current technologies employ disks in which the recording bits are magnetized perpendicularly to the disk plane and write-read head elements with lateral dimension of 10–20 nm, reaching storage areal density of the order of 100 Gbits/in^2. The development of magnetic recording technologies in the last two decades has enabled the increase of storage density of hard disk drives by several orders of magnitude. This, together with the high-density magnetic tapes described in the previous section, constitute essential hardware elements in today's information age for operation of data centers installed by the cloud computing industry.

9.5.4.2 Magnetic Random-Access Memory

Another important spintronic device for data storage, developed more recently, is the magnetic random-access memory (MRAM). The MRAM consists of a large number of magnetic memory cells arranged in a network with the form of a matrix, similar to the one with semiconductor memories shown in Fig. 7.37. There are several types of MRAM cells. We shall present here one type that is currently used in commercial devices, the so-called STT-MRAM. Its operation is based on two different mechanisms, the tunneling magnetoresistance (TMR) and the spin-transfer torque (STT).

Figure 9.31 shows the basic structure of a MRAM cell, consisting of two metallic ferromagnetic (FM) layers separated by a dielectric spacer layer, the tunnel barrier. The dielectric layer is very thin, typically less than 2 nm, so that electrons can flow from one FM layer to the other by means of a quantum mechanical tunneling, as presented in Sect. 3.3.3. For this reason, the element in Fig. 9.31 is also called magnetic tunnel junction (MTJ). The upper FM layer is the free layer, also called recording layer, that stores the information by means of the direction of its magnetization. The scheme shown in Fig. 9.31 employs in-plane magnetization, but it is common also to use perpendicular magnetization. The bottom FM layer is the pinned or reference layer, that has magnetization fixed in a certain direction by means of the exchange bias due to the contact with the antiferromagnetic (AF) layer. The lateral dimension of the MTJ in on the order of few tens of nanometers. The layers have elliptical shape so that the magnetized FM layers have only one domain.

The reading of the stored information is made by means of a small reading current to sense the device resistance. This is determined by the relative orientations of the magnetizations. If the two magnetizations are parallel, the resistance is low because the majority band electrons can tunnel into the majority band on the opposite side of the barrier. If the magnetizations are antiparallel the resistance is high since the majority band electrons have to tunnel into the minority band of the opposite FM layer. Thus, for a given reading current, a low voltage represents bit 0 and a higher voltage represents bit 1. The TMR parameter is defined just like the GMR parameter as in Eq. (9.73). The first MRAMs were made with GMR cells that have smaller resistances and require higher currents to have good voltage signals. The TMR cells have the advantage of quite larger resistances and much higher TMR parameters than

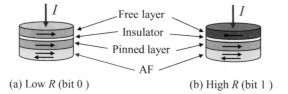

(a) Low R (bit 0)   (b) High R (bit 1)

Fig. 9.31 Sketch of a MRAM cell with in-plane magnetization. **a** State of low resistance with parallel magnetizations, bit 0. **b** State of high resistance with antiparallel magnetizations, bit 1

GMR cells. Typical MRAM cells using FeCoB for the FM layers and crystalline MgO for the dielectric layer have TMR as high as 500%.

The writing process in the MRAM also employs an electric current through the memory cell and is based on the mechanism of spin-transfer torque (STT). When a current pulse is injected into the upper FM layer, as in Fig. 9.31, electrons of the circuit current flow upward. As they pass through the FM pinned layer they become spin-polarized, with more electrons with spins in a certain direction than in the opposite one, so that they carry angular momentum into the free layer. Since the time-derivative of angular momentum is a torque, the pulsed spin-polarized current that is injected into the free layer exerts a torque on the magnetization and can reverse its orientation. Thus, in a cell with the two magnetizations in parallel, representing bit 0, the bit 1 can be written simply by passing a current pulse through the cell with intensity above a certain threshold value.

The MRAM device consists of an array of MTJs, or memory cells, in which each one is connected in series to a MOSFET used as a switch that controls the current through the MTJ. The array is similar to the one in a semiconductor RAM (SRAM), illustrated in Fig. 7.37, with the MOS capacitors replaced by MTJs. While in the SRAM the stored information is represented by the absence (bit 0) or presence (bit 1) of charge in the MOS capacitor, in the MRAM it is represented by the orientation of the magnetization in the free layer of the MTJ. In both types of RAMs the writing and reading operations are made by current pulses applied to the word line and bit line to select a specific cell address. One of the main advantages of the STT-MRAM over the SRAM is the nonvolatility. Since the MOS capacitors in the SRAM lose their charge over time, the device must refresh all cells periodically, typically every 50–70 ms, reading each one and re-writing its contents. In contrast, MRAM never requires a refresh. The nonvolatility also eliminates the need for rebooting when the computer is turned on, because all information stored in the RAM is saved before it is turned off.

As presented in Sect. 7.8.2, the semiconductor flash memories that employ floating-gate avalanche injection MOS, are also nonvolatile, like the MRAM. When used for reading, flash and MRAM are very similar in power requirements. However, flash is re-written using a large pulse of voltage (about 10 V) that is stored up over time in a charge pump, which is both power and time consuming. In addition, the current pulse physically degrades the flash cells, which means that flash can only be written to some finite number of times before it must be replaced. In contrast, MRAM requires only slightly more power to write than read, and no change in the voltage, eliminating the need for a charge pump. This leads to much faster operation, lower power consumption, and an indefinitely long lifetime. Currently, the main disadvantage of MRAM is the lower storage capacity than flash memories. However, the use of a new writing process based on the so-called spin–orbit torque promises to increase the storage capacity of MRAMs.

9.6 Magnetic Devices for Microwave Circuits

An important area for application of magnetic materials is that of non-reciprocal devices for microwave circuits. Electromagnetic waves with frequency in the microwave range, 1–30 GHz, are used in communications between ground stations and satellites, and in mobile telephony. They are also used in radar and in scientific and industrial equipment, as well as household appliances. Microwave circuits employ certain devices, such as isolators and circulators, in which the central element is made of ferrites. To understand the effect of a ferrite on the electromagnetic wave, it is necessary to study its susceptibility at high frequencies. For this, let us study initially the behavior of the magnetization in a material subjected to an external magnetic field.

9.6.1 The Magnetization Precession Motion

When an electromagnetic wave penetrates a magnetic medium, its *rf* magnetic field interacts with the microscopic magnetic moments. In case the medium is conductive and thick, the amplitude of the wave decays rapidly due to the skin effect, so that the magnetic effect is small. However, if the medium is insulating, the attenuation is small and the magnetic interaction produces a large effect on the polarization of the wave. This is why magnetic materials used in microwaves are insulating ferrites. To calculate the response at high frequencies, we initially consider an infinite ferrite material, subjected to a static magnetic induction field $\vec{B}$. In the equilibrium situation, the magnetic moment $\vec{\mu}$ is aligned with $\vec{B}$, because this is the situation in which the energy, given by Eq. (9.8), is minimum. However, if the moment deviates from the equilibrium direction, the field exerts a torque on it given by $\vec{\tau} = \vec{\mu} \times \vec{B}$. From this relation we can obtain the equation of motion for the magnetization subject to a magnetic field. First recall Newton's second law for an angular momentum $\vec{J}$ subject to a torque

$$\vec{\tau} = \frac{d\vec{J}}{dt}.$$ (9.74)

In the case of the atomic magnetic dipole, $\vec{\mu}$ and $\vec{J}$ are related by an expression obtained from Eqs. (9.12)–(9.15),

$$\vec{\mu} = -\gamma \mu_0 \vec{J},$$ (9.75)

where

$$\gamma = \frac{g \mu_B}{\hbar}.$$ (9.76)

is the **gyromagnetic ratio** of the magnetic atom, or ion. Substituting Eq. (9.75) into (9.74) and using the expression for the torque exerted by a field on the magnetic moment we obtain the equation of motion

$$\frac{d\vec{\mu}}{dt} = -\gamma\,\vec{\mu}\times\vec{B}.$$

Considering that the magnetization $\vec{M}$ is the magnetic moment per unit of volume, using Eq. (9.3) and the fact that $\vec{M}\times\vec{M} = 0$, we obtain the equation of motion for the magnetization in a magnetic field $\vec{H}$

$$\frac{d\vec{M}}{dt} = -\gamma\mu_0\,\vec{M}\times\vec{H}. \tag{9.77}$$

This is called **Landau-Lifshitz equation**. To understand the free motion of $\vec{M}$ when it is deviated from the direction of equilibrium, we choose a coordinate system in which the z-axis has the direction of the static field, that is $\vec{H} = \hat{z}H$. We can then write the magnetization in the form

$$\vec{M} = \hat{x}\,m_x(t) + \hat{y}\,m_y(t) + \hat{z}\,M_z, \tag{9.78}$$

where we use lowercase letters for the x and y components because they vary in time, while the z-component is static, and also because $m_x, m_y << M_z$. From Eq. (9.77) we obtain the equations for the transverse components of the magnetization

$$\frac{dm_x}{dt} = -\gamma\mu_0\,m_y H, \qquad \frac{dm_y}{dt} = \gamma\mu_0\,m_x H. \tag{9.79}$$

A solution for Eq. (9.79) is

$$m_x(t) = m_0\cos(\omega_0 t), \qquad m_y(t) = m_0\sin(\omega_0 t). \tag{9.80}$$

The motion of the magnetization described by Eq. (9.80) is illustrated in Fig. 9.32. The magnetization vector $\vec{M}$ precesses about the field $\vec{H}$, with its tip in a circular

Fig. 9.32 Precession motion of the magnetization of a ferrite about a static magnetic field H

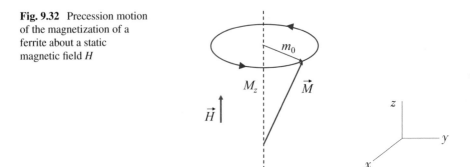

motion with amplitude m_0 defined by the initial condition. This motion is similar to that of a gyroscope precessing about the gravitational field. The angular frequency of precession, which is readily obtained by replacing one of the two equations in Eq. (9.80) into (9.79), is given by (in the SI).

$$\omega_0 = \gamma \mu_0 H = \gamma B. \tag{9.81}$$

In the CGS, since $\mu_0 = 1$, the precession frequency is given by $\omega_0 = \gamma H$. This is also the *magnetic resonance frequency* at which the magnetization response is maximum when driven by a *rf* field, as will be shown later. The resonance frequency is proportional to the magnetic field, with a factor that is the gyromagnetic ratio given by Eq. (9.76). For $g = 2$ its value is

$$\gamma = 2\pi \times 28 \text{ GHz/T} \quad \text{(SI)}, \qquad \gamma = 2\pi \times 2.8 \text{ GHz/kOe} \quad \text{(CGS)}.$$

Thus, for typical magnetic fields of magnets or electromagnets, $H = 1$ kOe, the precession frequency is 2.8 GHz, so that the natural response of a ferrite has a frequency situated in the microwave region. This is the reason for the importance of ferrites for microwave devices.

9.6.2 Dynamic Susceptibility of Ferrites

To calculate the response of a ferrite material to microwave radiation, we consider an alternating magnetic field, or *rf* field, with frequency ω, transverse to the static field. The total field is then

$$\vec{H} = (\hat{x}h_x + \hat{y}h_y)e^{-i\omega t} + \hat{z}H, \tag{9.82}$$

where h_x, $h_y \ll H$, since H is of the order of hundreds or thousands of Oe, while the *rf* field of a wave is of the order of fraction of Oe. Substituting Eqs. (9.78) and (9.82) into (9.77), we obtain the equations of motion for the transverse components of the magnetization

$$\frac{dm_x}{dt} = -\gamma \mu_0 m_y H + \gamma \mu_0 M_z h_y e^{-i\omega t}, \tag{9.83}$$

$$\frac{dm_y}{dt} = \gamma \mu_0 m_x H - \gamma \mu_0 M_z h_x e^{-i\omega t}. \tag{9.84}$$

Since we are only interested in the stationary response, we make

$$m_x(t) = m_x e^{-i\omega t}, \qquad m_y(t) = m_y e^{-i\omega t}.$$

Substituting these expressions in (9.83) and (9.84) and making $\omega_0 = \gamma \, \mu_0 H$ we obtain

$$-i\omega m_x = -\omega_0 m_y + \gamma \mu_0 M_z h_y, \tag{9.85}$$

$$-i\omega m_y = \omega_0 m_x - \gamma \mu_0 M_z h_x. \tag{9.86}$$

From these expressions we can write the relationship between the *rf* components of $\vec{M}$ and $\vec{H}$ in the form,

$$\vec{m} = \overline{\chi} \cdot \vec{h}, \tag{9.87}$$

where the vectors $\vec{m}$ and $\vec{h}$ are represented by the column matrices

$$\vec{m} = \begin{pmatrix} m_x \\ m_y \end{pmatrix}, \qquad \vec{h} = \begin{pmatrix} h_x \\ h_y \end{pmatrix}, \tag{9.88}$$

and $\overline{\chi}$ is the *rf* magnetic susceptibility tensor, represented by the square matrix

$$\overline{\chi} = \begin{bmatrix} \chi_{xx} & \chi_{xy} \\ \chi_{yx} & \chi_{yy} \end{bmatrix}, \tag{9.89}$$

where

$$\chi_{xx}(\omega) = \chi_{yy}(\omega) = \frac{\omega_M \omega_0}{\omega_0^2 - \omega^2}, \tag{9.90}$$

$$\chi_{yx}(\omega) = -\chi_{xy}(\omega) = i\frac{\omega_M \omega}{\omega_0^2 - \omega^2}, \tag{9.91}$$

where $\omega_M \equiv \gamma\mu_0 M_z \approx \gamma\mu_0 M$. Note that in the Gaussian system, $\omega_M = \gamma 4\pi M$, because $\mu_0 = 1$ and 4π is the factor that enters in the relationship between the permeability and the susceptibility, Eq. (9.7). This result shows that in a ferrite, the application of a *rf* field in the *x* direction, produces *rf* components of the magnetization in both *x* and *y* directions. Likewise, a h_y field produces m_x and m_y components. This is due to the fact that the natural motion of the magnetization is the precession about the *z*-axis. Thus, the application of a *rf* field either in the *x* or in the *y* direction, h_x or h_y, produces the precession motion, and consequently the components m_x and m_y. For this reason, the relationship between $\vec{m}$ and $\vec{h}$ is not a scalar, but a tensor.

Equations (9.90) and (9.91) indicate that the amplitude of the magnetization response increases rapidly as the driving frequency ω approaches ω_0, characteristic of a resonance process. In fact, the susceptibilities diverge at $\omega = \omega_0$, which is an unphysical situation, showing that relaxation cannot be neglected. Here we introduce the magnetic relaxation, or damping, in a phenomenological manner. This can be done by introducing an imaginary part in the resonance frequency, that corresponds to multiplying the amplitudes in Eq. (9.80) by an exponential function that decays in time. Thus, replacing ω_0 in Eqs. (9.90) and (9.91) by $\omega_0 - i\eta$, where η is the magnetic relaxation rate, the components of the *rf* susceptibility tensor become, approximately,

$$\chi_{xx}(\omega) = \chi_{yy}(\omega) = \frac{\omega_M \omega_0}{\omega_0^2 - \omega^2 - i2\omega_0\eta}, \tag{9.92}$$

$$\chi_{yx}(\omega) = -\chi_{xy}(\omega) = i\frac{\omega_M \omega}{\omega_0^2 - \omega^2 - i2\omega_0\eta}, \tag{9.93}$$

where we have assumed that $\eta << \omega_0$, that means small damping.

This result shows that the magnetic response of a ferrite behaves analogously to the electric susceptibility of an atom subjected to an electromagnetic radiation, studied in Sects. 8.2.2 and 8.3.2. The difference between the two situations is that, while in the atom the frequency ω_0 is in the optical region of the spectrum, in the magnetic case ω_0 is in the microwave region. The analogy between the two situations also allows for a quantum interpretation for the magnetic effect in ferrites. The motion of the magnetization corresponds to quantum transitions between two energy levels split by the magnetic field. In Eq. (9.18), we see that the energy separation between two neighboring levels is $\Delta E = g\mu_B B$. This energy corresponds to photons with frequency $\omega_0 = \Delta E/\hbar = \gamma B$, which is precisely the precession frequency (9.81).

An important aspect of the ferrite response to a microwave field, is that the precession frequency ω_0 varies linearly with the field H. This allows one to tune the ferrite response to a desired frequency. Figure 9.33 shows the real and imaginary parts of the diagonal susceptibility component χ_{xx} as a function of the applied field H, for a fixed frequency $\omega/2\pi = 9.8$ GHz. The other parameters used to make the plots in Fig. 9.33 are: $g = 2$; $\omega_M/\gamma = 0.3$ T (3 kOe); and $\eta/\omega = 0.04$. The reason for making the plot as a function of field, and not frequency, is that usually magnetic resonance experiments are carried out with the sample inside a microwave cavity to

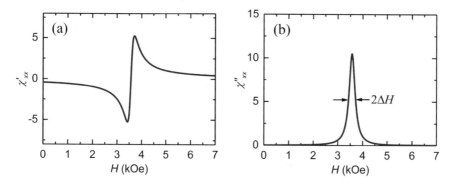

Fig. 9.33 Real (a) and imaginary (b) parts of the diagonal susceptibility tensor calculated with $\omega/2\pi = 9.8$ GHz, $g = 2$, $\omega_M/\gamma = 0.3$ T (3 kOe) and $\eta/\omega = 0.04$

enhance the microwave field. Thus, the frequency is kept fixed and the applied field is swept.

Note the similarity between this figure and Fig. 8.5 which shows the real and imaginary parts of the electric permittivity. The imaginary part $\chi_{xx}^{''}$ has the form of a Lorentzian function, and it is related to the microwave power absorbed by the ferrite. As shown in Fig. 9.33, when the field H is scanned and its value coincides with ω/γ, $\chi_{xx}^{''}$ is maximum and so is the absorbed microwave power. The phenomenon by which the absorbed power grows sharply and goes through a maximum at $H = \omega/\gamma$ is called **ferromagnetic resonance** (FMR). The resonance field is the one for which $\omega = \omega_0$. For a frequency of 9.8 GHz, the resonance field is 3.5 kOe.

The difference between the two field values for which $\chi_{xx}^{''}$ is half the peak value is called *full linewidth*, while one half of which is called *half-width at half maximum* (HWHM), or simply *linewidth*, denoted by ΔH. From Eq. (9.92) we can readily obtain a relation between the linewidth and the relaxation rate. For $\eta << \omega_0$ it is

$$\Delta H = \eta/\gamma. \tag{9.94}$$

Thus, the measurement of the linewidth in a magnetic resonance experiment gives information about the magnetic damping. The linewidth of the sample with susceptibility shown in Fig. 9.32 is $\Delta H = 140$ Oe, which is typical of polycrystalline ferrite materials.

In the ferromagnetic resonance mode, all spins in the medium precess about the magnetic field with the same phase, *i.e.*, all parallel to each other. Since the spins interact by means of the exchange interaction, they can also precess with a phase that varies in space, as in a wave. This constitutes a **spin wave**, that will be presented in the next section. Like other waves studied earlier, spin waves are quantized, and their quanta are called **magnons**. Thus, the FMR mode corresponds to a spin wave with zero wave number, or $k = 0$ magnons.

9.6.3 Electromagnetic Waves in Ferrites

The characteristics of an electromagnetic wave propagating in a ferrite medium subjected to a static magnetic field H are determined by Maxwell's Eqs. (2.1)–(2.4), with the magnetic permeability obtained with the results of the previous section. Since the susceptibility of a ferrite is a tensor, the permeability, defined by Eqs. (9.6) and (9.7), is also a tensor. In the SI we have

$$\bar{\mu} = \mu_0(\bar{1} + \bar{\chi}). \tag{9.95}$$

As a result, in addition to having a phase change and attenuation along space, an electromagnetic wave in a ferrite medium can have a change in polarization. The effects of the ferrite response on the wave depend on the directions of propagation and polarization, and also on the difference between the wave frequency and the FMR frequency.

A special important situation is that of propagation along the static field. In this case, the *rf* field $\vec{h}$ is perpendicular to the *z*-axis, and therefore has components h_x and h_y. Let us look at the behavior of circularly polarized waves in this region. With Eqs. (9.88) and (9.89) we obtain

$$m^{\pm} = m_x \pm i m_y = (\chi_{xx} \mp i \chi_{xy})(h_x \pm i h_y), \tag{9.96}$$

where m^+ and m^- represent the magnetizations of clockwise and counter-clockwise (right and left) circularly polarized waves, respectively. This result means that, although the relationship between $\vec{m}$ and $\vec{h}$ is tensorial, in the case of circularly polarized waves the relationship is scalar. Denoting by $b^{\pm}$ and $h^{\pm}$ the components of the circularly polarized fields, with Eqs. (9.3) and (9.96) we have

$$b^+ = \mu_+ h^+, \qquad b^- = \mu_- h^-, \tag{9.97}$$

where

$$\mu_{\pm} = \mu_0 = (1 + \chi_{xx} \mp i \chi_{xy}) \tag{9.98}$$

are the scalar permeabilities for circularly polarized waves. If ω is very different from ω_0, relaxation can be neglected, and from Eqs. (9.90) and (9.91) we obtain

$$\mu_+ = \mu_0 \left(1 + \frac{\omega_M}{\omega_0 + \omega} \right), \tag{9.99}$$

$$\mu_- = \mu_0 \left(1 + \frac{\omega_M}{\omega_0 - \omega} \right). \tag{9.100}$$

This result means that the relationships for waves in media with scalar permeability, as obtained in Chap. 8, apply to circularly polarized waves in ferrites. For example, the wave vectors of these waves have moduli

$$k^{\pm} = \frac{\omega \, \varepsilon^{1/2}}{c} \left(1 + \frac{\omega_M}{\omega_0 \pm \omega} \right)^{1/2}. \tag{9.101}$$

The fact that circularly polarized waves to the right and to the left have different propagation vectors, gives rise to the phenomenon of **Faraday rotation**, illustrated in Fig. 9.34. Consider a linearly polarized wave propagating in the direction of the *H* field, along the *z*-axis. We choose the *x*- axis as the direction of the *rf* field *h* in the plane at $z = 0$. Denoting by h_0 the amplitude of the field at $z = 0$, we have

$$\vec{h}^{\pm}(0, t) = \mathrm{Re} \left[\hat{x} \, h_0 \, e^{-i\omega t} \right] = \hat{x} \, h_0 \cos \omega t$$

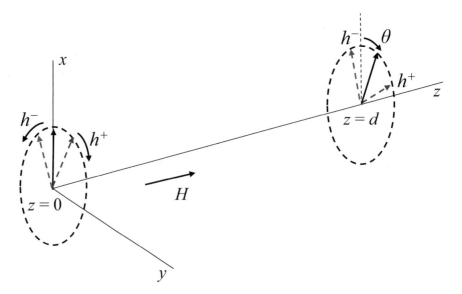

Fig. 9.34 Faraday rotation of an electromagnetic wave propagating in the direction of the magnetic field in a ferrite

It is easy to see that this field can be written as the sum of two circularly polarized components, with amplitudes $h_0/2$, rotating in opposite directions, $\vec{h} = \vec{h}^+ + \vec{h}^-$, where

$$\vec{h}^{\pm}(0, t) = \mathrm{Re}\left[(\hat{x}\, h_0/2 \pm \hat{y}\, i\, h_0/2)e^{-i\omega t}\right] = \hat{x}\,\frac{h_0}{2}\cos\omega t \pm \hat{y}\,\frac{h_0}{2}\sin\omega t. \quad (9.102)$$

Each circularly polarized wave propagates with a different wave vector, so that at the $z = d$ plane we have

$$\vec{h}^{\pm}(d, t) = \mathrm{Re}\left[(\hat{x}\, h_0/2 \pm \hat{y}\, i\, h_0/2)e^{ik^{\pm}d - i\omega t}\right]. \quad (9.103)$$

The field at $z = d$ is given by the sum of the two fields in Eq. (9.103). Its Cartesian components can be written in the form

$$\vec{h}_x(d, t) = \mathrm{Re}\left[h_0\cos\theta\, e^{ik_m d - i\omega t}\right], \quad (9.104)$$

$$\vec{h}_y(d, t) = \mathrm{Re}\left[h_0\sin\theta\, e^{ik_m d - i\omega t}\right], \quad (9.105)$$

where

$$k_m = (k^+ + k^-)/2, \quad (9.106)$$

and

$$\theta = (k^- - k^+) \, d/2. \tag{9.107}$$

Note that Eqs. (9.104) and (9.105) represent a **linearly polarized field**, in a direction at an angle θ with the x-axis. This shows that the composition of the two circularly polarized fields at $z = d$, with phases different than those at $z = 0$, results in another linearly polarized field, but with a polarization direction at an angle θ with the x-axis.

As a result of this process, the original wave propagates through the ferrite maintaining linear polarization, but in a direction that gradually rotates around the static field, in the sense from x to y (since $k^- > k^+$). This is the phenomenon of Faraday rotation. The rotation angle of the polarization direction, given by Eq. (9.107), is proportional to the distance travelled and to the difference between the wave numbers of polarizations $+$ and $-$. It is important to note that the direction of Faraday rotation is defined by the orientation of the H field, and does not depend on the orientation of the wave propagation.

Example 9.4 A microwave radiation with frequency 9.4 GHz propagates along the H field in a ferrite with $M = 250$ emu/cm^3, $g = 2$, $\Delta H = 25$ Oe and $\varepsilon = 4\,\varepsilon_0$. Calculate the wave absorption coefficient for $H = 2.5$ kOe, considering: (a) Circularly polarized wave in the $+$ direction; (b) Circularly polarized wave in the $-$direction.

(a) The wave numbers for the circular polarizations $+$ and $-$ including magnetic relaxation are given by Eq. (9.101), with the substitution of ω_0 by $\omega_0 - i\eta$,

$$k^{\pm} = \frac{\omega \, \varepsilon^{1/2}}{c} \left(1 + \frac{\omega_M}{\omega_0 \pm \omega - i\eta} \right)^{1/2}.$$

The introduction of the imaginary term in this equation results in an imaginary component of the wave number, which produces attenuation in the wave. To calculate the imaginary part, it is necessary to work with the complex number inside the square root, which leads to big expressions in the general case. To simplify the calculation, let us obtain the numerical values for the quantities of interest and compare them. Since the important quantities enter in the equation above in a ratio, let us make explicit the factor 2π that relates the angular frequency with the frequency. Using the CGS system we have

$$\omega_0 = \gamma H = 2\pi \times 2.8 \times 10^6 \times 2.5 \times 10^3 = 2\pi \times 7.0 \text{ GHz},$$

$$\eta = \gamma \Delta H = 2\pi \times 2.8 \times 10^6 \times 25 = 2\pi \times 0.07 \text{ GHz},$$

$$\omega_M = \gamma 4\pi M = 2\pi \times 2.8 \times 10^6 \times 4\pi \times 250 = 2\pi \times 8.8 \text{ GHz}.$$

We see then that the imaginary term in the denominator of the expression for $k^{\pm}$ is much smaller than the real term. We can then use a binomial expansion to obtain the real and imaginary parts of the square root. Thus, we have

$$k^{\pm} = \frac{\omega \varepsilon^{1/2}}{c} \left\{ 1 + \frac{\omega_M}{(\omega_0 \pm \omega)[1 - i\eta/(\omega_0 \pm \omega)]} \right\}^{1/2}$$

$$k^{\pm} \approx \frac{\omega \varepsilon^{1/2}}{c} \left\{ 1 + \frac{\omega_M}{(\omega_0 \pm \omega)} \left[1 + \frac{i\eta}{(\omega_0 \pm \omega)} \right] \right\}^{1/2}.$$

With the binomial expansion of the square-root term we have for the imaginary part of the wave number

$$k^{\pm \prime\prime} \approx \frac{\omega \varepsilon^{1/2}}{c} \frac{\omega_M \eta}{2(\omega_0 \pm \omega)^{3/2}(\omega_0 \pm \omega + \omega_M)^{1/2}}.$$

According to Eq. (8.13), the absorption coefficient is twice the imaginary part of k, therefore

$$\alpha^{\pm} = \frac{\omega \varepsilon^{1/2}}{c} \frac{\omega_M \eta}{(\omega_0 \pm \omega)^{3/2}(\omega_0 \pm \omega + \omega_M)^{1/2}}.$$

Substituting the values of the quantities, we have

$$\alpha^{\pm} = \frac{2\pi \times 6.0 \times 10^9 \times 4^{1/2}}{3 \times 10^{10}} \frac{8.8 \times 0.07}{(7.0 \pm 6.0)^{3/2}(7.0 \pm 6.0 + 8.8)^{1/2}}$$

So, for the $+$ wave we have

$$\alpha^+ = \frac{1.55}{(7.0 + 6.0)^{3/2}(7.0 + 6.0 + 8.8)^{1/2}} = 0.007 \text{ cm}^{-1}.$$

(b)   For the $-$ wave we have,

$$\alpha^- = \frac{1.55}{(7.0 - 6.0)^{3/2}(7.0 - 6.0 + 8.8)^{1/2}} = 0.49 \text{ cm}^{-1}.$$

We see then that the $-$ circularly polarized wave has a much larger absorption coefficient than the $+$ wave, because it is closer to the resonance condition, where the energy loss is much higher.

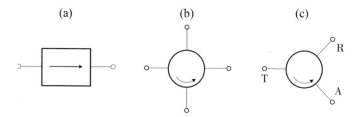

Fig. 9.35 Circuit symbols of non-reciprocal devices: **a** Isolator; **b** 4-port circulator; **c** 3-port circulator

9.6.4 Microwave Ferrite Devices

In this section we shall present, qualitatively, the ferrite devices most used in microwave circuits, namely, isolators, circulators, and YIG filters. These devices are used throughout the microwave region, from 1 to 100 GHz. Each unit operates efficiently only in a narrow range of frequencies, with a bandwidth that depends on the characteristics of the ferrite and the device geometry. Figure 9.35 shows the circuit symbols of the isolator and two circulators.

The isolator is a two-port device, which transmits radiation in one direction and fully absorbs radiation in the opposite direction. It is used in the output of microwave generators, to prevent the reflections from the external circuit to return to the generator and interfere in its operation. Figure 9.35a shows the circuit symbol of an isolator. Since isolators are non-reciprocal devices, their operation depends fundamentally on materials that have non-reciprocal properties. In the microwave region, the materials that have these properties are ferrites. Their origin is based on the gyroscopic behavior of the atomic magnetic moments, whose sense of precession is determined by the orientation of the applied static field.

One of the first built ferrite isolators for microwave waveguides employed the effect of Faraday rotation. Its structure, shown in Fig. 9.36, consists of the following elements: a cylinder of a ferrite material longitudinally magnetized by an external field, such as to produce Faraday rotation of 45°; two resistive plates with their planes at an angle of 45° to each other, placed close to the two ports and parallel to the directions of the larger dimension of the rectangular waveguides; a circular waveguide in which the ferrite cylinder is mounted, with transitions to the rectangular

Fig. 9.36 Schematic illustration of the structure of a waveguide Faraday rotation isolator

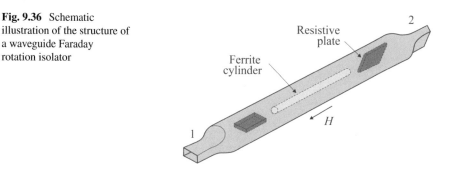

sections of the ports, at an angle of 45° between them. The operation of the device is described as follows.

An electromagnetic wave entering in port 1 passes through the resistive plate without attenuation, because it has electric field perpendicular to the plane of the plate (in the rectangular guide, the electric field has the direction of the smallest dimension), and therefore does not produce current in the plate. Since the Faraday rotation of the ferrite rod is 45° clockwise, as the wave reaches port 2 it is transmitted to the rectangular waveguide. However, a wave in the opposite direction, entering in port 2, has its polarization rotated 45° in the same sense, so that it is partially absorbed by the resistive plate near port 1 and also has polarization perpendicular to the one that is supported by the rectangular waveguide. Thus, the device transmits the microwave in the direction $1 \rightarrow 2$, but not in the opposite direction. The Faraday rotation isolator is not used in low microwave frequency ranges because the waveguides have sizeable dimensions. However, an analogous device, based on the magneto-optical Faraday effect, is used in the near infrared region in optical fiber communication systems.

The ferrite isolator most used in microwave equipment in ground-based stations has operation that relies on the ferromagnetic resonance phenomenon. The structure of the resonance isolator for waveguides, illustrated in Fig. 9.37, consists of a section of a waveguide, a ferrite plate located in a certain plane of the waveguide, and a permanent magnet used to magnetize the ferrite and determine the resonance frequency ω_0. In the rectangular waveguide the microwave magnetic field has two components in the x–y plane, 90° out of phase and with amplitudes that vary along the x-direction. As a result, there are two planes of the waveguide, P1 and P2 shown in Fig. 9.37, symmetric about to the middle plane, in which the fields are circularly polarized, one of them to the right and the other to the left. The distances from P1 and P2 to the side walls are determined by the frequency of the wave and the dimensions of the waveguide (Problem 9.16). The isolator operates in a frequency range around ω_0. For a wave propagating in a certain direction along the guide, the circular polarization in the plane P1, where the ferrite is located, has the sense opposite to magnetization precession. In this case there is no ferromagnetic resonance and the wave is transmitted with small attenuation. However, when wave propagates in the opposite direction, the senses of the circular polarizations in P1 and P2 are inverted, producing attenuation due to the resonance absorption in the ferrite. To increase the absorption, it is common to place a thin resistive plate next to the ferrite. The

Fig. 9.37 Structure of the resonance isolator for rectangular waveguides

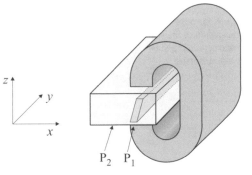

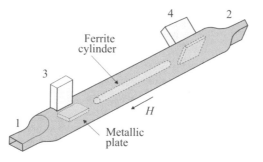

Fig. 9.38 Schematic illustration of the Faraday rotation 4-port circulator

resonance isolators have simple construction and can have a ratio between the transmissions in the two directions up to 10^4. They are also made with coaxial cables and miniaturized microstrips used in devices for portable equipment.

Another important microwave ferrite device is the **circulator**, also called gyrator. It is a device with three or more ports, which transmits the radiation that enters in a certain port to a neighboring one, in a "one-way" sense. Figure 9.35b and c show the circuit symbols of 3- and 4-port circulators, respectively. An important application of the 3-port circulator is in systems of transmission and reception with a single antenna. As shown in Fig. 9.35b, the circulator causes the radiation from the transmitter (T) to be directed to the antenna (A). On the other hand, the radiation received by the antenna is directed to the receiver (R). Since the direction of the gyrator is one-way, the signal from the transmitter is not carried to the receiver.

Like with the isolator, the first 4-port circulator was based on the Faraday rotation. Its components, shown in Fig. 9.38, are: a ferrite cylinder placed in a circular waveguide section, with 45° Faraday rotation; four ports made of sections of rectangular waveguides, with port 2 at an angle of 45° from port 1, port 3 perpendicular to 1, and port 4 at 45° of port 3. With this arrangement, radiation entering at any port is rotated by 45° and leaves from the next port, with the other ports maintained isolated. The circulator of Fig. 9.38 is no longer used due to the difficulty of its miniaturization. Figure 9.39 shows a 3-port microstrip line circulator, used in miniaturized microwave circuits. Its operation is also based on gyroscopic properties of the ferrite disc, but

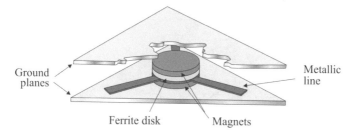

Fig. 9.39 Illustration of a 3-port microstrip line ferrite circulator

the configuration of the fields in the disk is complex and will not be analyzed here. Note that by placing a matched load in one port, the 3-port circulator behaves as an isolator. With the structure in Fig. 9.39, these non-reciprocal devices can be fabricated in thin-film batch-processed miniature microwave hybrid circuits that are use in a variety of microwave components and subsystems.

Another magnetic device used in microwave circuits is the **YIG filter**. As we saw in Sect. 9.4, YIG is a ferrimagnetic garnet, not a ferrite. However, its classification in the category of ferrites is justified by the similarity of their magnetic properties. The ferrites used in isolators and circulators are polycrystalline ceramics, with FMR linewidths of tens or hundreds of Oersted. The large linewidths are necessary for devices that operate in wide frequency bands. On the other hand, in filters one uses single-crystal YIG, in the form of highly polished spheres, that have linewidth of about 0.1 Oe. Since at $\omega = \omega_0$ the susceptibilities in Eqs. (9.92) and (9.93) have amplitude $\omega_M/2\eta = M/2\Delta H$, the small ΔH makes the YIG resonance very intense. Notice that in a ferrite with the curves shown in Fig. 9.33, the peak of the imaginary susceptibility is about 10, while in YIG the peak is $M/2\Delta H \approx 10^4$. These properties allow the construction of narrow-band transmission filters, electrically tunable, and other devices.

Figure 9.40 shows the basic structures of YIG filters with one and two stages. The electromagnetic structure of each stage consists only of two fine wire half-loops, or half-rings, perpendicular to each other. At the center of the loops there is small single-crystal YIG sphere, with diameter of the order 0.5 mm, subjected to the static magnetic field H of an electromagnet. The field value is adjusted by the current in the electromagnet. The microwave current in one of the half- loops creates a *rf* magnetic field on the sphere, perpendicular to the static field. If the microwave frequency ω is away from $\omega_0 = \gamma H$, the susceptibility is negligible, so that m_x and m_y are negligible. Since the half-loops are in planes perpendicular to each other, in this situation the signal transmitted from one to the other is also negligible. When $\omega \approx \omega_0$, the field h produced by one of the loops creates in the YIG sphere a *rf* magnetization in the x–y plane, given by Eqs. (9.88)–(9.91). The component of the *rf* magnetization perpendicular to the other loop induces an output signal, proportional to the input signal. Thus, the device operates as a narrowband transmission filter. Since $\omega_0 = \gamma H$, the filter tuning is made through the current in the electromagnet. The response curve of the one-stage filter, consisting of the amplitude of the output

Fig. 9.40 Schematic diagrams of YIG filters: **a** one stage; **b** two stages

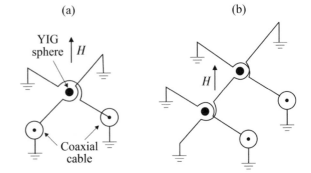

signal as a function of frequency, has the shape of the resonance curve. Other forms of the response curve appropriate for specific transmission filters, can be synthesized in devices using multiple stages.

Figure 9.40b shows the diagram of a two-stage filter. Two YIG spheres are mounted on the same structure and are subject to the same static field. In this way, when the static field is varied, the resonance frequencies of the two spheres vary equally. The synthesis of the filter response curve is possible because the resonance frequency of each sphere varies finely with the direction of its crystalline axes relative to the external field. Since the transmission curve of the filter is the product of the responses of the two stages, it is possible to vary the form of the curve by finely adjusting the crystal orientation of one sphere relative to another. Figure 9.41 shows the response curve of a two-stage filter built at the Physics Department of UFPE. The filter has a bandwidth of 15 MHz and can be tuned in the frequency range of 4 to 6 GHz.

YIG filters find various applications in microwave equipment, in functionalities that require electronic tuning. They are used in the input stages of simple tunable receivers and in intermediate stages of superheterodyne receivers. They are also employed to stabilize the frequency of microwave oscillators, such as the Gunn diode, or oscillators with GaAs or GaN MESFET transistors, with the advantage of allowing the electronic tuning of the frequency.

The interaction between a microwave magnetic field and the spin precession in a YIG sphere enables the coherent transfer of information between the electromagnetic and spin subsystems. This is particularly interesting if the microwave structure is a resonant cavity, because in this case one has two strongly coupled oscillators of quite different natures, constituting a hybrid photon-magnon system. Since the spin-wave resonance frequency can be easily tuned by the applied magnetic field, this system has a unique advantage over other hybrid coupled-oscillators systems. In recent years, the investigation of hybrid dynamic systems involving YIG resonators has attracted great attention due to their possible applications in quantum computation, information processing, and sensing. Figure 9.42a shows the image a photon-magnon

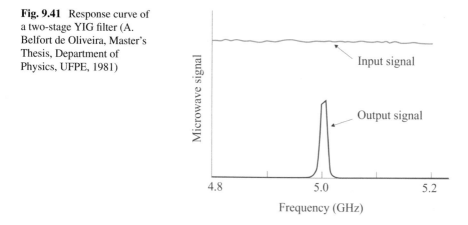

Fig. 9.41 Response curve of a two-stage YIG filter (A. Belfort de Oliveira, Master's Thesis, Department of Physics, UFPE, 1981)

(a)   (b)

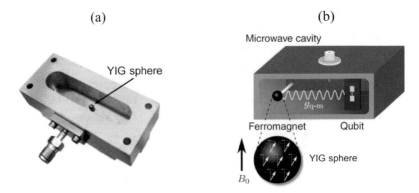

Fig. 9.42 Illustration of two hybrid systems employing a YIG-sphere magnon resonator: (a) Image of a microwave cavity, without the top cover, with a YIG sphere at the position of maximum *rf* magnetic field. Reproduced with permission from X. Zhang et al., Phys. Rev. Lett. 113, 156401 (2014). Copyright (2014) by the American Physical Society; (b) YIG sphere and a superconducting qubit inside a microwave cavity. The qubit and the FMR mode interact through virtual excitations in the cavity modes. Reproduced from D. Lachance-Quirion et al., Sci. Adv. **3**, 1603150 (2017), with permission of the American Association for the Advancement of Science

system employing a small YIG sphere. The microwave cavity made of copper is shown without the cover. The YIG sphere is placed in a position of maximum *rf* magnetic field of the operating mode, and is subject to an external DC magnetic field that can be varied by the current in the electromagnet. The cavity is excited through the coaxial cable connector, that is also used to measure the response of the photon-magnon system. This simple system has been shown to have great potential for several types of information processing operating at room temperature.

Figure 9.42b shows another interesting hybrid system that consists of a YIG sphere in a position of maximum *rf* magnetic field of a microwave cavity and a superconducting qubit in a position of maximum *rf* electric field. As will be shown in Sect. 10.4.4, the qubit is made of superconducting Josephson junctions that have quantized magnetic flux, which can be in any superposition of the 1 and 0 quantum states. In this device, that only operates at very low temperatures, the qubit and the FMR mode of the YIG sphere interact through virtual photon excitations in the cavity modes. The two systems in Fig. 9.42 provide a new paradigm for combining platforms and devices that can perform different tasks such as storing, processing, and transmitting coherent information.

9.7 Magnonics: Concepts and Perspectives for Device Applications

As mentioned in the previous section, the ferromagnetic resonance mode in which all spins in the medium precess about the magnetic field with the same phase corresponds to a spin wave with zero wave number. As we did in Sect. 2.2 to introduce elastic waves, we consider here a simplified one-dimensional model of a spin system, consisting of a linear chain of classical identical spins, illustrated in Fig. 9.43. Consider that the chain has N spins with magnitude S, uniformly spaced and separated by a distance a, coupled with the nearest neighbors by the exchange interaction in Eq. (9.31). The total exchange energy of the chain is given by

$$U_{exc} = -2J \sum_i \vec{S}_i \cdot \vec{S}_{i+1}, \tag{9.108}$$

where J is the nearest neighbor exchange parameter and $\vec{S}_i$ denotes the classical spin at the coordinate $x_i = ia$. In the ground state all spins are parallel, as in Fig. 9.43a, so that with $\vec{S}_i \cdot \vec{S}_{i+1} = S^2$ the exchange energy of the system becomes $U_0 = -2JNS^2$.

In the early studies of the magnetic properties of matter, Pierre Weiss considered that the first excited state consisted of one localized spin reversed, as in Fig. 9.43b. In this case, the exchange energy of the linear chain becomes $U_1 = U_0 + 8JS^2$. Although the Weiss molecular field model explained well the overall temperature dependence of the magnetization in ferromagnets, in 1930 Felix Bloch showed that the low-lying excitations of the spin system consisted of nonlocalized, collective spin deviations. Bloch called these excitations **spin waves**, and showed that they dominate the magnetic thermodynamics at low temperatures.

We treat spin waves in the linear chain of spins with motion governed by the classical mechanical equation for the torque. The torque acting on the spin $\vec{S}_i$ with its associated magnetic moment $\vec{\mu}_i = -g\mu_B \vec{S}_i$, given by Eq. (9.73), is written in the form $\vec{\tau} = \vec{\mu}_i \times \vec{B}_T$, where $\vec{B}_T$ is an effective field representing all interactions on the spin $\vec{S}_i$. This field can be found considering that the energy of the moment has the form $U_i = -\vec{\mu}_i \cdot \vec{B}_T$. Comparison with Eq. (9.108) shows that in the linear spin chain the effective field arising from the exchange interaction is

$$\vec{H}_{exc}^{eff} = -\frac{2J}{g\mu_B\mu_0} (\vec{S}_{i-1} + \vec{S}_{i+1}). \tag{9.109}$$

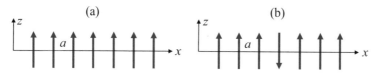

Fig. 9.43 a Linear chain of classical spins in the ground state. **b** Linear chain of spins in an excited state with one spin reversed

We consider that in addition to the exchange interactions with their neighbors, the spins are subject to an applied static magnetic field $\vec{H}$, so that the total field acting on the spins is $\vec{B}_T = \mu_0(\vec{H} + \vec{H}_{exc}^{eff})$. Using the torque equation $d\hbar\vec{S}/dt = \vec{\tau}$ we obtain the equation of motion for the spin $\vec{S}_i$

$$\frac{d\vec{S}_i}{dt} = -\gamma\mu_0 \vec{S}_i \times (\vec{H} + \vec{H}_{exc}^{eff}), \tag{9.110}$$

where γ is the gyromagnetic ratio, defined in Eq. (9.76). Since the magnetic moment has a direction opposite to the spin, to be consistent with the ground state depicted in Fig. 9.43a, we consider that the external field is applied in the $-\hat{z}$ direction, so that $\vec{H} = -\hat{z}H$. Then the equation for the spin component S_i^x becomes

$$\frac{dS_i^x}{dt} = \gamma\mu_0 S_i^y[H + \frac{2J}{g\mu_B\mu_0}(S_{i-1}^z + S_{i+1}^z)] - \gamma\mu_0\frac{2J}{g\mu_B\mu_0}(S_{i-1}^y + S_{i+1}^y) S_i^z.$$

Considering that the amplitude of the spin excitation is small, we linearize this equation using S_i^x, $S_i^y \ll S_i^z \approx S$. Then the equations for the two transverse spin components become

$$\frac{dS_i^x}{dt} = \gamma\mu_0 H S_i^y + \frac{2JS}{\hbar}(2 S_i^y - S_{i-1}^y - S_{i+1}^y), \tag{9.111}$$

$$\frac{dS_i^y}{dt} = -\gamma\mu_0 H S_i^x - \frac{2JS}{\hbar}(2 S_i^x - S_{i-1}^x - S_{i+1}^x). \tag{9.112}$$

Equations (9.111) and (9.112) show that the motion of the spin in any site is coupled to the motions of the neighboring spins, indicating that their solutions must be collective excitations. Consider for possible solutions spin excitations in the form of harmonic travelling waves

$$S_i^x = A_x e^{i(kx_i - \omega t)}, \qquad S_i^y = A_y e^{i(kx_i - \omega t)}, \tag{9.113}$$

where ω is the angular frequency and k is the wave number. Substitution of (9.113) into Eq. (9.111) leads to

$$-i\omega A_x = A_y [\gamma\mu_0 H + \frac{2JS}{\hbar}(2 - e^{-ika} - e^{ika})], \tag{9.114}$$

where we have cancelled the term $\exp i(kx_i - \omega t)$ on both sides. Equation (9.114) can be written as

$$-i\omega A_x = A_y[\gamma\mu_0 H + \frac{4JS}{\hbar}(1 - \cos ka)]. \tag{9.115}$$

Similarly, we obtain from Eq. (9.112)

$$-i\omega A_y = -A_x[\gamma H + \frac{4JS}{\hbar}(1 - \cos ka)]. \tag{9.116}$$

Equations (9.115) and (9.116) can be written in matrix form.

$$\begin{bmatrix} i\omega & [\gamma\mu_0 H + \frac{4JS}{\hbar}(1 - \cos ka)] \\ -[\gamma\mu_0 H + \frac{4JS}{\hbar}(1 - \cos ka)] & i\omega \end{bmatrix} \begin{pmatrix} A_x \\ A_y \end{pmatrix} = 0. \tag{9.117}$$

The solution of Eq. (9.117) is obtained by equating the main determinant to zero, which gives for the frequency

$$\omega_k = \gamma\mu_0 H + \frac{4JS}{\hbar}(1 - \cos ka). \tag{9.118}$$

This is the dispersion relation for spin waves in the linear spin chain. Substitution of Eq. (9.118) in either (9.115) or (9.116) gives the relation between the amplitudes of the spin components $A_y = -iA_x = -iA_0$. Thus, the real parts of the transverse spin components become

$$S_i^x = A_0 \cos(kx_i - \omega_k t), \qquad S_i^y = A_0 \sin(kx_i - \omega_k t), \tag{9.119}$$

while the longitudinal component is $S_i^z \approx S$. These equations show that the classical picture of a spin wave in one dimension consists of spins precessing circularly about the equilibrium direction, as illustrated in Fig. 9.44 at a certain instant of time. In a travelling wave propagating in the $+x$ direction, the spin precession has the same amplitude along the chain and has a phase that varies with the position as $\phi_i = kx_i$. The shortest distance between two spins precessing with the same phase corresponds to the wavelength, related to the wave number by $\lambda = 2\pi/k$.

A quantum formulation of this problem shows that spin waves are quantized. Their quanta are called **magnons**, that have energy $\hbar\omega_k$. For the one-dimensional chain just studied the magnon frequency is

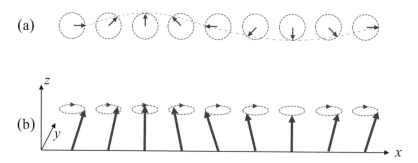

Fig. 9.44 Illustration of a spin wave in a linear chain of classical spins propagating in the $+x$ direction. The distance between the two spins at the ends corresponds to one wavelength. **a** Top view of the spins. **b** Side view

$$\omega_k = \gamma \mu_0 H + \frac{\omega_{ZB}}{2}(1 - \cos ka). \tag{9.120}$$

where $\omega_{ZB} = 4\,J\,S/\hbar$ is the zone boundary frequency in the absence of applied field. This equation shows that for $k = 0$ the magnon frequency is determined only by the magnetic field intensity $\omega_0 = \gamma \mu_0 H$, which is the FMR frequency in Eq. (9.81). This is due to the fact that if all spins precess in phase, there is no contribution from the exchange energy. As the wave number increases, the difference in the precession phase of neighboring spins increases and so does the exchange energy.

Figure 9.45a shows the magnon dispersion relation over the positive side of the first Brillouin zone. For magnetic field intensities typical of laboratories, the frequency gap at $k = 0$ is many orders of magnitude smaller than the frequency for $ka = \pi$, which is situated in the terahertz range, and cannot be seen in the plot. Spin waves with small wave numbers are important for many magnonic phenomena. Using the binomial expansion of the exchange term in Eq. (9.120) for $ka \ll 1$ we obtain the dispersion relation for small wave number magnons

$$\omega_k \approx \gamma \mu_0 (H + Dk^2), \tag{9.121}$$

where

$$D = \frac{2J\,Sa^2}{\gamma\,\hbar\mu_0} \tag{9.122}$$

is called exchange parameter. Equation (9.121) obtained for a linear spin chain, is actually valid for a 3-dimensional spin system with only the exchange interaction. It reveals that magnons have a quadratic dispersion relation, similar to electrons in an uniform potential, Eq. (3.33), as shown in Fig. 9.45b. The unique feature of magnons, that is important for many applications, is the fact that their frequencies can be easily tuned by the applied magnetic field.

The field of science and technology that employs spin waves for information transport and processing is known as **magnonics, or magnon spintronics**. Besides the frequency tuneability, spin waves have other unique features that make them

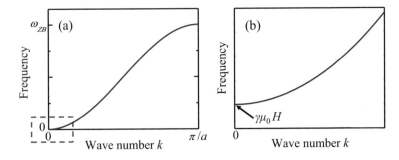

Fig. 9.45 Dispersion relation for magnons in a linear chain. **a** View of the full Brillouin zone. **b** Zoom near the origin ($ka \ll 1$)

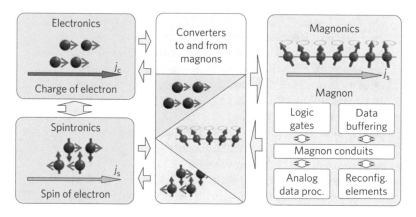

Fig. 9.46 Building blocks of magnon spintronics. Information encoded into charge or spin currents is converted into magnon packets that are processed in the magnonic system and converted back. Reproduced from A. V. Chumak, V. I. Vasyuchka, A. A. Serga, and B. Hillebrands, Nat. Phys. **11**, 453 (2015), with permission by Springer-Nature

suitable for information processing. One of them is that spin-wave packets can be easily excited in magnetic films by microwave current pulses in a simple wire antenna and propagate over distances of hundreds nanometers or several micrometers. The most commonly used materials are single-crystal YIG films that have very small damping, and metallic permalloy films that have higher damping than YIG but are suitable for microsized patterning. The scheme in Fig. 9.46 shows a few magnonic analog and digital devices and the interfaces with converters to and from electronic and spintronic carriers. Several magnonic devices have been demonstrated in research laboratories but they are still not available in commercial products.

Problems

9.1 (a) From the definition (9.9) of the angular momentum operator, obtain the expressions for the L^2 and L_z operators. b) Show that the orbitals ψ_{110} and ψ_{111} of the hydrogen atom are eigenstates of the L^2 and L_z operators and that the eigenvalue equations satisfy the relations (9.10) and (9.11).

9.2 Apply Hund's rules to obtain the ground states of the ions Ni^{2+} and Eu^{2+} and calculate the corresponding g-factors.

9.3 An atom with $S = 1/2$ and $L = 0$ is placed in a magnetic field $H = 2$ kOe. Calculate the frequency, in GHz, of the photon emitted in a magnetic dipole transition between the two spin states in the field.

9.4 A paramagnetic sample has 4×10^{22} cm^{-3} ions with $S = 2$ and $L = 0$. Calculate the magnetic susceptibility of the sample at $T = 4$ K and $T = 300$ K.

9.5 Iron crystallizes in the bcc structure, with a lattice parameter $a = 2.87$ Å. Considering that the magnetic moment of iron is 2.22 μ_B per atom, calculate its saturation magnetization and compare with the value in Table 9.2.

9.6   Iron has a Curie temperature $T_c = 1{,}043$ K. Calculate its molecular field and the J_1 exchange constant.

9.7   A permanent magnet has the shape of a cylindrical rod with diameter 2 cm and length 10 cm, with a uniform magnetization, parallel to the axis, with value $4\pi M = 15$ kG. Calculate the field created by the magnet at three points on its axis, 1 mm, 10 mm, and 50 mm apart from one of the cylinder bases.

9.8   Consider a permanent magnet in the form of a thin disk, with uniform magnetization M perpendicular to the plane. Calculate the fields B and H at a point external to the disk, close to the surface of the north pole. Locate the operating point of the magnet in the curve of Fig. 9.21.

9.9   An electromagnet has a magnetic circuit like the one in Fig. 9.22, with a cylindrical core made of iron, with diameter 10 cm, length 120 cm, and air gap distance 5 cm. Considering a winding with 200 turns, calculate the magnetic field at a point close to the center of the air gap surface, produced by a current of 10 A.

9.10   Show that the functions ψ_1, ψ_2, and ψ_3, given by Eqs. (9.63)-(9.65), satisfy the Poisson Eq. (9.53) for the magnetic potential in the geometry of Fig. 9.25, which represents a magnetic tape with a recorded sinusoidal signal.

9.11   (a) Calculate the wavelength of a 1 kHz signal, recorded on a magnetic tape with speed 20 cm/s.
         (b) Calculate the distance of a point to the tape at which the magnetic field created by the tape is 5% of the value on its surface.

9.12   A magnetic tape with magnetization $M = 500$ emu/cm^3, thickness 15 μm, and track width 1 mm, slides at a speed of 20 cm/s under an induction reading head with 20 turns, efficiency 0.8, at a distance of 2 μm. Calculate:

   (a)   The signal frequency for which the reproduction amplitude is maximum;
   (b)   The value of the output signal in this condition.

9.13   From Eqs. (9.85) and (9.86), show that the *rf* magnetic susceptibility tensor of a ferrite is given by Eqs. (9.90)–(9.91).

9.14   An electromagnetic wave with frequency 9.8 GHz propagates along the field H, in a ferrite with parameters $M = 300$ emu/cm^3, $g = 2$, $\Delta H = 50$ Oe, $\varepsilon = 4\varepsilon_0$. (a) Considering that the wave is circularly polarized in the $-$ direction, calculate its absorption coefficient for $H = 1$ kOe and $H = 3.5$ kOe.
         (b) Calculate the absorption coefficient for the same fields as in item a), for a wave circularly polarized in the $+$ direction.

9.15   If the wave in the previous problem is linearly polarized, calculate the Faraday rotation angle, in rad/cm, for the two field values given.

9.16   The microwave magnetic field propagating in a rectangular waveguide, in the fundamental mode, has two components:

$$h_x = -\frac{ik_g a}{\pi} h_0 \sin\left(\frac{\pi x}{a}\right) e^{ik_g z - i\omega t},$$

$$h_z = h_0 \cos\left(\frac{\pi x}{a}\right) e^{ik_g z - i\omega t},$$

where z is the direction along the guide, x is the large transverse direction, a is the width of the waveguide (in the x direction), h_0 is the amplitude of the longitudinal field and k_g is the wave number in the longitudinal direction, given by

$$k_g = \frac{1}{c}[\omega^2 - (\pi c/a)^2]^{1/2},$$

where c is the speed of light. Considering that a X-band waveguide has dimension $a = 2.3$ cm, determine the distance from the side wall where a ferrite plate should be placed for the device to operate as an isolator for 9.4 GHz.

9.17 Calculate the group velocity of a spin-wave packet propagating in a single-crystal YIG sample with a wave number peak at $k = 6 \times 10^5$ cm^{-1}, considering that YIG has an exchange parameter $D = 5.4 \times 10^{-9}$ Oe cm^2 and gyromagnetic ratio $\gamma = 2\pi \times 2.8 \times 10^6$ s^{-1} Oe^{-1}.

Further Reading

H.N. Bertram, *Theory of Magnetic Recording* (Cambridge University Press, Cambridge, 1994)

S. Blundell, *Magnetism: A Very Short Introduction* (Oxford Univ. Press, Oxford, 2012)

R.L. Comstock, *Introduction to Magnetism and Magnetic Recording* (Wiley, New York, 1999)

R.E. Hummel, *Electronic Properties of Materials* (Springer-Verlag, Berlin, 2011)

C. Kittel, *Introduction to Solid State Physics* (Wiley, New York, 2004)

B. Lax, K. Button, *Microwave Ferrites and Ferrimagnetics* (McGraw-Hill Book Co., New York, 1962)

D. Jiles, *Introduction to Magnetism and Magnetic Materials* (CRC, Boca Raton, 2015)

A.H. Morrish, *The Physical Principles of Magnetism* (IEEE Press, New York, 2001)

D.M. Pozar, *Microwave Engineering* (Wiley, New York, 2012)

S.M. Rezende, *Fundamentals of Magnonics. Lecture Notes in Physics,* Vol. 969 (Springer, Cham, 2020).

R.M. White, *Introduction to Magnetic Recording* (IEEE Press, New York, 1985)

Chapter 10
Other Important Materials for Electronics

In this chapter we present the basic physical properties and some applications of certain materials important for electronics, not studied in the previous chapters. Dielectric materials find a variety of applications in electronics since its emergence in the beginning of the twentieth century. In the last decades these applications have become more diverse and sophisticated with the discovery of new materials and phenomena and the development of optoelectronics and photonics. These dielectric materials and their applications are the subject of the first two sections. Then we describe basic properties of phosphorescent ceramics, liquid crystals and organic conducting materials, that find increasingly sophisticated applications in displays and screens of video monitors. The last section is devoted to superconductivity, the main properties and basic physics. Superconducting materials already have important applications and a great potential for many more, but the full realization of this potential still depends on the development of new materials and technologies. In the last section we briefly introduce their application in quantum computers.

10.1 Dielectric Materials

As we saw in Chap. 4, materials with a large energy gap between the valence and conduction bands do not have electrons in this band and, therefore, they are electrical insulators. Insulators are of great importance for electronics, since they are necessary to assemble or electrically isolate wires and parts of devices and circuits. The most used materials in these applications are ceramics of inorganic oxides, resins, and a wide variety of polymeric materials commonly called plastics. However, free electrons are not the only ones responsible for the response the materials to external electric fields. In general, insulators have ions or molecules that, under the action of an external field, undergo small displacements or reorientations. In this way, even without producing electric currents, these materials present a response to an electric

© The Author(s), under exclusive license to Springer Nature Switzerland AG 2022
S. M. Rezende, *Introduction to Electronic Materials and Devices*,
https://doi.org/10.1007/978-3-030-81772-5_10

field. They are called **dielectric materials**, and find various specific applications in electronics.

10.1.1 Polarization of Materials

The behavior of dielectric materials in an external electric field is determined by the properties of their microscopic **electric dipoles**. These dipoles can be permanent, or induced by the external electric field. They are produced by the separation between the positively charged nuclei and the negative electrons in the atoms, ions, or molecules that form the material. Materials that have permanent microscopic electric dipoles are called **polar**, while those that do not have permanent dipoles are **nonpolar**. When the material is submitted to an external electric field, this exerts opposite forces on positive and negative charges, so that the dipoles are oriented as illustrated in Fig. 10.1. As a result, the dipoles create a field that superimposes to the external field and determines the dielectric response of the material. The electric dipole created by separating two charges of opposite signs, $\pm q$, distant from each other by a displacement vector $\vec{d}$, has a dipole moment.

$$\vec{p} = q\vec{d}. \tag{10.1}$$

Macroscopically, the quantity that represents the dielectric state of a material is the polarization vector $\vec{P}$, defined analogously to the magnetization vector, that is, as the electric dipole moment per unit volume

$$\vec{P} = \frac{1}{V} \sum_i \vec{p}_i, \tag{10.2}$$

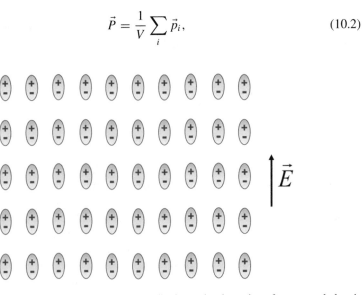

Fig. 10.1 Orientation of the microscopic electric dipoles under the action of an external electric field

where the summation is carried out over all points i where the microscopic dipoles are located inside a volume V. As in the magnetic case, the volume should be large enough for a good macroscopic media, but small relative to the sample size, so that $\vec{P}$ represents a local property. Recall that $\vec{P}$ is related to the electric field vector $\vec{E}$ and the displacement vector $\vec{D}$ through relations that depend on the system of units. In the SI

$$\vec{D} = \varepsilon_0 \vec{E} + \vec{P}, \tag{10.3}$$

where $\varepsilon_0 = (4\pi \times 9 \times 10^9)^{-1}$ C^2/Nm^2 is the permittivity of vacuum. Note that the unit C^2/Nm^2 is equivalent to farad/meter. The unit of E is V/m, while the unit of D and P is C/m^2. In the CGS $\varepsilon_0 = 1$, so that the relationship between the fields is

$$\vec{D} = \vec{E} + 4\pi \vec{P}. \tag{10.4}$$

Contrary to the magnetic case, where the SI and CGS systems are equally used, in the electric case the most used system is the SI. For this reason, we will not make much use of the CGS in this section. In vacuum, since there are no dipoles, $P = 0$ and therefore $D = \varepsilon_0 E$. The response of a dielectric to an electric field can be expressed by the electric susceptibility χ, or by the permittivity ε. In the case of simple dielectrics, the field $\vec{E}$ produces a polarization $\vec{P}$ in the same direction, so that χ is a scalar. By definition

$$\chi = \frac{P}{\varepsilon_0 E}, \qquad \varepsilon = \frac{D}{E}. \tag{10.5}$$

The relationship between these quantities, obtained by substituting (10.5) into Eq. (10.3), is

$$\varepsilon = \varepsilon_0 (1 + \chi). \tag{10.6}$$

It is also common to use the relative permittivity, or dielectric constant, defined by $\varepsilon_r = \varepsilon/\varepsilon_0$. The response of a dielectric to an external field varies with the frequency of the field. A typical form of variation of the susceptibility $\chi(\omega)$ with frequency is shown in Fig. 10.2. In the near infrared, visible, and ultraviolet regions, the response is dominated by electron transitions in atoms, as studied in Sects. 8.2.2 and 8.3.1.

In the infrared region, the main contribution to $\chi(\omega)$ comes from the interaction between the field and the ions that form the material. This contribution is illustrated in Fig. 10.3a, that shows the effect of an electric field on the ions of a crystal lattice, represented by a linear chain. The field displaces the ions of charges $+$ and $-$ in opposite directions, producing a typical vibration motion in the optical mode, studied in Sect. 2.2. The susceptibility can be calculated using a model similar to the one in Sect. 8.2.2, and considering that the field interacts with the charge ions $+q$ and $-q$.

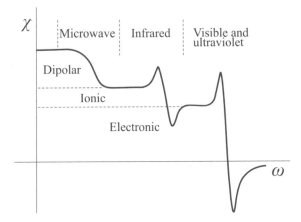

Fig. 10.2 Variation of the susceptibility of a dielectric with the frequency of the applied electric field

With this model it can be shown that the contribution of the ions to the susceptibility is

$$\chi_{\text{ion}}(\omega) = \frac{Nq^2/m_r\varepsilon_0}{\omega_0^2 - \omega^2 - i\omega\Gamma},\tag{10.7}$$

where $m_r = M_1 M_2/(M_1 + M_2)$ is the reduced mass of the ions with masses M_1 and M_2, N is the number of unit cells per unit volume, ω_0 is the angular frequency of the optical vibration mode at $k = 0$ and Γ is the damping rate. The ionic contribution to the response of dielectric materials is important in the infrared region, because optical vibration modes of the crystal lattice have frequencies in this region. The amplitude of this response is smaller than the contribution of electrons in the visible region because the mass of the ions is larger than that of electrons. It is important to note, however, that although the contribution of electrons, given by Eq. (8.36), is larger in the visible region, it is still significant at lower frequencies. Since the

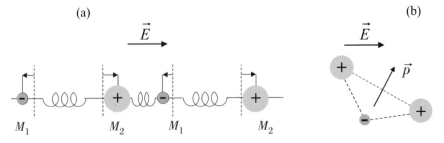

Fig. 10.3 Illustration of two mechanisms of dielectric response to an applied electric field: **a** Ionic; **b** Dipolar

ionic contribution adds to that of electrons, χ does not have a zero in the infrared region, as shown in Fig. 10.2. At frequencies below the infrared region, the dielectric response of certain materials contains a dipolar contribution that adds to the ionic and electronic components. This occurs in dielectrics that have molecules with permanent dipoles, as illustrated in Fig. 10.3b. The application of the field tends to produce a rotation of the dipole moments in its direction. Although this trend is partly cancelled by the effect of thermal agitation, there is a resulting moment in the direction of the field.

10.1.2   Capacitors

One of the most traditional applications of dielectric materials in electronics is in the fabrication of capacitors. Figure 10.4 shows a simple capacitor made of a dielectric layer between two parallel metallic plates. One of the basic functions of the capacitor is to store charge, and therefore electric energy. When a voltage V is applied between the plates, an electric field is created in the direction of the plate $+$ to the plate $-$. Away from the edges the field is uniform, with intensity $E = V/d$, where d is the distance between the plates. Since the capacitance of the capacitor is $C = Q/V$, where Q is the modulus of the charge on each plate, to calculate C it is necessary to relate the electric field with the charge. For this we use the integral form of Eq. (2.1),

$$\oint_S \vec{D} \cdot d\vec{a} = \int \rho \, dV = q, \qquad (10.8)$$

where ρ is the density of free charges, and q is the total free charge inside the volume limited by the closed surface S. Application of Eq. (10.8) to a cylinder containing a base inside one of the metal plates and the other in the dielectric, where there are no free charges, gives

$$D = \sigma = \frac{Q}{A}, \qquad (10.9)$$

Fig. 10.4 Parallel plate capacitor

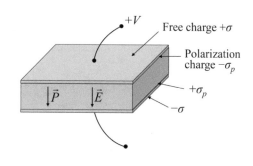

where σ is the surface density of free charges on the inner surface of the positive metallic plate, whose area is A. From this result and with $V = Ed$, we obtain the capacitance as a function of the capacitor dimensions and the permittivity of the dielectric

$$C = \frac{\sigma A}{V} = \frac{\varepsilon E A}{V} = \varepsilon \frac{A}{d}. \tag{10.10}$$

Note that since as $\varepsilon > \varepsilon_0$, the presence of the dielectric increases the capacitance relative to its value with air between the plates. To better understand the role of the dielectric in the capacitor, let us look at the behavior of the polarization in the dielectric. The polarization vector $\vec{P}$ is created by the electric field $\vec{E}$, and therefore it is directed from the $+$ plate to the $-$ plate. In this situation the microscopic dipoles induced by the field are uniformly distributed and directed downwards as in Fig. 10.4. As a result, the charges that form the dipoles cancel each other inside the dielectric. However, on the two surfaces this cancellation does not occur, resulting in the formation of surface charges. They are called **polarization charges** and result from the discontinuity of $\vec{P}$ on the surface. Note that the charges are negative at the top surface of the dielectric and positive at the bottom surface, due to the direction from top to bottom of the non-compensated dipoles. For this reason, they are also called **depolarization charges**.

Formally these charges can be introduced using an equation similar to (2.1), or its integral form. Denoting by ρ_p the volumetric density of polarization charges, we have

$$\nabla \cdot \vec{P} = -\rho_p, \qquad \oint_S \vec{P} \cdot d\vec{a} = -\int \rho_p \, dV. \tag{10.11}$$

Application of the integral form of (10.11) to a cylinder with a base inside the dielectric layer and the other outside, shows that the modulus of the surface density of polarization charge is $\sigma_p = P$. Finally, the relationship between the polarization charge and the electric field, obtained by replacing Eqs. (10.3) and (10.11) in (2.1) gives

$$\nabla \cdot \varepsilon_0 \vec{E} = \rho + \rho_p = \rho_t, \tag{10.12}$$

where ρ_t is the total charge density, resulting from the sum of free and polarization charges. The integral of (10.12) leads to

$$\varepsilon_0 \oint_S \vec{E} \cdot d\vec{a} = \int (\rho + \rho_p) \, dV = q_t. \tag{10.13}$$

This is Gauss's law for the electric field in the presence of dielectric materials. The electric field is created by the sum of the free and polarizing charges, as if they were in vacuum. The importance of this result for the capacitor comes from fact that the polarization charges have sign opposite to the free charges. As a result, for a certain charge Q in the capacitor, the presence of the dielectric results in an electric field smaller than there would be without it. This produces a smaller potential difference V and, thus, larger capacitance.

Several dielectric materials are used to make capacitors. As presented in Chap. 7, in integrated circuits the oxides of the own semiconductors are used to manufacture capacitors. A common type of capacitor used in the past was the paper capacitor. It was made by two aluminum sheets intercalated with sheets of wax paper. The set was rolled up to form a small cylinder and encapsulated after welding the wire terminals on the aluminum foils.

A very common type of capacitor used today is the electrolytic one. In the past the electrolytic capacitor used a liquid or a paste of a dielectric electrolyte solution. Subsequently, they were replaced by an oxide film, deposited on a sheet of aluminum, or tantalum, through the electrolysis of an electrolytic solution. In this technique, after the formation of the film with the desired thickness, the liquid solution is removed. The surface of the film is then covered with a metallic layer, forming the second plate of the capacitor. The set is finally rolled up into a cylinder. The film can be made with very reduced thicknesses, in the range of 1–10 nm, making possible to obtain capacitances in the range $1–10^5$ μF.

Two dielectric materials widely used in electrolytic capacitors are aluminum oxide and tantalum oxide. These oxides are easily formed on sheets of the corresponding metals. These materials have relatively high permittivity and also high **dielectric strength**, or **breakdown field** (E_b). This is the maximum electric field that a material can sustain without undergoing breakdown and becoming electric conductive. This field limits the maximum voltage that can be applied to a capacitor. In the case of tantalum oxide $\varepsilon_r \approx 28$ and $E_b \approx 10^8$ V/m.

Finally, there are several ceramics used as dielectrics in capacitors, making possible to obtain capacitances in a wide range of values. One advantage of ceramics over oxides is their much higher resistivity. As a result, ceramic capacitors have lower losses than electrolytic capacitors. Table 10.1 presents the main parameters of some important dielectrics for electronics.

Table 10.1 Relative permittivity $\varepsilon_r = \varepsilon/\varepsilon_0$ at low frequencies and dielectric strength E_b of some dielectric materials at room temperature

Material	ε_r	E_b (10^6 V/m)
Bakelite	4.8	12
Mica	5.4	160
Aluminum oxide (Al_2O_3)	10	5
Tantalum oxide (Ta_2O_5)	28	100
Titanium oxide (TiO_3)	94	6
Paper	3.5	14
Porcelain	6.5	4
Fused quartz (SiO_2)	3.8	8
Teflon (PFTE)	1.9	60

10.1.3   Piezoelectric Materials

Piezoelectricity is the property that some dielectrics have of developing a polarization when subjected to a mechanical stress. As shown in the previous section, the polarization produced by a voltage creates polarization charges and, therefore, an electric field. Conversely, the application of an electric field to a piezoelectric material results in a mechanical deformation (called reverse piezoelectric effect). In both cases, the change in the direction of the disturbance produces an inversion in the direction of the effect. These phenomena were discovered at the end of the nineteenth century by Pierre Curie, who coined the name piezoelectricity to the effect (piezo means pressure).

Figure 10.5 shows in a two-dimensional model how the compression of a crystal induces an electric dipole moment in the direction of the deformation. In the crystal without deformation, as in (a), the three dipoles formed by ion A and its neighbors (each charge is divided into three) have total moment null. However, when the crystal is deformed as indicated in (b), the angles between the dipoles produce a resulting moment in the direction of deformation.

It is important to note that there can be no piezoelectricity in crystals with a center of inversion symmetry. This property can be demonstrated generically from symmetry relations between fields. Physically it can be understood with the two-dimensional model in Fig. 10.6. Note that in the square lattice with inversion symmetry, the total electric dipole moment is null, both in the equilibrium situation in (a), and in the deformed lattice in (b). Piezoelectricity is not restricted to insulators. It also occurs in several semiconductors, such as CdS and ZnO.

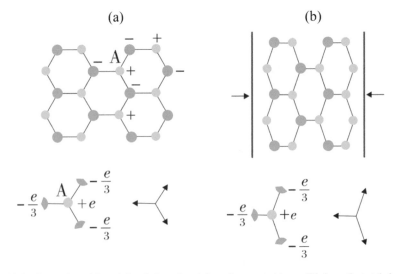

Fig. 10.5 Illustration of the origin of piezoelectricity. **a** In a crystal in equilibrium, the total electric dipole moment is zero. **b** Electric dipole resulting from mechanical deformation is not zero

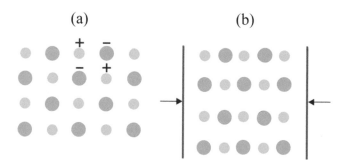

Fig. 10.6 Demonstration of the absence of piezoelectricity in crystals with a center of inversion symmetry

The application of a mechanical stress in a certain direction of the crystal results, in general, in a polarization at a different direction. Thus, the relationships between various quantities involved in piezoelectricity are tensorial. However, in some particular symmetry directions of the crystals, the vectors are in the same direction. In this case the relations are scalar and can be written in the form

$$P = dT + \varepsilon_0 \chi E, \tag{10.14}$$

$$R = sT + dE, \tag{10.15}$$

where T is the stress applied to the material (force per area unit), E is the applied electric field, P is the induced polarization, and R is the resulting deformation per unit length. The constants d, s and χ are characteristic parameters of each material. The constant d is the one that characterizes piezoelectricity, because it relates the induced polarization with the applied mechanical stress, or the deformation produced by an applied electric field. Since R is dimensionless, d has the inverse unit of the electric field, m/V in the SI, or cm/statvolt in the CGS. Actually, each material has several $d_{\alpha\beta\gamma}$ piezoelectric constants, relating the polarization induced in the α direction with the $\beta\gamma$ component of the tensor that characterizes the mechanical stress. Since α can assume 3 values and $\beta\gamma$ can assume 3×3 values, the piezoelectric tensor can have 27 components. However, due to the symmetry of the crystal, several components are equal to others and several are null, so that only a few are relevant.

Currently, about one thousand piezoelectric materials are known, but practical applications are dominated by only a few of them. The piezoelectric and dielectric constants of the most important materials are presented in Table 10.2. One of the most traditional piezoelectric crystals is quartz (SiO_2), whose longitudinal piezoelectric constant is $d_{11} = 2.3 \times 10^{-12}$ m/V. To have an idea of the meaning of this value, consider a quartz disc of thickness $l = 1$ mm, subjected to a voltage $V = 100$ V. The variation Δl in the disk thickness, given by (10.15), is

Table 10.2 Values of the largest components of the piezoelectric constant and dielectric constant tensors of important piezoelectric materials

Material	d (10^{-12} m/V)	ε_r
Piezoelectric		
Quartz (SiO$_2$)	−2.3	4.5
Turmaline	−3.7	6.3
Aluminum oxide (Al$_2$O$_3$)	21	40
Ferroelectric		
Barium titanate (BaTiO$_3$)	390	2900
PZT (Pb$_{0.5}$Zr$_{0.5}$ TiO$_3$)	370	1700

$$\frac{\Delta l}{l} = d \frac{V}{l}, \tag{10.16}$$

which is only $\Delta l = 2.3 \times 10^{-10}$ m $= 2.3$ Å.

In genuinely piezoelectric materials, the polarization is zero in the absence of mechanical stress or external electric field, as shown in Eq. (10.14). There is another class of materials, which will be presented in the next section, in which there is a spontaneous polarization in the absence of external fields. They are called **ferroelectric materials** and, as will be shown in the next section they have a piezoelectric effect. The ferroelectric and piezoelectric materials most important for application in electronics are lithium niobate, barium titanate, and lead and zirconium titanate, this one known as PZT, whose parameters are in Table 10.2. PZT is generally used in the form of polycrystalline ceramics, sintered under an external electric field. The application of the field during the cooling process produces an alignment of the crystalline grains along a certain crystallographic axis, so that the material exhibits a macroscopic piezoelectric effect. The PZT ceramic is widely used today because of its high piezoelectric constant, $d_{33} = 3.7 \times 10^{-10}$ m/V, about two hundred times larger than in quartz.

An important application of piezoelectric materials is in the fabrication of electromechanical transducers for the generation of elastic waves, as illustrated in Fig. 10.7. In applications with low frequencies (up to tens of kHz) the most used material in transducers is PZT, while at higher frequencies (≥ 1 MHz) crystalline quartz is the most used. The transducer is formed by a disc, or a rectangular plate, of PZT or quartz, with the two faces covered by metallic films. The metallic cover of one of the faces is extended to the side edge to allow electric contact with an external wire. A voltage applied between the electrodes creates an electric field in the piezoelectric material, resulting in a mechanical deformation. When the transducer is placed in contact with another material, the application of an *ac* voltage generates an elastic wave in the material. This technique is used to generate ultrasonic waves, used in medical, scientific, and industrial equipment. The reflected ultrasonic waves carrying information on the probed material are converted into an electrical signal by another receiving transducer, or by the transmission transducer itself.

(a)   (b)

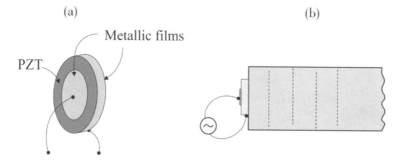

Fig. 10.7 **a** PZT piezoelectric transducer. **b** Use of the transducer to generate ultrasonic (elastic) waves

Although quartz has a much smaller piezoelectric effect than PZT, it finds several important applications in electronics due to its low acoustic loss. This property makes a block of crystalline quartz to be an excellent mechanical resonator, with a very low damping rate. The application of a voltage pulse to the block causes a mechanical vibration, which in turn creates an oscillating electric voltage through the piezo-electric effect. Thus, the quartz block with metallic films on two opposite sides is electrically equivalent to a parallel *RLC* resonant circuit. The vibration frequencies depend on the dimensions and shape of the block. In the case of a thin plate with parallel faces, the fundamental resonance mode corresponds to a stationary acoustic wave, successively reflecting at the two faces, having a wavelength equal to twice the thickness of the plate. Thus, for a plate of thickness l and acoustic wave velocity v in the direction perpendicular to the face, the oscillation frequency is

$$f = \frac{v}{2l}. \tag{10.17}$$

The wave velocity depends on the crystallographic direction of the cut of the crystal. For the so-called X-cut in quartz, $v = 5.4 \times 10^3$ m/s. So, a crystal with thickness 1 mm in this cut, oscillates with a frequency of 2.7 MHz. Due to the resonant oscillation, the variation in thickness is much larger than in the DC situation, given by (10.16).

Quartz crystals are used to synchronize electronic oscillators for watches, computers, radio and TV transmitters and receivers. The oscillator consists of an amplifier circuit with feedback, using a small quartz plate with metallic contacts on the two opposite faces instead of a *RLC* resonant circuit. Figure 10.8 shows the circuit symbol for the quartz crystal and its equivalent electric circuit. The quartz oscillators have two advantages relative to *RLC* circuits: at frequencies of few MHz, they have much smaller losses, so that their resonance quality factor is much higher (in a *RLC* circuit the quality factor is $Q = \omega L/R$); the stability of the resonance frequency in a quartz crystal is much higher than in a *RLC* circuit.

Fig. 10.8 a Circuit symbol
of a quartz crystal oscillator.
b Equivalent electric circuit

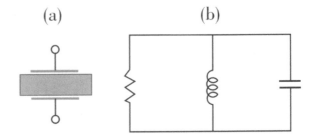

Finally, another important application of piezoelectric materials is in surface acoustic wave (SAW) devices. The simplest SAW device, shown in Fig. 10.9, is formed basically by a substrate slab of quartz or lithium niobate, having one polished surface, on which two interdigital transducers (IDT) are made. Each transducer consists of a metallic film, deposited on the substrate, having the shape of two combs with intercalated fingers. An *ac* voltage applied between the two transducer terminals, produces an elastic deformation in the region near the surface. This generates an acoustic wave that propagates in the slab, confined to a superficial layer, with velocity similar to that of volume waves. This acoustic surface wave is detected by the second transducer, which converts the acoustic signal into an electric signal. Since two neighboring fingers have opposite polarities, the efficiency of the transducer is maximum for a frequency whose acoustic wavelength is twice the distance between them. The SAW technology is used to manufacture several signal processing devices with frequency in the range of tens or hundreds of MHz, such as delay lines and filters. Before the advent of this technology, these devices were bulky, made by series of tuned circuits of discrete capacitors and inductors. The development of SAW devices made possible the miniaturization and the integration of circuits for the VHF and UHF frequency bands.

Fig. 10.9 Surface acoustic
wave (SAW) device

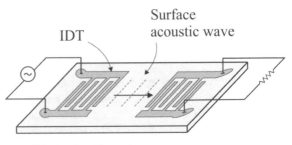

10.1.4 Ferroelectric Materials

Ferroelectric materials are those that have a spontaneous polarization in the absence of external fields. This spontaneous polarization originates from the electric dipole moment that appears in the unit cell due to a displacement of the center of the positive charges relative to the center of the negative charges. This displacement results from a small distortion in the crystal structure, which occurs below a certain critical temperature, in order to minimize the energy of the system. Figure 10.10 shows the unit cell of barium titanate ($BaTiO_3$), indicating the displacement of positive ions that produces the electric dipole moment.

Example 10.1 Calculate the displacement of the Ti^{4+} ion relative to the center of the unit cell in $BaTiO_3$, considering that at $T = 300$ K its spontaneous polarization is $P_s = 0.26$ C/m^2 and that the lattice parameter is $a = 4.0$ Å.

Since the polarization is the electric dipole moment per unit volume, the moment of the unit cell is

$$p = P_s\, a^3 = 0.26 \times \left(4.0 \times 10^{-10}\right)^3 = 1.66 \times 10^{-29} \text{ Cm.}$$

Considering that the total charge of the Ba^{2+} and Ti^{4+} ions inside the cell is $6e$, this dipole moment results from a displacement given by

$$\delta = \frac{p}{6e} = \frac{1.66 \times 10^{-29}}{6 \times 1.6 \times 10^{-19}} = 0.17 \times 10^{-10} \text{ m} = 0.17 \text{ Å.}$$

Note that the displacement is much smaller than the dimensions of the cell.

The spontaneous polarization in ferroelectric materials disappears above a critical temperature T_c, analogously to the magnetization in ferromagnets. Here T_c is also called Curie temperature. In the case of $BaTiO_3$, $T_c = 393$ K. Another important ferroelectric material, lithium niobate, $LiNbO_3$, has $T_c = 1470$ K. The polarization of ferroelectric materials can be changed by the application of an external electric field

Fig. 10.10 Unit cell of $BaTiO_3$ showing the displacements of the positive ions that produce the electric dipole moment

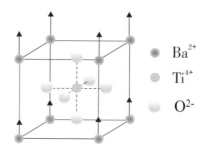

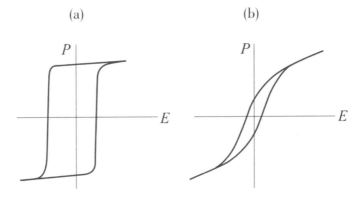

Fig. 10.11 Hysteresis cycles of ferroelectric materials. **a** Rectangular cycle observed in crystals. **b** Elongated cycle in aligned polycrystalline ceramics

E. Here there is also an analogy with the ferromagnetic case, since the variation of P with E follows a hysteresis cycle like that of M vs. H. In the case of single crystals, the curve has a rectangular shape, shown in Fig. 10.11a, which resembles the hysteresis cycle of permanent magnets. The material maintains a remanent polarization P_r, with value close to that of saturation, after the electric field is removed. For applications in which the E field varies in time the rectangular loop is undesirable, because the variation of P is discrete and because the energy loss is large. This can be avoided with the preparation of ceramic materials made of aligned crystalline grains. The alignment is obtained through the application of an external E field during the cooling process. This process results in a thin and elongated hysteresis cycle, as shown in Fig. 10.11b.

Ferroelectric materials are used in three different situations: applications that require high permittivity dielectrics; applications based on the rectangular hysteresis cycle; and as piezoelectric materials. The property that all ferroelectric materials are piezoelectric can be understood by means of Fig. 10.12. In (a), a two-dimensional model of a ferroelectric material shows that with no external mechanical stress, there

Fig. 10.12 Illustration of the piezoelectric effect in ferroelectric crystals. **a** Crystal under no stress. **b** Crystal under mechanical stress showing the variation Δp in the electric dipole moment

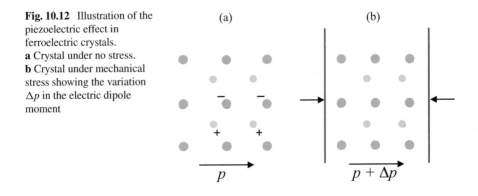

is an electric dipole moment due to the displacement between the centers of positive and negative charges. Figure 10.12b shows the crystal deformed by the application of an external stress. We see that the deformation results in a variation Δp in the electric dipole moment of the unit cell, producing a polarization in the material.

10.1.5 Electrets

A special class of dielectric materials, with properties that resemble those of ferro-electrics, is that of **electrets**. The electret is formed by a layer of dielectric material on which positive and negative electric charges are deposited in the fabrication process. The charges remain trapped, close to the surfaces or inside the volume, generating a macroscopic polarization and, therefore, an electric field. Figure 10.13 illustrates various forms of charge trapping in electrets. In (a) the negative charges trapped on the top surface of the layer induce compensation charges on the metallic film under the layer. These compensation charges remain in the film because they cannot pass through the potential barrier between the metal and the dielectric. The situation in (b) is similar to (a), but the negative charges are trapped on the surface and inside the dielectric layer. In (c) the $+$ and $-$ charges are inside the material, forming domains that behave like electric dipoles.

A basic difference of an electret to a ferroelectric material, is that its polarization gradually decays in time, that is, it is not permanent. This results from the fact that the charges are placed artificially, generating a metastable state. The charges remain trapped in local potential wells, but they can be released through thermal activation. The charge decay time depends on the host material, the conditions of preparation of the electret, and the temperature. Denoting by ΔE the average height of the potential barriers that trap the charges, the charge decay time is

$$\tau = \tau_0\, e^{-\Delta E/k_B T}, \tag{10.18}$$

where τ_0 is a characteristic time of the material and the preparation conditions. We see then that the charge decay time decreases with the increasing temperature. This is so because as T increases, the polarization gradually decreases, without a phase

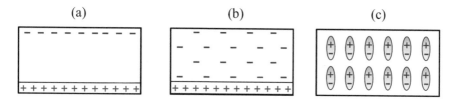

Fig. 10.13 Illustration of some types of electrets. In **a** and **b** the bottom surface of the dielectric layer is metallized

transition as in ferroelectric materials. The decay time at room temperature can vary from a few seconds to tens or hundreds of years. Therefore, although the polarization of electrets is not permanent, for practical purposes materials with $\tau \sim 100$ years behave as if they are quasi-permanent.

Electrets can be made from a wide range of materials, prepared using various loading techniques. The first electrets studied were vegetable waxes. One of the earliest recipes consisted of 45% carnauba wax, 45% white rosin, and 10% white beeswax, melted, mixed together, and left to cool in a static electric field of several kilovolts/cm. The thermo-dielectric effect, related to this process, was first described by the Brazilian physicist Joaquim da Costa Ribeiro, a pioneer of solid-state physics in Brazil. Many materials can be used to make electrets, organic substances such as anthracene, naphthalene, and several polymers, as well as inorganic compounds, such as quartz, sulfur, and ionic crystals. There are also several electrets of biological materials, called bioelectrets, such as bones, teeth, tissues, proteins, etc. Among the charging methods, the most important are: electric discharge in air from a high voltage metallic plate near the surface of the material; various ionizing radiations, such as X-rays, gamma rays, alpha particles, and ultraviolet.

The important electrets for electronics are polymer films, mainly of two types of teflon, polyfluorethylene propylene (FEP) and polytetrafluoroethylene (PTFE). They are made in the form of films with thickness in the range of 10–50 μm, metallized by evaporation on one or both surfaces. The charges are produced by high voltage discharge, with charge densities in the range 10^{-4}–10^{-2} C/m^2 and decay time $\tau \sim 10^9$ s ($\sim$32 years).

One of the main applications of electrets in electronics is in electrostatic transducers for microphones. The electret microphones are small, very sensitive and are fabricated at very low cost. For these reasons, they have replaced the traditional magnetic microphones in most applications. Figure 10.14 shows the cross section of a simple electret microphone. It consists of a diaphragm formed by a teflon film (FEP or PTFE) with thickness of the order of 20 μm, with the upper surface covered by a metallic film (thickness $\sim$50–100 nm). The teflon film, containing a surface charge as in Fig. 10.13a, is mounted on a metallic plate, supported on spacers that leave an air layer with a certain thickness. When a sound wave hits the diaphragm, it produces a deflection that varies the thickness of the air layer. Thus, the electric field between the film and the metal plate produces a variation in the output voltage proportional to the diaphragm displacement. With some modifications relative to the

Fig. 10.14 Illustration of the cross section of a simple electret microphone

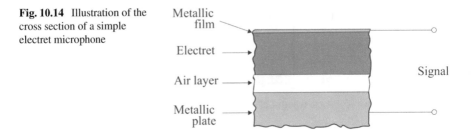

Metallic film

Electret

Air layer

Metallic plate

Signal

scheme of Fig. 10.14, the electret microphones acquire higher stability and better frequency response.

10.2 Dielectric Materials for Optoelectronics and Photonics

The optoelectronic devices presented in Chap. 8 have the main function of converting an electric signal into an optical signal, or vice versa. The development of optoelectronics has resulted in devices for processing information directly on the optical signal, avoiding the need for its conversion into an electronic signal, its processing, and the conversion back into the optical signal. These devices have a much faster response time than those that employ conversion to electronic signal and also less insertion loss, and form the basis of the active field of **Photonics**. The operation of several of these devices is based on the optical properties of dielectric materials that we shall present in this section.

10.2.1 Electro-optic and Elasto-optic Effects

These two effects have a strong analogy with the piezoelectric effect, studied in the previous section. When a macroscopic electric field is applied to a dielectric material, it acts on the electric dipole moments and produces macroscopic effects. The action of the field on the ionic dipoles results in a deformation of the crystalline lattice, and therefore in the inverse piezoelectric effect. On the other hand, the action of the field on the electronic dipoles, produces a change in the optical dielectric constant of the material, giving rise to the **electro-optic effect**. This change comes mainly from the variation of the electronic energy levels produced by the external field, known as the Stark effect. This variation results in a change in the optical dielectric constant, because as we saw in Chap. 8, this depends directly on the energies of the states involved in the electronic transitions.

Like piezoelectricity, the electro-optic effect requires that the material does not have an inversion symmetry. For this reason, piezoelectric crystals are also electro-optic active. Since the dielectric constant of a crystal is characterized by a tensor with 9 components, while the applied electric field is a vector with 3 components, the relationship between them involves a tensor with 27 components. To avoid algebraic complications, let us assume that the electro-optic tensor is dominated by one of its components. In this case, the relationship between the applied electric field E and the variation in the relevant optic dielectric constant ε_r takes the form

$$\Delta\left(\frac{1}{\varepsilon_r}\right) = rE, \tag{10.19}$$

Table 10.3 Ordinary refractive index n and main electro-optic constant r in some materials, measured at the wavelength λ indicated

Material	λ (nm)	n	r (10^{-12} m/V)
BaTiO$_3$	514	2.44	820
CdTe	1000	2.84	4.5
GaAs	1150	3.43	1.43
KDP	514	1.51	10.6
LiNbO$_3$	633	2.29	32.6
Quartz	514	1.54	0.53
Ti:LiNbO$_3$	1500	2.20	31.0

where r is the electro-optic constant. Note that since ε_r is dimensionless, r has the inverse unit of the electric field, which is m/V in the SI. The change in the dielectric constant produced by the electric field produces a variation in the refractive index n of the material. Since $\varepsilon_r = n^2$, the variation of n with the applied field is given by

$$\Delta n = -\frac{1}{2}n^3 r E. \tag{10.20}$$

The main electro-optic crystals are also the main piezoelectric materials, due to the absence of inversion symmetry. Table 10.3 presents the refractive indices and the values of the main component of the electro-optical tensor of some of these materials. In the case of LiNbO$_3$, $n = 2.29$ and $r = 3.26 \times 10^{-11}$ m/V for visible light at the wavelength $\lambda = 633$ nm. Therefore, an electric field of $E = 10$ V/m applied to this material produces a change in the refractive index of only $\Delta n = 1.96 \times 10^{-4}$. Although small, this variation is sufficient to produce macroscopic effects that enable the construction of electro-optic devices.

The **elasto-optic effect**, also called **photoelastic**, or **acousto-optic**, is the phenomenon by which the elastic deformation of a material results in a variation of the optic dielectric constant, and therefore in the refractive index. This effect results from changes in the electronic energy levels resulting from the change in the crystalline electric field, produced by the deformation of the lattice. Although this effect is also characterized by a tensor, for simplicity we shall consider just the simple case of the relationship involving the largest variation

$$\Delta\left(\frac{1}{\varepsilon_r}\right) = p\,R, \tag{10.21}$$

where p is the photoelastic constant and R is a component of the deformation tensor, defined as the variation in the dimension of the material in a certain direction, per unit length. Since ε_r and R are dimensionless quantities, the photoelastic constant is also dimensionless. Table 10.4 shows the values of the main photoelastic constant and the refractive index of some materials. Note that fused quartz, lithium fluoride, rutile, and other materials in Table 10.4 have inversion symmetry and have a photoelastic effect.

Table 10.4 Average refractive index and main photoelastic constant in the visible region in some dielectrics

Material	n	p
LiNbO$_3$	2.25	0.15
LiF	1.39	0.13
Rutile (TiO$_2$)	2.60	0.05
Sapphire (Al$_2$O$_3$)	1.76	0.17
Fused quartz	1.46	0.2

The reason for this is that the photoelastic tensor is characterized by components with four indices, $p_{\alpha\beta\gamma\delta}$. In this case, it is not necessary the absence of inversion symmetry in the material for some components of $p_{\alpha\beta\gamma\delta}$ to be nonzero.

Like the electro-optic effect, the elasto-optic effect results in a variation of the refractive index n of light produced by an elastic deformation, given by

$$\Delta n = -\frac{1}{2}n^3 p R. \tag{10.22}$$

In a piezoelectric material, the elasto-optic effect can give rise to an electro-optic effect. The application of a field E produces a deformation R, given by Eq. (10.15), which in turn results in a variation in the refractive index given by (10.22). Combining these equations, we see that the electromagnetic optical constant resulting from this indirect effect is $r = dp$. In the case of quartz, the values d and p in Tables 10.2 and 10.4 give $r = 4.6 \times 10^{-13}$ m/V, a value smaller than the electro-optic constant in Table 10.3. This result, obtained here for quartz, is valid for other materials. This means that the direct electro-optic effect produced by the variations of the electronic structure caused by the electric field, is larger than the indirect effect, resulting from the combination of piezoelectricity and photoelasticity.

The variation of the refractive index produced by the electro- and elasto- optic effects gives rise to various phenomena of interest, both scientific and technological. One of the most evident phenomenon is the birefringence induced by an electric field, or by a mechanical deformation. When an electromagnetic wave propagates in a material, having polarization components in two perpendicular directions, its behavior is influenced by the indices of refraction in these two directions. Since the direction of the refractive index change depends on the direction of the external disturbance and the characteristics of the material, the electro- or elasto-optic effects can produce variations in the refractive index in just one direction. Thus, if the material in equilibrium is isotropic, the disturbance results in different indices in the two directions. This birefringence causes a variation in the polarization of the wave, which can be controlled by the external disturbance, either an electric field or a mechanical deformation in the material. This phenomenon finds several applications in photonics. In Sect. 10.2.3 we shall present some photonic devices whose operation is based on the electro- and elasto-optic effects.

10.2.2   Nonlinear Optical Materials

In the presentation of the electro-optic effect, we considered that the electric field in the material was created by an external source. Actually, the field of an electro-magnetic wave propagating in the material can produce an electro-optic effect. In this case, the refractive index that determines the velocity of the wave depends on the field amplitude in the wave. To quantify this phenomenon, consider the variation in the polarization $P^{(2)}$ resulting from the electro-optic effect created by the electric field E of the wave. Using Eqs. (10.5) and (10.19) we obtain

$$P^{(2)} = \varepsilon_0 \, \Delta\chi \, E = \varepsilon_0 \, \Delta\varepsilon_r \, E = -\varepsilon_0 \, \varepsilon_r^2 \, r \, E^2. \tag{10.23}$$

This result shows that the contribution of the electro-optic effect to the polarization varies with the square of the field, while the usual contribution is linear in the field. Materials that have this property are called nonlinear optical materials. Only crystals without inversion symmetry exhibit non-linear responses of the type in Eq. (10.23). Actually, in addition to the quadratic nonlinearity of Eq. (10.23), it is possible to have higher order contributions. The total polarization created by an electric field can then be written in the form

$$P = \varepsilon_0 \big(\chi^{(1)} \, E + \chi^{(2)} \, E^2 + \chi^{(3)} \, E^3 + \cdots \big), \tag{10.24}$$

where $\chi^{(1)}$ is linear susceptibility, which we previously represented only by χ, while $\chi^{(2)}$ and $\chi^{(3)}$ are the quadratic and cubic susceptibilities. Actually, the quantities E^2 and E^3 in Eq. (10.24) can be products of components of fields in different directions, such as $E_\beta E_\gamma$ and $E_\beta E_\gamma E_\delta$, while P is the P_α component of the polarization vector. Thus, in the most general case, the susceptibilities appearing in (10.24) are components of the tensors $\chi^{(1)}_{\alpha\beta}$, $\chi^{(2)}_{\alpha\beta\gamma}$, and $\chi^{(3)}_{\alpha\beta\gamma\delta}$. While the tensor $\chi^{(2)}_{\alpha\beta\gamma}$ is null in crystals with inversion symmetry, the tensor $\chi^{(3)}_{\alpha\beta\gamma\delta}$ is not necessarily null whatever the crystal symmetry is.

Since the nonlinear effects vary with the square and the cube of the electric field, they are important only for high field intensities, as illustrated in Fig. 10.15. The nonlinear effects are manifested in high power waves, typically on the order or above 1 MW/cm^2. For this reason, the field of nonlinear optics was developed only after the invention of the laser. The nonlinear optical effects have a wide variety of applications in optics and photonics. One of the most evident is the mixture of waves. When two waves with frequencies ω_1 and ω_2, with amplitudes E_1 and E_2, respectively, propagate in a nonlinear medium, they generate a polarization $P^{(2)}$ given by

$$P^{(2)} = \varepsilon_0 \, \chi^{(2)} \, E_1 \, E_2 \, e^{i(\pm\omega_1 \pm\omega_2)t}. \tag{10.25}$$

Fig. 10.15 Illustration of the deviation from linearity of the polarization for intense electric fields in crystals without inversion symmetry

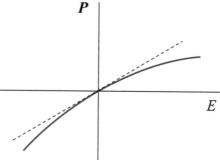

This polarization results in a third wave, with frequency given by the sum or difference of the original frequencies. Another simple non-linear effect is the doubling of the frequency resulting from the term $\chi^{(2)}E^2$ in Eq. (10.24). This so-called **second-harmonic generation** occurs naturally when a high-power wave passes through a crystal without inversion symmetry, such as KDP, LiNbO₃, or quartz. However, to have an efficient second-harmonic generation, it is necessary to cut the crystal in certain crystallographic directions, with a certain thickness, to avoid destructive interference due to the dispersion. Since some important commercial lasers operate in the infrared, second- and third- harmonic generators are widely used to convert their radiation into visible light. This is the case of Nd:YAG lasers, which operate at $\lambda = 1060$ nm. KDP frequency doublers converts this radiation into green light, with $\lambda = 530$ nm, with efficiency higher than 60%.

10.2.3 Electro-optical Waveguide Devices

The dissemination of optical communications led to the development of several devices for processing optical signals, and many of them are based on optical waveguide technology. The basic idea of the waveguide is the same as in the optical fiber, shown in Sect. 8.8. When a wave propagates along a medium with a certain refractive index, surrounded by other media with a smaller refractive index, it may suffer internal reflections and be confined to the region of the first medium. It is possible to manufacture waveguides for visible or infrared light wave on substrates of a dielectric material or a semiconductor, using photolithographic and diffusion techniques, similar to those used to make integrated electronic circuits. These waveguides can be used to conduct the wave from one device to another, in a collection of devices manufactured in the same substrate, constituting an integrated optical circuit.

Figure 10.16 shows a simple light waveguide used in optical devices. It is made of a plate of dielectric material, such as quartz or LiNbO₃, or a semiconductor, such as GaAs or Si. The waveguide is made by means of the interdiffusion of a certain impurity along a strip on the surface of the plate. The diffusion of the impurity produces a

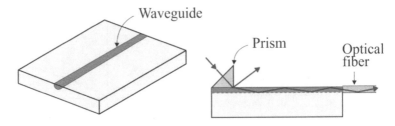

Fig. 10.16 a Light waveguide in a plate of a dielectric material or semiconductor. **b** Illustration of methods to couple external light to the waveguide

channel of a material with refractive index larger than of the substrate, constituting a light waveguide, as illustrated in Fig. 10.16a. The waveguide has a typical width of few μm and the impurity normally used in LiNbO$_3$ is titanium. Figure 10.16b illustrates two methods used to couple external light to the waveguide. In the first, the light from a collimated laser beam falls on a prism and undergoes total internal reflection on the surface in contact with the plate. This results in an evanescent wave that penetrates the plate, reproducing approximately the mode profile in the waveguide. This method allows an efficient coupling of the waveguide with an external light beam, both for input and output. In the case of coupling with optical fibers, the situation is simpler, since the fiber mode has a configuration similar to that of the guide. Thus, the coupling can be done simply by gluing the end of the fiber to the front surface of the device, as shown in Fig. 10.16b.

A simple electro-optical device based on the guide in Fig. 10.16 is the **phase modulator**. It consists of a waveguide in an electro-optic material, between two electrodes made by deposition of metallic strips on the plate surface, illustrated in Fig. 10.17. When a voltage V is applied between the electrodes, an electric field is created through the guide. The field strength is given, approximately, by $E = V/d$, where d is the distance between the electrodes. The field produces a variation Δn in the refractive index, resulting in a change in the lag suffered by the wave in the guide along the electrodes, given by

Fig. 10.17 Electro-optical phase modulator

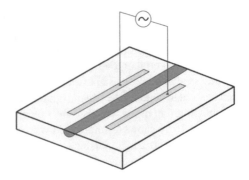

$$\Delta\phi = L\,\Delta k = \frac{2\pi L}{\lambda}\Delta n, \qquad (10.26)$$

where L is the length of the electrodes. Using Eq. (10.20), one can express the phase change as a function of the applied voltage

$$\Delta\phi = -\frac{\pi n^3 r}{\lambda\,d}L\,V. \qquad (10.27)$$

We see that the phase change is proportional to the length of the electrodes and the applied voltage. It is easy to see that the value of the LV product necessary to have $\Delta\phi = \pi$ is

$$(LV)_\pi = \frac{\lambda\,d}{n^3\,r}. \qquad (10.28)$$

Example 10.2 Calculate the voltage that must be applied between two electrodes distant $d = 7.0\ \mu m$, in a $LiNbO_3$ electro-optic phase modulator, with a Ti:$LiNbO_3$ waveguide, for light with $\lambda = 1.5\ \mu m$.

Using the parameters for Ti:$LiNbO_3$ given in Table 10.3 in Eq. (10.28) we have

$$(LV)_\pi = \frac{1.5 \times 10^{-6} \times 7.0 \times 10^{-6}}{2.2^3 \times 31 \times 10^{-12}} \approx 0.032\ \text{Vm} = 32\ \text{Vmm}.$$

This result means that for a modulator with an electrode of length 8 mm, a voltage of 4 V is sufficient to produce a phase change of π.

Based on the electro-optic phase modulation, it is possible to build very efficient light **amplitude modulators**. Figure 10.18 shows the basic scheme of an amplitude modulator that uses a Mach-Zehnder interferometer, made of two Y-junctions connected in opposite directions to two waveguides. A pair of electrodes is deposited around one of the guides, so that a voltage applied to it produces a phase lag $\Delta\phi$ in the wave that propagates in this waveguide relative to the wave in the other guide. The

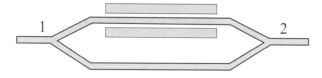

Fig. 10.18 Top view of an electro-optic amplitude modulator with a Mach-Zehnder interferometer

wave incident in the input terminal 1 is divided equally between the two legs of the Y-junction, giving rise to two waves that propagate independently in the two waveguides with the same amplitude E_1. At the output junction the two waves superimpose, giving rise to a wave whose amplitude is

$$E_2 = E_1 + E_1 e^{i\Delta\phi}.$$

Since the output power is proportional to $E_2 E_2^*$, it can be shown (Problem 10.5) that the device transmission, defined as the ratio between the output and input powers, $T = P_2/P_1$, is given by

$$T = \frac{1 + \cos \Delta\phi}{2}, \tag{10.29}$$

where $\Delta\phi$ is proportional to the voltage applied to the electrodes, according to Eq. (10.27). Figure 10.19 shows the transmission curve as a function of $\Delta\phi$. We see that the transmission is maximum at $\Delta\phi = 0$, that is, when the applied voltage is zero. In the ideal device the maximum value is 1, but in real devices there are losses due to reflections in the connections and at the junctions, reducing the maximum transmission to about 0.5 (corresponding to an insertion loss of 3 dB). The transmission drops to zero at $\Delta\phi = \pi$, 3π, etc., which allows digital on/off modulation for V varying between 0 and the value given by Eq. (10.28). The device can also be used for analog signal modulation. For this, it is necessary to superimpose the *ac* signal with a DC polarization voltage so that the operation point lies in the linear region around $\Delta\phi = \pi/4$ (point A in Fig. 10.19). Commercial electro-optic modulators operate with voltages of few volts and have a modulation bandwidth larger than 1 GHz, in the case of analog signals, or modulation rate above 1 Gbit/s, for digital signals. The waveguide technology is used for the fabrication of a variety of other electro-optic devices for optical communications, such as switches, directional couplers, multiplexers, etc.

Fig. 10.19 Transmission of the amplitude modulator as a function of the offset angle $\Delta\phi$, which is proportional to the applied voltage. A is the operating point for modulation of analog signals

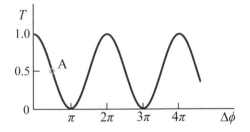

10.3 Materials for Video Displays

One of the most important functions in electronics is the conversion of information contained in electric signals into visual information. In the early days of electronics, the main purpose of this function was to indicate the status of an equipment to the external user. Then, with the invention of television, it became an essencial function for the presentation of sequences of images forming a motion picture. Today, this is a basic component for television displays, video equipment, computer monitors, and a wide variety of mass consumption mobile devices, as well as in vehicles, aircrafts, appliances, and many equipment for scientific, medical, and industrial applications.

Until the 1970s, light displays were made with incandescent or gas bulb lamps, of different sizes, while television displays, computer, and scientific equipment monitors employed a cathode-ray picture tube, called kinescope. Then, light indicators began to be replaced by solid state devices, using electroluminescent displays, liquid crystals, and light emitting diodes, which are much more efficient, resistant, and economical. Only in the 1990s video monitors using kinescope tubes began to be replaced by solid state displays. In this section we shall present materials and devices used in the manufacture of video displays that were not studied in Chap. 8, phosphorescent ceramics, liquid crystals, and organic conductors.

10.3.1 Phosphorescent Ceramic Materials

As we saw in Chap. 8, luminescence is the property that some materials have of emitting light in radiative transitions from excited states to lower energy states. Two common forms of luminescence are photoluminescence and electroluminescence. In the first, electrons are excited to higher energy states by means of photon absorption, while in the second the excitation results from an electric stimulus. Materials that have these properties are called, respectively, photoluminescent and electroluminescent.

An important class of electroluminescent materials is that of semiconductors, that emit light with the passage of an electric current. As we saw in Chap. 8, the current in a forward-biased junction diode produces injection of electrons and holes, which emit light in the recombination process. Another class of technological importance is that of phosphorescent ceramics, which can be excited electrically or optically in many ways. One electric excitation process is the bombardment by electrons at high speed, as in an electron beam. When they collide with the atoms of the material, they take some bound electrons to excited states. Then, they quickly relax to states with longer lifetimes and subsequently undergo radiative transitions to lower energy states, emitting photons with energy determined by the difference between the energies of the states involved, as studied in Sect. 8.3. Another form of electric excitation of phosphorescent ceramics is the application of an intense electric field. When the field exceeds the value of dielectric strength, which is of the order of 10^6 V/cm, certain atoms are ionized and the ejected electrons are accelerated. Then they collide with other atoms, taking them to excited states and producing luminescence by the process described above.

Two important properties of luminescent materials that determine their application is the decay time of the radiative transition and the spectrum of the emitted light. The decay time determines the duration of the light pulse produced after the external excitation. When this time is on the order of nanoseconds or less, the process is called **fluorescence**. In this category are the light emission by interband transitions in direct gap semiconductors and the emission by atomic transitions in ionized gases of fluorescent lamps. On the other hand, when the duration of the emission is on the order of milliseconds or longer, the process is called **phosphorescence**. Phosphorescent materials were very important in electronics for their application in the manufacture of video screens. A motion picture is formed by a series of still images, each one differing slightly from the preceding one. If the duration of each image is of the order of tens of milliseconds, the observer has a visual sensation of a continuous change in the picture. Thus, it is important that the duration of the luminescence, also called **persistence**, be of tens of milliseconds. Times shorter than this give the sensation that the image blinks, as in a strobe, while longer times result in a blur, because a new image is formed before the previous one disappears. Phosphorescent ceramic materials employed in video screens are generically called **phosphors**, to account for the literal meaning of the word, they light up when excited. They are phosphate compounds, oxides, tungstates, sulphates, and sulphites of various metals, such as zinc and cadmium, insulators or semiconductors, doped with impurities of iron transition elements or rare earths. The impurities used are those that have metastable states adequate for the luminescent transitions. The chemical composition determines the persistence and the light emission spectrum, and therefore the application of the materials in phosphors.

The most important application of phosphors in electronics is in video screens. The oldest technology employed cathode ray tubes (CRT), also called kinescopes, which were used extensively for several decades in television receivers, computer monitors, and many other equipment. Figure 10.20 shows the external view of a kinescope, consisting of a tube of reinforced glass, evacuated and sealed, with pyramidal shape and having an elongated neck at the end. Inside the neck there is an electron gun, consisting of a cathode and several electrodes, that produces the electron beam. This is directed to the front face by means of a high voltage applied between it and the cathode. The front face is internally covered by a phosphor layer, that when bombarded by electrons emits light that is transmitted through the glass and seen externally.

The electron gun is made of a heated cathode, a grid and some electrodes, all mounted in a socket with pins for the external connections. When heated by a tungsten filament with an electric current, the cathode emits electrons that are accelerated by the voltages applied to the electrodes, forming a monoenergetic beam, that is, with little velocity dispersion. When focusing on the phosphor layer, the electron beam produces a bright spot on the tube screen.

Fig. 10.20 Illustration of a
kinescope, or picture tube

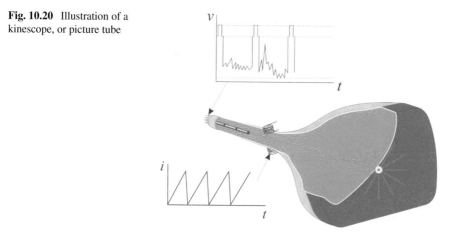

The formation of the image on the screen requires two processes: scanning the
beam, so that the lighted spots produce a picture frame on the entire area of the
screen; modulation of the beam intensity, that is, variation of the number of electrons
per unit time, so as to change the brighness of the spot according to the electric
video signal. The beam is scanned by the magnetic fields created by two pairs of
coils placed outside the kinescope neck, as illustrated schematically in Fig. 10.20.
Only one pair of coils is shown to facilitate the view. The coils create a magnetic
field in the vertical direction that deflects the beam horizontally. Another pair of
coils, in the vertical plane, creates a horizontal field for the vertical deflection. The
coils are fed by voltage signals that produce currents that vary in time in the form
of a saw-tooth periodic wave, shown in Fig. 10.20. This waveform makes the beam
scan the screen in one direction, and quickly return in the opposite direction. The
simultaneous application of the currents in the two pairs of coils causes the lighted
spot to scan the screen in the horizontal and vertical directions, however, each vertical
scan corresponds to many horizontal sweeps, as illustrated in Fig. 10.21a. The lighted
spot describes a zig-zag motion, from left to right in the horizontal direction, and
from top to bottom in the vertical direction. In the standard-definition TV system
(SD-TV), a picture frame consists of 525 horizontal lines, and the motion picture
has 60 frames per second. Thus, the frequency of the saw-tooth wave of the current
in the vertical deflection coils is 60 Hz, while in the horizontal deflection it is $525 \times$
60 Hz $= 31.5$ kHz.

For the formation of a black-and-white image, it is necessary to vary the intensity
of the beam as it scans the screen, so that the brightness of each spot corresponds
to that of the image. The variation in beam intensity is made by means of a voltage
signal applied to the grid of the electron gun, which controls the number of the
electrons in the beam. Figure 10.20 shows the waveform of an analog video signal
during a time interval of two periods of the horizontal scan. The video signal is such
that higher amplitudes produce darker spots. Thus, as the beam scans the screen, the
black-and-white picture frame is formed by the horizontal lines scanned from left to

(a)   (b)

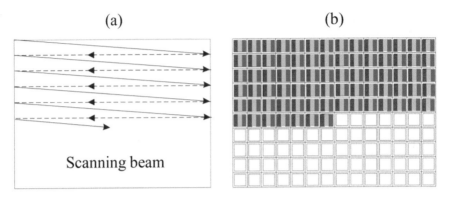

Scanning beam

Fig. 10.21 Illustration of the scanning process in video displays. **a** In a kinescope screen the light spot describes a zig-zag motion to form a picture frame. **b** In solid-state displays, the pixels are lit in sequence, from left to right, from top to bottom

right, and displayed from top to bottom. The spikes at the ends of the scans are used as synchronization pulses. They serve to trigger the return in the oscillator circuit that generates the saw-tooth current, so as to synchronize the screen scanning with that of the camera that produced the video signal. During one horizontal scan there are about 200 variations in the video signal, such that its frequency is on the order of 6 MHz. This is the bandwidth needed for the transmission of analog video signals.

The color CRT display works basically with three systems of monochrome images in the same tube, with three identical electron-gun devices, that produce three independent parallel electron beams. The three beams are emitted with the same kinetic energy so that they are equally deflected by the deflection coils. The intensity of each beam is determined by the video signal corresponding to each color. There are two technologies for the layout of the electron guns. They can be one above the other, in a vertical alignment, or arranged at the vertices of a triangle. In the first, the phosphors of the three colors are deposited in horizontal lines on the screen, so that each electron beam hits a different line. In the second, the three phosphors are deposited in small adjacent circles, with the centers at the vertices of a triangle, so that each circle is hit by one of the three beams. The information on the intensities of the three colors is carried by the video signal through a time-multiplexing process. The time interval of the signal corresponding to one pixel of the image is subdivided into three shorter intervals, one for each color, red, green and blue. This signal is decoded at the receiver, so that the information on the intensity of each color is processed, generating a voltage signal that acts on the corresponding electron gun. Despite several drawbacks, such as fragility, size, weight, and large power consumption, cathode-ray tubes were used for image displays in TV receivers, computers, and a variety of equipment for many decades. Only in the 1990s they began to give way to solid-state displays, initially with electroluminescent ceramics in flat-screen

TV receivers and with liquid-crystal displays in notebooks. Then, a variety of technologies were developed, most based on light emitting diodes, either with inorganic semiconductors or with organic conductors.

The first solid-state technology used in video displays that replaced the CRT screens in various applications employed electroluminescent ceramic devices. In an electroluminescent display, or ELD, the luminescence is produced directly in a piece of a ceramic material by an intense electric field, as described in the beginning of the section. This technology made possible the manufacture of displays with flat screen, with less weight, less power consumption, and longer lifetimes than CRT kinescopes. Figure 10.22 illustrates an electroluminescent device that produces light when subjected to an adequate electric voltage. It consists basically of five layers, deposited on a substrate using thin film preparation techniques. The phosphor is a layer of an electroluminescent ceramic with thickness of the order 500–1000 nm. It is excited by the electric field created by the voltage applied between the two electrodes, which are isolated from the phosphor by means of two layers (about 300 nm) of insulating material. The most used insulators are aluminum oxide, Al_2O_3, and an alloy of Al, Ti, and O, known as ATO. One of the electrodes is a thick metallic layer, for example Al, that reflects the light emitted by the phosphor. The other electrode is a layer (about 300 nm) of a transparent conductor, such as indium-tin oxide, known as ITO. For the light emission by the phosphor it is necessary that the electric field exceeds the dielectric strength of the material, so in general the applied voltage is pulsed and alternating, with amplitude in the range 120–200 V.

Each piece of the image, called **pixel**, is formed by three different devices of the type in Fig. 10.22, each with a different color phosphor. Among the most used phosphor materials are ZnS:Mn, ZnS:Cl, and CaS:Eu for the red color, ZnS:Tb and SrS:Ce for green, and Ga_2S_3:Ce, SrS:Eu, and SrS:Ag for blue. The intensity of the light emitted by each device is controlled by the repetition rate of the pulses of the applied voltage during the corresponding interval in the video signal. Combining the emission intensities in the devices of the three colors, it is possible to obtain any color of the visible spectrum. The screen is made up of hundreds of thousands of

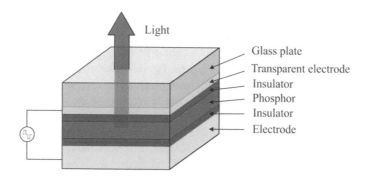

Fig. 10.22 Illustration of an electroluminescent device used in ELD displays

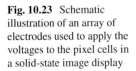

Fig. 10.23 Schematic
illustration of an array of
electrodes used to apply the
voltages to the pixel cells in
a solid-state image display

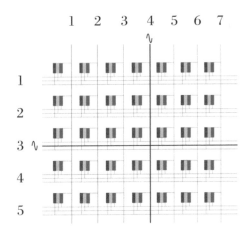

pixels, arranged side by side in rows and in columns, forming an array as illustrated in Fig. 10.21b. The image is produced by lighting the pixels in sequence, through a process of horizontal and vertical scans, following the same pattern previously described for kinescopes. The application of the voltage in each device is made through a mesh of addressing electrodes, schematically illustrated in Fig. 10.23, also manufactured using thin films techniques with appropriate masks.

The voltage in the form of pulses is applied between a line electrode and a column electrode, so that the device connected to these electrodes lights up. For example, in the case of Fig. 10.23, the device that is lit is the one corresponding to the green cell of the pixel in row 3 and column 4. The pixels are lit one at a time, in sequence, and with the intensity determined by the video signal, in a scanning process similar to that of the electronic beam at the cathode-ray tube.

ELD screens had several advantages over CRT kinescopes, such as less weight, higher mechanical resistance, thinner display, longer lifetime, and lower power consumption, so they gradually replaced the tubes in TV receivers, desktop computer monitors, medical and industrial equipment, etc. However, due mainly to the need of hundreds volts to light the phosphor, which is not convenient for portable equipment, in few years they gave way to LED displays in TV receivers and other equipment. At the same time, liquid crystal displays became the best technology for the use in notebooks, tablets, mobile phones, and a variety of other applications.

One technology that was also used in large flat video screens is the plasma display panel (PDP). The plasma screen is also based on emitting devices that employ luminescent materials, such as the phosphorescent ceramics used in CRT tubes and ELD screens. The difference to the others is that the light emission occurs by photoluminescence, and the excitation is produced by ultraviolet radiation emitted by a plasma. The screen is made of hundred thousands glass cells with the format illustrated in Fig. 10.24. The cells are hermetically sealed and contain a mixture of xenon-helium gas at low pressure. When an AC voltage of the order of 100 V is applied between the addressing electrodes, there is a discharge in the gas and the emission of ultraviolet

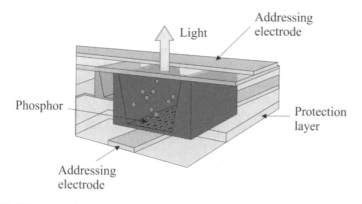

Light

Addressing electrode

Phosphor

Protection layer

Addressing electrode

Fig. 10.24 Illustration of a plasma cell

radiation. This radiation excites the phosphor layer deposited on the bottom of the cell, causing it to emit visible light. Figure 10.25 illustrates the arrangement of cells in a color PDP screen, which is powered by a mesh of addressing electrodes, like the one in Fig. 10.23. Plasma screens are durable, display bright images that can be seen in a large range of angles, and for several years were the best choice for large TV displays. However, due to the need of high voltages and the high fabrication costs, in the last decade they lost market and were gradually replaced by other technologies.

Currently, the most used technologies for computer monitors and small TV receivers are liquid crystals display (LCD) with LED backlight and thin-film transistor liquid-crystal display (TFT-LCD). The main technologies for large TV screens employ semiconductor light emitting diodes (LED), organic light emitting diodes (OLED), and quantum dot display (QLED). In all of them, the screen is made of

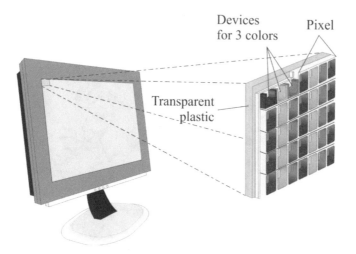

Devices for 3 colors

Pixel

Transparent plastic

Fig. 10.25 Schematic illustration of a solid-state color video screen

a large number of pixels, each with three light emitting devices arranged side-by-side in rows, such as in the schematic illustration of Fig. 10.25. In one standard of high-definition television (HD-TV), a picture frame has 1080 lines, instead of 525 as in the SD-TV, and each line has 1920 pixels. Thus, the number of pixels in one frame is 1920 × 1080, which is approximately two million pixels. Since the video information for one pixel is carried by several pulses, and a motion picture has 60 frames per second, the bandwidth required for a simple video signal exceeds several hundred Mbit/s, much larger than the 4 MHz bandwidth of SD-TV. The transmission of HD-TV video signals by carrier waves with frequencies in the usual TV bands was only made possible by the techniques for digital signal compression developed in recent times.

10.3.2   Liquid Crystals

As presented in Sect. 1.4.4, liquid crystals consist of elongated molecules, oriented approximately along the same direction, with no fixed positional order, so they can flow as in a liquid. One characteristic of liquid crystals that distinguish them from common liquids, is that their molecules are long and relatively rigid. It is the "contact" between long molecules that prevent them from occupying random directions, as in an isotropic liquid. A layer of molecules in a liquid crystal is, to a certain extent, similar to a set of wood logs, floating on the surface of a river.

Actually, the liquid crystal is a phase that certain substances exhibit in a temperature range between the solid phase and the liquid phase. Figure 10.26 illustrates the positions and orientations of molecules of a certain substance in three temperature ranges characterizing the solid, liquid, and liquid crystal phases. At $T < T_1$ the binding energy of the molecules dominates, so that they occupy fixed positions and exhibit positional and orientational order, characterizing the solid crystalline phase. At $T > T_2$ the thermal energy dominates, breaking the bonds between the molecules and making them have random positions and orientations, characteristics

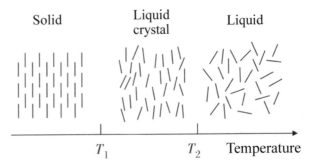

Fig. 10.26 Illustration of the positions and orientations of the molecules of a certain substance, characterizing the solid, liquid, and liquid crystal phases

of the liquid phase. The intermediate temperature range, $T_2 < T < T_1$, corresponds to the liquid crystal, where the thermal energy is sufficient to overcome the molecular bond without destroying completely the orientational order of the molecules.

The average direction of orientation of the molecules in the liquid crystal phase is called **director**, and represents a preferred alignment direction. The orientation of the molecules varies randomly over time, maintaining an average along the director. This situation resembles the configuration of the magnetic moments in a ferromagnetic material at high temperatures but still below T_c. However, in a liquid crystal the molecules also have a random displacement motion and can flow as in a liquid. It is customary to characterize the ordered phase of a physical system by an **order parameter**, that usually varies with temperature. The order parameter in magnetic materials is the magnetization, while in ferroelectrics is the polarization. In liquid crystals, the order parameter in a region with N molecules is defined by

$$S = \frac{1}{N} \sum_i (3\cos^2 \theta_i - 1)/2, \qquad (10.30)$$

where θ_i is the angle between the direction of each molecule and the director. Note that S is an angular average whose value is 1 for a system with a perfect orientational order ($\theta_i = 0$) and 0 for an isotropic system (Problem 10.7). Figure 10.27 shows the typical behavior of the order parameter of a substance with a liquid crystal phase. The order parameter gradually decreases with temperature in the liquid crystal phase, dropping sharply to zero in the transition to the liquid phase, that occurs at the critical temperature T_c.

The main application of liquid crystals is in various types of displays, simple ones for signs and alphanumeric symbols, and more sophisticated ones for images and picture motion. The liquid crystal display, known by the acronym LCD, is of the passive type, that is, it does not generate its own light. For this reason, its energy consumption is very low, which gives it a huge advantage over the emissive displays in applications that use small batteries, such as wristwatches and hand calculators.

There are two basic types of LCDs, reflection and transmission. In both types the liquid crystal has the role of changing the polarization of the light provided by other

Fig. 10.27 Variation with temperature of the order parameter in a substance in the liquid crystal phase

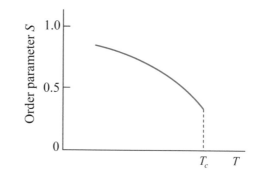

sources, such as external ambient light, or internal produced by a lamp or a LED. The LCD reflection mode uses front light, while the transmission mode uses backlight. The device consists of a liquid crystal layer, with thickness of the order 10 μm or less, placed between two transparent glass slides or plastic sheets, sealed at the ends, forming a closed cell. The surfaces of the layers have transparent conducting films, such as ITO, used to create an electric field by an applied voltage, that acts in the liquid crystal. The effect of LCD on the external light is produced by the liquid crystal molecules, oriented in a direction that can be altered by the electric field.

Among the types of liquid crystals presented in Sect. 1.4, the most used in LCDs is the nematic one. A simple LCD cell is made with a nematic liquid crystal between two plates with internal surfaces that are treated in order to force the molecules of the closest layers to orient themselves in the plane of the surfaces. If the two plates have alignment directions perpendicular to each other, with no applied electric field the director gradually changes from one surface to the other, as shown in Fig. 10.28a. When a voltage V is applied between the plates, the electric field tends to reorient the molecules in its direction, which is perpendicular to the surfaces. The reorientation occurs for fields larger than a threshold value E_T, corresponding to a voltage V_T on the order of a few volts.

The action of the liquid crystal on the external light is due to the strong polarization produced by the organic molecules. When polarized light passes through the cell with no electric field, as in Fig. 10.28a, its polarization follows the orientation of the molecules and undergoes a 90° rotation. However, for voltages in the range $V_T < V < V_b$, the electric field changes the orientation of the molecules so that the angle α of the polarization of the light that passes through the cell can be controlled by the applied voltage, as in Fig. 10.28b.

Figure 10.29 illustrates the operation of an LCD reflection device. It consists of the liquid crystal cell, two crossed polarizer sheets P_1 and P_2, and a mirror to reflect the incident light. The external light, initially depolarized, incident from left to right, becomes vertically polarized going through the polarizer P_1 and enters the liquid

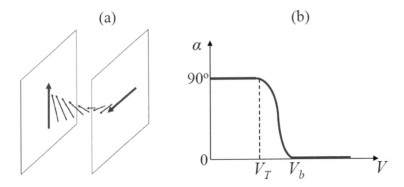

Fig. 10.28 Illustration of the orientation of the molecules of a liquid crystal in an LCD cell with the applied voltage V. **a** $V = 0$. **b** Variation with the applied voltage of the angle α of the polarization vector of the light after passing through the LCD cell

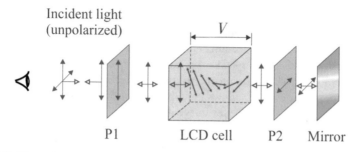

Fig. 10.29 Illustration of the operation of a LCD reflection device

crystal cell. If no voltage is applied to the cell, the polarization rotates by 90°, so that the light passes through the polarizer P_2, reflects on the mirror and returns, as shown in Fig. 10.29. Thus, with $V = 0$, the incident radiation is reflected by the device, which looks bright to an external observer on the left side. On other hand, when the applied voltage is such that $V > V_T$, the polarization angle of the light passing by the LCD cell is controlled by the voltage, so that the intensity of the light transmitted by the polarizer P_2 and reflected the mirror varies with the voltage. Thus, the brightness of the LCD device seen by the external observer is controlled by the applied voltage.

An LCD display consists of an array of LCD devices with a mesh of addressing electrodes, such as the one in Fig. 10.23. Each device for operation in the reflection mode is made of an LCD cell, two polarizers, and one mirror, all glued together, as shown in Fig. 10.30. It has total thickness of few μm and lateral dimensions of few mm. The electrodes are connected to the transparent metallic films on the two outer surfaces of the cell in each LCD device. The addressing mesh feeds the frame of pixels, so that each one has brightness determined by the applied voltage. Since the liquid crystal is insulating, the current through the cell is extremely small, so that the power consumption is very small. Actually, the use of DC voltages tends to shorten the lifetime of the LCD because after a number of cycles there are electromechanical reactions that hinder the motion of the molecules. For this reason, alternating voltages with square-wave form are used, with frequency in the range of 25 Hz to 1 kHz. This produces capacitive currents that result in a small increase in power consumption.

After dominating the market for displays in wrist watches, hand calculators and electronic equipment in general, liquid crystals entered the segment of video screens for computers monitors and TV receivers, initially black and white, and then color. Currently, liquid crystal video displays operate mainly in the transmission mode, in which cells such as the ones in Fig. 10.30 without the mirror are illuminated

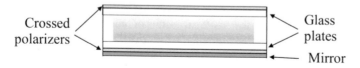

Fig. 10.30 Illustration of the cell of a liquid crystal reflection display (LCD)

from the back side by light emitting diodes. The performance of LCD screens is greatly improved by means of active-matrix techniques. These techniques consist of incorporating to each pixel a semiconductor device, a diode or a transistor. Since the activation of the liquid crystal pixel has a non-linear response, and it requires voltages $V > V_T$, the incorporation of a nonlinear semiconductor device increases the addressing possibilities. This allows the fabrication of screens with higher contrast and brightness, and therefore better image quality.

The liquid crystals used in LCD displays are organic compounds, with molecules formed by two or three benzene rings bound directly to each other. These compounds, synthesized in the last decades, have great chemical stability and exhibit liquid crystal phase in extensive ranges of working temperatures. As mentioned earlier, the only function of the liquid crystal is to vary the polarization of light, making possible to control its intensity by means of crossed polarizers. The colors of the cells that form the pixels on the color video displays are created by optical filters made of dielectric layers. The techniques for the fabrication of video screens employ the deposition of successive layers of thin films, so as to manufacture the semiconductor devices integrated with the liquid crystal cells and the color filters.

10.3.3   Organic Conducting Materials

The main characteristic of organic materials is that their molecules are hydrocarbons, that is, composed of carbon and hydrogen atoms with covalent bonds. The plant world and the animal world are formed by organic compounds produced by nature. In the twentieth century, the technology for the manufacture of artificial organic materials was developed, making possible the commercial production of a wide variety of materials for different applications. Currently, more than two million organic materials are known. They can be grouped into two broad categories, polymeric materials and non-polymeric materials.

Polymeric materials, commonly called plastics, have a huge variety of applications in our daily life. As shown in Sect. 1.4.3, polymers consist of molecules with long chain structure, formed by the repetition of simpler units, called monomers. These chains are easily formed by C and H atoms, so polymers are generally organic materials. The wealth of polymers stems from the fact that small changes in the constitution of monomers result in profound changes in its physical-chemical properties. Although polymers can be synthesized from a wide variety of raw materials, the most economical manufacturing processes are based on the transformation of petroleum derivatives. This is why the continuous emergence of new plastic materials after the Second World War is associated with the evolution of the petrochemical industry.

Polymeric materials used in traditional sectors of industry are electrical insulators. In electronics, they are essential for manufacturing different parts and pieces, such as, wire jackets of electric cables, insulating supports, equipment cases, buttons, knobs, and other pieces. Since traditional plastics are insulating, it caused a big surprise in the 1970s the discovery of new electrically conducting polymers, having electrical properties that resemble those of metals, semiconductors, or even superconductors. These materials are also known as non-conventional polymers. Recently these materials have found unusual applications in electronics, and several organic conductor devices are already commercially manufactured. The possibility of obtaining materials for practical use, combining electrical properties typical of inorganic materials with certain features of plastics, such as mechanical flexibility and optical transparency, has motivated intense research activity in the field of conducting polymers. Several of recent developments in this area are due to discoveries and scientific contributions made by Alan J. Heeger, Alan G. MacDiarmid, and Hideki Shirakawa, who received the Nobel Prize in Chemistry in the year 2000.

The binding of the atoms that form the polymer chains is of the covalent type, in which the valence electrons are shared by neighboring atoms. This binding is of the same type that exists in most inorganic semiconductors, but it is much stronger than metallic and molecular bonds. Each C atom, like Si and Ge, has four valence electrons, that are shared with the neighboring atoms. It is the strong covalent bond of atoms along the chain that gives cohesion to polymers. This enables the fabrication of thin sheets of plastic, with thicknesses of the order of some μm, with malleability not found in sheets made of other types of materials.

In contrast to the strong cohesion along the chains, the binding between neighboring chains is of the molecular type, so it is weak. For this reason, plastics commonly used are made with interlaced chains, in order to produce uniform resistance in all directions. However, for application in electronics, it is important that the material has the highest structural order as possible. This can be achieved through polycrystalline structures, like the one illustrated in Fig. 10.31. The material consists of regions of ordered polymer chains, separated by amorphous regions.

One of the most studied conducting polymers is **polyacetylene**. It consists of a chain of monomers containing only C and H atoms, represented by $(CH)_x$. It is a **conjugate polymer**, a name given to polymers that have carbons along the chain with alternating bonds, one single bond with one neighbor, and one double bond with another neighbor. Polyacetylene can be synthesized with two distinct structures, called **cis** and **trans**, shown in Fig. 10.32. Since the two structures have identical chemicals formulas, they are called isomers. In the cis structure, the H atoms bound

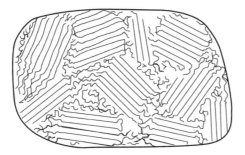

Fig. 10.31 Illustration of a polycrystalline polymer

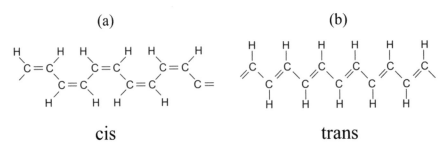

Fig. 10.32 Two polyacetylene isomers: **a** cis-$(CH)_x$; **b** trans-$(CH)_x$

to neighboring C atoms with double bonds are on the same side of the carbon chain, while in the trans structures, the H atoms are bound to C atoms alternately, on opposite sides of the chain. Thus, the neighboring H atoms are closer to each other in the cis structure than in the trans configuration. Polyacetylene is normally synthesized in cis form. Heating at 150 °C for some minutes produces the isomerization and transforms the cis form into the trans structure.

The different configurations of the H atoms in the cis-$(CH)_x$ and trans-$(CH)_x$ monomers result in very distinct electronic band structures, and therefore in different electric properties. While cis-$(CH)_x$ is electrically insulating, trans-$(CH)_x$ is a semi-conductor. Figure 10.33 shows the band structures of trans-$(CH)_x$, calculated for different distances of the carbon-carbon bonds. Since in conjugate polymers there are two types of bonds along the chain, it is necessary to consider two distances between neighboring carbons, d_1 for the C–C bond, and d_2 for C $=$ C. Figure 10.33 shows that the energy gap E_g between the valence and conduction bands depends on the distance of the bonds. In (a), with equal distances, $d_1 = d_2 = 1.39$ Å, the gap is zero, and therefore the polymer behaves as a metal. Figure 10.33b shows that slightly different distances, $d_1 = 1.43$ Å and $d_2 = 1.36$ Å, are already sufficient to produce

Fig. 10.33 Energy bands in trans-$(CH)_x$ polyacetylene for different distances of the bonds C–C (d_1) and C $=$ C (d_2): **a** $d_1 = d_2 = 1.39$ Å; **b** $d_1 = 1.43$ Å, $d_2 = 1.36$ Å; **c** Actual values $d_1 = 1.54$ Å and $d_2 = 1.34$ Å. Reproduced from P. M. Grant and I. P. Batra, Solid State Communications **29**, 225 (1979), with permission from Elsevier

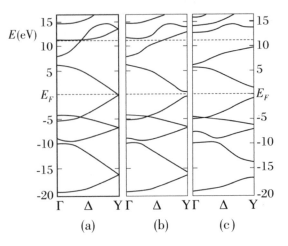

an energy gap. As can be seen in Fig. 10.33c, a larger difference between the bond distances, $d_1 = 1.54$ Å and $d_2 = 1.34$ Å, results in a larger gap. This is the reason why conjugate polymers, that necessarily have different bond distances d_1 and d_2, are the ones that have semiconductor properties interesting for electronics.

The values of d_1 and d_2 used in Fig. 10.33c are the actual bond distances of trans-polyacetylene at room temperature. They result in a band structure with direct gap, with energy $E_g = 1.5$ eV. This value is small enough for electrons to go from the valence to the conduction band by thermal excitation at room temperature. The chemical consequence of the electron transfer from the valence to the conduction band corresponds to the breaking of a double bond between the carbon atoms, that when changing to a single bond releases one electron to conduct the electric current. Electrons at the minimum of the conduction band have wave number $k = \pi/a$, effective mass $m^* = 0.1 \, m_0$, and collision time $\tau_e \sim 10^{-14}$ s. These values result in a mobility along the chain, given by Eq. (5.49), of $\mu_n \sim 200 \, \text{cm}^2/\text{Vs}$. Comparing this value with the data in Table 5.2, we see that it is of the same order of magnitude as the hole mobility in the traditional Si and GaAs semiconductors. For this reason, a sheet of trans-polyacetylene presents an optical brightness similar to that of silicon, however with mechanical flexibility typical of plastics.

The electronic properties of trans-polyacetylene can be changed by doping with donor or acceptor impurities, as in inorganic semiconductors. The p-type semiconductor can be obtained with impurities of arsenic pentafluoride (AsF_5) or iodine (I_2), diffused in $(CH)_x$ by means of vapor phase techniques or electrochemical processes. With doping, electron transfer occurs from the atoms of the polymer chains to the molecules of impurities, producing holes in the chains and consequently p-type semiconductor behavior. Figure 10.34 shows the increase in the conductivity of trans-$(CH)_x$ with the concentration of AsF_5 and I_2, expressed as a fraction of impurity molecules relative to those of the polymer. It can be seen that the conductivity varies almost seven orders of magnitude with doping by AsF_5. Similar conductivity

Fig. 10.34 Variation of the conductivity of trans-polyacetylene with the impurity concentration

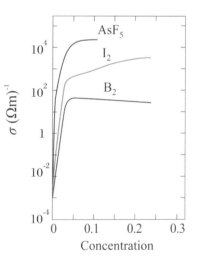

behavior is obtained by doping with alkali metal atoms, which produces a *n*-type semiconductor. It is important to draw attention to the fact that the charge transport mechanisms in conducting polymers is more complex than in metals and inorganic semiconductors. These mechanisms involve the motion of conformational defects of the type "soliton, or "polaron, that occur in the alternate bonds of conjugate polymers and which have no analogous in traditional materials. Thus, quite generally, one anticipates that solitons, polarons, and bipolarons will be the excitations of major importance in this class of one-dimensional polymer semiconductors.

Note that polyacetylene is conceptually important, because the discovery of its high conductivity upon doping was key to launch the field of organic conducting polymers and its study helped in the understanding of other organic materials. However, currently polyacetylene has no application in commercial devices. Other important conjugate polymers with semiconductor properties are polyaniline (PANI) and polyphenylene vinylene (PPV), whose chemical structures are shown in Fig. 10.35. Polyaniline looks similar to the plastics used in photographic films. It was one of the first polymers to be synthesized. Its manufacture is simple, low cost, and it is stable in the air. It has well-known physical-chemical properties, and can be synthesized with controlled impurities to produce suitable conductivities for different applications. Polyaniline is also widely used for printed circuit board manufacturing, in the final finish, for protecting copper from corrosion and preventing its solderability.

Figure 10.35b shows the chemical structure of PPV, that is stable up to temperatures of 400 °C. It has mechanical properties that allow its manufacture and processing in the form of thin films, with thicknesses in the range 0.02–1 μm. One of its most important property for application in electronics is electroluminescence. Similarly to direct gap inorganic semiconductors, electrons and holes in PPV can undergo recombination with the emission of photons with energy approximately equal to E_g. An advantage of PPV over many inorganic semiconductors and that it produces light in the visible spectrum, with wavelength that can be tuned by varying the distances of chemical bonds in the chain.

In addition to non-conventional polymers, there is a wide variety of organic materials with properties of electric conductors. Some are even superconductors at very

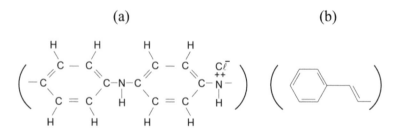

Fig. 10.35 Chemical structures of the monomers of two important semiconductor polymers: **a** polyaniline; **b** polyphenylene vinylene (PPV). In (**b**) the letters C and H indicative of the atoms are omitted to simplify the notation

low temperatures. A class of these materials which has been extensively investigated is that of charge transfer salts. One of these most studied salts is TTF-TCNQ, whose basic unit consists of a tetrathiafulvalene (TTF) molecule attached to a tetracyanoquinodimethane molecule (TCNQ). These molecules have a flat structure that aggregate on top of each other, forming layers of molecules, arranged along flat sheets. In the binding mechanism there is an electron transfer from the TTF molecule, called donor, to the TCNQ molecule, called acceptor. The overlap of the wave functions along the stack produces a conduction band that is partially filled with electrons from the charge transfer. As a result, the conductivity along the stack is reasonably high, of the order of $2 \times 10^3 \, \Omega^{-1} \, cm^{-1}$, while the conductivity along the planes is low, because the interaction between the layers of molecules is small. For this reason, TTF-TCNQ has conductivity predominantly in one dimension.

Another important conducting organic material is aluminum hydroxyquinoline, known as AlQ. It belongs to a class of compounds known as small molecule, because it contains a number of atoms much smaller than in most organic compounds. Its molecule is formed by an O_3N_3 group, surrounded by six benzene rings, some incomplete. AlQ is prepared in the form of small crystals, arranged in layers, that exhibit conduction and electroluminescence properties similar to those of PPV. One of the advantages of organic materials over inorganic ones is that they can be deposited in the form of films with an ordered structure on a wide variety of substrates. The manufacture of devices with organic materials has relatively low cost and can be made on polymeric substrate sheets, which can be rolled up and used in unusual applications. A disadvantage of organic materials is the low electron mobility. It is of the order of 1 cm^2/Vs in the best organic films, which is very low compared to the values 10^3–10^6 cm^2/Vs characteristic of inorganic semiconductors. This results in low response speed of devices with organic conductors.

The main applications of electronic devices made of organic conductors are biochemical sensors, thin polymer film transistors, and the organic light emitting diode (OLED). The polymer transistor has low response speed compared to those of silicon. For this reason, its use is restricted to low frequency applications, as is the case of video displays. They are used in active-matrix liquid crystal displays, in which each LCD device is activated by a polymer transistor. The advantage of polymer transistor over silicon is its lower processing cost and the ease of its direct deposition on the liquid crystal.

One of the most important uses in commercial products of organic material devices is the OLED, used in optical displays and image screens. Figure 10.36 shows the basic structure of an OLED. It consists of a transparent substrate, glass or plastic, on which five films are successively deposited: a positive metal electrode or anode; three layers of conductive organic materials; and one negative metallic electrode. The positive electrode is made of a transparent conductor, like ITO. The negative electrode, or cathode, is a common metal film, such as aluminum, which reflects visible light. The most used organic materials placed between the two electrodes are PPV and AlQ, which with the addition of dyes emit light at any wavelength in the visible range. The film that emits light, made of an intrinsic semiconductor, is located

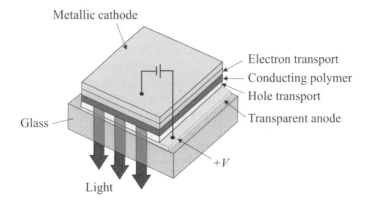

Fig. 10.36   Basic structure of an organic light emitting diode (OLED)

between a film doped with donor impurities and one with acceptor impurities, called layers of electron and hole transport, respectively.

When a voltage is applied between the electrodes, in the direction shown in Fig. 10.36, electrons are injected into the middle film by one layer, while holes are injected by the other layer. The recombination of electrons and holes produces light that is reflected by the aluminum film and is emitted by the front side through the glass plate. A major advantage of this LED is exactly the fact that the light comes out frontally, in a wide area, instead of the lateral emission confined to the junction region, as in inorganic semiconductor diodes. Currently, OLEDs made with PPV and AlQ operate with voltages less than 10 V and have conversion efficiency around 10%. They are employed in mobile phones, tablets, digital cameras, and a variety of video devices. In this segment they have supplanted liquid crystal displays because they emit light frontally, with higher brightness and larger viewing angle. Recently, more sophisticated OLED structures began to be used in large screens of TV receivers.

10.3.4   Touch-Sensitive Screens

A very important innovation introduced in the last two decades in electronic equipment with image displays is the touch-sensitive screen, or simply touchscreen. The image display can employ LCD or OLED, as used in smartphones, tablets, and laptops. The touchscreen allows the user to give input or control the information processing system through simple or multi-touch gestures by touching the screen with the fingers or a special stylus. The user can react to what is displayed and, if the software allows, to control how it is displayed, for example, zooming to increase the image or text size. The touchscreen enables the user to interact directly with what is displayed, rather than using a keyboard, mouse, or joystick. There are several technologies for operating touchscreens, the two most important are called resistive and capacitive.

The resistive touchscreen has the simplest structure and was the first to be manufactured commercially. It is made of a few layers of transparent plastic, placed above the screen that displays the image. The processing of the touch information can be done in several ways. One of them employs two active layers, each having one of its surfaces with thin parallel connecting lines, made with a resistive film. The two layers, separated by a thin gap, have the surfaces with the lines facing each other and oriented so that the lines of one are perpendicular to lines on the other. The gap is maintained by means of protrusions in one of the plastic layers. In this way the two layers form a matrix with rows and columns that define the coordinates of each point. A voltage is applied to the lines of one of the layers in such a way that when the outer layer is pressed by a finger, or a stylus, the two inner surfaces touch at that point and the voltage is transmitted to the other surface. This allows the circuit to identify the coordinates of the touch point on the screen. This process is done by pulsed signals that scan the lines and columns of the screen and are controlled by a software for processing the information. Resistive technology was introduced in electronic game screens, bank ATM displays, industrial and medical equipment, and dominated the market of touchscreens until about 2010.

The capacitive technology is the most used today in touchscreens of smart phones, tablets, and laptop computers. Like the resistive technology, there are several ways to detect and process touch information with capacitive screens. In one of them, the screen assembly consists of a rigid and insulating glass plate glued to the screen that displays the image and on which two plastic layers are placed. On the top surface of the glass plate, parallel metallic lines are made that form the sensing lines. The first layer of insulating plastic is glued to the glass and has on the top surface metallic lines oriented perpendicular to the lines on the glass, forming a matrix. The second layer is also a plastic one and serves to protect the system. The intersection of the lines in the glass and in the plastic forms parallel plate capacitors in which the insulator is the plastic layer itself. Capacitors are kept charged by a voltage applied to the plastic lines. When the user touches the screen with the finger, another capacitor is formed in the touched region, since the finger conducts electricity. This produces a variation in the capacitor charge in the region, resulting in an electric signal that is processed by the sensor line circuit indicating the position and characteristics of the touch. An advantage of the capacitive touchscreen is that it transmits about 90% of the light of the image produced on the screen, while the resistive screen transmits about 75%. Note that the operation of the capacitive screen requires the user to touch it with the finger or some object that conducts electricity. It does not work if the user is wearing a rubber glove or using an insulating touch object, such as an ordinary plastic pen.

10.4 Superconducting Materials

Superconducting materials are those with negligible resistance to electric currents. Superconductivity is observed in certain metal elements or alloys, at temperatures below a critical value T_c. This phenomenon was discovered in 1911 by the

Fig. 10.37 Variation of the
resistance of a sample of
mercury with temperature,
measured by Kamerlingh
Onnes in 1911

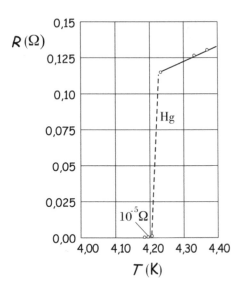

Dutch physicist Kamerlingh Onnes, who had managed to liquefy helium three years earlier. When making measurements of the electric resistance of materials around the temperature at which helium undergoes a transition from the gas to the liquid phase (4.2 K), he observed that the resistance of mercury dropped sharply to negligible values at a certain temperature $T_c \approx 4.2$ K. A reproduction of the original plot made by Kamerlingh Onnes is shown in Fig. 10.37.

In the following years, Onnes discovered that, even at $T < T_c$, superconductivity was destroyed and the resistance had normal values when the material was subjected to a magnetic field of intensity above a critical value H_c. He also observed that superconductivity was destroyed with the passage of an electric current with a density above a critical value J_c. From then on, countless laboratories and researchers all over the world began to investigate the electric and magnetic properties of materials, in search for new superconductors with higher critical temperatures. On the other hand, theoretical physicists started to seek an explanation for the unusual phenomenon. Onnes was awarded the Physics Nobel Prize in 1913 for the discovery of superconductivity.

Early studies in the field revealed that several simple metals were superconducting, all with low T_c values. Onnes himself observed the superconductivity in lead (Pb) in 1913, with $T_c = 7.2$ K. The highest critical temperature in a simple metal, niobium (Nb), with $T_c = 9.2$ K, was observed in 1930. Then, the investigations turned to alloys and intermetallic compounds, and various Nb compounds were discovered with higher T_c. However, until 1986, the highest known critical temperature was 23.2 K, observed in Nb_3Ge. That year, Alex Müller and Georg Bednorz, researchers at the IBM laboratory in Zurich, observed superconductivity in ceramic LaBaCuO, with critical temperature $T_c \approx 30$ K. This discovery revolutionized the field of superconductivity and stimulated researchers to look for superconductivity in new classes

Fig. 10.38 Variation of $YBa_2Cu_3O_7$ resistivity with temperature measured by Chu. Reproduced with permission from P. Chu et al., Physical Review Letters **58**, 908 (1987). Copyright (1987) by the American Physical Society

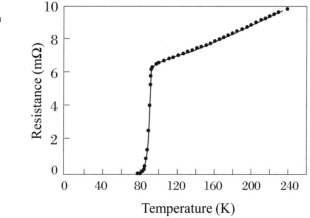

of materials still unexplored. Müller and Bednorz were awarded the Physics Nobel Prize in 1987 for their seminal discovery.

Shortly after Müller and Bednorz's discovery, Paul Chu observed superconductivity in a ceramic material with chemical formula $YBa_2Cu_3O_7$, known by the acronym YBaCuO, with $T_c = 92$ K. Figure 10.38 shows the measurement of the electric resistivity of this compound as a function of temperature, measured by Chu. The importance of Chu's discovery lies in the fact that YBaCuO was the first material to exhibit superconductivity at a temperature above 77 K. This is the temperature of liquefaction of nitrogen, much higher than that of helium. Helium and nitrogen are the most used cryogenic liquids to keep materials at temperatures much lower than 300 K.

Since it is much easier and more economical to work with liquid nitrogen than with liquid helium, the discovery of superconductivity in YBaCuO aroused the hopes of practical application of superconductors. Since 1987, several other superconducting cuprous oxides have been synthesized with critical temperatures above 77 K. The stable material of highest known T_c at normal pressure is $HgCa_2Ba_2Cu_3O_8$, that has $T_c = 134$ K. These materials are called high T_c superconductors. Table 10.5 presents the critical temperatures, critical fields and two important lengths that will be explained in Sect. 10.4.2, for superconducting materials of different classes.

10.4.1 Magnetic Properties of Superconductors

Superconducting materials exhibit strong magnetic behavior at temperatures below T_c. This was first observed by Meissner and Ochsenfeld, in 1933, who discovered that simple metals which are superconductors have **perfect diamagnetism** at $T < T_c$. They observed that when a superconductor is subjected to an external magnetic field H, the induction field B is expelled from its interior as the temperature is lowered

Fig. 10.39 Illustration of the Meissner effect in a superconducting sphere. The induction field B is expelled from inside the sphere at $T < T_c$

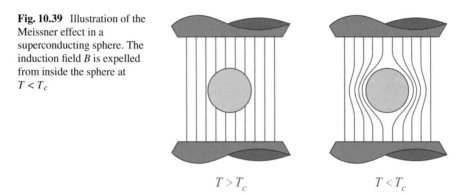

$T > T_c$ $T < T_c$

below T_c. This phenomenon, illustrated in Fig. 10.39, known as **Meissner effect**, occurs only for fields with intensity $H < H_c$, because above H_c the material is in the normal state at any temperature. Since $\vec{B} = \mu_0(\vec{H} + \vec{M})$, the Meissner effect implies that, at $T < T_c$ and $H < H_c$,

$$\vec{B} = -\vec{M}, \tag{10.31}$$

inside the superconductor. The magnetization in the superconductor does not originate from atomic magnetic dipoles, as in magnetic materials. It results from macroscopic currents, induced in the superconductor by the application of the magnetic field, called **supercurrents**. Supercurrents are induced by the Faraday effect, and since the material resistance is negligible, they persist for a long time. In pure materials they can last up to thousands of years. Due to Lenz's law, supercurrents have a sense such as to counteract the magnetic field, and for that reason they create an effective magnetization in opposition to the field. Actually, only superconductors made of simple metals have magnetization given by Eq. (10.31) in the entire range $H < H_c$. These materials, called **type I superconductors**, have magnetization M that varies with H as in Fig. 10.40a.

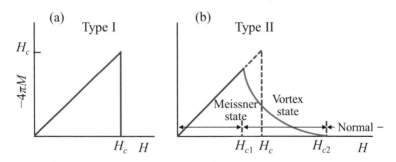

Fig. 10.40 Variation of the magnetization with the applied magnetic field H in superconducting materials. **a** Type I. **b** Type II

There is another class of materials, called **type II superconductors**, in which $M = -H$ only for H fields smaller than a value H_{c1}. In these materials there are two critical fields, H_{c1} and H_{c2}. The field H_{c2} is the value above which the material is no longer superconducting, that is, it has normal resistance, while H_{c1} is the field below the which the material is perfectly diamagnetic.

In type II superconductors, the variation of M with H has the shape shown in Fig. 10.40b. Thus, the magnetic behavior is characterized by three distinct phases: $H < H_{c1}$, the Meissner phase, the material is completely diamagnetic ($M = -H$); For $H > H_{c2}$, we have the normal phase, in which $M = 0$ and the resistance is normal; For intermediate fields, $H_{c1} < H < H_{c2}$, there is a mixed phase, in which the magnetic behavior is more complex. In this phase the material is diamagnetic, but the diamagnetism is not perfect, that is, $|M| < |H|$ because the expulsion of the B field from the interior of the material is not complete. As shown in Fig. 10.41, some induction lines remain inside the material, confined to tiny filaments with diameter less that 100 nm, called **vortices**. In the filamentary regions with the induction lines the material is in the normal state, while in the rest it is in the superconducting phase. In the superconducting regions there are supercurrents circulating around the filaments, in order to maintain the field of the vortices. For this reason the material is also said to be in a **vortex state**. Using concepts of quantum mechanics applied to superconductors, it is possible show that the magnetic flux in each vortex, called a **fluxoid**, is given by

$$\Phi_0 = \frac{h}{2e} = 2.067 \times 10^{-7} \text{ gauss cm}^2. \tag{10.32}$$

Thus, the magnetic flux through a superconducting the material is quantized, and is equal to a multiple of Φ_0.

Type I superconductors are simple metals, consisting only of one chemical element. As shown in Table 10.5, they have critical temperatures below 10 K and critical fields of a few hundred oersteds. On the other hand, intermetallic compounds

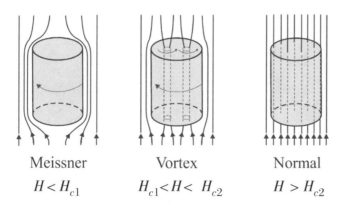

| Meissner | Vortex | Normal |
| $H < H_{c1}$ | $H_{c1} < H < H_{c2}$ | $H > H_{c2}$ |

Fig. 10.41 Behavior of the magnetic induction lines in the three magnetic phases of type II superconductors

Table 10.5 Parameters of some superconducting materials

Material	T_c (K)	H_c (kOe)	λ_L (Å)	ξ (Å)
Simple metals				
Al	1.1	0.1	160	16,000
Sn	3.9	0.3	340	2,300
Pb	7.2	0.8	370	830
Nb	9.5	2.0	400	380
Alloys and binary compounds				
$Nb_{0.3}Ti_{0.7}$	9.2	140	600	450
Nb_3Al	18.5	325	–	–
Nb_3Sn	18.1	240	800	35
Nb_3Ge	23.2	380	–	
High T_c cuprous oxides				
$YBa_2Cu_3O_7$	92	~1,500	4,000	~10
$Bi_2Ca_2Sr_2Cu_3O_{10}$	110	~2,500	~6,000	~10
$Tl_2Ca_2Ba_2Cu_3O_{10}$	125	>1,300	–	~13
$HgCa_2Ba_2Cu_3O_8$	134	>1,500	–	~13

and cuprous oxides are type II superconductors, with higher critical temperatures. In this case, the critical fields presented in Table 10.5 are those in which superconductivity is destroyed, that is, H_{c2}. We see that the critical fields in type II superconductors are considerably higher than in type I superconductors. This is the main reason why type II superconductors are more important technologically than type I.

The critical field values presented in Table 10.5 are valid at $T = 0$ K. Actually, the fields H_c, H_{c1}, and H_{c2} vary with temperature. As shown in Fig. 10.42, the critical fields decrease with increasing temperature, so that the fields that destroy the superconductivity decrease at larger temperatures. The phase diagrams shown in

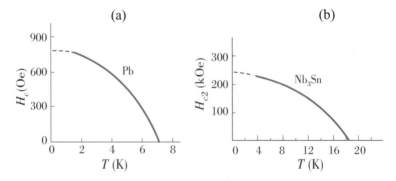

Fig. 10.42 Variation of critical fields with temperature in typical superconductors. **a** Pb, type I. **b** Nb_3Sn, type II

Fig. 10.42 for typical superconductors, Pb and Nb_3Sn, have shapes that are similar to all type I and type II superconductors.

10.4.2 The Physics of Superconductivity

The superconductivity of materials was one of the most intriguing and challenging physical phenomenon of the twentieth century. Since its casual discovery in 1911, it attracted great interest of experimental and theoretical physicists, in the search for new superconducting materials and the theoretical explanations for the observed phenomena. In 1934 Hans and Fritz London announced a phenomenological theory that explained the Meissner effect. However, two more decades of work were necessary until the formulation of a convincing microscopic theory was formulated, as announced in 1957 by John Bardeen, Leon Cooper, and Robert Schrieffer. The so-called BCS theory was very successful in explaining various aspects of the superconductivity observed in several materials, and Bardeen, Cooper, and Schrieffer received the Physics Nobel Prize in 1972 for their groundbreaking contribution.

The BCS theory seemed to solve the mysteries of superconductivity. However, with the discovery of high T_c superconductors in 1986, it was realized that the BCS theory did not explain the superconductivity of these materials, and, so far, they still do not have a convincing microscopic theory. In this section we shall present the most important results of London's theory and some notions about the basic mechanism of the BCS theory.

The London theory for the behavior of the magnetic field is based on the equations of electromagnetism and the basic property of superconductors, that is, zero resistance. The model assumes that the electric current in the material is carried by two types of particles, normal electrons, that are scattered by impurities or by phonons, and superconducting particles, that do not suffer collisions. The component of the current carried by the superconducting particles is called **supercurrent**. The equation of motion of these particles in an electric field $\vec{E}$ is

$$m \frac{d\vec{v}_s}{dt} = q \vec{E}, \qquad (10.33)$$

where m, $\vec{v}_s$, and q are, respectively the mass, velocity and charge of the superconducting particles. Denoting by n_s the concentration of these particles, the current density $\vec{J} = n_s q \vec{v}_s$ obtained with Eq. (10.33) satisfies the following equation

$$\frac{d\vec{J}}{dt} = \frac{n_s q^2}{m} \vec{E}, \qquad (10.34)$$

Substitution of this expression for the electric field in Maxwell's Eq. (2.3) leads to

$$\frac{\partial}{\partial t}\left(\nabla \times \vec{J} + \frac{n_s q^2}{m}\vec{B}\right) = 0. \tag{10.35}$$

Integrating this equation in time and considering that with $\vec{B} = 0$ there is no current in the superconductor, we obtain.

$$\nabla \times \vec{J} + \frac{n_s q^2}{m}\vec{B} = 0. \tag{10.36}$$

This is the London equation, that relates the current with the magnetic field in a superconductor. To obtain the field equation, we substitute (10.36) in Maxwell's Eq. (2.4). Considering that the fields do not vary in time ($\partial/\partial t = 0$) and the relation $\vec{B} = \mu_0 \vec{H}$, valid for the microscopic field, we obtain

$$\nabla \times \nabla \times \vec{B} + \mu_0 \frac{n_s q^2}{m}\vec{B} = 0. \tag{10.37}$$

Using known relations between differential operators and Eq. (2.2), we obtain the equation that describes the variation of the $\vec{B}$ field in a superconductor

$$\nabla^2 \vec{B} = \frac{1}{\lambda_L^2}\vec{B}, \tag{10.38}$$

where

$$\lambda_L = \left(\frac{m}{\mu_0 n_s q^2}\right)^{1/2} \tag{10.39}$$

is the **London penetration length, or London length**. Table 10.5 presents the values of λ_L for some superconductors. Let us use Eq. (10.38) to calculate the variation of the magnetic field in a semi-infinite superconductor, with a flat surface, illustrated in Fig. 10.43. We assume that the field is uniform outside the material, $x < 0$, and parallel to the surface, $\vec{B} = \hat{z}\, B_0$. Since B_z only varies in the x direction, Eq. (10.38) at $x > 0$ reduces to

$$\frac{d^2 B_z(x)}{dx^2} = \frac{1}{\lambda_L^2}B_z(x). \tag{10.40}$$

The solution of this equation is

$$B_z(x) = C_1\, e^{-x/\lambda_L} + C_2\, e^{x/\lambda_L}. \tag{10.41}$$

Since the field must be finite at $x \to \infty$, it is necessary that $C_2 = 0$. Due to continuity on the surface at $x = 0$, $C_1 = B_0$. Therefore,

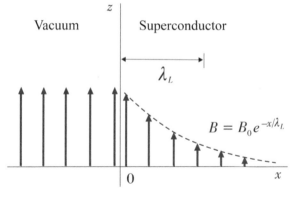

Fig. 10.43 Illustration of the variation of the magnetic induction field B in a superconductor. The field is practically confined to a superficial layer of thickness λ_L

$$B_z(x) = B_0\, e^{-x/\lambda_L}. \tag{10.42}$$

Thus, the external magnetic field applied to the superconductor penetrates only in a layer of thickness λ_L on the surface, and decays exponentially to zero inside the material. Since λ_L is in the range 50–500 nm, the layer is very thin, so that the field barely penetrates inside. This result explains the Meissner effect.

At the time of its publication in 1934, London theory was greeted with enthusiasm, but soon it was realized that it had a phenomenological character that explained only one aspect of superconductivity, the Meissner effect. However, the BCS theory announced two decades later, explains microscopically the zero resistance of super-conductors and is entirely quantum. Understanding the BCS theory requires advanced knowledge of quantum mechanics and statistical mechanics, which is beyond the level of this book. However, some quite elementary notions of the mechanism of superconductivity can be understood qualitatively.

The first important concept in the BCS theory is that of **Cooper pairs**. Under certain conditions, in a crystal lattice, two electrons forming a bound pair have less energy than they would have if they were independent. Since electrons have charges with the same sign, they suffer electrostatic repulsion. Thus, the formation of a pair requires the existence of an attractive interaction by some other mechanism. Using quantum theory, Cooper showed in 1956 that the interaction between electrons and phonons in a crystalline lattice can produce an attractive interaction between electrons and result in the formation of pairs. Figure 10.44 illustrates qualitatively how this is possible. When an electron travels in a lattice in equilibrium ($T = 0$ K), the ions of the lattice around it are disturbed slightly, due to the electrostatic interaction. Thus, upon reaching a certain point, the electron e_1 momentarily attracts neighboring ions. This produces a vibration wave in the lattice, that is, a phonon. This wave propagates in the lattice and can produce, at another point, a displacement of ions in the sense of creating an attractive potential for another electron e_2. If the energy of this pair is lower than that of the two independent electrons, they will form a bound state, called Cooper pair. The size of this pair is characterized by a distance ξ, called coherence length. The electrons that participate in this process have states close to the Fermi

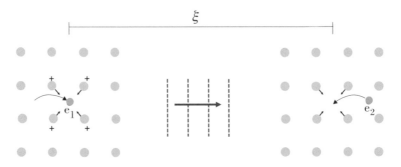

Fig. 10.44 Simple illustration of the attractive interaction between two electrons by means of perturbations in the crystal lattice. This the basic mechanism for the formation Cooper pairs

surface, and have the Fermi velocity v_F. It can be shown that the coherence length is given by

$$\xi = \frac{\hbar v_F}{\pi \Delta}, \tag{10.43}$$

where Δ is the energy reduction that an electron undergoes in the formation of the Cooper pair. This energy is of the order of some meV, which is a typical phonon energy. Table 10.5 shows the values of ξ for some superconductors. We see that in traditional materials, ξ is much larger than the distance between neighboring atoms in a crystal lattice. This means that two electrons can establish an attractive interaction and form a pair, having a large number of ions between them.

Actually, this view of the formation of Cooper pairs is extremely simplified. As mentioned earlier, the interaction between electrons through the phonons is an eminently quantum phenomenon. Its description is made in momentum space, and it can be shown that the two electrons in the pair have opposite wavevectors, $\vec{k}$ and $-\vec{k}$, as well as opposite spins. The Cooper pairs with charge $q = -2e$ and mass $m = 2m_0$, are the particles that produce the supercurrent. In the supercurrent, the Cooper pairs have a collective drift motion. So, while normal electrons move individually undergoing scattering by phonons and impurities, the pairs move collectively, without collisions. Therefore, the superconducting state results from the ordering of conduction electrons in pairs that are formed to decrease the total energy of the system.

The BCS theory explains why the superconducting state can be destroyed by an increase in temperature, or by the application of a magnetic field. The thermal energy resulting from an increase in temperature causes the effective binding energy of a Cooper pair to decrease with temperature, as shown in Fig. 10.45. Note the similarity between the curve in this figure and that of the variation with temperature of the magnetization in a ferromagnet shown in Fig. 9.4. This similarity is not accidental. The binding energy is the order parameter of the superconductor, and therefore has a certain analogy with the spontaneous magnetization of the ferromagnet. In both cases the thermal energy is equal to the ordering energy at the critical temperature

Fig. 10.45 Variation with temperature of the binding energy of Cooper pairs in superconductors

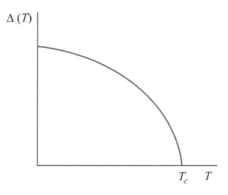

T_c, where a phase transition occurs. Below this temperature the order dominates and the collective phenomenon is established.

The destruction of superconductivity with the application of a magnetic field higher than a critical value is explained by the nature of the electrons in the Cooper pairs. As mentioned earlier, the two electrons in the pair have opposites spins. Thus, since the magnetic field tends to align the spins in its direction, the increase in H tends to break the pair, which occurs for $H > H_c$. Thus, it is expected that the field H_c varies with temperature in a way similar to the binding energy. This is the reason for the similarity between Figs. 10.42 and 10.45. The BCS theory for the zero resistance applies to both type I and type II superconductors. However the mechanisms for the magnetic behavior are different in the two types. The full theory for type II superconductors was developed by the soviet physicicts, Vitaly Ginzburg, Lev Landau, Alexei Abrikosov, and Lev Gorkov. Landau received the Physics Nobel Prize in 1962 for the theory of superfluidity, while Ginzburg and Abrikosov were awarded the Physics Nobel Prize in 2003 for their work in superconductivity.

The magnetic behavior of superconductors, and therefore their classification as type I or II, is directly associated with the relationship between the two relevant lengths in the theory, λ_L and ξ. Type I superconductors have $\xi \geq \lambda_L$, because they must have a distance between the electrons in the Cooper pairs (ξ), and therefore the spatial length of the superconducting state, larger that the characteristic distance of the magnetic field variation (λ_L). Notice in Table 10.5 that this is the case of simple metals Al, Hg and Pb. On the other hand, type II superconductors have $\xi \leq \lambda_L$, because in this case the field penetrates the material at distances larger than the length of the superconducting state. Thus, the material is characterized by normal regions, in the form of filaments of radius ξ, crossed by field lines, which are the vortices. This is the case of binary compounds and high-T_c superconductors, listed in Table 10.5. Note that despite being a simple metal, Nb has a behavior closer to type II superconductors.

To conclude this section, it is important to mention that the mechanisms responsible for the superconductivity in high-T_c materials are still not completely understood. It is known that the supercurrent is produced by particles of charge $q = -2e$, and therefore the superconducting state is formed by pairs of electrons, as in traditional materials. However, there are several experimental evidences that the formation of Cooper pairs is not mediated by phonons. This is consistent with the fact that these materials have a coherence length comparable to the lattice parameter, as shown in Table 10.5. In this situation it is expected that the attractive interaction between electrons is mediated by some mechanism of local interaction, which is not the case with vibration waves. So far, this mechanism has not been identified in all details.

10.4.3   Junctions with Superconductors

Chapters 5 and 6 presented various types of junctions of different materials, involving semiconductors, metals and insulators. In all cases the behavior of the current at the junction as a function of the applied voltage is determined by the properties of the particles responsible for the current. Since in superconductors these particles are electron pairs, it is to be expected that the junctions involving these materials have different properties than those previously studied.

To analyze the junctions with superconducting materials, it is necessary initially to understand certain properties of the conduction electrons. To form the Cooper pairs in the superconducting state, the energy of the electrons is reduced by a value Δ. Since the electrons that form pairs are those that are close to the Fermi surface in momentum space, this reduction produces an opening in the curve of the density of states around the energy E_F. Figure 10.46 shows the density of electronic states $D(E)$ as a function of the energy in a superconductor. The dashed line represents $D(E)$ in the normal metal at $T > T_c$, as in Fig. 4.10. At $T < T_c$ there is a reduction of Δ in the energy of the pairs and a corresponding increase Δ in the energy of the states with $E > E_F$, so that the energy gap becomes $E_g = 2\Delta$. At $T = 0$ only states with energy less than $E_F - \Delta$ are occupied. At $T > 0$, some electrons have enough

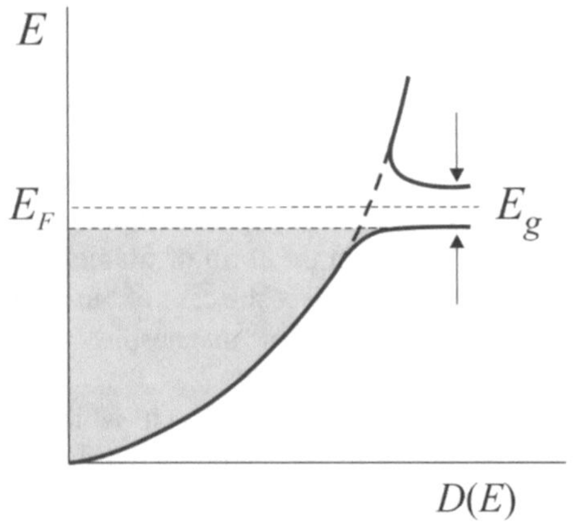

Fig. 10.46 Density of electronic states $D(E)$ in a superconductor

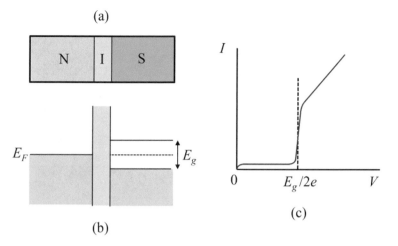

Fig. 10.47 NIS junction. **a** One-dimensional model. **b** Energy diagrams in the metal and in the superconductor. **c** I-V characteristics of the junction

thermal energy to go to the upper branch of the curve, breaking the corresponding Cooper pairs. Notice that the origin of this energy gap is entirely different than the one in insulators. While in insulators the gap is due to the interaction of electrons with the atoms or ions of the lattice, in a superconductor it originates in the attractive interaction of electrons above the Fermi level mediated by phonons.

Now consider a NIS junction formed by a normal metal (N), separated from a superconductor (S) by means of a thin insulating layer (I), as illustrated in Fig. 10.47a. If the thickness of the insulating layer is of the order of 10 nm or larger, the potential barrier prevents the flow of electrons from the N side to the S side, and vice-versa. However, if the layer is sufficiently thin ($\sim$1–2 nm), there is a significant probability that electrons on one side will go to the other side by means of the **tunnel effect**. For this to occur, it is necessary to have occupied states on one side and unoccupied states on the other side with the same energy.

As can be seen in Fig. 10.47b, this does not happen in equilibrium. One needs to apply a voltage V to the junction, in either direction, to make the energy diagram on one side go up, or go down, relative to the other side, by a value of eV. Thus, only if $V \geq E_g/e$, the tunneling current I will increase significantly. The variation of I with V in the NIS junction is shown in Fig. 10.47c.

Another important junction is the SIS, made of two superconductors separated by a thin insulating layer. In this case, if the superconducting material is the same on both sides, the energy diagram has the shape shown in Fig. 10.48a. In order for isolated electrons to go from one side to the other, there must be occupied states on one side, with the same energy as unoccupied states in the other side. Thus, when the voltage applied to the junction is $V \geq E_g/e$, there is a tunneling current of isolated electrons, as indicated in Fig. 10.48b. However, even with $V = 0$, there is a current with maximum value I_0, produced by tunneling of Cooper pairs. This phenomenon

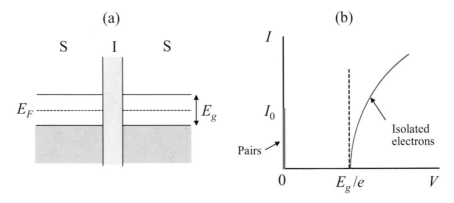

Fig. 10.48 Superconductor-insulator-superconductor (SIS) junction. **a** Energy diagram. **b** *I-V* characteristic, showing the dc Josephson effect at $V = 0$

is called **dc Josephson effect**, in honor of the British physicist Brian Josephson, who theoretically predicted it in 1962 in his Ph.D. thesis. The current at $V = 0$ is of a quantum nature, and can flow in either direction. Its value depends on the phase between the wavefunctions of the superconducting state in the two sides. The SIS junction is also known as **Josephson junction**.

At the SIS junction there is another important phenomenon, called **ac Josephson effect**. The application of a constant voltage V at the junction produces an alternating current, with frequency

$$f = \frac{2e}{h} V = \frac{V}{\Phi_0}, \tag{10.44}$$

where Φ_0 is the quantum of magnetic flux, given by Eq. (10.32). This phenomenon, also of quantum nature, results in an oscillation of the Cooper pairs current due to the variation in the phase of the wavefunction on one side of the junction relative to the other. For $V = 0.1$ mV, the frequency given by Eq. (10.44) is 48.36 GHz, located in the microwave region.

10.4.4 Some Established Applications of Superconductors

The most important technological applications of superconducting materials at the moment are concentrated in equipment that use intense magnetic fields. These fields are generated by coils made of superconducting wires with a large number of turns. Since the resistance of the wire is very small, it can carry a high current to generate an intense field, with very small heating. The superconducting coils are routinely used in laboratory electromagnets, in medical equipment for nuclear magnetic resonance tomography, and in high-power motors and generators. In general, they are made with multifilament thin wires of Nb-Ti or Nb_3Sn, that have critical field H_{c2} of 150 and 240 kOe, respectively, and critical currents on the order of 10^5 A/cm^2. The superconductor coils are commercialy made for fields in the range 100–200 kOe. They operate immersed in a liquid helium bath to maintain the low temperature and ensure that the wire remains in the superconducting phase. For this reason, equipment using superconducting coils are bulky and have high cost. High T_c superconductors are not yet used in these applications because they are brittle, and therefore difficult to handle for make windings. Furthermore, in the ceramic form they do not have sufficiently high critical currents.

Superconducting materials still do not have routine applications in electronic devices, mainly because of the need to operate at low temperatures. A possible potential application is in high integration circuits of electronic devices, in which the reduction of the component physical dimensions limits the dissipation of thermal energy. In this case, the replacement of the metallic films of the contacts and interconnections between the components by superconducting films would allow a further reduction in the device dimensions. In some situations, the use of superconducting films in these devices can be advantageous, even with the need to keep them at low temperature.

The junctions of superconductors also have potential application in specific electronic equipment. The Josephson junction, with the I-V characteristic in Fig. 10.48, presents a behavior with two different current states: $I \approx 0$ for $V < E_g/e$; $I > 0$ for $V > E_g/e$. In superconductor junctions, the switching from one state to the other is very fast, with picosecond time intervals (10^{-12} s), and with power dissipation of the order of pW. These features make Josephson junctions very attractive for digital applications, in logic circuits and in fast computer memories. Again, the main difficulty with this technology is the need to operate at low temperatures.

The ac Josephson effect has an important application in metrology. The traditional standard for voltage is an electrochemical battery, known as Weston cell. This cell has a voltage of 1.018 V and stability around 1 ppm. With this effect, it is possible to convert voltage into frequency, and vice versa, with great precision in the measurement of frequency. This is used in a voltage standard with precision and stability about 100 times higher than the Weston cell.

Another application of the Josephson junctions is in devices known as SQUID, a word formed by the initial letters of Superconducting Quantum Interference Device. The SQUID device is made of two Josephson junctions in parallel, as in Fig. 10.49. The current I that enters the device is divided into two components, which flow through the two Josephson junctions in the form of Cooper-pair currents. In this case, it can be shown that the dependence of each current on the phases of the wavefunctions on both sides results in a current that varies with the magnetic flux Φ across the circuit contour in the form

$$I = I_0 \, |\cos(\pi \, \Phi/\Phi_0)|, \tag{10.45}$$

(a)   (b)

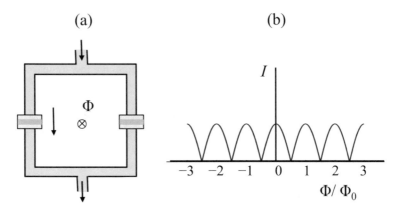

Fig. 10.49 SQUID device. **a** Schematic of the connection of the Josephson junctions. **b** Variation
of the current with the magnetic flux through the device

where Φ_0 is the quantum of magnetic flux, given by Eq. (10.32). This result shows
that when the SQUID is subjected to a magnetic field, the current varies periodically,
going through consecutive maxima as the flow passes by multiples of the quantum
Φ_0, as illustraded in Fig. 10.49b. Then, by means of a digital counter circuit, it
is possible to count the number of maxima that the current goes through, so as to
determine the final flux. We see from Eq. (10.45) that if the circuit has an area 1 cm^2,
the field corresponding to a quantum Φ_0 is $B \approx 2 \times 10^{-7}$ gauss. This extremely small
value (the Earth's magnetic field is about 0.5 gauss) allows the SQUID device to be
used to measure magnetic fields with great sensitivity and precision. The SQUID
magnetometers are routinely used in scientific, medical, and industrial equipment.

10.4.5   Superconducting Devices for Quantum Computing

To conclude this chapter, and the book, we present a promising application of superconducting devices in quantum computers. Quantum computing is the use of quantum
phenomena such as superposition and entanglement of quantum states to store and
process information. Quantum computers are expected to be able to solve certain
specific problems much faster than classical computers that operate with classical
bits. There are several models of quantum computing systems, but the most widely
used is the quantum circuit, based on the **quantum bit**, known as **qubit**.

In classical digital computers, all information is represented by a sequence of
binary digits, characterized as 0 or 1. These bits are stored by means of a charge,
or its absence, in a semiconductor memory device, or by the magnetization in one
direction, or in the opposite, in a magnetic medium. The bits are processed by means
of a low voltage signal with two levels, that is sent from one logic circuit to another
in the central processing unit (CPU), or from the CPU to a memory unit. It turns out

that a set with N bits (a register) has 2^N possible states in the range $\{0\ldots00, 0\ldots01, 0\ldots10,\ldots,1\ldots11\}$. Thus, a 200-bit memory can store any of 2^{200} possible bit strings, a set so large that it is physically impossible to deal with. However, while a 200-bit register can assume any of its 2^{200} possible states, we know that each bit is either zero or one, so at any given time the register can store only one of the bit strings.

In a quantum computer, the basic information is represented by a qubit. Just like a regular bit, a qubit has a zero (ground) state and a one (excited) state. However, what distinguishes a qubit from a classical bit is that it can be in a superposition of its 0 and 1 states, so that the number of possible states is very large. Mathematically, the state of a qubit is described by a 2D complex state vector of unit amplitude, that can be written as

$$|\psi\rangle = \begin{pmatrix} \alpha \\ \beta \end{pmatrix} = \alpha|0\rangle + \beta|1\rangle. \tag{10.46}$$

This plays the role of a wavefunction, where the first notation is a matrix representation, and the second is in the Dirac notation, where the symbol $|\ \rangle$ represents a *ket vector*, or simply *ket*. The coefficients α and β are complex numbers corresponding to the probability amplitudes of finding the state in one of the basis vectors, 0 or 1. Since the vector has unit amplitude, the coefficients are constrained by

$$|\alpha|^2 + |\beta|^2 = 1. \tag{10.47}$$

Note that a qubit in the superposition state given by (10.46), does not have a value between 0 and 1. Rather, when measured, the qubit has a probability $|\alpha|^2$ of having the value 0, and a probability $|\beta|^2$ of the value 1. In other words, superposition means that there is no way, even in principle, to tell which of the two possible states forming the superposition state actually pertains. Furthermore, the probability amplitudes, α and β, encode more than just the probabilities of the outcomes of a measurement. They also have relative phases, that are responsible for quantum interference. The great advance of using qubit instead of binary digits, is that with two bits and two probability densities, it is possible to represent a very large number of states. Besides the huge advantage in the representation of the information, the processing on a quantum computer is typically carried out as an in-memory operation. That is, rather than moving data from the memory to a CPU and back, the quantum register is operated upon directly. These two features give quantum computing a massive computation capacity.

In a quantum computer the set of qubits are produced by some initialization method, that depends on the physical circuits, and are processed in circuits consisting of quantum logic gates. Mathematically, the qubits undergo a (reversible) unitary transformation described by a matrix operation that preserves the qubits unit amplitudes. These gate operations must be fast relative to the qubit coherence time. Since a quantum computation consists of a series of gate operations, it is critical that the individual processes be carried out on a timescale orders of magnitude shorter than the coherence time of the individual qubits. Another requirement for a quantum computer is to have a method to measure the state of individual qubits to determine the solution of some computation in a reliable manner. The measurement is an irreversible operation that contributes to the loss of qubit coherence, so it must employ a very efficient mechanism to minimize interference in the processing.

Several qubit technologies have been used to manufacture the circuits for quantum computers. The main requirements of these technologies are: have a physical two-level quantum-mechanical system that can be used to represent qubits with sufficiently long coherence times; have the ability to control large arrays of qubits and an efficient manner for the qubits to interact with external controls; be scalable. The main physical systems used for implementation of qubits are nuclear spins of some isotopes, such as ^{31}P, electronic spins in cold atoms, in doped semiconductors, in trapped ions, and in quantum dots, single-photon polarization in integrated photonic devices, as well as in superconducting circuits.

Quantum computing with superconducting circuits are made mainly with **flux qubits**, also known as **persistent current qubits**. They consist of micrometer sized loops of a superconducting metal that is interrupted by one or more Josephson junctions. During fabrication, the junction parameters are engineered so that a persistent current will flow continuously when an external magnetic flux is applied. Only an integer number of flux quanta are allowed to penetrate the superconducting ring, resulting in clockwise or counter-clockwise mesoscopic supercurrents (typically 300 nA) in the loop, representing bits 0 or 1. When the applied flux through the loop area is close to a half-integer number of flux quanta, the two lowest energy eigenstates of the loop will be a quantum superposition of the clockwise and counter-clockwise currents, that is, a superposition of the two bits. The two lowest energy eigenstates differ only by the relative quantum phase between the composing current-direction states. Higher energy eigenstates correspond to much larger (macroscopic) persistent currents, that induce an additional flux quantum to the qubit loop, thus are well separated energetically from the lowest two eigenstates. This separation, known as the qubit nonlinearity criteria, allows operations with the two lowest eigenstates only, effectively creating a two-level system. Usually, the two lowest eigenstates will serve as the computational basis for the logical qubit.

Initializations and computational operations are implemented by pulsing the qubit with microwave pulses of frequency corresponding to the energy difference between the two basis states. Properly selected pulse duration and strength can put the qubit into a quantum superposition of the two basis states, while subsequent pulses can manipulate the probability weighting that the qubit will be measured in either of the two basis states, thus performing a computational operation.

Flux qubits are fabricated using techniques similar to those used to make semiconductor integrated circuits. The devices are usually made on silicon or sapphire wafers using electron beam lithography and metallic thin film evaporation processes. To create Josephson junctions, a technique known as shadow evaporation is normally used. This involves evaporating the source metal alternately at two angles through the lithographically defined mask in the electron beam resist. The process results in two overlapping layers of the superconducting metal, such as aluminum, between which a thin insulating layer of aluminum oxide is deposited, forming a Josephson junction.

The efficient coupling between qubits is essential for implementing many qubit gates. The coupling can be made in several ways. An example of a device with inductive coupling between two flux qubits is shown by the images in Fig. 10.50. In each qubit there are three Josephson junctions made as described in the previous paragraph, in a circuit closed by means of a superconducting Nb microstructure. The device is placed near the shorted end of a $\lambda/4$ microwave resonator used for readout. The opposite open end of the resonator is capacitively coupled to an on-chip coplanar waveguide. The circulating currents of the qubits inductively affect one another, clockwise current in one qubit induces counter-clockwise current in the other. This enables the implementation of a NOT gate operation.

Several circuits with Josephson junctions have been used to implement qubits in some successfully developed prototypes of quantum computers. This technology is attractive because the low dissipation inherent to superconductors enables, in principle, long coherence times. In addition, because complex superconducting circuits can be microfabricated using integrated-circuit processing techniques, scaling to a large number of qubits is relatively straightforward. Currently there is a worldwide race to manufacture quantum computers, and some first models have been announced. The reason for the race is that quantum computers are believed to be able to quickly solve certain problems that no classical computer could solve *in any feasible amount of time,* a feat known as "quantum supremacy."

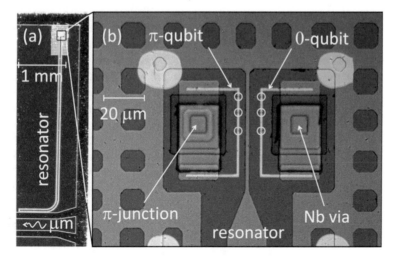

Fig. 10.50 Images of a quantum computer device with two coupled flux qubits. **a** $\lambda/4$ microwave resonator capacitively coupled to the transmission line. **b** Optical picture of the two composite Nb/Al flux qubits placed near the shorted end of $\lambda/4$ resonator. The Nb part of the left qubit contains the π-junction. Right qubit has Nb "via" structure forming a superconducting short. The yellow circles mark the positions of aluminum Josephson junctions. Reproduced with permission from A. V. Shcherbakova et al., Supercond. Sci. Technol. **28**, 025,009 (2015)

Problems

10.1   A parallel plate capacitor with a tantalum oxide insulator of thickness 1 μm
         has capacitance C = 1.0 μF:

(a)   Calculate the maximum voltage that can be applied to the capacitor;
(b)   Calculate the free charge density and the polarization charge density
         when the applied voltage is 10 V.

10.2   Ten PZT disks of thickness 1 mm are stacked to form a micropositioner.
         The disks are placed on top of each other, with alternating polarities, and
         separated by a copper sheet for application of the voltage. The terminals
         of the sheets are interconnected so that all disks are subjected to the same
         external voltage, alternately, so that the stack expansion is the sum of the
         disk expansions. Calculate the variation in the length of the micropositioner
         when subjected to a voltage of 100 V.

10.3   Calculate the thickness of a X-cut quartz plate, used to stabilize the oscillator
         of the transmitter in a radio station with frequency 720 kHz.

10.4   The unit cell of $BaTiO_3$ has a lattice parameter of 4.0 Å and electric dipole
         moment $p = 1.66 \times 10^{-29}$ Cm due to a small spontaneous displacement of
         the Ti^{4+} ions. Estimate the value of the piezoelectric constant of this material
         and compare with the data in Table 10.2.

10.5   Show that in a Mach-Zehnder type electro-optic modulator, the transmission
         is given by Eq. (10.29).

10.6   An electro-optic modulator of the Mach-Zehnder type with waveguides of
         $Ti:LiNbO_3$ has electrodes 5 mm long and 5 μm apart. Calculate the voltage
         required to produce the cut-off in a modulation of the type on-off.

10.7   In an isotropic liquid, the molecules can assume any direction in space with
         equal probability. Show that the integral in three dimensions of the angular
         factor in Eq. 10.30 is zero in this situation.

10.8   In the superconducting vortex state, each vortex has a flux Φ_0. Calculate the
         number of vortices per cm^2 in a material when the magnetic field through it
         is 5 T.

10.9   Calculate the London penetration length for a simple metal with $n_s = 10^{23}$
         cm^{-3}, $m = 2m_0$, and $q = -2e$, and compare with the data in Table 10.5.

10.10   What voltage is needed to produce an oscillation with frequency 100 GHz
         in a Josephson junction?

10.11   A SQUID magnetometer has a detector with area 2 cm^2. What is, in gauss
         and in tesla, the smallest variation in the magnetic field that can be measured
         by the magnetometer?

Further Reading

P.J. Collings, *Liquid Crystals* (Princeton University Press, Princeton, 1990)

M. Tinkham, *Introduction to Superconductivity* (Dover Publications, Mineola, N. Y., 2004)

R.E. Hummel, *Electronic Properties of Materials* (Springer-Verlag, Berlin, 2011)

I.C. Khoo, *Introduction to Liquid Crystals* (Wiley, New York, 2006)

C. Kittel, *Introduction to Solid State Physics* (Wiley, New York, 2004)

Y.A. Ono, *Electroluminescent Displays* (World Scientific, Singapore, 1995)

T. Ikeda, *Fundamentals of Piezoelectricity* (Oxford University Press, Oxford, 1997)

M.A. Nielsen, I.L. Chuang, *Quantum Computation and Quantum Information* (Cambridge University Press, Cambridge, 2000)

K.M. Rabe, C.H. Ahn, J.M. Triscone, *Physics of Ferroelectrics: A Modern Perspective, Topics in Applied Physics 107* (Springer-Verlag, Berlin, 2007)

L. Solymar, D. Walsh, *Lectures on the Electrical Properties of Materials* (Oxford University Press, Oxford, 2018)

R. Syms, J. Cozens, *Optical Guided Waves and Devices* (Mc Graw-Hill Book Co., New York, 1992)

A. Yariv, P. Yeh, *Photonics: Optical Electronics in Modern Communications* (Oxford University Press, Oxford, 2007)

Appendix A

Perturbation Theory

Calculation of the Transition Probability

In this appendix we present the calculation of the transition probability per unit time for a quantum system, initially in a state n, to go to another state m, due to some perturbation. The calculation is based on the theory of time-dependent perturbation studied in Sect. 8.3.1. As shown in that section, the quantum state of a system with Hamiltonian $H = H_0 + H(t)$ is described by a wave function $\Psi(t)$, which can be expanded as in (8.48)

$$\Psi(t) = \sum_n a_n(t)\psi_n e^{-iE_n t/\hbar}, \tag{A.1}$$

where ψ_n are the eigenfunctions of the constant part of the Hamiltonian, H_0, with energies E_n. Since the wavefunctions ψ_n are known, to determine the evolution of the system subjected to an excitation variable in time represented by the Hamiltonian $H'(t) = H' \exp(-i\omega t)$, it is necessary to obtain the coefficients $a_n(t)$. The starting point is Eq. (8.51)

$$\frac{da_m}{dt} = \frac{1}{i\hbar}\sum_n a_n(t)\,H'_{mn}\,e^{i(\omega_{mn}-\omega)t} \tag{A2}$$

where $\omega_{mn} = (E_m - E_n)/\hbar$. Note that this result is exact, since no approximation has been made so far. The problem is that Eq. (A.2) cannot be resolved analytically exactly for a general perturbation. To solve it approximately, we employ perturbation theory. For this we consider that the Hamiltonian of the excitation in time is small compared to the static Hamiltonian, that is, $H' \ll H_0$. Thus, the coefficients $a_n(t)$ can

S. M. Rezende, *Introduction to Electronic Materials and Devices*,
https://doi.org/10.1007/978-3-030-81772-5

be expanded in power series

$$a_n = a_n^{(0)} + a_n^{(1)} + a_n^{(2)} + \cdots , \tag{A.3}$$

where $a_n^{(0)}$ is the value that a_n would have if $H' = 0$, $a_n^{(1)}$ is the first-order term in H', $a_n^{(2)}$ is the second-order term, etc. Substitution of Eq. (A.3) into (A.2) gives

$$\dot{a}_m = \dot{a}_m^{(0)} + \dot{a}_m^{(1)} + \dot{a}_m^{(2)} + \cdots = \frac{1}{i\hbar} \sum_n \left(a_n^{(0)} + a_n^{(1)} + a_n^{(2)} + \cdots \right) H'_{mn} \, e^{i(\omega_{mn} - \omega)t}. \tag{A.4}$$

Equating the terms of the same order on the right- and left-hand sides of this equation we obtain

$$\dot{a}_m^{(0)} = 0$$

$$\dot{a}_m^{(1)} = -\frac{i}{\hbar} \sum_n a_n^{(0)} H'_{mn} \, e^{i(\omega_{mn} - \omega)t}$$

$$\dot{a}_m^{(2)} = -\frac{i}{\hbar} \sum_n a_n^{(1)} H'_{mn} \, e^{i(\omega_{mn} - \omega)t}$$

$$\vdots$$

$$\dot{a}_m^{(s)} = -\frac{i}{\hbar} \sum_n a_n^{(s-1)} H'_{mn} \, e^{i(\omega_{mn} - \omega)t} \tag{A.5}$$

The zero-order solution is obtained from the first equation, $a_m^{(0)} = $ constant. This means that if there is no perturbation, the system remains in the initial stationary state indefinitely. Assuming it is initially in state n we have

$$a_n^{(0)} = 1,$$
$$a_m^{(0)} = 0 \quad m \neq n \tag{A.6}$$

The first-order solution is obtained from the second equation in (A.5), which can be written in the form

$$\dot{a}_m^{(1)} = -\frac{i}{\hbar} H'_{mn} \, e^{i(\omega_{mn} - \omega)t}. \tag{A.7}$$

Let us now consider that the excitation of the system is turned on only at a time $t = 0$, that is, $H' = 0$ for $t < 0$. Integration of (A.7) leads to

$$a_m^{(1)}(t) = -\frac{i}{\hbar} \int_0^t H'_{mn} e^{i(\omega_{mn}-\omega)t'} dt' = -\frac{1}{\hbar} \left[H'_{mn} \frac{e^{i(\omega_{mn}-\omega)t} - 1}{\omega_{mn} - \omega} \right]. \qquad (A.8)$$

Since $\Psi_m^* \Psi_m$ is the probability density of finding the system in the state m, it can be seen that the probability of the system to undergo a transition from state n for another state m is given by

$$\left|a_m^{(1)}\right|^2 = \frac{4\left|H'_{mn}\right|^2}{\hbar^2} \frac{\sin^2[(\omega_{mn} - \omega)t/2]}{(\omega_{mn} - \omega)^2}. \qquad (A.9)$$

As we know, the linewidth of the transition cannot be zero. So we shall consider that n and m are actually two groups of states. Thus, the probability of the system to be found in the group of m states is given by

$$\left|a_m^{(1)}\right|^2 = \frac{4}{\hbar^2} \int_{-\infty}^{\infty} \left|H'_{mn}\right|^2 \left\{ \frac{\sin^2[(\omega_{mn} - \omega)t/2]}{(\omega_{mn} - \omega)^2} \right\} D(\omega_{mn}) d\omega_{mn}, \qquad (A.10)$$

where $D(\omega_{mn})$ is the joint density of states associated with the two groups of states m and n. Note that the function of ω_{mn} between the brackets has a value $t^2/4$ for $\omega_{mn} = \omega$. When ω_{mn} goes away from ω, this function oscillates with decreasing amplitude due to the increase in the denominator. It is easy to see that the linewidth of this function around $\omega_{mn} = \omega$ is approximately $2\pi/t$. Thus, after a relatively large time t, the function between the brackets has a small width in the region $\omega_{mn} \approx \omega$. Thus, the density of states can be considered approximately constant with the value $D(\omega_{mn} = \omega)$ in this region, so that it can be removed from the integral. Using the definite integral

$$\int_{-\infty}^{\infty} \frac{\sin^2(xt/2)}{x^2} dx = \frac{\pi t}{2}, \qquad (A.11)$$

we obtain

$$\left|a_m^{(1)}\right|^2 = \frac{2\pi}{\hbar^2} \left|H'_{mn}\right|^2 D(\omega_{mn} = \omega) t. \qquad (A.12)$$

Therefore, the probability per unit time for the system to undergo a transition from the group of states n to the group of states m, given by $\left|a_m^{(1)}\right|^2/t$, becomes

$$W_{n \to m} = \frac{2\pi}{\hbar} \left|H'_{mn}\right|^2 D(E_m = E_n + \hbar\omega) \qquad (A.13)$$

where $D(E)\, dE = D(\omega)\, d\omega$ is the number of states with energy between E and $E + dE$. This result is known as the **Fermi golden rule**, Eq. (8.53).

Appendix B

Physical Constants and Table for Conversion of Energy Units

B1-Physical Constants

Quantity	Symbol	Value	CGS	SI
Electron mass	m_0	9.10956	10^{-28} g	10^{-31} kg
Electron charge	e	1.60219	–	10^{-19} C
(modulus)		4.80325	10^{-10} esu	–
Planck constant	h	6.62620	10^{-27} erg.s	10^{-34} J.s
	$\hbar = h/2\pi$	1.05459	10^{-27} erg.s	10^{-34} J.s
Speed of light	c	2.99792	10^{10} cm/s	10^8 m/s
Proton mass	M_p	1.67261	10^{-24} g	10^{-27} kg
Boltzmann constant	k_B	1.38062	10^{-16} erg/K	10^{-23} J/K
Bohr magneton	μ_B	9.27410	10^{-21} erg/G	10^{-24} J/T
Permittivity of vacuum	ε_0	–	1	$10^7/4\pi c^2 = 8.85 \times 10^{-12}$ C^2/Nm^{-2}
Permeability of vacuum	μ_0	–	1	$4\pi \times 10^{-7}$ T.m/A

© The Editor(s) (if applicable) and The Author(s), under exclusive license
to Springer Nature Switzerland AG 2022
S. M. Rezende, *Introduction to Electronic Materials and Devices*,
https://doi.org/10.1007/978-3-030-81772-5

B2-Conversion of Energy/Frequency Units

	Hz	cm^{-1}	eV	J	K	Oe*
Hz	1	3.3357×10^{-11}	4.1357×10^{-15}	6.6262×10^{-34}	4.7994×10^{-11}	3.5714×10^{-7}
cm^{-1}	29.979×10^9	1	1.2398×10^{-4}	1.9865×10^{-23}	1.4388	1.0707×10^4
eV	2.4180×10^{14}	8.0655×10^3	1	1.6022×10^{-19}	1.1605×10^4	8.6355×10^7
J	1.5092×10^{33}	5.0341×10^{22}	6.2414×10^{18}	1	7.2431×10^{22}	5.3898×10^{26}
K	20.836×10^9	0.69502	8.6170×10^{-5}	1.3806×10^{-23}	1	7.4413×10^3
Oe*	2.80×10^6	9.3399×10^{-5}	1.1580×10^{-8}	1.8554×10^{-27}	1.3438×10^{-4}	1

To convert the value of a quantity expressed in the column unit on the left, to the corresponding unit to one of the columns, multiply by the value in the corresponding row and column.

*Calculated with $\gamma = 2.8$ MHz/Oe

Appendix C

Periodic Table of the Elements

Legend (example cell):

Fe 26	← Atomic number
55,847	← Atomic mass
bcc	← Crystal structure
7,87	← Specific mass (g/cm³) in the solid phase

Periodic Table

I A	II A	III B	IV B	V B	VI B	VII B	VIII	VIII	VIII	I B	II B	III A	IV A	V A	VI A	VII A	VIII A
H¹ 1,00797 hcp 0,088																	He² 4,00260 hcp 0,205
Li³ 6,941 bcc 0,542	Be⁴ 9,01218 hcp 1,82											B⁵ 10,81 rombo 2,47	C⁶ 12,01115 diamante 3,516	N⁷ 14,0067 cúbico 1,03	O⁸ 15,9994 complexa	F⁹ 18,99840	Ne¹⁰ 20,179 fcc 1,51
Na¹¹ 22,98977 bcc 1,013	Mg¹² 24,305 hcp 1,74											Al¹³ 26,98154 fcc 2,70	Si¹⁴ 28,0867 diamante 2,33	P¹⁵ 30,97376 complexa	S¹⁶ 32,064 complexa 2,03	Cl¹⁷ 35,453 complexa 2,03	Ar¹⁸ 39,948 fcc 1,77
K¹⁹ 39,098 bcc 0,91	Ca²⁰ 40,08 fcc 1,53	Sc²¹ 44,9559 hcp 2,99	Ti²² 47,90 hcp 4,51	V²³ 50,9414 bcc 6,09	Cr²⁴ 51,996 bcc 7,19	Mn²⁵ 54,9380 cúbico 7,47	Fe²⁶ 55,847 bcc 7,87	Co²⁷ 58,9332 hcp 8,9	Ni²⁸ 58,71 fcc 58,71	Cu²⁹ 63,546 fcc 8,93	Zn³⁰ 65,38 hcp 7,13	Ga³¹ 69,72 complexa 5,91	Ge³² 72,59 diamante 5,32	As³³ 74,9216 rombo 5,77	Se³⁴ 78,96 hex. 4,81	Br³⁵ 79,904 complexa 4,05	Kr³⁶ 83,80 fcc 3,09
Rb³⁷ 85,4678 bcc 1,629	Sr³⁸ 87,62 fcc 2,58	Y³⁹ 88,9059 hcp 4,48	Zr⁴⁰ 91,22 hcp 6,51	Nb⁴¹ 92,9064 bcc 8,58	Mo⁴² 95,44 bcc 10,22	Tc⁴³ 98,9062 hcp 11,5	Ru⁴⁴ 101,07 hcp 12,36	Rh⁴⁵ 102,9055 fcc 12,42	Pd⁴⁶ 106,4 fcc 12,00	Ag⁴⁷ 107,868 fcc 10,50	Cd⁴⁸ 112,40 hcp 8,65	In⁴⁹ 114,82 tret. 7,29	Sn⁵⁰ 118,69 diamante 5,76	Sb⁵¹ 121,75 rombo 6,69	Te⁵² 127,60 hex. 6,25	I⁵³ 126,9045 complexa 4,95	Xe⁵⁴ 131,30 fcc 3,78
Cs⁵⁵ 132,905 bcc 1,997	Ba⁵⁶ 137,34 bcc 3,59	La⁵⁷ * 138,9055 hex. 6,17	Hf⁷² 178,49 hcp 13,20	Ta⁷³ 180,9479 bcc 16,66	W⁷⁴ 183,85 bcc 19,25	Re⁷⁵ 186,2 hcp 21,03	Os⁷⁶ 190,2 hcp 22,58	Ir⁷⁷ 192,22 fcc 22,55	Pt⁷⁸ 195,09 fcc 21,47	Au⁷⁹ 196,9665 fcc 19,28	Hg⁸⁰ 200,59 rombo 14,26	Tl⁸¹ 204,37 hcp 11,87	Pb⁸² 207,19 fcc 11,34	Bi⁸³ 208,9804 rombo 9,8	Po⁸⁴ (210) sc 9,31	At⁸⁵ (210)	Rn⁸⁶ (222)
Fr⁸⁷ (223)	Ra⁸⁸ 226,0254	Ac⁸⁹ ** (227) fcc 10,07															

* Lanthanides

Ce⁵⁸	Pr⁵⁹	Nd⁶⁰	Pm⁶¹	Sm⁶²	Eu⁶³	Gd⁶⁴	Tb⁶⁵	Dy⁶⁶	Ho⁶⁷	Er⁶⁸	Tm⁶⁹	Yb⁷⁰	Lu⁷¹
140,12 fcc 6,77	140,9077 hex. 6,78	144,24 hex. 7,00	(147)	150,4 complexa 7,54	151,96 bcc 5,25	157,25 hcp 7,89	158,9254 hcp 8,27	162,50 hcp 8,53	164,9304 hcp 8,80	167,26 hcp 9,04	168,9342 hcp 9,32	173,04 fcc 6,97	174,97 hcp 9,84

** Actinides

Th⁹⁰	Pa⁹¹	U⁹²	Np⁹³	Pu⁹⁴	Am⁹⁵	Cm⁹⁶	Bk⁹⁷	Cf⁹⁸	Es⁹⁹	Fm¹⁰⁰	Md¹⁰¹	No¹⁰²	Lr¹⁰³
232,0381 fcc 11,72	231,0359 tret. 15,37	238,029 complexa 19,05	237,0482 complexa 20,45	(244) complexa 19,81	(243) hex. 11,87	(247)	(247)	(251)	(254)	(257)	(258)	(255) fcc	(256)

Index

Printed in the United States
by Baker & Taylor Publisher Services